Photosynthesis during leaf development

# Tasks for vegetation science 11

Series Editors

HELMUT LIETH

*University of Osnabrück, F.R.G.*

HAROLD A. MOONEY

*Stanford University, Stanford CA, U.S.A.*

# Photosynthesis during leaf development

edited by

ZDENĚK ŠESTÁK

Contributors:

*Zdeněk Šesták, Jiří Čatský,
Ingrid Tichá, Danuše Hodáňová, Jaromír Kutík,
Jarmila Solárová, Jana Pospíšilová,
Jan Zima, Miloslav Kaše*

1985 **DR W. JUNK PUBLISHERS**
a member of the KLUWER ACADEMIC PUBLISHERS GROUP
DORDRECHT / BOSTON / LANCASTER

**Distributors**

*for the United States and Canada:* Kluwer Boston, Inc., 190 Old Derby Street, Hingham, MA 02043, USA

*for Czechoslovakia, Albania, Bulgaria, Cuba, China, German Democratic Republic, Hungary, Mongolia, Northern Korea, Poland, Rumania, U.S.S.R., Vietnam, and Yugoslavia:* Academia, Publishing House of the Czechoslovak Academy of Sciences, P.O.B. 896, 112 29 Prague 1, Czechoslovakia

*for all other countries:* Kluwer Academic Publishers Group, Distribution Center, P.O. Box 322, 3300 AH Dordrecht, The Netherlands

**Library of Congress Cataloging in Publication Data**

Main entry under title:

Photosynthesis during leaf development.

  (Tasks for vegetation science; 11)
  Includes bibliographical references and indexes.
  1. Photosynthesis. 2. Leaves-Development.
I. Šesták, Zdeněk, 1932 -     . II. Series.
QK882.P554 1984     581.1'3342     83-14992

ISBN-13: 978-94-010-8941-8          e-ISBN-13: 978-94-009-5530-1
DOI: 10.1007/978-94-009-5530-1

Scientific Editor: Prof. RNDr. PhMr. Miroslav Penka, DrSc.
Scientific Adviser: RNDr. Bohdan Slavík, DrSc.

Joint edition published by
Dr W. Junk Publishers,
P.O. Box 163, 3300 AD Dordrecht, The Netherlands
and Academia, Publishing House of the Czechoslovak Academy of   Sciences,
P.O.B. 896, 112 29 Prague 1, Czechoslovakia

# CONTENTS

# 0 LEAF DEVELOPMENT - TERMS AND PHOTOSYNTHETIC ASPECTS

*Z. Šesták and J. Čatský*

Photosynthesis is the primary metabolic process in nature: without photosynthesis life on our planet would cease. Green plants together with photosynthetic bacteria are the only primary producers on Earth: they provide energy and biomass for animals and men and keep the oxygen stable in the atmosphere.

The most important producers of biomass and oxygen are higher plants (mainly forming tropical forests) which produce about 70% of the total biomass, *i.e.* about $12 \times 10^{13}$ kg per year (*cf.* Lieth and Whittaker 1975, Whittaker and Likens 1975), a great deal of it being of immediate significance for man as nutrition or raw materials. Crops on cultivated land produce about 5% of the total biomass, *i.e.* less than $10^{10}$ t per year.

Even if we take into account the important contribution of photosynthesis of ears, stems and other green plant parts to the total plant biomass production, the major role in most plants is played by the leaves, the total surface of which is in the biosphere about $65 \times 10^7$ km². During its life, however, the leaf morphology, structure of its tissues and cells, and its composition change, and physiological and biochemical activities (including those embraced by the concept of photosynthesis) develop in a manner similar to that of any creature.

## 0.1  PHASES OF LEAF DEVELOPMENT AND INSERTION PROFILE

During leaf development, or leaf life span, generally three phases may be distinguished: (1) leaf formation related to leaf area increase, (2) the period of leaf maturity after maximum leaf area is reached, and (3) senescence associated with a decrease in leaf area.

Leaf formation begins with the origin of leaf primordia at the shoot apex. Cell division is replaced by cell enlargement as the chief means of leaf expansion when leaves have attained about one third or half of maximum leaf area (*cf.* Avery 1933, Hannan 1968). The duration of leaf maturity is different in herbs, deciduous and evergreen trees. The start of senescence is usually characterized by the appearance of signs of cell and tissue degradation.

With respect to the carbon economy of the leaf, three more or less corresponding phases may be distinguished (Stoddard and Thomas 1982): (1) Initially, the leaf is a net carbon-importing structure and remains so until the full development of photosynthetic capacity and the subsidence of the peak demand for photosynthates for the assembly of cells. (2) There then follows a period, of variable duration, when the leaf becomes an asset to the carbon economy of the plant. This period may be called a period of "photosynthetic maturity" (Šesták and Čatský 1962), during which peak values of the most important photosynthetic parameters are reached. It coincides with the maximum rate of leaf area expansion (cf. Rawson and Hackett 1974). The period ends with the advent of internal or environmental conditions which initiate senescence. (3) From this point the leaf progresses into a period of massive mobilization and export of carbon and minerals. This period is accompanied by a gradual decline in photosynthetic capacity.

Even if we analyse the ontogeny of one leaf, the ontogenetic diversity of different regions of its blade has to be taken into account. Meristematic activity takes place in the base of the blade, the cells, chloroplasts, mitochondria, etc. of which are thus young, while the leaf tip region contains the oldest tissues.

Subsequently formed leaves develop at different phases of plant ontogeny and in various microenvironmental conditions, especially under increasing irradiance. The phyllotactic arrangement originated in the phase of leaf primordia is maintained during the whole life of a plant. The vertical displacement of leaf primordia during stem development leads to the formation of the so-called insertion gradient (insertion profile). Leaves reaching various insertion levels on the stem attain not only different final area of leaf blades, but also different values of maximum photosynthetic (and other physiological) activities. These peak values are reached in subsequent leaves at different time intervals after their unfolding. Hence, the development of leaves of different insertion has to be completed with the ontogenetic changes in the character of the leaf insertion gradient at various phases of plant ontogeny.

At this point, some problems of terminology have to be cleared up. The term "leaf development" is probably the most correct one, as it covers the whole life span of the leaf from the formation of primordium to leaf death. "Leaf ontogeny" has the same meaning. Nevertheless, ontogeny is usually defined as "the history of development and growth of an individual", even if the Greek words forming the term (on = being, genesis = origin, descent) do not fully condition that. Hence, one can explain the term leaf ontogeny in the sense of the well-known Timiryazev motto "Plant — that is a leaf", or as the shortened explanation of "the development of an individual leaf during development of an individual plant". The term leaf ageing means virtually the same, as everything ages from the very moment of its conception. Leopold (1980) defines aging as "processes of accruing maturity with the passage of time". Nevertheless, ageing is subconsciously connected mainly with processes after reaching leaf maturity, and thus incorrectly connected with senescence. Senescence is better referred to as the deteriorative processes that are natural causes of death, and that provide for the endogenous regulation of death (Leopold 1980). For the insertion gradient the terms lower, middle, and upper leaves may be recommended, in contrast to the terms young, mature (adult), and old leaves, characterizing leaf ontogeny.

## 0.2 COMMON IMPERFECTIONS OF LEAF ONTOGENETIC STUDIES

The amount of papers in which changes of one or another parameter of photosynthesis have been described under a more or less pronounced aspect of leaf and plant ontogeny, is rather vast. Nevertheless, this literature has only rarely been reviewed and analysed, and the extent of research tenacious of purpose is rather limited. Summarizing the data existing in the literature is difficult for several reasons:

(1) Only correctly planned experiments can be taken into account. The studies in which e.g. always the third leaf from the top is analysed throughout the plant ontogeny are worthless from the point of view of leaf development.

(2) Unfortunately, many authors recall the well-known sigmoid growth curves and Gaussian distribution and take them as universal for all life processes. Accordingly, very often two points on the scale of the life span of the leaf are randomly taken, and the "young and mature", "young and adult", "young and senescent", etc. leaves are compared. Nevertheless, even three points ("young — mature — old") are not satisfactory for the evaluation of changes of many leaf characteristics, whose ontogenetic course often does not follow any simple smooth curve. It is always necessary to pursue the whole leaf ontogeny and take into account leaf insertion in order to obtain reliable data.

(3) The ontogenetic formation of leaf structure, composition, and metabolic functions is genetically inherent, but their proper expression is dependent on optimum microenvironmental factors (composition of atmosphere, irradiance, temperature, water supply, etc.) during leaf growth at different insertion levels. And even the actual capacity of the photosynthetic apparatus depends on environmental conditions to the combination of which some components of the apparatus may acclimate rather easily. Every researcher knows that a sudden transient change of the conditions in a growth chamber (e.g. a drop in temperature) may induce a days-long shock reflected in the physiological activities of the plant. Nevertheless, many authors describe ontogenetic changes in a plant grown in uncontrolled conditions of environment, or do not specify them. Very often the cultivar is not

given, and the basic parameter of leaf growth — area of its blade — is not shown for comparison.

(4) Different ontogenetic courses may be obtained, when the given parameter is related to leaf area unit, dry matter, or per one leaf. The best papers relate the photosynthetic characteristics to all these basic units. The worst ones use fresh matter for the basic unit. In special cases (activities of photosystems, photophosphorylation, enzymes) special relation units (chlorophyll amount, protein, *etc.*) are used.

(5) Sometimes misunderstandings originate by the use of incorrect terms, *e.g.* speaking of enzyme activity when only the amount was determined. In insertion gradients the plant age should always be given, and a clear statement should show whether the youngest or more correctly (as in this volume) the oldest leaf bears the number one.

## 0.3  SPECIALIZED MONOGRAPHS AND THE SCOPE OF THIS BOOK

The above mentioned objections are responsible for the present situation, when leaf ontogeny is the factor only rarely taken into account in chapters on photosynthesis in textbooks and monographs on physiology, biochemistry, or biophysics of plants. Also the amount of summarizing reviews or monographs on this topic is rather limited.

As concerns monographs dealing with photosynthetic characteristics during leaf ontogeny, a pioneering book edited by Sironval (1967) collected 31 papers which reviewed various aspects of chloroplast development; the monograph edited by Marcelle *et al.* (1979) contains 31 papers dealing with photosynthesis during plant development. Some information may be found in the monograph on chloroplasts by Kirk and Tilney-Bassett (1978). Recently, two books appeared dealing with plant ontogeny, *i.e.* the monograph focused on various senescence phenomena in plants (Thimann 1980), and the monograph of Mokronosov (1981) where the analysis of ontogenetic problems is based mainly on his own detailed experiments with potato plants. From the specialized reviews, those of Treffry (1978) and Bradbeer (1981) deal with the formation of the photosynthetic apparatus, while those of Woolhouse (1967, 1974, 1982), Woolhouse and Batt (1976) and Noodén (1980) with the senescence phenomena. The largest complex of literature was analysed in the series of specialized reviews on individual parameters of photosynthesis by Šesták and Čatský (1966, 1967c), Tichá (1966, 1982), Šesták (1977a, b, 1978, 1983, 1984), Zima and Šesták (1979), Tichá and Čatský (1981), Čatský and Tichá (1982), and Solárová and Pospíšilová (1983).

The present monograph is intended to show, with results selected from the literature and the experience of the authors themselves, the peculiarities of changes of individual photosynthetic parameters during leaf ontogeny. The authors have taken into account that the photosynthetic activity of a leaf measured as the $CO_2$ uptake rate, dry matter or energy accumulation (see Chapter 7) is a result of the formation of leaf structure (Chapter 1) and ultrastructure, especially that of chloroplasts (Chapter 2), its chemical composition (with special emphasis on the location of compounds in thylakoids), its physical and photophysical properties (Chapter 4), and biochemical activities. The contents of photosynthetic pigments (Chapter 3), components of the electron transport chain (Chapter 5), and enzymes of the carbon fixation cycles (Chapter 6), their states *in vivo* and activities may change in accordance or in contrast with the changing leaf morphology. They together determine the absorption, transport and utilization of radiant energy on the one hand (Chapter 4), and of carbon dioxide and water on the other (see Chapter 8). The cooperation of processes under various climatic conditions lead to ecological adaptations which were in the course of plant phylogeny genetically fixed, as are the $C_3$, $C_4$ and CAM pathways of carbon fixation (see Chapters 1, 6, and 7) and the related removal of undesirable products in the process of photorespiration (see Chapter 9). The final photosynthetic balance evident from the rates of gas exchange or dry matter accumulation is the result of interaction of the whole set of photosynthetic parameters and reactions (see Chapter 10).

In the field of photosynthesis, ontogenetic changes are especially important as they should to be included into models of plant and canopy productivity. Up till now, only a few models of canopy productivity have taken into account the great changes in leaf optical properties based mainly on pigment amounts and leaf anatomy, photosynthetic, photorespiration and respiration rates, *etc.* The selection of new cultivars for crops of higher productivity and of the composition of biomass more advantageous for mankind needs ideotypes of appropriate leaves with optimum ontogeny of the photosynthetic apparatus. The aim of this monograph was to collect, analyse and synthesize the ontogenetic aspects of leaf photosynthesis for their dissemination not only in basic research but also in agriculture and forestry.

# 1 ONTOGENY OF LEAF MORPHOLOGY AND ANATOMY

*Ingrid Tichá*

## 1.1 LEAF AS A PHOTOSYNTHETIC STRUCTURE

Because of their role in intercepting and absorbing radiant energy, and transforming it to energy of organic substances through the complex process of photosynthesis, leaves are the most important organs for plant production, and hence for agriculture.

Leaf structure developed as an adaptation of plants to terrestrial environment as a compromise between photosynthetic gains and inevitable transpirational losses. The anatomical structure of the leaf represents a skeleton, in which physiological processes take place including photosynthesis with its photophysical, photochemical and biochemical reactions; furthermore, $CO_2$ and $O_2$ transfers are located there. Thus, studying photosynthesis during leaf development requires also the study of the interrelationships between photosynthesis and leaf structure, and to evaluate how changes in leaf structure with leaf age may limit the photosynthetic performance of the leaf.

In this chapter values on changes in morphological characteristics and anatomical parameters during leaf ontogeny, in leaves of different insertion levels, and on the leaf blade in leaves of higher plants are summarized and evaluated. Mainly those anatomical parameters which are connected with photosynthesis are treated here. Leaf initiation or development of leaf primordia in the apex are not included (for reviews see, *e.g.*, Milthorpe 1956, Maksymowych 1973, Napp-Zinn 1973, 1974, Williams 1975, Dale 1982). We shall start with a leaf just at leaf unfolding when the photosynthetic rate can be measured (for reviews see also Milthorpe 1976, Mokronosov 1978, 1981, Scott and Possingham 1982).

## 1.2 LEAF GROWTH, LEAF AREA AND SIZE, LEAF THICKNESS

### 1.2.1 Leaf Area

Rates of photosynthesis are commonly expressed as $CO_2$ uptake per unit time per unit leaf area. Leaf area is often used as the reference unit also in other physiological studies. The total leaf area of a plant is sometimes called the photosynthetic

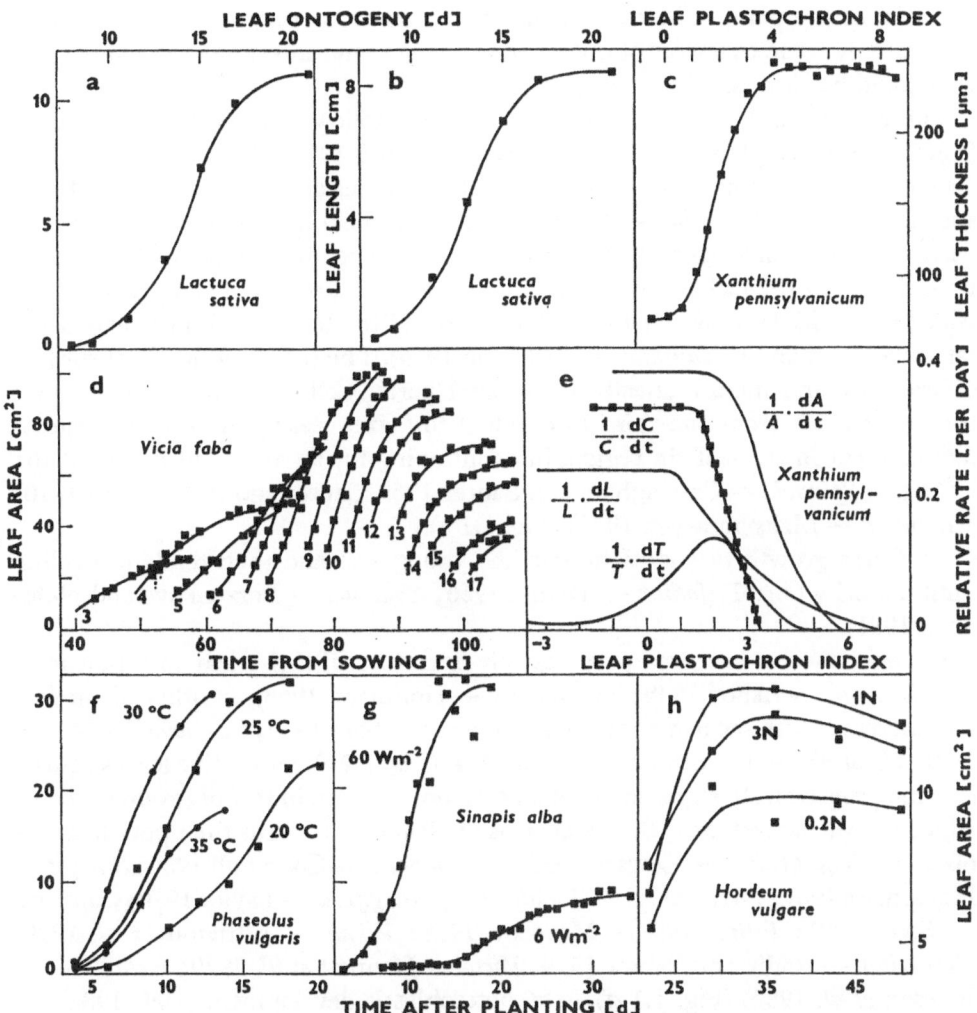

**Fig. 1.1.** During leaf ontogeny indicated in days (d) or as leaf plastochron index, a nearly parallel increase in leaf area (a) and lenght (b) (*Lactuca sativa* L., from Bourdu *et al.* 1975), and leaf thickness (c — *Xanthium pennsylvanicum* Walbr., from Maksymowych 1973) is found. The relative rates of cell formation (C), expansion in area (A) and leaf length (L) are high in the unfolding leaf but the relative rate of lamina thickness growth (T) is negligible; during further leaf development cell division stops first, followed by increase in lamina thickness, leaf area and, finally, leaf length (e — the same plant and reference as c). Leaf area expansion in leaves of different positions on the plant varies in its rate and in maximum leaf area reached (d — *Vicia faba* L., from Dennett *et al.* 1979). Leaf area expansion is modified by various environmental factors, *e. g.* temperature (f — *Phaseolus vulgaris* L., from Wilson and Ludlow 1968), irradiance (g — *Sinapis alba* L., from Wild and Wolf 1980) and nitrogen supply (h — *Hordeum vulgare* L., from Mader *et al.* 1981).

potential of the plant. Therefore, leaf area is estimated very frequently; in this chapter only some examples of ontogenetic changes and insertion gradients in leaf area could be chosen.

The increase in leaf area during leaf ontogeny follows a typical sigmoid curve (*e.g. Cucumis* – Hopkinson 1964; *Capsicum* — Schoch 1972a; *Vigna* — Schoch and Candelario 1973; *Phaseolus* — Čatský *et al.* 1976, Tichá and Čatský 1977, Verbelen and de Greef 1979; *Malus* — Kennedy and Johnson 1981) (Fig. 1.1 a). Various parts of the lamina expand at different rates, depending upon their distance from the tip of the leaf and increase in the basal direction (basipetal expansion — *Xanthium* — Maksymowych 1973). The centre of the leaf has a higher expansion rate than the margin (*Xanthium* — Erickson 1966). The tissues at the tip of the leaf differentiate and mature ahead of the basal parts. Cell division continues until the leaves reach from one-third to one-half full size, ceasing at an early stage of development in the leaf tip region, but continuing for an extended period at the leaf base (*Spinacia* — Possingham and Saurer 1969, Saurer and Possingham 1970; *Xanthium* — Maksymowych 1973) (Fig. 1.1 e).

Leaf area growth rate is highest in the young leaf and then tends to decline until maturity (*e.g. Trifolium* — Denne 1966; *Solanum* — Borzenkova and Nefedova 1981).

In some species leaf area progressively decreases with the leaf insertion level (*e.g. Ipomoea* — Ashby 1948; *Arabidopsis* — Hoffmann 1968), in other plants leaf area increases with the leaf position (*e.g. Lolium* — Sant 1969; *Capsicum* — Schoch 1971; *Nicotiana* — Hackett and Rawson 1974) (Fig. 1.2 a). Successive leaves usually become progressively larger in the final area up to a certain leaf or group of leaves which has the largest area; then a decrease in leaf area towards the upper leaves is observed (*e.g. Malus* — Cowart 1936; *Callistephus* — Cockshull 1966; *Sinapis* — Humphries 1967; *Plectranthus* — Tichá 1968; *Mangifera* — Taylor 1970; *Capsicum* — Steer 1971; *Filipendula* — Morozov 1978; *Vigna* — Littleton *et al.* 1979; *Gossypium* — Radin and Parker 1979; *Vicia* — Dennett *et al.* 1979; *Helianthus* — Rawson *et al.* 1980) (Fig. 1.1 d) — *cf.* heteroblastic development (Dale 1982).

Area growth of leaves and thus ontogenetic and insertion changes in leaf area are strongly modified by temperature (*Phaseolus* — Wilson and Ludlow 1968; *Vicia* — Dennett *et al.* 1979, Dennett and Auld 1980; *Glycine, Gossypium* — Jones and Hesketh 1980; *Helianthus* — Rawson and Hindmarsh 1982) (Fig. 1.1 f), radiation (*Tropaeolum* — Rumi and Carpinetti 1977; *Vicia* — Dennett *et al.* 1979; *Sinapis* — Wild and Wolf 1980) (Fig. 1.1 g), shading (*Cucumis* — Hopkinson 1966) which hastens leaf senescence, photoperiod (*Nicotiana* — Hackett and Rawson 1974) (Fig. 1.2 b), air humidity (*Capsicum* — Schoch 1971), water supply (*Ipomoea* — Ashby 1948; *Sorghum* — McCree and Davis 1974; *Zea, Glycine* — Boyer 1970; *Helianthus* — Rawson *et al.* 1980, Takami *et al.* 1981, Rawson and Turner 1982) (Fig. 1.2 c), salinity (*Phaseolus* — Wignarajah *et al.* 1975), and nitrogen nutrition (*Lolium* — Robson and Deacon 1978; *Gossypium* — Radin and Parker 1979;

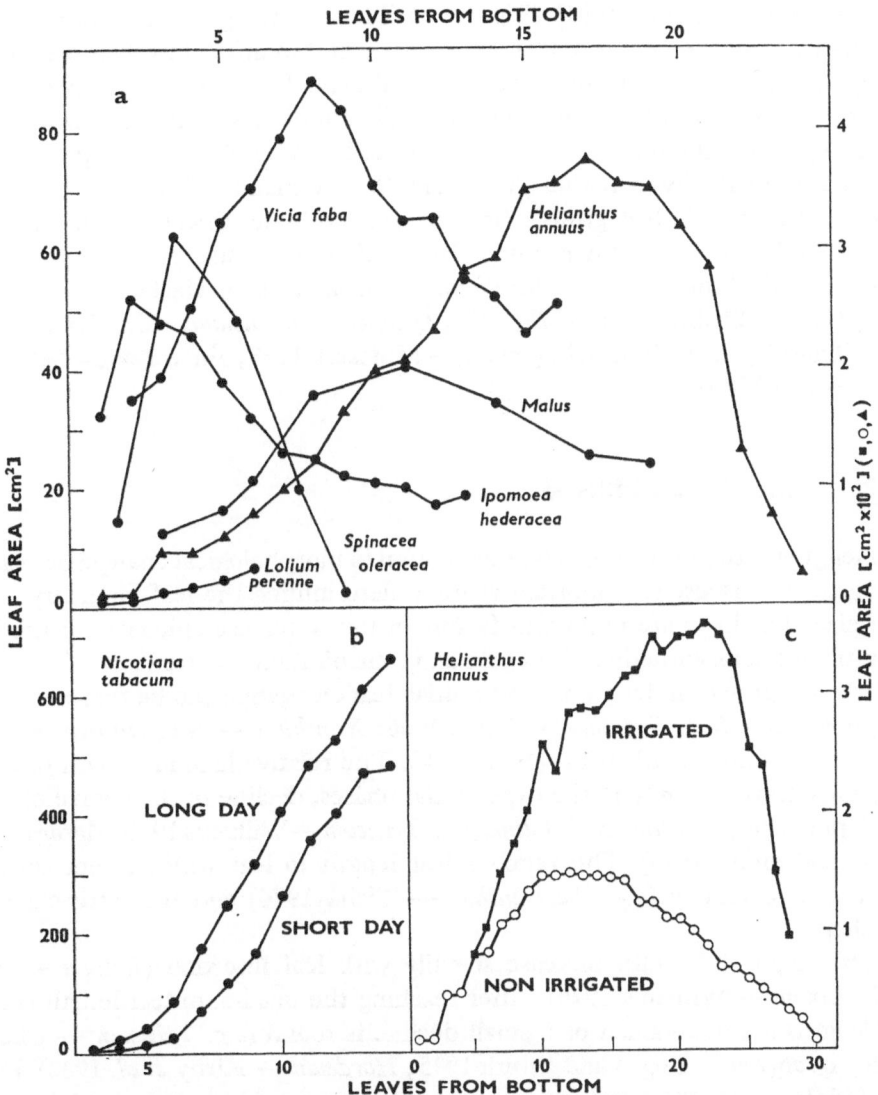

**Fig. 1.2.** Insertion patterns of leaf area in different plants. Maximum leaf area is reached in lower leaves (a — *Ipomoea hederacea* Jacq., from Ashby 1948), in middle leaves (a — *Helianthus annuus* L., from Rawson *et al.* 1980; *Malus domestica* Borkh., from Cowart 1936; *Vicia faba* L., from Dennett *et al.* 1979; *Spinacea oleracea* L., from Possingham and Saurer 1969) or upper leaves (a – *Lolium perenne* L., from Sant 1969). Insertion gradients in maximum leaf area are modified in their absolute values by environmental factors, *e. g.* photoperiod (b — *Nicotiana tabacum* L., from Hackett and Rawson 1974) or water supply (c — *Helianthus annuus* L., from Rawson *et al.* 1980), but the trend of the gradient is maintained.

*Hordeum* — Mader *et al.* 1981) (Fig. 1.1 h). The ascending and descending insertion gradients in the final leaf area are maintained, but absolute values are shifted as well as the nodus with the maximum final area. In successive sowings (April to June) of *Ipomoea* plants maximum growth rate of leaves, maximum leaf area at full expansion, the rate of leaf production, and also number of cells per leaf and the cell area markedly increased (Ashby and Wangermann 1950).

In the last decade leaf growth (in area, size, volume, thickness, and also dry matter) under different environmental conditions was also analysed by means of mathematical models (*e.g. Nicotiana* — Hackett 1973, Hackett and Rawson 1974; *Vicia* — Dennett *et al.* 1978, 1979; *Lycopersicon, Cucumis* – Charles-Edwards 1979, Thornley *et al.* 1981; *Gossypium* — Mutsaers 1983; for a review see Jones and Hesketh 1980).

## 1.2.2 Leaf Size and Shape

Leaf length represents an important dimension in morphological changes associated with growth, it is also an important value in determining the leaf boundary layer. Leaf size and shape are important factors in the water use efficiency of the leaf, the ratio of photosynthetic carbon gain and transpirational water loss.

Early leaf growth in length is exponential, leaf elongation can be represented by a sigmoid curve (*e.g. Trifolium* – Denne 1966; *Xanthium* — Maksymowych 1973; *Lactuca* — Bourdu *et al.* 1975) (Fig. 1.1 b). The relative lamina growth rates for successive leaves, mainly in the exponential phases, decline in the course of plant development (*e.g. Nicotiana, Trifolium,* and *Triticum*—Williams 1975; *Agropyron*—Rogan and Smith 1975). The ratio of leaf length to leaf width is not constant during leaf ontogeny (*e.g. Plectranthus* — Tichá 1974) and indicates changing leaf shape.

Leaf length and width increase steadily with leaf insertion (*Lolium* — Sant 1969; *Linum* — Williams 1975). After reaching the maximum leaf length similar leaf lengths are maintained or a small decline is found (*e.g. Triticum* — Chonan 1965; *Agropyron* — Rogan and Smith 1975; *Hordeum* — Kirby *et al.* 1982). Up to the middle leaves, the points of inflection of the sigmoid curves are high on the curves, with curvatures greater above than below that point. This condition rapidly reverses for later leaves, as the point of inflection falls gradually lower (*Linum* — Williams 1975).

Various environmental factors influence final lamina size, *e.g.* radiation (*Cucumis*— Horie *et al.* 1979), temperature (*Phaseolus* — Dale 1965), nitrogen supply (*Hordeum* — Mader *et al.* 1981), *etc.* The development of optimal size in relation to the environment was analysed by means of models (*e.g.* Parkhurst and Loucks 1972, Givnish 1978).

## 1.2.3 Leaf Thickness

Leaf thickness increases, in parallel to leaf area, only slightly at first followed by a period of rapid expansion in the thickness of the lamina, which involves both cell division and cell enlargement. This period of rapid expansion in the thickness of the lamina with the expansion of the palisade cells and development of intercellular spaces in the spongy parenchyma occurs at about one-third of the mature length of the leaf (*e.g. Nicotiana* — Avery 1933; *Xanthium* — Maksymowych 1973 — see Figs. 1.1 c and 1.6). During this period, the number of cell layers in the leaf increases rather quickly from the six basic layers up to eight or more (Maksymowych 1973). This expansion is completed when the mature thickness of the lamina is reached. Further increase in leaf thickness is very slow (*Quercus* — Pieters 1962; *Arabidopsis* — Hoffmann 1968; *Lolium* — Sant 1969; *Mangifera* — Taylor 1970; *Phaseolus* — Gausman *et al.* 1970, Čatský *et al.* 1976, Verbelen and de Greef 1979). On the other hand, in wheat a decrease in leaf thickness from the eighth to the fifteenth day of leaf ontogeny was found (Hoffmann 1968).

In some plant species no clear tendency in the relatively small changes in thickness in leaves of different insertion levels was found (*e.g. Oryza* – Kataoka and Oohara 1980). In poplar leaves grown under constant conditions, leaf thickness increased a little from the lower to middle leaves, and then decreased towards the upper leaves (Pieters 1962). Sometimes only an increase towards the top of the plant was found (*e.g. Nerium* at low irradiance — Turrell 1965; *Eucalyptus* — Pereira and Kozlowski 1976). Leaf thickness may, however, be also highest in lower leaves and then decrease towards the plant apex (*e.g., Vinca* — Turrell 1965; *Gossypium* — Gausman *et al.* 1971b; *Populus* — Isebrands and Larson 1973; *Betula* — Öquist *et al.* 1982). In some plant species the lowest and highest leaves were the thickest on the plant (*Cucumis* — Guretskaya 1954; *Malus* — Cowart 1936).

## 1.2.4 Leaf Plastochron Index

For studying leaf development a numerical index, the leaf plastochron index (LPI) was developed (Erickson and Michelini 1957; for reviews see Maksymowych 1973, Lamoreaux *et al.* 1978).

A plastochron is the period of time between initiation of successive leaf primordia at the shoot apex of a higher plant or the interval between the corresponding stages of development of successive leaves. Erickson and Michelini (1957) recognized that when successive plastochrons are of equal duration the plastochron can serve as the elemental unit of a quantitative scale of shoot development instead of chronological age. They developed a numerical index of the developmental age of plants, the plastochron index (PI); for characterizing leaf development, a similar leaf plastochron index (LPI) was derived.

Plastochron index, PI, can be written as:

$$PI = n + \frac{\log L_n - \log \lambda}{\log L_n - \log L_{n+1}}$$

where $n$ denotes the serial number of the youngest leaf whose length exceeds that of the reference value $\lambda$, leaf $n + 1$ is the following leaf, $L_n$ and $L_{n+1}$ are lamina lengths [mm] of leaves $n$ and $n + 1$, respectively, $\lambda$ is chosen according to the plant material and selected so that the index leaf of that length is still in an early stage of development; usually it is 10 or 20 mm long.

Leaf plastochron index of the $i^{th}$ leaf on a shoot is defined as $LPI_i = PI - i$, where $i$ is the serial number of the leaf in question (always calculated from the oldest one). Leaves shorter than the index leaf have negative LPIs and leaves longer than the index leaf have positive LPIs. A leaf, the length of which is exactly $\lambda$, is of zero LPI.

The derivation of the plastochron index is based on the asumptions that (a) early leaf growth occurs at an exponential rate, (b) early growth of successive leaves on a single plant occurs at the same relative rate, and (c) successive plastochrons are of the same length for a particular plant. Before applying the PI and LPI, one should therefore always verify that these assumptions are reasonably met.

For characterization of *Hevea brasiliensis* leaf age the Leaf Blade Class Concept which is based on measurement of the angle formed between the middle leaflet and the leaf stalk, was developed by Samsuddin *et al.* (1978).

## 1.3    LEAF EPIDERMIS

The epidermis of leaves of land plants protects the plant against desiccation in the air (environment). The epidermis controls not only loss of water vapour but also entry of the $CO_2$ essential for plant photosynthesis. Stomata on the epidermis enable communication between the outer atmosphere and that inside the leaf (see Section 1.3.3). The epidermis is covered by a more or less impermeable outer layer — the cuticle, various wax overlays or secretions, and trichomes (see Section 1.3.2).

### 1.3.1    Number of Epidermal Cells, their Sizes, and Thickness of Epidermis

The number of epidermal cells per unit leaf area decreases on both epidermes with leaf age, as the cells grow larger (*Arabidopsis* — Hoffmann 1968). Area growth of epidermal cells during leaf ontogeny follows a sigmoid curve, the shape of which is modified by air temperature and humidity (*Capsicum* — Schoch 1971) or irradiance (*Vigna* — Schoch and Candelario 1973).

Linear sizes of both upper and lower epidermal cells decrease with increasing leaf insertion level (*Populus* — Isebrands and Larson 1973) as does the area of epidermal cells (*Ipomoea* – Ashby and Wangermann 1950; *Capsicum* — Schoch 1971). Consequently, the number of epidermal cells in a leaf in terrestrial plants increases with insertion level (*Ipomoea* — Ashby and Wangermann 1950; *Capsicum* — Schoch 1971). In aquatic species no consistent differences in leaves of different insertion levels in size or number of epidermal cells per unit leaf area were found (*Potamogeton* — Lehner 1946, Tichá 1964).

The thickness of the epidermis either slightly increases (*e.g. Xiphium* – Pazourek 1973c) or declines (*Eucalyptus* — Pereira and Kozlowski 1976) with increasing insertion level of the leaf.

The pattern of changes of epidermal cell length and width along the maize leaf is clearly different for the adaxial and abaxial surfaces: at the leaf base an increase in both parameters is found on both leaf surfaces followed by a sharp decrease on the adaxial side at 5 cm from the base (Miranda *et al.* 1981a). The area of epidermal cells and the thickness of epidermal cells on abaxial epidermes of *Magnolia* and *Ligustrum* vary very little on leaf lamina (Kutík 1973).

During leaf development leaf temperature affects epidermal cell size but not epidermal cell number (*Vicia* – Auld *et al.* 1978). Insertion gradients may be substantially modified by environmental factors: *e.g.* growth irradiance changes not only absolute values but also trend of the gradient in thickness of the epidermis (*Xiphium* – Pazourek 1973c), while nitrogen nutrition changes only the absolute values of the gradient in epidermal cell area, but the trend is maintained (*Gossypium* — Radin and Parker 1979).

## 1.3.2 Cuticle, Waxes, Trichomes

Both leaf epidermes are usually covered by cuticle, wax particles and trichomes or emergences which are important mainly as protection against water loss by the leaf. On the other hand, the quality of the epidermal surface determines the thickness of the leaf boundary layer and thus the conductance for $CO_2$ and water vapour transfer (*cf.* Section 8.1.1).

Cuticle thickness increases during leaf ontogeny, *e.g.* in *Panicum* four fold from the lower to upper leaves (Wilson 1976) and about three fold within the 3 cm from the leaf base on both surfaces of maize leaf, but there are no further changes towards the leaf tip (Miranda *et al.* 1981a).

During leaf ontogeny the number of trichomes per leaf area unit increases (*Hordeum* — Dutzmann *et al.* 1981). The number of trichomes increases also from the lower to the upper leaves of the plant (numerous herbaceous plants — Zalenskiï

1904; *Nicotiana* — Barrera and Wernsman 1966; *Lolium* — Sant 1969; *Hordeum* — Pazourek 1966, Dutzmann *et al.* 1981). The frequency of trichomes may be different on adaxial and abaxial leaf surfaces but the described gradient holds for both leaf surfaces (*Hordeum* — Pazourek 1966).

### 1.3.3    Stomata

Stomata, which enable communication in the gaseous phase between the inside of the leaf and the ambient air, optimize the balance of the $CO_2$ influx during leaf photosynthesis and water vapour efflux in the course of leaf transpiration. Stomata density and sizes are thus important anatomical parameters contributing to the resistance (conductance) of the leaf to $CO_2$ and water vapour transport (*cf.* Chapter 8) and thus determining the water use efficiency of the leaf. Ontogenetic changes in stomata density and sizes were recently reviewed by Tichá (1982).

#### 1.3.3.1    *Stomata Density*

According to stomata distribution on leaf surface amphistomatous (with equally distributed stomata on the abaxial and adaxial leaf surfaces), hypostomatous (stomata mainly on the abaxial side), and epistomatous (stomata prevailing on the adaxial side) leaves can be distinguished.

Leaf ontogeny: Stomata density on both leaf surfaces (if stomata are present) of herbaceous plants increases in the expanding leaf when stomata are initiated (for examples see Table 1.1), and then decreases at first rapidly and then gradually more slowly towards the end of leaf life. At a certain point of leaf development stomata division ceases but maturation of stomata continues (*Phaseolus* — Evans and Ting 1974; *Vicia* — Lurie 1977). Peak values of stomata density were found at a leaf area expansion of 10 to 60% of their final size (Fig. 1.3). Very often the period of stomata formation was not followed and thus only a decrease in stomata density with leaf age was found (*e.g. Citrus* — Reed and Hirano 1931; *Plectranthus* — Tichá 1970a; *Phaseolus* — Solárová 1973, Tschakalova 1976; *Nicotiana* — Rawson and Craven 1975; *Triticum* — Tschakalova and Hoffmann 1976) (Fig. 1.4 *top*).

The pattern of development of stomata density is the same in individual leaves of different insertion levels on plants, but the absolute values of stomata density are often different (*Plectranthus* — Tichá 1970a, 1982 — Fig. 1.4 *top*; *Phaseolus* — Tschakalova 1976). In contrast to etiolated plants the pattern of stomata density development may be greatly compressed in time in light-grown plants (*Vicia* – Lurie 1977).

The ratio of stomata density in the adaxial/abaxial leaf epidermes increases during leaf ontogeny, *e.g.* in tobacco (Rawson and Craven 1975) or primary leaves of *Phaseolus* (Šesták *et al.* 1978a), which is due to a more pronounced decrease in stomata density on the abaxial leaf surface than the adaxial surface. But, in other plants an increase and then a decrease in this ratio during leaf ontogeny (*Triticum* — Tschakalova and Hoffmann 1976; *Sinapis* — Wild and Wolf 1980) or only a decrease (*Phaseolus* — Solárová 1973, Tschakalova 1976) may be found. This indicates the uneven development of stomata density on abaxial and adaxial leaf surfaces in leaves of different plants.

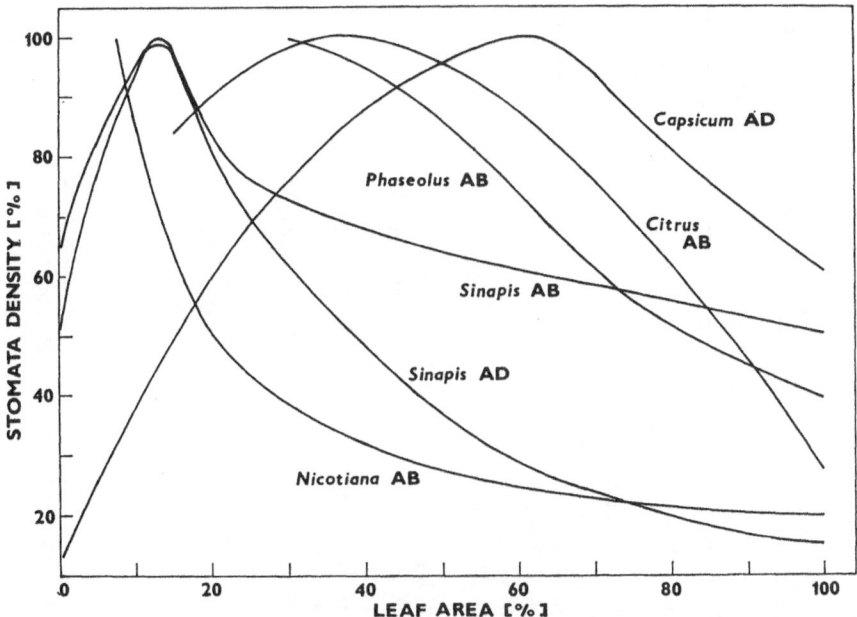

**Fig. 1.3.** Ontogenetic changes in abaxial (AB) and adaxial (AD) stomata density as related to leaf area development. 100 % leaf area is the maximum leaf area reached. Maximum stomata density is reached at 8 to 60 % of maximum leaf area. *Citrus sinensis* Osbeck, redrawn from Reed and Hirano (1931); *Capsicum annuum* L., from Schoch (1972a); *Phaseolus vulgaris* L., from Solárová (1973); *Nicotiana tabacum* L., from Rawson and Craven (1975); *Sinapis alba* L., from Wild and Wolf (1980).

The pattern of ontogenetic changes in stomata density is mostly maintained in leaves grown under various irradiance. Nevertheless, the primary leaves of *Sinapis alba* grown at 60 W m$^{-2}$ had at the end of leaf life span more than twice as many stomata per unit leaf area and the stomata were longer by 2 μm than at low irradiance (6 W m$^{-2}$) (Wild and Wolf 1980). Higher mean daily irradiance during leaf ontogeny increased stomata density but reduced stomata length and area per stoma in tobacco and sunflower (Rawson and Craven 1975). Higher growth irradiance increased the ratio of frequencies of stomata in the upper/lower

**Table 1.1** Changes in stomata density and sizes during leaf ontogeny.
Abbreviations and symbols: $I$ — irradiance; — not studied; 0 — no changes in columns (5) and (6); max — maximum; ↑ increase; ↓ decrease.

| Plant (and cultivation) (1) | Age of leaf (numbered from the oldest one) (2) | Adaxial/abaxial stomata density [mm⁻²] (3) | Adaxial/abaxial stomata sizes [μm] (4) | Changes induced by leaf ageing in stomata density (5) | size (6) | Author(s) and year of publication (7) |
|---|---|---|---|---|---|---|
| *Capsicum annuum* L. cv. Yolo Wonder (pots, field) | leaf 3, 20 to 55–60 d after sowing | (1) 50/--...320/--...180/-- <br> (2) 83 W m⁻² global $I$: <br> 25/--...250/--...10/-- <br> 136 W m⁻²: <br> 40/--...200/--...35/-- <br> 273 W m⁻²: <br> 90/--...180/--...110/-- | | ↑, ↓/-- (max d 32) <br> ↑, ↓/-- (max d 34) | ↑/↑ <br> ↑/↑ | Schoch 1972a (1), 1972b (2) |
| *Nicotiana tabacum* L. cv. Mammoth (phytotron, glasshouse) | leaves 11 and 18 from 10 % of final to maximum leaf area | leaf 11, low $I$: <br> 150/450...38/91 <br> leaf 11, high $I$: <br> 125/310...48/108 <br> leaf 18, high $I$: <br> 165/560...47/110 | area [μm²]: <br> 280/260...976/1000 <br> 390/420...620/837 <br> 200/320...695/820 | ↓/↓ <br> ↓/↓ <br> ↓/↓ | ↑/↑ <br> ↑/↑ <br> ↑/↑ | Rawson and Craven 1975 |
| *Phaseolus vulgaris* L. cv. Jantar (growth chamber) | primary leaves, 8 to 30 d | 170/990...60/500 | length: <br> 16/12...28/24 | ↓/↓ | ↑/↑ | Solárová 1973 |
| *Sinapis alba* L. (growth chamber) | primary leaves, 3 to 19 d after potting | 60 W m⁻² (400–700 nm): <br> 105/190...208/290...40/145 <br> 6 W m⁻² (400–700 nm): <br> 10/120...67/175...15/65 | mean length of pore for both leaf surfaces: <br> 11...21 <br> 7...19 | ↑, ↓/↑, ↓ (max d 6) <br> ↑, ↓/↑, ↓ (max d 15) | ↑ <br> ↑ | Wild and Wolf 1980 |

| Triticum aestivum L. cv. Ramses (laboratory) | leaves 1 to 9 (= flag leaf), 35 to 110 d after sowing | length: | | | | Tschakalova and Hoffmann 1976 |
|---|---|---|---|---|---|---|
| | | leaf 1: 43/50...43/50 | 68/68...74/76 | ↑/↑ | 0/0 | |
| | | 2: 60/51...36/44 | 63/68...74/69 | ↑/↑ | ↓/↑ | |
| | | 4: 53/36...43/40 | 73/71...62/66 | ↓/↑ | ↓/↑ | |
| | | 6: 51/50...44/44 | 58/54...68/70 | ↑/↑ | ↓/↑ | |
| | | 8: 46/47...54/43 | 64/62...66/64 | ↑/↑ | ↑/↑ | |
| | | 9: 58/47...58/47 | 61/59...58/63 | ↓/↑ | 0/0 | |
| | | width: | | | | |
| | | leaf 1: 15/15...10/16 | | ↓/↑ | | |
| | | 2: 15/16...11/14 | | ↓/↑ | | |
| | | 4: 14/12...14/12 | | 0/0 | | |
| | | 6: 13/12...14/10 | | ↑/↑ | | |
| | | 8: 14/11...11/12 | | ↓/↑ | | |
| | | 9: 14/12...10/10 | | ↓/↑ | | |
| Vicia faba L., etiolated plants (dark room) | lowest and uppermost leaf, 6 to 15 d after sowing | lowest leaf: −/18...−/58...−/52 | | −/↑,↓ | | Lurie 1977 |
| | | uppermost leaf: −/20...−/58...−/54 | | −/↑,↓ | | |

epidermes during leaf ontogeny (*Phaseolus* – Šesták *et al.* 1978a; *Sinapis* — Wild and Wolf 1980). Irradiance was more effective in young than senescing leaves (*Sinapis* — Wild and Wolf 1980). Similarly, when transferring plants from high to low irradiance or *vice versa*, the smaller the leaf at the time of transfer the greater the effect of that transfer on stomata density (*Lycopersicon* — Gay and Hurd 1975).

Addition of sodium chloride into an aerated Hoagland solution in which plants of *Phaseolus* were cultivated resulted in reduced cell extension, whereas the number of cells was not or only slightly affected; this was concluded from a three to four times increase in stomata density and only small effects of the treatment on the total number of stomata per leaf during leaf ontogeny (Brouwer 1963).

Leaf insertion: Stomata density on leaves of different insertion levels was studied as early as more than 100 years ago (Weiss 1865); therefore an enormous quantity of values is available (for recent reviews see Tichá 1966, 1970a, 1982, Napp-Zinn 1973, 1974).

There is a regular increase in the number of stomata per unit area from the basal to the apical leaves on both leaf surfaces (if stomata are present) (*e.g.* 50 plant species — Zalenskiï 1904; 11 species — Salisbury 1927; *Citrus* — Reed and Hirano 1931; *Hordeum, Lycopersicum, Linum, Avena, Phaseolus* — Farkas and Rajháthy 1955; *Nicotiana* — Glater *et al.* 1962; *Hordeum* — Pazourek 1966, 1969; *Phragmites* — Pazourek 1973b; *Beta* — Tretyak and Okanenko 1966; *Plectranthus* — Tichá 1970a, 1982; *Lolium* — Sant 1969; *Gossypium*

27

— Nagarajah 1975b, Ackerson 1980; and Table 1.2 and Fig. 1.4, *bottom left*).

A similar trend, *i.e.* an increase in stomata density towards the apex of the crown, was observed in trees (*e.g. Abies* — Nestsyarovich *et. al.* 1963; *Pinus* — Żelawski and Gowin 1967a, b; *Acer* — Eliáš and Huzulák 1975, Huzulák and Eliáš 1975; *Carpinus* — Huzulák and Eliáš 1975; *Fagus* and *Quercus* — Aussenac and Ducrey 1977). Stomata density on conifer needles of different ages (current year and 1, 2, 3 years old) usually increased on both needle surfaces with increasing age (*e.g. Abies* — Nestsyarovich *et al.* 1963; *Picea* — Watts *et al.* 1976).

When all leaves on a plant including immature ones are examined, or in some plant species, an increase followed by a decrease in stomata density in successive leaves on the plant is often found (*e.g. Petunia, Chrysanthemum, Antirrhinum, Medicago, Imperata* — Migahid and Abu Raya 1952a; *Lolium* — Sant 1969; *Eucalyptus* — Cameron 1970a; *Iris* — Pazourek 1970; *Zea* — Heichel 1971a, b; *Capsicum* — Schoch 1972a; *Sorghum* — Liang *et al.* 1975).

In addition to the continuous increase in stomata density from lower to upper leaves, a somewhat higher stomata density for the lowest leaves was sometimes found (*e.g. Phaseolus* — Tschakalova 1976; *Triticum* — Tschakalova and Hoffmann 1976). In plants where leaves are arranged in whorls a great variability in the insertion gradients in stomata density was found (*e.g. Antirrhinum* — Harte and Hansen 1971); leaves of whorl 1 of *Asperula odorata* had a higher stomata density than leaves of whorls 2 and 3 (Eliáš and Kozinka 1976). Juvenile leaves of *Hedera helix* had a lower stomata density than the adult ones (Bauer and Bauer 1980).

When soybean leaves at each position are sampled after they have reached full expansion (from June to September) an increase in stomata density towards leaves of middle insertion followed by a decrease towards the uppermost leaves is found (Lugg and Sinclair 1979).

The insertion gradients in stomata density usually have the same trend on abaxial and adaxial epidermes, however, Kisser (1927) reports opposite trends in wheat and Rübel (1920) in sunflower; differences for both leaf surfaces were also found in *Bromus* and *Epilobium* (Salisbury 1927), tobacco (Turner and Begg 1973), *Phaseolus vulgaris* (Tschakalova 1976), wheat (Tschakalova and Hoffmann 1976), and barley (Yoshida 1978). The mean ratio of adaxial/abaxial stomata density for field-grown soybeans is 0.50 to 0.55; this ratio is higher for leaves at the middle insertion level than for leaves in upper or lower positions (Lugg and Sinclair 1979).

---

→

**Fig. 1.4.** Comparison of ontogenetic changes in abaxial stomata density in seven successive leaves (*top*) and insertion gradients in abaxial stomata density and absolute stomata number per abaxial leaf surface during plant development indicated in days (*bottom*). Leaves numbered from the oldest one. Plants of *Plectranthus fructicosus* L'Hérit grown in warmer and drier [day/night 27/22 °C, and 50/80 % relative humidity (RH) — *top left* and ◆], and in colder and more humid [20/15 °C and 80/98 % RH — *top right* and ●] controlled environments. (From Tichá 1970a, 1982.)

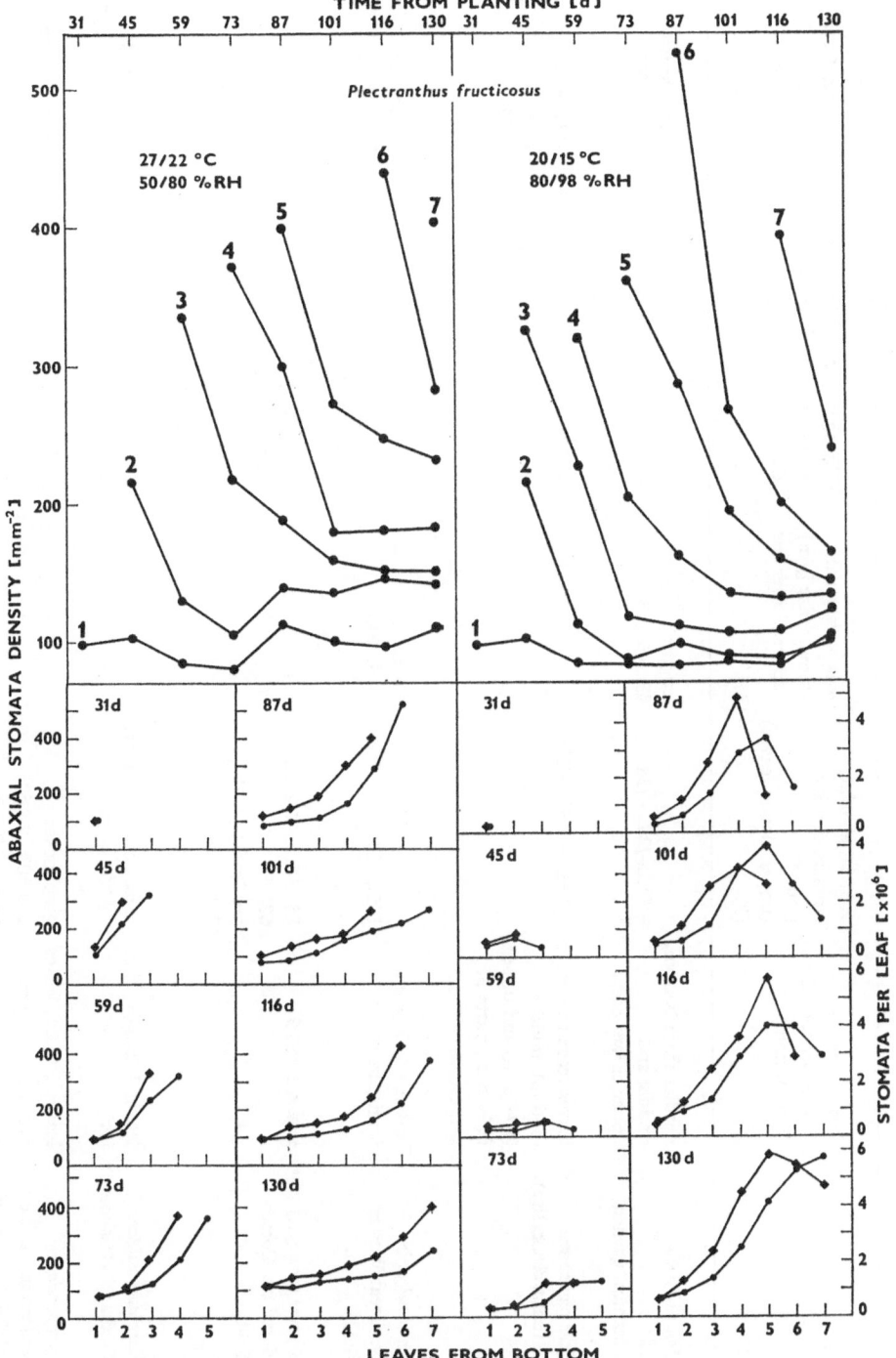

TIME FROM PLANTING [d]

*Plectranthus fructicosus*

27/22 °C
50/80 %RH

20/15 °C
80/98 %RH

ABAXIAL STOMATA DENSITY [mm⁻²]

STOMATA PER LEAF [×10⁶]

LEAVES FROM BOTTOM

**Table 1.2** Stomata density and size as affected by leaf insertion level. For abbreviations and symbols see Table 1.1.

| Plant (and cultivation) (1) | Leaf insertion (leaves numbered from bottom) (2) | Adaxial/abaxial stomata density [mm⁻²] (upper...lower leaves) (3) | Adaxial/abaxial stomata size [μm] (upper...lower leaves) (4) | Changes induced by decreasing insertion level in stomata density (5) | size (6) | Author(s) and year of publication (7) |
|---|---|---|---|---|---|---|
| *Abies sibirica* Ldb. (botanical garden) | needles from bottom, middle and upper crown parts | –/31...–/28...–/24 | length: 43...43...43 | –/↓ | 0 | Nestsyarovich *et al.* 1963 |
| *Acer campestre* L. (natural locality) | leaves from the lower (5–6 m), middle (10–11 m) and upper (15–16 m) parts of tree crown | –/363...–/289...–/206 | | –/↓ | | Eliáš and Huzulák 1975 |
| *Beta vulgaris* cv. Great Wester (field) | leaves 1 to 7 (6 and 7 immature) | 105...113...60 | length: 10.6...24.8 | ↑, ↓ (max leaf 6) | ↑ | Brown and Rosenberg 1970 |
| *Glycine max* (L.) Merril cv. Corsoy (field) | leaves 1 to 15 | 88/185...147/223...62/157 per leaflet [×10⁵]: 1.2/2.6...10.8/16.3 ...0.7/1.4 | | ↑, ↓ / ↑, ↓ (max leaf 7) ↑, ↓ / ↑, ↓, (max leaf 7) | | Lugg and Sinclair 1979 |
| *Hordeum vulgare* L. (field and glasshouse) | leaves 1 to 6 (= flag leaf) | means of five cultivars: 71/76...–/35 | | –/↓ | | Miskin and Rasmusson 1970 |
| *Lycopersicon esculentum* Mill. cv. Minibelle (glasshouse or growth chamber) | leaves 1 to 24; leaves 4, 5, 6, 7 (fully expanded) | glasshouse: 140/500...20/60 growth chamber: 100 W m⁻² (400–700 nm): | | ↓ / ↓ | | Gay and Hurd 1975 |

| | | | | | |
|---|---|---|---|---|---|
| *Medicago sativa* L. (garden) | leaves 1 to 7 | 32/108...30/111..34/95 20 W m⁻² (400–700 nm): 1/82...2/95...1/78 | | 0/↑,↓ | Migahid and Abu Raya 1952 a |
| *Panicum maximum* var. *trichoglume* (growth chamber) | leaves 5, 7, 9, 11, 13, 15 and flag leaf | 255/216...264/216 ...162/116 | | 0/↑,↓ | Wilson 1977 |
| | | 287/–...91/– | ↑,↓/↓ (max. leaf 4) | ↓/– | Frimmel 1977 |
| *Triticum aestivum* L., 15 cultivars (field) | leaves 1 to 6 (= flag leaf) | length: –/54...–/21...–/22 –/53...–/80...–/7? | | ↓,↑ | ↑,↓ |
| *Zea mays* L. cv. Pioneer 395 (glasshouse) | leaves 5 to 10 | 53/80...37/54 | | ↓/↑ | Domes and Bertsch 1969 |

The humidity of the external and internal environment of a leaf affects the formation of insertion gradients in stomata density and sizes (*e.g.* 50 species of herbaceous plants — Zalenskiï 1904; 11 plant species — Salisbury 1927; *Lycopersicum, Linum, Avena, Hordeum* — Farkas and Rajháthy 1955; *Vicia, Petunia, Delphinium, Hyoscyamus* — Migahid and Abu Raya 1952b, c, d, e). The more arid the conditions of plant growth or cultivation the higher is usually the stomata density and the steeper the insertion gradients in stomata density between successive leaves on a plant (*e.g. Trifolium* — Simonis 1947; *Vicia, Petunia, Delphinium, Hyoscyamus* — Migahid and Abu Raya 1952b; *Lycopersicum, Linum, Avena, Hordeum* — Farkas and Rajháthy 1955; *Pisum* — Manning *et al.* 1977). On the other hand, the more humid the environment the smaller are the differences between successive leaves, *e.g.* in water plants no or very small differences were reported (Salisbury 1927). The effect of leaf water content or water supply on stomata density is more pronounced in upper leaves than in the lower ones: the lower leaves of a shoot being more turgid than the upper ones, are less sensitive to changes in water supply (*e.g. Vicia, Petunia, Delphinium, Hyoscyamus* — Migahid and Abu Raya 1952b). The duration of water stress and time of application are, however, also important: insertion gradients in stomata density were not substantially altered by short-time wilting cycles in sunflower (Tumanow 1927) or cotton plants (Ackerson 1980).

The insertion gradients in stomata density were analogous in fully-expanded leaves of tomato plants grown at 100 and 20 W m$^{-2}$ (Gay and Hurd 1975) or in leaves of *Iris* grown at 12, 37, 75 and 100% sunlight but the absolute numbers of stomata in one leaf increased with increasing irradiance (Pazourek 1970). The higher the irradiance the steeper may be the insertion gradient in stomata density (*e.g. Helianthus* — Rübel 1920; *Iris* — Pazourek 1970; *Capsicum* — Schoch 1972a), at very low irradiance the maxima of the gradients may flatten (*Iris* — Pazourek 1970). Successive leaves of tomato plants grown under high irradiance were amphistomatous with about three times as many stomata on their lower as on their upper surface; those grown under low irradiance were hypostomatous and had a slightly lower stomata density on the lower epidermis than leaves under high irradiance (Gay and Hurd 1975). Leaves of *Circea, Scilla, Ficaria* and *Lepidium* plants grown in light and darkness, or sun and shade, but under a similar humidity, showed no significant differences in stomata density (Salisbury 1927).

Growing wheat plants at 10, 18 and 27 °C in growth chambers led to changes in stomata density in successive leaves, but the character of the insertion gradients remained similar with the exception of low stomata density in flag leaves at 27 °C (Frank *et al.* 1973). The combination of the effects of temperature and air humidity (colder and more humid "spring" conditions and warmer and drier "summer" conditions) showed similar results; nevertheless, in the warmer and drier conditions the insertion gradients were steeper and maxima in the absolute stomata number per leaf were reached somewhat earlier (*Plectranthus* — Tichá 1970a, 1982) (Fig. 1.4, *bottom*).

Also other environmental or biological factors (*e.g.* salinity: *Gossypium* — Gausman and Cardenas 1968; ploidy: *Beta* — Tretyak and Okanenko 1966, *Triticum* — Dunstone *et al.* 1973; winter hardiness: *Ilex* — Knecht and Orton 1970) may change the absolute values of stomata density but the character of the insertion gradients is maintained.

Differences on leaf blade: Stomata density varies also within the same leaf blade — for review see Tichá (1982). In a thorough analysis of 700 sites on one abaxial half leaf of *Vinca minor* differences were found in stomata density from 1 to 450 mm$^{-2}$ (Pazourek 1966). Nevertheless, stomatal development in young leaves seems to occur all over the blade and not especially in specific areas (*Vicia* — Lurie 1977).

The stomata density usually declines from the leaf tip to its base (*e.g. Alisma, Stellaria, Adoxa, Paris, Statice, Scabiosa* — Salisbury 1927; *Petunia, Chrysanthemum, Antirrhinum, Medicago* — Migahid and Abu Raya 1952a; *Kalanchoë* — Sharma and Dunn 1968; *Iris* — Pazourek 1970; *Dactylis* — Schäfer and Tirtapradja 1970; *Acer* — Eliáš and Huzulák 1975; *Asperula* — Eliáš and Kozinka 1976; *Hordeum* — Yoshida 1978; *Zea* — Miranda *et al.* 1981a) but opposite gradients are also found (*e.g. Setaria* — Migahid and Abu Raya 1952a; *Nicotiana* — Glater *et al.* 1962, Slavík 1963; *Sorghum* — Liang *et al.* 1975; *Corchorus* — Majid *et al.* 1978). In elongated leaves there is a fall in stomata density in the apical region, so that maximum values of stomata density are found in the middle of the leaf (*e.g. Iris, Carex, Luzula* — Salisbury 1927; *Imperata, Eragrostis* — Migahid and Abu Raya 1952a; *Phragmites* — Pazourek 1973a, b). Negligible or very small differences only were found in leaf blades of *Vitis* (Düring 1980), *Bromus* (Hunt and Christie 1969), *Magnolia* and *Ligustrum* (Kutík 1973).

Differences in stomata density on the edge, in the centre and near the midrib of the leaf blade also exist, but are usually not significant (*e.g. Nicotiana* — Slavík 1963; *Bromus* — Hunt and Christie 1969; *Datura* — Sharma and Dunn 1969; *Vitis* — Düring 1980).

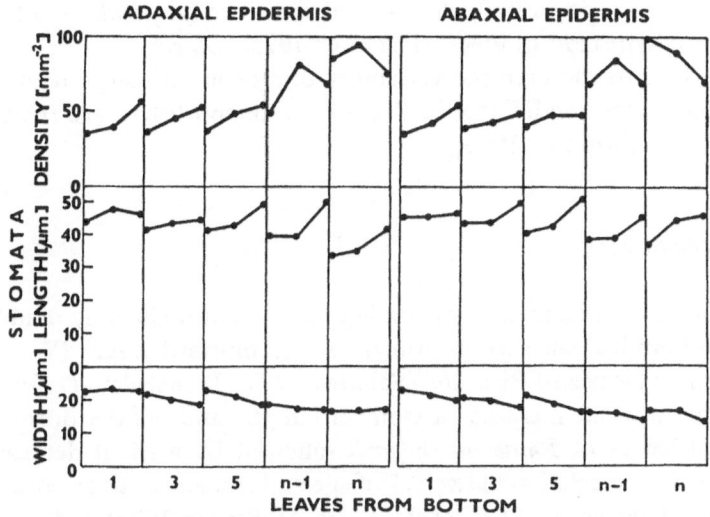

**Fig. 1.5.** Gradients of stomata density, length and width on both leaf surfaces of individual *Hordeum distichon* L. leaves. Each of the gradients indicates values on the base, in the middle and at the top of the leaf blade (from *left* to *right*). Leaf *n* is the uppermost one and leaf *n-1* is the leaf below leaf *n*. (Redrawn after Pazourek 1969.)

The differences in stomata density were analogous on abaxial and adaxial leaf surfaces (*e.g.* seven herbaceous species — Migahid and Abu Raya 1952a; *Nicotiana* — Glater *et al.* 1962, Slavík 1963; *Corchorus* — Majid *et al.* 1978) but also contradictory or different trends are described (*Triticum* — Milthorpe and Penman 1967; *Datura* — Sharma and Dunn 1969; *Phragmites* — Pazourek 1973a; *Sorghum* — McCree and Davis 1974).

The patterns in stomata density on the leaf blade differ for leaves of different insertion levels (*e.g.* Migahid and Abu Raya 1952a; *Hordeum* — Pazourek 1966, 1969; *Iris* — Pazourek 1970; *Phragmites* — Pazourek 1973a; *Plectranthus* — Tichá 1970a, 1982) or for leaves from individual crown layers of trees (*e.g. Carpinus, Acer* — Huzulák and Eliáš 1975).

In barley leaves stomata density increases on both leaf surfaces from the base to the top of the leaf blade in the lower leaves, and in the upper leaves the maximum values are in the middle region or at the base of the leaf (Pazourek 1966, 1969 — Fig. 1.5). On the other hand, in upper, middle and bottom leaves of *Vaccinium macrocarpon* the same course of gradients in stomata density on the abaxial leaf surface is found (Sawyer 1932).

The gradients in stomata density on the same leaf are quantitatively and qualitatively modified by various environmental and cultivation factors, *e.g.* irradiance (*Iris* — Pazourek 1970), water supply (*Gossypium* — Gindel 1969), soil salinity level (*Gossypium* — Gausman and Cardenas 1968), type of cultivation — glasshouse, field (*Vitis* — Düring 1980), closeness of the stand (*Asperula* — Eliáš and Kozinka 1976), ecotype (*Phragmites* — Pazourek 1973a, b), *etc.*

In comparison to the extreme variability of stomata density on the leaf blade the differences among individual leaflets of compound leaves are relatively small (*e.g. Medicago* — Pazourek 1965).

### 1.3.3.2 *Stomata Size*

Leaf ontogeny: The mean stomata length was approximately doubled during ontogeny on both leaf surfaces in primary white mustard leaves (Wild and Wolf 1980) and primary leaves of *Phaseolus* (Solárová 1973). Tschakalova (1976) described, however, only a small increase in stomata lengths and widths during ontogeny of individual leaves of *Phaseolus vulgaris* followed by a small decrease in both stomata sizes. No correlation between leaf age and stomata sizes or pore sizes could be found in wheat leaves (Tschakalova and Hoffmann 1976) and no significant changes in stomata lengths during ontogeny of individual oat leaves (Frommhold 1971, 1972).

The mean area of a single stoma calculated for tobacco leaves from stoma length × breadth × 0.785 (the standard formula for an ellipse) increased during leaf ontogeny (Rawson and Craven 1975).

Abaxial stomata are on an average larger than those on the adaxial surface (*Nicotiana* — Rawson and Craven 1975; *Phaseolus* — Tschakalova 1976), the adaxial stomata reach full size earlier in leaf ontogeny (Rawson and Craven 1975).

Leaf insertion: In stomata length generally a regular decrease in size from the basal to the apical leaves is found (Table 1.2, Fig. 1.5). In some plants only very small variations in stomata length with leaf insertion level are reported (*e.g.* deciduous trees — Nestsyarovich *et al.* 1963; *Avena* — Frommhold 1971). Juvenile leaves of *Hedera helix* had longer guard cells than the adult ones (Bauer and Bauer 1980).

Differences on leaf blade: Changes in stomata length on the leaf blade are usually the opposite to changes in stomata density, *i.e.* the more stomata per unit leaf area the smaller they are (*e.g. Nicotiana* — Slavík 1963; *Hordeum* — Pazourek 1969; *Bromus* — Tan and Dunn 1975; *Asperula* — Eliáš and Kozinka 1976) (Fig. 1.5). As a result of this, the relative index of the actual area of the stomata pores may be similar throughout the blade due to the compensation of the smaller number of stomata by their larger size (*e.g.* Slavík 1963). In some cases the differences in stomata or guard cell length on the same leaf are very small (*Magnolia, Ligustrum* — Kutík 1973; *Corchorus* — Majid *et al.* 1978).

### 1.3.3.3 *Absolute Number of Stomata per Leaf*

The absolute number of stomata per leaf (density × leaf area) on both leaf epidermes increases only in the first days after leaf unfolding and stabilizes then to a more or less constant value (*Phaseolus* — Brouwer 1963; *Lycopersicon* — Gay and Hurd 1975). In tomato 90% of the stomata on the upper epidermis were initiated by the time the leaf had expanded to 35% of its final area, whilst this was not achieved on the lower epidermis until the leaf reached 50% of its final area suggesting that initiation started earlier on the adaxial epidermis (Gay and Hurd 1975).

There may be a limit to the number of stomata present on a leaf under given conditions, and the presence of existing stomata may inhibit the production of further stomata: Stoma initials inhibit adjacent cells from becoming stoma initials (*Bougainvillea, Sambucus, Impatiens, Vriesea* — Bünning and Sagromsky 1948; *Sedum* — Sagromsky 1949, Korn 1972; *Pelargonium* — Korn 1972; *Lycopersicon* — Gay and Hurd 1975).

Absolute stomata number per leaf in leaves that had finished stomata differentiation was assumed to be constant (*Nicotiana, Solanum, Whitania* — Gupta 1961). However, regular changes in insertion gradients in stomata number per leaf, *i.e.* first an increase from the bottom towards leaves of middle insertion level followed by an decrease towards the top of the plant, are mostly reported (*e.g. Citrus* — Reed and Hirano 1931; *Iris* — Pazourek 1970; *Plectranthus* — Tichá 1970a, 1982; *Hordeum* — Yoshida 1978; *Glycine* — Lugg and Sinclair 1979) (Fig 1.4, *bottom right*).

Water stress increases stomata density but reduces stomata sizes and area so that the area of the stomata apparatus per unit leaf area or the stomata number per leaf may remain unchanged (*e.g. Helianthus* — Rawson *et al.* 1980).

### 1.3.3.4 *Stomatal Index*

As the stomata density on plant leaves is usually variable and influenced by environmental factors (see Section 1.3.3.1), the correlation between the number of stomata per unit area (S) and the number of epidermal cells (E) in the same area unit is often used. This stomatal index $SI = 100 \, S/(E + S)$ (Salisbury 1927) shows that the amount of stomata formed in the epidermis is not larger for sun than shade leaves. Similarly, the increment in stomata density in plants grown on dry rather than wet soil, or in small rather than large leaves is due chiefly to differences in the spacing of stomata and not to differences in the amounts of stomata developed. This holds also for the variations in stomata density in different parts of the same leaf. On the other hand, high humidity tends to reduce the proportion of stomata formed, and aquatic plants have a low SI. Variations in SI are mainly due to internal factors (Salisbury 1927).

## 1.4    LEAF MESOPHYLL

### 1.4.1    Leaf Anatomy of $C_3$, $C_4$, and CAM Plants

The main types of photosynthetic $CO_2$ fixation [$C_3$ plants, $C_4$ plants, Crassulacean Acid Metabolism (CAM) plants] are connected with a characteristic leaf anatomy (for their photosynthetic performance see Chapters 6 and 7, Table 7.1).

Leaf mesophyll of $C_3$ p l a n t s is usually divided into layers of palisade parenchyma and spongy parenchyma. The cells of the palisade parenchyma are elongated and regularly oriented with their long axes at right angles to the leaf surface. The spongy parenchyma appears to be less regular and has conspicuous intercellular spaces (Fig. 1.6). Palisade parenchyma is presumed to be the major site of $CO_2$ fixation for its large surface area (1.6 to 3.5 times that of spongy cells: succulent, mesomorphic and xeromorphic species — Turrell 1936), many chloroplasts (1.5 to 3 times more than in spongy parenchyma: *Spinacia* — Possingham and Saurer 1969), and for the proximity to the upper surface of the leaf (more radiant energy). The spongy parenchyma is thought to act as an intermediate conducting tissue, for the movement of photosynthates to the veins, as well as a photosynthetic tissue (*Solanum, Nicotiana, Tussilago* — Mokronosov *et al.* 1973a; *Vicia* — Outlaw and Fisher 1975a, b, Outlaw *et al.* 1975; *Acer, Betula* — Malkina 1976a). Palisade parenchyma cells contain more photosynthetic pigments (*Acer* – Malkina 1976a; *Betula, Hedera, Sophora, Nuphar, Nymphaea, Zea* — Maróti 1976) and they are enriched in Photosystem (PS) 1, spongy parenchyma cells are enriched in PS 2 (*Hedera, Nuphar, Nonea* — Maróti and Gábor 1976). The kinetics of photosynthetic carbon metabolism is quantitatively similar in the two tissues, quantitative differences in relative photosynthetic rates are according to Outlaw and Fisher (1975b) and Outlaw *et al.* (1976) due largely to an irradiance gradient through the leaf (*cf.* Chapter 4).

Most $C_4$ p l a n t s possess a wreath-like or "Kranz" leaf anatomy (for an exception see, *e.g.*, *Suaeda* — Shomer-Ilan *et al.* 1975). Vascular bundles are surrounded by concentric layers of two types of photosynthetic tissue: the bundle sheath cells,

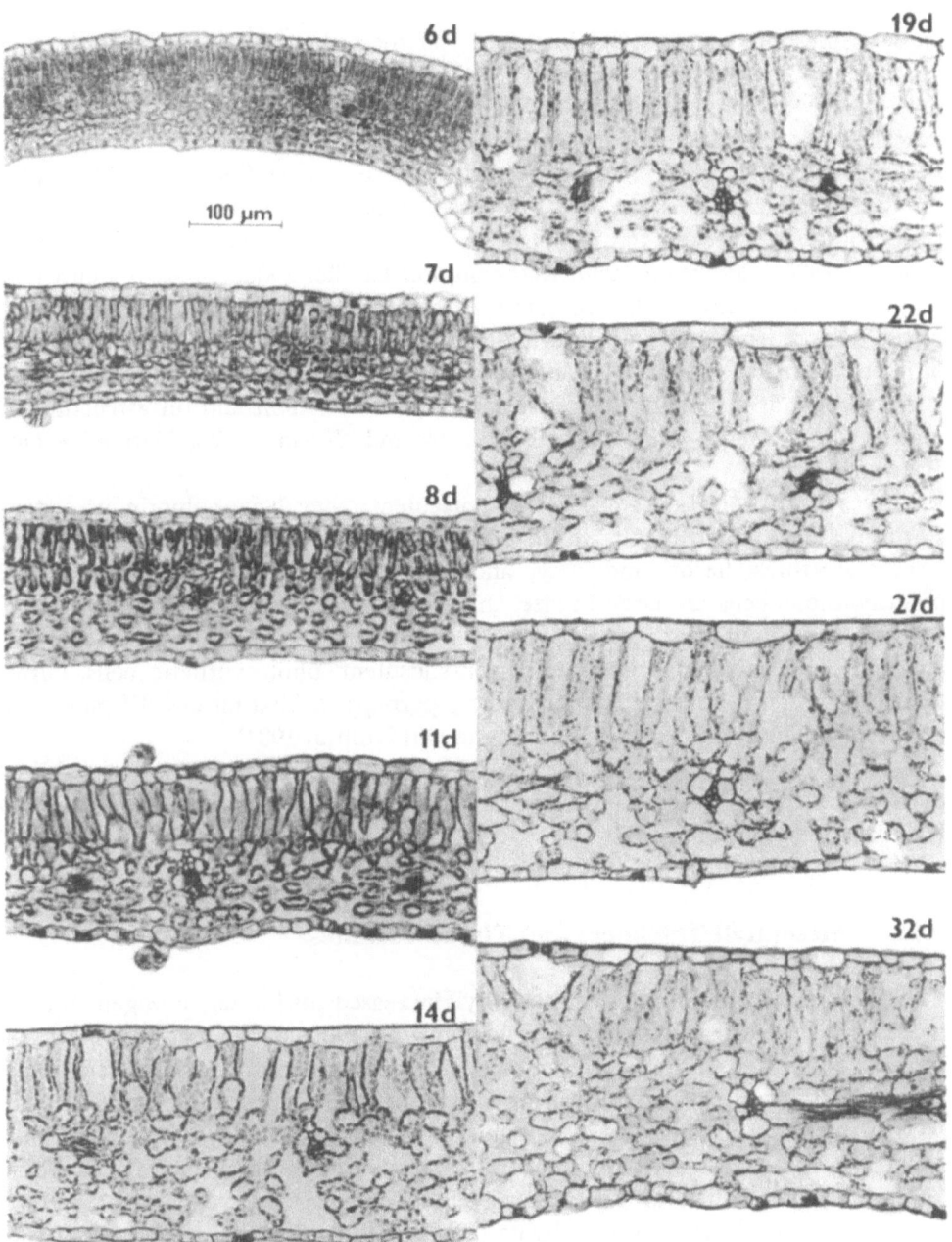

**Fig 1.6.** Leaf structure in primary leaves of *Phaseolus vulgaris* L. cv. Harzgruss during leaf development (from d 6 to d 32 after sowing). The originally tightly packed mesophyll cells enlarge dramatically during leaf development and large intercellular spaces and substomatal cavities are built. Mesophyll cells are full of chloroplasts, in the old leaf they are destroyed. Leaf cross sections (8 μm), fixed in FAA, stained with safranine and Ehrlich's haematoxyline. (Photo J. Pazourek.)

and the mesophyll cells. (Originally, as leaf mesophyll all leaf tissues between both the upper and lower epidermes were described.) Chloroplast dimorphism may appear: in some groups of $C_4$ plants chloroplasts of bundle sheath cells are granaless but contain starch, whereas mesophyll cells have grana. The distances between vascular bundles and between the leaf substomatal cavities and vascular bundles are relatively short, and the presence of a less extensive leaf air-space system implies a greater tissue density per unit leaf volume than in $C_3$ plants (Laetsch 1974, Crookston 1980, Rathnam and Chollet 1980). Stomata density in $C_4$ grasses is 1.5 times higher than in $C_3$ grasses (Apel 1979). Three groups of $C_4$ plants distinguished by their differing $C_4$ acid decarboxylating systems show also differences in the intracellular location of chloroplasts (centrifugal, centripetal) and mitochondria in bundle sheath cells, and the content and ultrastructure of mitochondria (Hatch et al. 1975, Hattersley and Watson 1976, Hattersley and Browning 1981).

Leaves of CAM plants are usually succulent: they have voluminous water-storing tissues, which results in an increase in volume relative to surface area. Leaves are thick, fleshy, and juicy, and tend to have a spherical shape. CAM photosynthetic cells are large in size, thin-walled with a narrow peripheral band of cytosol and huge water-storing vacuoles. The number of chloroplasts per cell appears small in comparison with non-succulent photosynthetic cells. CAM tissues have large intercellular air spaces, perhaps facilitating gas diffusion (for reviews see Kluge and Ting 1978, Osmond and Holtum 1981).

Leaf anatomy of $C_3$ — $C_4$ intermediate species is intermediate, resembling more or less the $C_3$ or the $C_4$ leaf anatomy (for reviews see Apel and Peisker 1979, Rathnam and Chollet 1980).

## 1.4.2 Mesophyll Thickness and Tissue Volumes

The thickness of the assimilating mesophyll increased during leaf ontogeny nearly twice (primary leaves of *Sinapis* — Wild and Wolf 1980, and *Phaseolus vulgaris* — Tichá, unpublished; Fig. 1.7a). A tenfold increase in the incoming radiation doubles the thickness of the mesophyll during leaf ontogeny (Wild and Wolf 1980). The ratio of thickness of palisade parenchyma to thickness of spongy parenchyma was defined as the "mesophyll quotient"; this index increased with leaf ontogeny (Fig. 1.7 a) and insertion level as the thickness of palisade parenchyma enlarged more than that of spongy parenchyma (Neese 1917). Thus the thickness of the palisade layer increased more than twofold in upper leaves (Pereira and Kozlowski 1976) and the spongy parenchyma layer in upper leaves was about one third or more thinner than in lower leaves (*Malus* — Cowart 1936, *Eucalyptus* — Pereira and Kozlowski 1976).

Recently, stereological methods for measuring internal leaf structure, mainly tissue volumes, surface areas per unit volume, *etc.* were developed (Parkhurst 1982). Unfortunately, they have not yet been used in ontogenetic studies of plant leaves but only for leaves of different insertions: in the fully expanded first and second

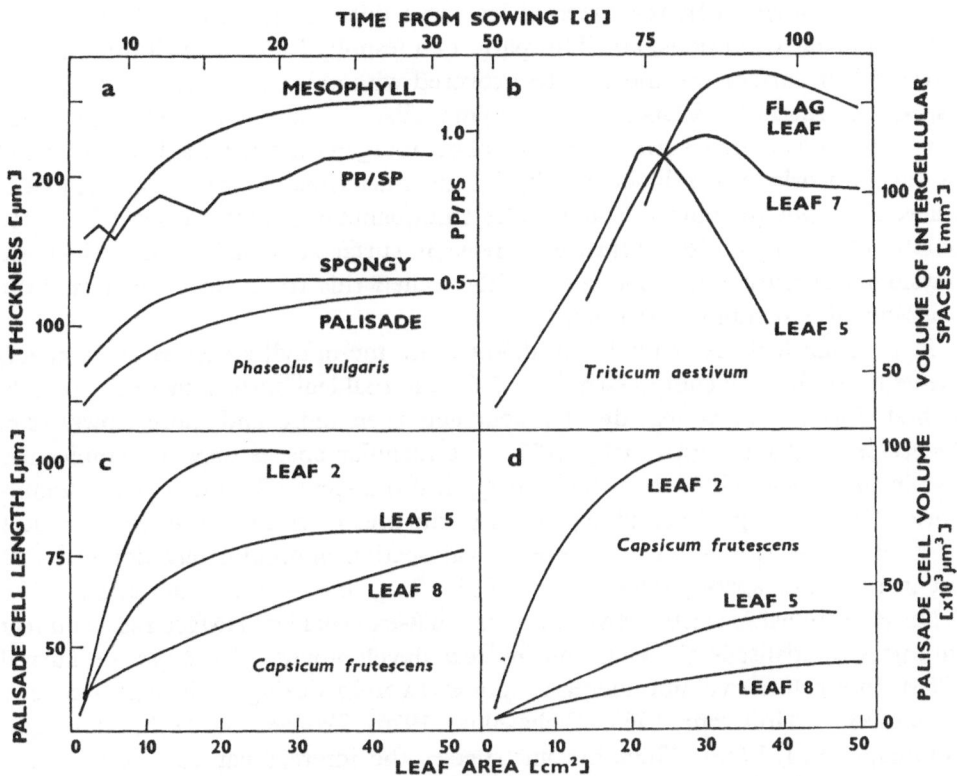

**Fig. 1.7.** During leaf development the thickness of the whole leaf mesophyll increases due to increasing thickness of palisade (PP) and spongy (SP) parenchyma layers. SP stops thickness growth earlier but PP continues to thicken; as the result the "mesophyll quotient" PP/SP slowly increases during leaf development (a: primary leaf of *Phaseolus vulgaris* L. — Tichá, unpublished). Contemporarily the volume of intercellular spaces in the leaf rises (b: *Triticum aestivum* L. — Tschakalova and Hoffmann 1976). Palisade cell length (c) and volume (d) (*Capsicum frutescens* L. — Steer 1971) increase during leaf development (indicated as leaf area increase) differently in leaves of different positions on the plant (leaves 2, 5, 8 from bottom).

barley leaves the relative volume of the photosynthetic tissue increased from 53.4 to 54.8% of leaf volume, the volume of photosynthetic tissues per leaf area unit increased from 59.4 to 73.5 × $10^{-3}$ mm³ mm⁻² (Pazourek and Nátr 1981). These studies are very prospective for physiological leaf anatomical investigations.

### 1.4.3 Intercellular Air Spaces and Internal Leaf Surface

Through intercellular air spaces, $CO_2$ is transported in the gaseous phase from the stomata to the mesophyll cell surface. The internal leaf surface is usually 10 to 40 times larger than the external leaf area depending on the plant species, environmental conditions, *etc*. The parts of mesophyll cells contributing to the internal leaf surface are more or less covered with chloroplasts (80% in palisade parenchyma and 20% in spongy parenchyma cells, *cf*. Esau 1977). The significance of internal and external surface exposure of plant organs and cells has been discussed by several workers in relation to physiological activities. Surfaces are the places where the input and output of molecules and quanta of radiation essential also for photosynthesis take place. Thus, an increase in surface exposure tends to increase physiological activities. Haberlandt (1924) called this trend on a cell level "the principle of maximum exposure".

In a young leaf just after leaf unfolding the mesophyll structure is compact, the cells are densely compressed (Fig. 1.6). Internal leaf surface increases rapidly at first during further leaf development and then more and more slowly (*e.g. Phaseolus* — Tichá and Čatský 1977). Intercellular spaces expand significantly also in the mesophyll of sugar beet during leaf ontogeny (Kursanov and Paramonova 1976). During a fivefold expansion in the area of cotton leaves the numbers of intercellular spaces on a transverse leaf section approximately doubled and the intercellular space increased threefold (*Gossypium* — Gausman *et al*. 1970). A linear relationship exists between the internal-external leaf surface ratio and leaf thickness, or palisade thickness during leaf development (*Medicago* — Turrell 1965). Intercellular volume increases almost twofold during early leaf ontogeny (*Phaseolus* — Hoffmann 1968, Tschakalova 1976; *Triticum* — Tschakalova and Hoffmann 1974, 1976). Then, in some species the intercellular volume stabilizes for a long period or decreases slightly (*Phaseolus* — Tschakalova and Hoffmann 1974, Tschakalova 1976; *Triticum* — Tschakalova and Hoffmann 1976) (Fig. 1.7b).

With increasing leaf insertion level the volume of intercellular space usually decreases (*e.g. Malus* — Cowart 1936; *Nicotiana* — Glater *et al*. 1962; *Nerium, Vinca* — Turrell 1965; *Hedera* — Klee and Steubing 1967); in grasses increases (*Triticum* — Tschakalova and Hoffmann 1976). With leaf insertion the ratio leaf volume/leaf area decreases in birch leaves which is most likely related to the decrease in leaf thickness (Öquist *et al*. 1982). The cell surface to cell volume ratio increases in higher leaves (*Brassica* — Sasahara and Tsunoda 1971). The tips of leaves contain larger intercellular air spaces than the middle or base (*Nicotiana* — Glater *et al*. 1962). In a sunny locality or at high irradiance, leaves have a small intercellular air space system but the insertion gradient is maintained (*Hedera* — Klee and Steubing 1967).

## 1.4.4 Mesophyll Cells

The development of leaf mesophyll was recently reviewed by Gamaleï and Kulikov (1978). Cell division of mesophyll cells is greatest in young leaves and virtually ceases at further leaf development; cell size increases in proportion to leaf length (*Medicago* — Koehler 1973; *Tropaeolum* — Rumi and Carpinetti 1981). Division ceases earlier in spongy parenchyma, whereas the palisade cells continue to divide. Spongy parenchyma cells enlarge only (*Nicotiana* — Glater *et al.* 1962).

During leaf ontogeny the number of palisade cells and spongy cells per unit leaf area declines twice to three times (*e.g. Arabidopsis, Phaseolus* — Hoffmann 1968; *Mangifera* — Taylor 1970; *Solanum* — Mokronosov *et al.* 1973a; *Phaseolus* — Verbelen and de Greef 1979), but the number of palisade cells per leaf increases somewhat (*Arabidopsis, Phaseolus* — Hoffmann 1968; *Mangifera* — Taylor 1970; *Phaseolus* — Verbelen and de Greef 1979; *Tropaeolum* — Rumi and Carpinetti 1981). Palisade cell number per tobacco leaf increases at an exponential rate until the leaf is about one-tenth of its final size; the rate of increase then decreases sharply and becomes zero when the leaf is about half-grown (Clough and Milthorpe 1975). The length of palisade cells increases progressively with leaf ontogeny (*Capsicum* — Steer 1971; *Cucumis* — Chugunova *et al.* 1980; Fig. 1.7c); their surface increases with a small decrease in old leaves, and their volume expands twice to ten times (*Arabidopsis, Phaseolus* — Hoffmann 1968; *Capsicum* — Steer 1971; *Sinapis* — Wild and Wolf 1980) (Fig. 1.7d, Fig. 10.2).

Cell size in the palisade and spongy parenchyma declines also with the leaf order (*e.g. Nicotiana* — Glater *et al.* 1962; *Brassica* — Sasahara and Tsunoda 1971; *Populus* — Isebrands and Larson 1973). Thus palisade cell length decreased from 97 μm in lower leaves to 15 μm in the upper leaf (*Lycopersicon* — Briant 1974). Also the diameter of the protuberances of the arm-palisade cells in wheat and rice decreased from the lower to the upper leaves, while the number of protuberances increased (Chonan 1965, 1967). Spongy parenchyma cells became more undulated (6 to 8 armed) in upper leaves (*Nicotiana* — Glater *et al.* 1962). The number of cells per unit leaf area increased from the lower to the upper leaves (*Triticum, Oryza* — Chonan 1965, 1967; *Lolium* — Sant 1969). The cells in the upper leaves had a smaller surface and volume than lower leaves, the ratio of the cell surface to the cell volume increased from the lower to the upper leaves (*Triticum, Oryza* — Chonan 1965, 1967).

The expansion of palisade parenchyma cells and spongy parenchyma cells had already stopped in the leaf tip, whereas base and lateral margins continued to expand (*Populus* — Isebrands and Larson 1973). In the lower wheat leaves, the protuberance diameter of arm-palisade cells decreased from the base to the tip of leaf, while such a relation was not seen in the upper leaves (*Triticum* — Chonan 1965).

Radiant energy seemed to be very important for differentiation of palisade and spongy parenchymas: in shade no palisades or only one layer of palisade cells were formed, and large differences in ontogeny and insertion gradients in sun and shade leaves were described (*e.g. Sinapis* — Wild and Wolf 1980; *Tropaeolum* — Rumi and Carpinetti 1981; for review see Boardman 1977c, Lichtenthaler *et al.* 1981b). Irradiance during leaf ontogeny modified the number of palisade cells per unit leaf area and the number of palisade cells per leaf (*Phaseolus* — Verbelen and de Greef 1979); at high irradiance the volume of palisade parenchyma cells was four times as large as that at low irradiance (*Sinapis* — Wild and Wolf 1980).

Furthermore, development of palisade cells is modified by water supply: cell expansion is rapidly reduced by small water deficits and ceases early, but cell division is much less sensitive, it continues, although at a reduced rate, even after cell expansion has ceased (*Nicotiana* — Clough and Milthorpe 1975).

The thickness of cell walls of mesophyll cells increased slightly during leaf ontogeny: those from young leaves were 1.5 to 2 times thinner than those of mature leaves (*Beta* — Kursanov and Paramonova 1976).

In very young leaves mesophyll cells are filled with cytosol. With proceeding leaf ontogeny the cells vacuolize and cytosol with chloroplasts may be found only in a thin layer near the cell wall. This fact compensates to some extent the larger diffusion pathway of $CO_2$ in liquid phase due to the thickening of cell walls, and the enlargement of cells and chloroplasts during leaf ontogeny. On the other hand, in larger mesophyll cells and larger chloroplasts more parallel diffusion pathways for $CO_2$ can be expected.

## 1.4.5 Chloroplasts

Chloroplasts are special semi-autonomous cell organelles of eukaryotes. The mature chloroplasts are usually discoid or lens shaped, they are surrounded by a double membrane (envelope) and contain thylakoid membranes which are regularly stacked as grana within the matrix (stroma) of the organelle (for details see Chapter 2). In the thylakoids the photochemical reactions of photosynthesis (see Chapter 5) and in the matrix the biochemical reactions of photosynthesis (see Chapter 6) take place. The chloroplast and its envelope are also compartments of the $CO_2$ diffusion pathway in the liquid phase from cytosol to carboxylation centres in the chloroplasts (see Chapter 8). In this section the anatomical parameters of chloroplasts as chloroplast numbers per cell, chloroplast sizes, covering of the internal leaf surface by chloroplasts, *etc.* are treated during leaf ontogeny and on leaves of different order. For recent reviews on chloroplast anatomy see, *e.g.*, Kirk and Tilney-Bassett (1978), Reinert (1980), Schnepf (1980), Bradbeer (1981), Öquist (1981), and Robards (1981).

### 1.4.5.1  *Chloroplast Numbers*

A young leaf contains numerous chloroplasts small in diameter. With increasing leaf age the numbers of chloroplasts per palisade parenchyma cell and spongy parenchyma cell increase substantially (Figs. 1.8 a and 10.2, Table 1.3). The number of plastids in palisade parenchyma cells increases until the leaf growth ceases,

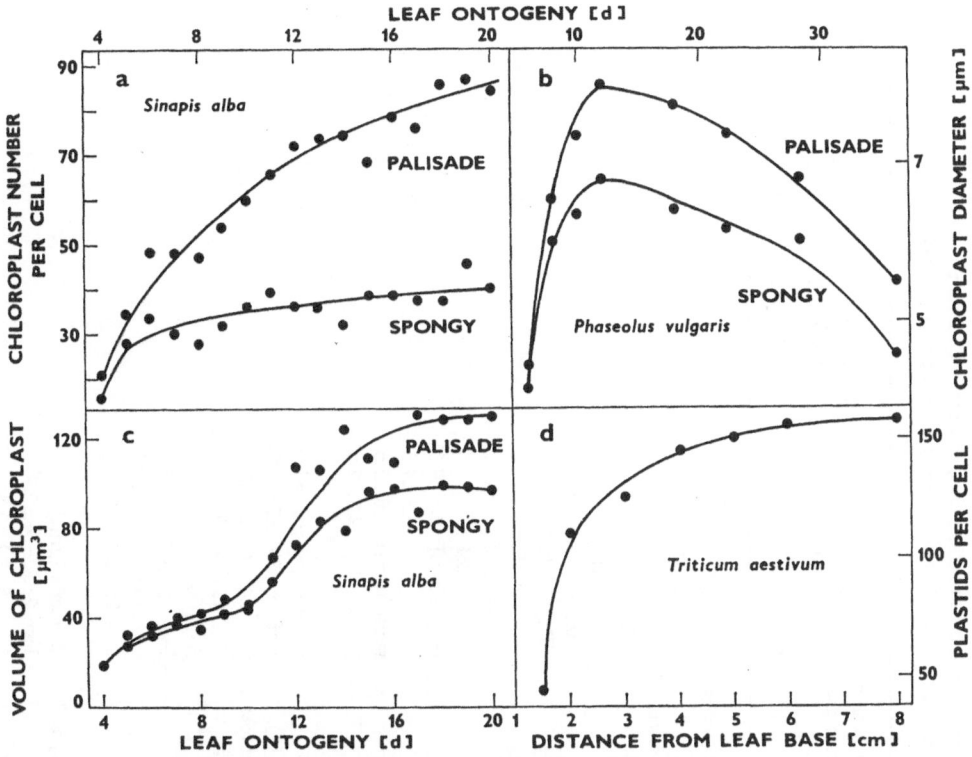

**Fig. 1.8.** Chloroplast number per palisade and spongy parenchyma cell increases during leaf development (a — *Sinapis alba* L., Wild and Wolf 1980). Chloroplast size increase is followed by a decline after a maximum leaf size is reached (b — *Phaseolus vulgaris* L., Naito *et al.* 1979) which is reflected in changes in chloroplast volume (c — the same plant and reference as a). Number of plastids per cell varies substantially along the leaf lamina, especially in the long grass leaves (d — *Triticum aestivum* L., Boffey *et al.* 1979) .

in spongy parenchyma cells when the leaf area of 10 to 15% of the maximum size is reached (*Solanum* — Mokronosov and Bagautdinova 1974). The rate of increase in plastids in palisade cells is highest before the period of maximum leaf expansion; by contrast the largest increase in the total numbers of plastids occurs during the period of the greatest leaf expansion when cell division has stopped and when the plastids have all developed into chloroplasts (*Phaseolus* — Whatley 1980).

**Table 1.3** Chloroplast number and size in mesophyll cells as affected by leaf age and insertion level. Abbreviations and symbols: max — maximum; PP — palisade parenchyma; SP — spongy parenchyma; ↑ increase; ↓ decrease.

| Plant species and cultivation (1) | Age of leaf or leaf insertion (leaves numbered from the oldest one) (2) | Chloroplast parameter (from young to old or from upper to lower leaves) (3) | Changes in chloroplast parameter induced by ageing (4) | Author(s) and year of publication (5) |
|---|---|---|---|---|
| *Acer platanoides* L. (natural locality) | second leaf pair on terminal shoot of 3 years-old seedlings, 0 to 82 d after breaking of buds | number [cell⁻¹]: | | Tsel' niker 1973, 1978 |
| | | PP: 16...29 | ↑ | |
| | | 8...31 | ↑ | |
| | | SP: 10...14 | ↑ | |
| | | 9...13 | ↑ | |
| | | number [×10⁷ leaf⁻¹]: | | |
| | | PP: 17...158 | ↑ | |
| | | 0.6...69 | ↑ | |
| | | SP: 24...62 | ↑ | |
| | | 2...24 | ↑ | |
| | | length [μm]: 2.3...4.3 | ↑ | |
| | | 3.3...4.3 | ↑ | |
| | | width [μm]: 1.8...2.6 | ↑ | |
| | | 2.2...2.6 | ↑ | |
| *Beta vulgaris* L. var. *saccharifera* | ontogeny of leaves 5 (June 11 to August 14), 14 (July 18 to September 16) and 25 (August 14 to September 16) | number per palisade cell: | | Smelyanskaya 1965 |
| | | leaf 5: 89...230...192 | ↑, ↓ | |
| | | 14: 30...58...53 | ↑, ↓ | |
| | | 25: 24...56 | ↑ | |
| | | size [μm]: | | |
| | | leaf 5: 3.7...5.9...5.3 | ↑, ↓ | |
| | | 14: 3.7...5.3...4.7 | ↑, ↓ | |
| | | 25: 3.4...4.9 | ↑ | |
| *Capsicum frutescens* L. cv. California Wonder (growth chamber) | ontogeny of leaves 2, 5, 8 | diameter [μm:] | | Steer 1972 |
| | | leaf 2: 2.2...4...2.9 | ↑, ↓ | |
| | | 5: 2.1...3.3 | ↑ | |
| | | 8: 2.2...5 | ↑ | |

44

| Species (conditions) | Developmental stage | Measurement | Trend | Reference |
|---|---|---|---|---|
| *Helianthus annuus* L. (field) | leaves 8, 12, 16, 20 | surface per leaf area [cm⁻² cm⁻²]: 32.7...21 | ↓ | Shul'gin *et al.* 1977 |
| *Lactuca sativa* L. cv. Amanda (growth room) | leaves 8 to 15 d after sowing | number per leaf area [×10³ mm⁻²]: 12...34...10 | ↑, ↓ (max d 11) | Bourdu *et al.* 1975 |
| *Nicotiana rustica* L. | leaf 7, 0 to 80 d | number [cell⁻¹]: PP: 40...410 / SP: 30...250; number per leaf area [×10⁶ cm⁻²]: 10...21 | ↑ ↑ ↑ | Mokronosov 1981 |
| | leaf 5, 3 to 65 d (leaf area 4 to 605 cm²) | number [cell⁻¹]: PP: 34...355 / SP: 26...214; number per leaf area [×10⁶ cm⁻²]: PP: 1.1...10.8...9.2 / SP: 9.5...10.4...7.0; volume [μm³]: PP: 19...62 / SP: 27...69 | ↑ / ↑ / ↑,↓ / ↑,↓ / ↑ / ↑ | Nekrasova 1978 |
| *Phaseolus vulgaris* L. cv. Canadian Wonder (pots, growth chamber) | primary leaf, 1 to 14 d after germination | number [cell⁻¹]: PP: 10...44; number per PP layer [×10⁶]: 5...650; long axis [μm]: 1.3...3.3; percentage cover: 3.5...14.3...9.9 | ↑ / ↑ / ↑ / ↑,↓ | Whatley 1980 |
| *Raphanus sativus* L. cv. Saxa Treib. (nutrient solution, growth chamber) | cotyledons, 1 to 5 d | length [μm]: 5.1...6.0; width [μm]: 2.1...3.8; volume [μm³]: 29...72 | ↑ / ↑ / ↑ | Meier and Lichtenthaler 1981 |

**Table 1.3** (continued)

| Plant species and cultivation (1) | Age of leaf or leaf insertion (leaves numbered from the oldest one) (2) | Chloroplast parameter (from young to old or from upper to lower leaves) (3) | Changes in chloroplast parameter induced by ageing (4) | Author(s) and year of publication (5) |
|---|---|---|---|---|
| *Solanum tuberosum* L. (field) | leaf 5, 0 to 50 d | number [cell⁻¹]: PP: 40...215...150 | ↑, ↓ | Mokronosov and Bagautdinova 1974 |
| | | SP: 50...85...65 | ↑, ↓ | |
| | leaves 5 and 13, 3 to 35 d | leaf 5: | | |
| | | PP: 90...350...280 | ↑, ↓ | |
| | | SP: 80...65 | ↓ | |
| | | leaf 13: | | |
| | | PP: 60...140...130 | ↑, ↓ | |
| | | SP: 40...60...40 | ↑, ↓ | |
| | | number per mm² [×10³]: | | |
| | | leaf 5: 580...400 | ↓ | |
| | | leaf 13: 600...380 | ↓ | |
| | | length [μm]: | | |
| | | leaf 5: | | |
| | | PP: 3.44...4.13...3.82 | ↑, ↓ | |
| | | SP: 2.66...3.50 | ↑ | |
| *Spinacea oleracea* L. (nutrient solutions, growth chamber) | leaves 4 and 5, 10 to 24 d | number per palisade cell: leaf 4: 60...330 | ↑ | Possingham and Saurer 1969 |
| | | leaf 5: 35...330 | ↑ | |
| | leaves 1 to 9 of a 21-d-old plant | 40...540 | ↑ | |

The number of chloroplasts in palisade parenchyma cells is usually higher than in spongy parenchyma cells (*e.g.* 2.5 to 3 times in *Solanum* — Mokronosov and Bagautdinova 1974, twice in *Sinapis* — Wild and Wolf 1980). The increase in chloroplasts per palisade cell during leaf growth is more rapid than in spongy mesophyll cells. The relative amount of chloroplasts in spongy cells decreased with leaf age from 22 to 16 % of the whole number per leaf (*Solanum* — Mokronosov *et al.* 1973a).

During leaf ontogeny the numbers of chloroplasts per cell in both the palisade and spongy parenchyma cells increases two to ten times (*e.g.* in *Phaseolus* from 10 to 45 — Więckowski 1967a, Whatley 1980; in *Spinacia* from 50 to 500 — Possingham and Saurer 1969, Saurer and Possingham 1970; in *Nicotiana* from 34 to 355 in leaf 5 and from 40 to 410 in leaf 7 — Nekrasova 1978, Mokronosov 1981; in *Solanum* from 42 to 118 in palisade cells and from 37 to 80 in spongy cells — Borzenkova and Nefedova 1981; see also Table 1.3). The number of chloroplasts per leaf area unit and per unit cell volume decreases during leaf ontogeny (*Solanum* — Mokronosov *et al.* 1973a, Mokronosov and Bagautdinova 1974; understory herbs — Goryshina 1980b; *Sinapis* — Wild and Wolf 1980). There were about 1.4 times more chloroplasts per unit cell volume in the palisade cells than in the spongy tissue (*Sinapis* — Wild and Wolf 1980). The number of chloroplasts per transverse leaf section in deciduous and evergreen plants increased in young leaves, then maintained for a long period (2—3 years) a relatively constant value (Gamaleï 1975). The number of chloroplasts per leaf declined with leaf age (in *Solanum* from $45 \times 10^6$ to $10 \times 10^6$ — Borzenkova and Nefedova 1981).

The seasonal dynamics of chloroplast density was due to leaf cell growth and, as a result, to a decrease in cell number per leaf surface unit. The chloroplast amount in the leaves is thus diluted, especially in spring ephemerals (a 3 to 6 times decrease), while in leaves of summer-vegetation plants it is more stable (Goryshina 1980a). In some species with seasonal leaf dimorphism, spring leaves with high chloroplast amounts were replaced by summer leaves with fewer chloroplasts per cell (Goryshina 1980a).

The lowest maize ($C_4$) leaf contained fewer chloroplasts per bundle sheath and mesophyll cells than the fifth leaf; however, bundle sheath cells contained more chloroplasts than mesophyll cells. The ratio number of bundle sheath chloroplasts to number of mesophyll cell chloroplasts was almost identical for leaves 1 and 5 (Crespo *et al.* 1979).

Chloroplasts per unit of cell volume and the total chloroplast surface per leaf area increased from leaves in the lower part of tree crown towards leaves in the middle and upper crown parts (*Acer, Quercus, Tilia, Ulmus* — Goryshina 1980a).

There is a continuous increase in chloroplasts per mesophyll cell from the base to the tip of the leaf (*Spinacia* — Possingham and Smith 1972; *Hordeum* — Robertson and Laetsch 1974; *Triticum* — Boffey *et al.* 1979; see Fig. 1.8 d).

Chloroplast replication is temperature dependent and stimulated by irradiance; cytokinin does not affect chloroplast replication (*Spinacia* — Possingham and Smith 1972). Under low irradiance there was only a small change from 20 to 25 chloroplasts in the palisade cells during leaf ontogeny; palisade cells from expanded high-irradiated leaves had 3.5 × more chloroplasts than cells from low-irradiated leaves (*Sinapis* — Wild and Wolf 1980). Number and size of chloroplasts can be controlled by shading during different phases of leaf development (*Acer* — Tsel'-niker 1973). Chloroplast numbers in mesophyll cells were influenced by pruning (*Vitis* — Golinka 1966), mineral nutrition (*Linum* — Porokhnevich 1972, *Glycine* — Weiland *et al.* 1975), and by environmental conditions in natural localities (*Anemone* — Goryshina *et al.* 1981), etc.

### 1.4.5.2 *Chloroplast Sizes*

Average diameter of chloroplasts increases during leaf ontogeny (from *ca.* 1.6 μm in very young leaves to about 6 μm in mature leaves of *Xanthium* — Holowinsky *et al.* 1965; for further examples see Table 1.3), a change in the shape of the plastids from circular to ellipsoid being concomitant with their growth. The rate of increase in the size of the chloroplast is at first low followed by a very rapid (70%) increase in the chloroplast diameter. This increase in the period of cellular differentiation coincides with the maximum rate of increase in leaf area and cell enlargement (*Xanthium* — Maksymowych 1973). A definite relationship of chloroplast size to leaf area was established (*Xanthium* — Holowinsky *et al.* 1965). In darkness the mean chloroplast diameter increased up to 10 d after which it remained steady (*Phaseolus* — Bradbeer *et al.* 1974).

During leaf ontogeny chloroplasts in cells of both palisade and spongy parenchymas enlarged, *e.g.* from 5.74 to 6.54 μm and 6.11 to 6.74 μm, respectively; for the small diameter the values were 3.07 to 3.12 and 3.09 to 3.26 μm, respectively (*Arabidopsis* during 22 d of leaf development — Hoffmann 1968). After reaching a maximum size, chloroplast diameter in palisade and spongy cells declines again (*Phaseolus* — Naito *et al.* 1979; *Capsicum* — Steer 1972) (Fig. 1.8 b). Similarly, the surface and volume of chloroplasts increased (*Arabidopsis* — Hoffmann 1968; *Sinapis* — Wild and Wolf 1980) (Figs. 1.8 c and 10.2). In potato the chloroplast volume fell a little with leaf senescence (Borzenkova and Nefedova 1981), in wheat, the maximum chloroplast volume was found at 80% of maximum leaf area (Tschakalova and Hoffmann 1976).

Total chloroplast surface per unit leaf area increases from lower to upper leaves, but the ratio between the free mesophyll cell surface and chloroplast surface remains rather constant (*Helianthus* — Shul'gin *et al.* 1977). Chloroplast size decreases from leaves in the lower part of the tree crown to those in the middle and upper crown parts (Goryshina 1980a).

Plastid size and volume increase from the base to the tip of the leaf (*Zea* — Tageeva *et al.* 1969; *Hordeum* — Robertson and Laetsch 1974).

Increase in chloroplast diameter in mesophyll cells was very strongly modified by temperature (*Xanthium* — Holowinsky *et al.* 1965) and irradiance (*Xanthium* — Holowinsky *et al.* 1965; *Sinapis* — Wild and Wolf 1980).

The changes during leaf ontogeny in covering the leaf internal surface by chloroplasts are only little known. In a young *Phaseolus* leaf, when cell expansion and plastid division are nearly completed and the cell vacuole is established, the total plastid face area occupies about 10% of the palisade parenchyma cell surface. During 14 d of early leaf development the relation between total plastid face area and cell surface area in palisade cells changes from 3.5 to 14.3 (at day 5) and then to 9.9% (*Phaseolus* — Whatley 1980). In the lowest maize leaf, 19% of mesophyll and 23% of bundle sheath internal cell area, whereas in the fifth leaf 36 and 48%, respectively, were covered by chloroplasts (Crespo *et al.* 1979).

Chloroplast movement in mesophyll cells which is primarily a response to irradiance, certainly may contribute to the $CO_2$ supply to carboxylation centres, as much as changes in cytosol streaming, viscosity, *etc.* But, unfortunately, nothing is known about the ontogenetic trends of these phenomena.

### 1.4.6 Vascular System

The midrib appears shortly after leaf initiation but lateral veins appear later, first at the tip of the leaf, then progressively towards the base. The total length of the vascular system increases during leaf ontogeny, the relative rate of increase in vascular length declines and the number of mesophyll cells in one layer decreases per unit length of vascular system throughout the growth of the leaf (*e.g. Trifolium* — Denne 1966).

With increasing insertion level an increase in density of the vascular system per leaf area unit in terrestrial plants is referred to by Zalenskii (1904); in aquatic plants, practically no differences could be found (*Potamogeton* — Tichá 1964).

## 1.5 ŽALENSKIĬ LAW

The anatomical characteristics of leaves from different insertion levels on a plant were thoroughly investigated by V. R. Zalenskii (1904) on nearly 50 plant species. According to Zalenskii, the upper leaves on a plant have more smaller stomata per unit leaf area, a more dense leaf nervature, smaller epidermal cells and trichomes, more trichomes per unit leaf area on both leaf surfaces, thicker cell walls and more developed wax layers in both epidermes, smaller mesophyll cells, more typical palisade and less typical spongy parenchyma, smaller intercellular spaces and more developed mechanical tissue than the lower leaves. This means that upper leaves show a more xeromorphic structure than lower leaves: this difference is due to the fact that leaves of individual insertion levels develop under different environmental conditions, from different primordia with different genetic properties, and mainly, leaves at the extremities are less supplied with water. The observations of Zalenskiĭ were supported by many authors (see Tichá 1982). Maximov (1929) formulated this concept into what he called the "Zalenskiĭ law": "The anatomical structure of individual leaves of the same shoot is the function of their distance from the root system".

## 1.6    CONCLUSIONS

The complex of anatomical and morphological structures of the leaf is the place where processes and reactions of photosynthesis take place. The marked changes in leaf anatomy and morphology during leaf ontogeny and with leaf insertion summarized in this chapter may substantially contribute to the respective changes in the course of leaf photosynthesis.

Some of these changes are in favour of a higher photosynthesis, some act in the opposite way. So, increase in leaf size and area during leaf ontogeny or with leaf insertion means a greater area also for interception of radiant energy and photosynthesis, thicker leaves mean more photosynthetic mesophyll cells or larger cells. On the other hand, in thicker leaves diffusion pathways for $CO_2$ are longer, but there might be more parallel diffusion pathways on the larger inner leaf surface. Continuing thickening of leaf epidermis, of cell walls, cuticle and other surface structures contribute less to variations in photosynthesis. What is important for the gas exchange of the leaves is the increase in stomata numbers per leaf and stomata sizes during ontogeny; stomata density during leaf ontogeny decreases due to the growth of leaf in area. The increase in chloroplasts per cell and chloroplast sizes during leaf ontogeny supports photosynthesis as more carboxylation centres, and larger amounts of thylakoids and photosynthetic enzymes may be expected. However, chloroplast numbers per leaf area unit decrease due to leaf growth. Larger chloroplasts also mean longer diffusion pathways for $CO_2$, but continuing vacuolization of photosynthetic cells, and thus replacing the chloroplasts in a more peripheral layer near the cell wall may compensate this to a certain extent.

The values of anatomical parameters are modified by various environmental factors but the ontogenetic or insertion pattern of the changes is usually maintained.

# 2 CHLOROPLAST DEVELOPMENT

*J. Kutík*

## 2.1 CHLOROPLAST STRUCTURE

Photosynthetic apparatus of higher plant leaves takes place (as in all photosynthesizing *Eukaryonta*) in cell organelles, called chloroplasts, first recognized as green globules of plant cells by Leeuwenhoek in the second half on the 17th century.

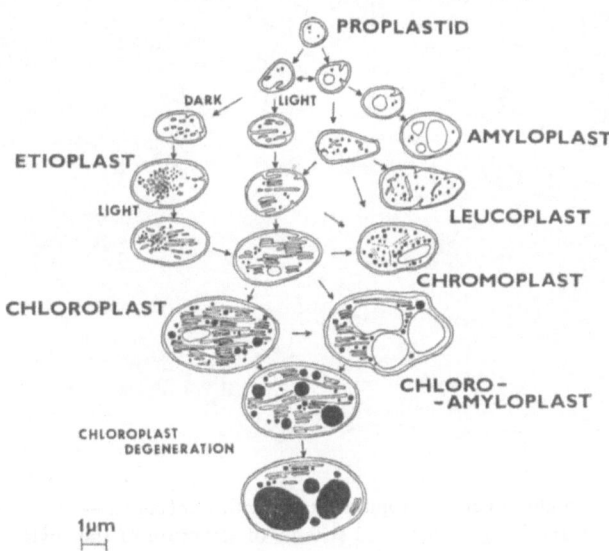

**Fig. 2.1.** Scheme of plastid development and mutual relationships in leaf cells (after Lichtenthaler 1968b).

Chloroplasts are the most important representatives of the more general group of plant cell organelles, plastids, which are double membrane enveloped, self reduplicating organelles bearing various plant pigments. Several plastid types or plastid stages (proplastids, leucoplasts, chloroplasts, chromoplasts) are currently encountered in higher plant cells (see Fig. 2.1).

Concerning the origin of plastids in the cells and mutual relationships of various types of plastids, it seems certain today that (1) plastids are not formed *de novo* but arise only from division, and (2) all different stages (types) of plastids in higher plants are readily interconverted (Schnepf 1980). However, in the specific plant cell at a certain stage in its development there are usually plastids of one type only (Woolhouse and Batt 1976).

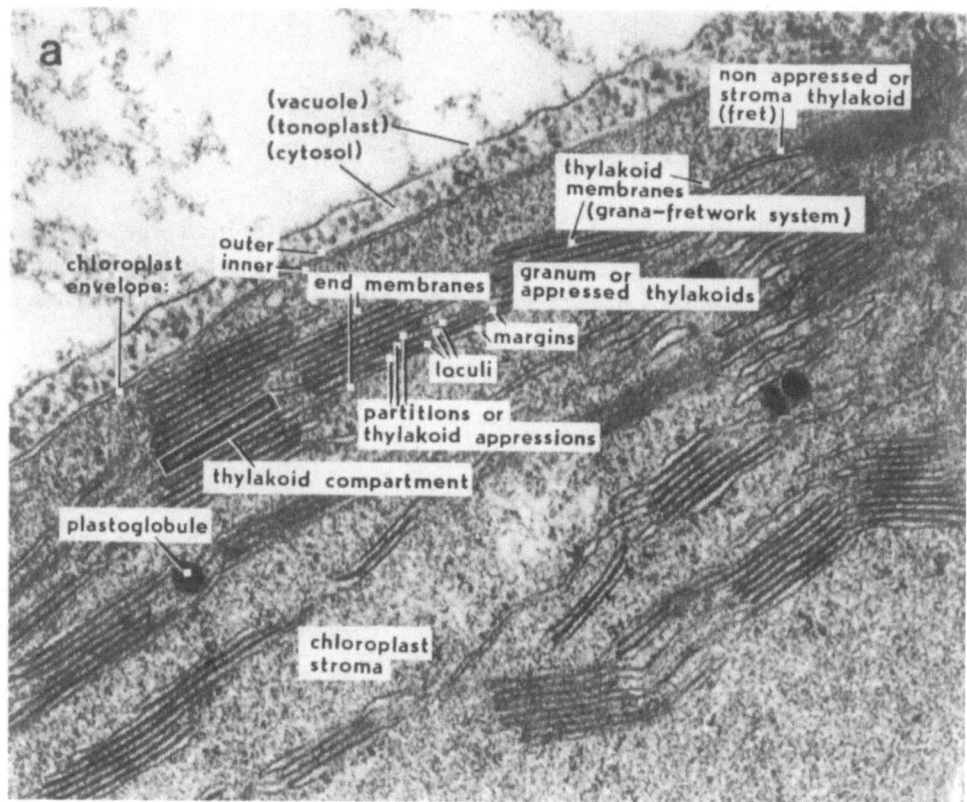

**Fig. 2.2.** Outline of higher plant chloroplast ultrastructure (*above:* a — chloroplast from primary leaf of *Phaseolus vulgaris* — orig. Kutík) and scheme of structure of the thylakoid membrane (*on p. 53:* b — after Staehelin 1981).

The evolutionary origin of plastids has not been fully explained up to now. However, it is very probable that they evolved from some free living unicellular photo-autotrophic *Prokaryonta*, by the way of endosymbiosis (see, *e.g.*, Whatley *et al.* 1979, Gibbs 1981, Schiff and Lyman 1982 for review). The alternative origin of these organelles, by endogenous compartmentation of cells, seems to be less probable. The questions on the origin of eucaryotic cells are being extensively studied at present, especially by the methods of molecular genetics. In this volume, we shall leave the question of the phylogenetic origin of chloroplasts aside.

Plastids have been studied for a very long time (see, *e.g.*, Schimper 1885). However, only the modern methodology of the last thirty years, especially electron microscopy, using the methods of ultrathin sections and freeze-fracturing or freeze-etching (see, *e.g.*, Hall 1978 for review), brought a considerable advance in the understanding of their structure and functions.

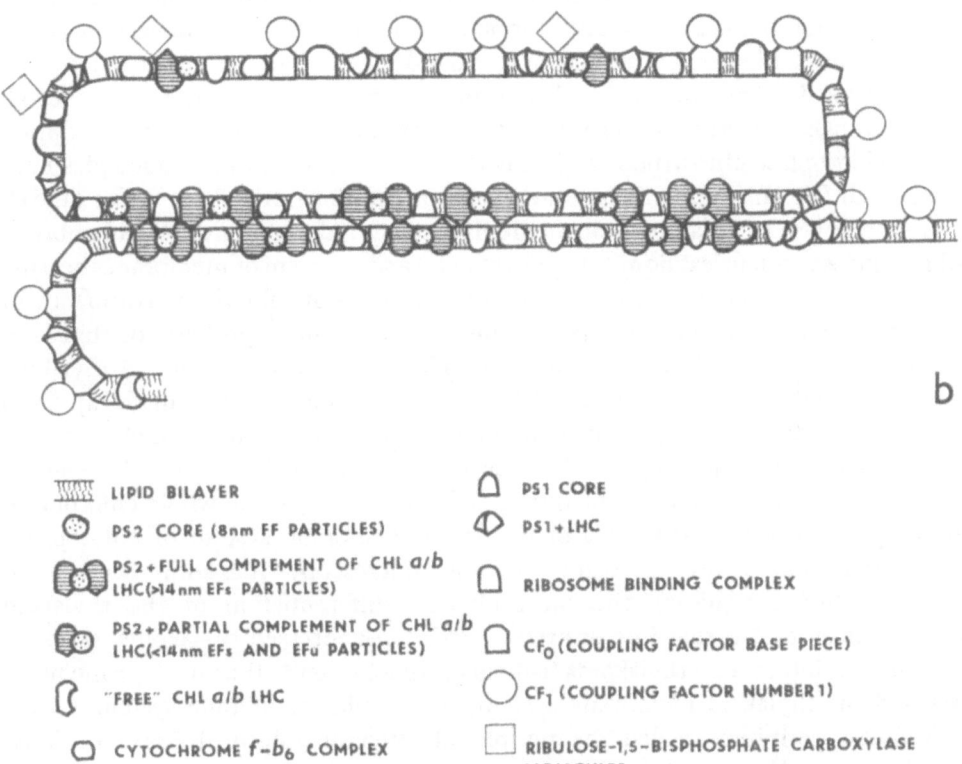

b

LIPID BILAYER

PS2 CORE (8nm FF PARTICLES)

PS2 + FULL COMPLEMENT OF CHL *a/b* LHC (>14nm EFs PARTICLES)

PS2 + PARTIAL COMPLEMENT OF CHL *a/b* LHC (<14nm EFs AND EFu PARTICLES)

"FREE" CHL *a/b* LHC

CYTOCHROME *f–b₆* COMPLEX

PS1 CORE

PS1 + LHC

RIBOSOME BINDING COMPLEX

CF₀ (COUPLING FACTOR BASE PIECE)

CF₁ (COUPLING FACTOR NUMBER 1)

RIBULOSE-1,5-BISPHOSPHATE CARBOXYLASE MOLECULES

We shall not follow the historical development of the knowledge on chloroplast structure and functions. This subject is reviewed comprehensively in a basic handbook of plant physiology (Granick 1955, Mühlethaler 1977, Sane 1977), and from various aspects in many reviews and monographs, *e.g.* Boardman (1977a), Kowallik and Herrmann (1977), Arntzen (1978), Kirk and Tilney-Bassett (1978), Kreutz (1978), Shiryaev (1978), Reinert (1980), Forti (1981), Robards (1981), Staehelin (1981), and Wettstein (1981).

The present state of the knowledge on chloroplast structure may be summarized as follows: Higher plant leaf cell chloroplasts are usually lens-shaped bodies having most frequently 4 to 8 μm in diameter. Their number in a mesophyll cell varies from several to several tens (for review see Butterfass 1979). In mature mesophyll cells, the chloroplasts are appressed to the cell walls, being embedded only by a tiny cytoplasmic layer. However, they may perform some movements in the cells (see Section 2.2.1).

In the study of chloroplast structure, two organization levels may be recognized, namely, the ultrastructure and substructure (see, *e.g.*, Kirk and Tilney-Bassett 1978, Shiryaev 1978). Under the term "chloroplast ultrastructure", we mean the arrangement of photosynthetic (thylakoid) membranes and other chloroplast parts whereas "chloroplast substructure" (or molecular structure, supramolecular structure), means the internal structure of thylakoid membranes and other chloroplast components. In a broader sense, "chloroplast ultrastructure" may be used for all the chloroplast structure not visible under a light microscope. For the description of the chloroplast ultrastructure (in a narrower sense), various terminologies have been used. We shall apply here that shown in Fig. 2.2a, based on Weier (1961).

A chloroplast is a plant cell organelle limited by a double envelope membrane which encloses colourless liquid stroma (matrix) and a system of membranes bearing chlorophylls, carotenoids and other substances necessary for the performance of primary reactions of photosynthesis. This system of photosynthetic or thylakoid membranes is derived from the inner chloroplast envelope membrane. A thylakoid was originally meant by Menke (1960) to be a separate flat bag made up from photosynthetically active plastid membrane. At present, the most widely accepted model of the chloroplast thylakoid membrane system (Paolillo *et al.* 1969, Paolillo 1970) supposes the continuity of intrathylakoidal spaces in the whole chloroplast. A complicated helical structure of the inner membrane system of chloroplasts was discovered. However, using electron microscopy (ultrathin sections or freeze-fracturing replicas), the most obvious differentiation of the thylakoid membrane system is that into appressed and non-appressed thylakoids, or grana and stroma (intergrana) thylakoids (lamellae). In addition to the envelope membranes and the thylakoid membrane system, the tubular membrane system on the chloroplast periphery, called the peripheral reticulum, derived from the inner chloroplast envelope seems to be currently present in chloroplasts; its role is not yet clear. In close relation to thylakoid membranes, there are sometimes present paracrystalline lattice structures called prolamellar bodies. They develop in the plastids under low irradiance; resulting plastids are usually called etioplasts. Under higher irradiance, prolamellar bodies give rise to the thylakoid membranes. Thus, chloroplasts are formed from etioplasts *via* some intermediary plastid stages (chloroetioplasts, etiochloroplasts). The reverse transformation (chloroplasts to etioplasts) also occurs. Prolamellar bodies develop from vesicles derived also from the inner plastid envelope.

The chloroplast stroma is the site of carbon dioxide fixation. Consequently, it contains mainly ribulose-1,5-bisphosphate carboxylase (RuBPC) and other enzymes of the carbon cycle. Moreover, plastid deoxyribonucleic acid (more electron transparent regions) and plastid ribosomes (smaller than cytoplasmic ones) are present in the stroma. The stroma contains also various inclusions, namely, starch grains which are abundant in chloroamyloplasts, and lipoid inclusions called plastoglobules, which are identical with the "star bodies" observed in plastids after permanganate fixation. Other inclusions are seen in chloroplasts only occasionally: stromacentres (semicrystalline arrays of granules and fibrils) are probably formed by RuBPC ("fraction 1 protein"); phytoferritin particles (electron opaque particles of about 10 nm in size) store iron bound to protein; crystalline bodies of various size store excess proteins; microtubules are present in plants with the CAM type of photosynthesis (see Section 2.2.3.3).

A recent concept of the substructure of chloroplasts (see Fig. 2.2b) is based on the fluid mosaic model of the biological membrane (Singer and Nicolson 1972). Similar models of the thylakoid membrane structure are based mainly on results obtained by the freeze-fracturing method and the "fracture faces" terminology (Branton 1966, Branton et al. 1975) is therefore used here for their description. The presented structure of the thylakoid membrane allows vectorial electron flow in accordance with Mitchell's chemiosmotic hypothesis of phosphorylation (see, e.g., Trebst 1978, for review; for the electron transport chain arrangement in the membrane see Fig. 5.1). The smaller particles revealed by freeze-fracturing on the outer fracture face (protoplasmic fracture face, stacked or unstacked; PFs or PFu) of the thylakoid membrane (having a diameter of about 12 nm) have been related to Photosystem 1 chlorophyll-protein complexes whereas the larger ones on the inner fracture face (exoplasmic fracture face, stacked or unstacked; EFs or EFu; with a diameter of about 18 nm) to Photosystem 2 complexes. The latter particles are more densely distributed in appressed thylakoid regions (partition regions) where also the activity of Photosystem 2 was proved. The distribution of both these particles in the plane of the thylakoid membrane and the degree of their submersion in the lipid bilayer of the membrane are variable (Ojakian and Satir 1974, Popov et al. 1980). The work by Izawa and Good (1966), on the influence of salt concentration on thylakoid stacking in isolated chloroplasts showed a reversible, dynamic nature of appressed thylakoid (grana) regions.

The plasticity of the thylakoid membrane system structure and the continuous turnover of its building elements are necessary conditions for substantial structural changes occurring in chloroplasts during their ontogeny.


## 2.2    ONTOGENY OF CHLOROPLAST STRUCTURE

### 2.2.1    Outlines of Chloroplast Development

In connection with the study of photosynthesis during leaf development, the conversion of proplastids (i.e., small non-green plastids possessing, at most, several simple membrane vesicles) to mature chloroplasts (green plastids with a well developed system of thylakoid membranes, differentiated usually into grana and intergrana regions), and later to senescent chloroplasts (plastids with a more or less destroyed system of thylakoid membranes and with predominating plastoglobules) is the most worthy of attention, especially as concerns the chloroplasts of leaf mesophyll cells (chlorenchyma cells).

The ontogeny of chloroplasts may be studied in three different ways (see, *e.g.*, review by Khavkin 1977), namely: (1) light induced transformation of isolated proplastids or etioplasts (*i.e.*, plastids with prolamellar bodies), (2) light induced transformation of plastids of intact etiolated plants, (3) normal chloroplast development during leaf ontogeny (under day/night light regime; eventually, under continuous illumination).

From our point of view, the third possibility is the most important as it enables to confront the structural and functional changes of photosynthetic apparatus during leaf development. Instead of successive studies on leaves of increasing age, leaves in the insertion gradient on a plant or various parts of the same leaf, representing also leaf tissues of increasing age, may be evaluated at a certain moment. Of course, there are some differences between structural development of greening etioplasts and normally developing chloroplasts (*e.g.* Baker *et al.* 1975, Leech 1976).

In all studies on the ontogeny of the photosynthetic apparatus it must be clear whether the development of a leaf, a tissue, a cell, or a chloroplast is being investigated. However, there is usually only one generation of chloroplasts in leaves and so the life span of the chloroplasts and that of the leaf is roughly equal (Gamaleï and Kulikov 1978), although some minor differences can be observed (see, *e.g.*, Hudák 1981).

Before treating ontogenetic changes in the chloroplast structure, spatial changes in chloroplast arrangement in a cell, *i.e.* chloroplast movements, must be briefly mentioned (for review see Haupt and Feinleib 1979). There are two kinds of chloroplast movements in higher plants: passive movement with cell cytosol streaming ("cyclosis") and chloroplast movement as the answer to irradiance, *i.e.* larger exposure to weak than to strong radiant energy ("epistrophe" versus "parastrophe"), while in darkness, chloroplasts are randomly distributed in cells ("apostrophe"). We have not yet found any work dealing with possible changes in the character and intensity of these movements during leaf ontogeny.

The results of studies on structural and functional development of chloroplasts are summarized in many reviews and monographs. However, most attention is usually paid there to the transformation of etioplasts into chloroplasts, see Sironval (1967), Kirk (1971), Rosinski and Rosen (1972), Leech (1976), Boardman (1977b), Bradbeer (1977, 1981), Mohr (1977, 1981), Akoyunoglou and Argyroudi-Akoyunoglou (1978), Boschetti (1978), Parthier (1978), Silaeva (1978b), Treffry (1978), Kasemir (1979), Voskresenskaya (1979), Schiff (1980), Sundqvist *et al.* (1980), Akoyunoglou (1981), Klein (1982).

Reviews dealing mainly with chloroplast ontogeny during normal leaf development are less numerous (*e.g.* in the monographs of Gamaleï and Kulikov 1978, Mokronosov 1981). Some of them, moreover, concentrate on the stage of leaf and chloroplast senescence, *e.g.* Lichtenthaler (1968b), Woolhouse and Batt (1976), Thomas and Stoddart (1980). The reviews by Khavkin (1977) and Šesták (1977a, b, 1978, 1981) cover the whole leaf ontogeny, but they concern mainly biochemical, not structural aspects of the ontogeny of the photosynthetic apparatus. The development of the ultrastructure of chloroplasts during the ontogeny of various organs of higher plants is reviewed, *e.g.*, by Abdullaev *et al.* (1979). Morphometric (quantitative) methods of evaluation of chloroplast structure development are reviewed by Silaeva and Silaev (1979). Otherwise, the observations and remarks on this subject are scattered throughout the literature on chloroplasts. Unfortunately, in many relevant papers, the tissue from which the chloroplasts have originated and the leaf developmental phase are not precisely defined. Also chloroplast developmental phases are not distinguished uniformly (see Section 2.2.3.1); however, at least young, mature, and senescent chloroplasts are usually recognized.

## 2.2.2    Non-ontogenetic Changes in Chloroplast Structure

The chloroplast ultra- and substructure is continuously subjected to various changes, some of which may interfere with the ontogenetic changes. Thus, periodical oscillations in chloroplast volume (recorded by microcamera) are evident within seconds; they are related to irradiance (Rudenko *et al.* 1979). After several minutes or less of exposure, irradiation induces chloroplast volume diminution and flattening of the shape of thylakoids as well as of whole chloroplasts (Nobel 1968, Miller and Nobel 1972, Zurzycki and Gierka 1978). This may increase photosynthetic efficiency of chloroplasts. Ivanchenko *et al.* (1979) have described volume oscillations of chloroplasts (from young pea leaves) within *ca.* 2 h period and with an amplitude exceeding 50%. Photophosphorylation and photosynthetic electron transport rates change in the same material within periods roughly half as long. Thus the volume changes of chloroplasts are a characteristic of the state of their structure. Diurnal changes in chloroplast structure (see, *e. g.*, Więckowski 1968, Lott 1970, Bartels 1971, Mark and Sweeney 1975) include night lysis of chloroplast starch inclusions, their restoration after irradiation and thylakoid system cyclic rebuilding. Especially the latter diurnal changes must be taken into account in all ontogenetic studies, *i.e.*, plant material must always be sampled at the same time in the diurnal cycle.

Irradiance influences also chloroplast ultrastructure: Under strong irradiance, there are more chloroplasts per leaf area unit, the grana are more numerous but smaller, the chloroplasts contain more stroma, *i.e.* higher amount of carbon cycle enzymes, less chlorophyll per chloroplast, and exert higher activities of photosystems than under weak irradiance (*e.g.* Osipova and Ashur 1965, Brangeon 1973,

Prioul 1973, Boardman *et al.* 1975, Lichtenthaler 1981, Lichtenthaler *et al.* 1981a). Goryshina and Laverycheva (1980) found differences in the chloroplast ultrastructure resembling those between the sun and the shade chloroplasts when they compared spring and summer leaves of two woodland herbs, *Pulmonaria obscura* and *Aegopodium podagraria*. Chloroplasts in 9 to 11 d old *Vicia faba* leaves have under low irradiance numerous small grana, however, in 13 d old leaves, larger grana arise probably by fusion, *i.e.* the "shade type" chloroplast ultrastructure is formed (Vlasova and Osipova 1973). Thus the ultrastructure of already formed chloroplasts can be changed after a change of irradiance, "shade type" chloroplasts having been less stable than "sun type" ones.

Even the thickness of thylakoid membranes can be changed under the influence of various qualities and quantities of the irradiation (Zurzycki and Gierka 1978).

### 2.2.3 Chloroplast Ultrastructure during Leaf Development

**2.2.3.1** *Leaves of Herbs and Deciduous Woody Plants*

The oldest papers on the development of chloroplast structure during leaf ontogeny appeared in the 1950s, when the development of electron microscopy enabled the study of chloroplast ultrastructure. Howevere, Tabentskiï (1953) stated, by means of mere light microscopy, that chloroplasts in sugar beet leaves reached their maximal dimensions approximately half way through the leaf's ontogeny and that the size of the grana increased during the whole ontogeny of the leaves. Genkel' and Morozova (1957, 1959) followed electron microscopically (in chloroplast suspensions) the ontogeny of chloroplasts during the development of the leaves of *Bellis perennis*. "Summer" chloroplasts (June) are oval, with thin stroma, conspicuous grana and some minute plastoglobules. During autumn (September, October) transition from this state to the "winter" state occurs. In winter (November to February), the chloroplasts are small, round, with dense stroma, inconspicuous grana, and numerous plastoglobules. In March, distinct grana appear again in the chloroplasts of the overwintering leaves but later (April, May) destruction of these chloroplasts occurs. Using the same technique in potato leaves, Shakhov and Golubkova (1964, 1966) described "lipoid granulation", *i.e.* accumulation of plastoglobules in chloroplasts of senescing leaves, as one of the most characteristic structural features of chloroplast senescence.

Rhodes and Yemm (1966) emphasized that in developing chloroplasts (in the first true leaf of barley) the formation of partitions (appressions of the thylakoids) seriously affected photosynthetic pigment system organization and so enabled full development of chloroplast functions. They proposed the model of grana formation by "primary lamellae" (primary thylakoids) evagination (see also Mühlethaler and Frey-Wyssling 1959). Röbbelen (1966) supposed, on the basis of the study of thylakoid formation in palisade parenchyma cell chloroplasts of the leaves of "*xantha*" mutant of *Arabidopsis thaliana*, that aggregation of fibrils (7.5 nm in diameter) connected with prolamellar body tubules was the possible mechanism of thylakoid growth.

The most complicated structural organization of chloroplasts (thickly packed thylakoids, grana in whole chloroplast section) was found in mature potato leaves where also the highest activity of adenosine triphosphatase and the highest $CO_2$ fixation rate were recorded (Kislyakova *et al.* 1967). Plastoglobules (the term proposed by Lichtenthaler and Peveling 1967; formerly the term "osmiophillic globules" and some others were used) occur in chloroplasts in all phases of their development (for review see Lichtenthaler 1968b). They are most conspicuous in senescent chloroplasts where either a great number of tiny plastoglobules (in annual herbs) or several large ones (in perennials and woody plants) are present. Plastoglobules probably serve as a pool of thylakoid lipids when thylakoids are not fully developed. Paolillo *et al.* (1969) assume that during chloroplast ontogeny, the thylakoid system of chloroplasts is continuously rebuilt; in mature chloroplasts, either the state of dynamic equilibrium or decrease in the rate of these processes is reached. Butler (1967) described the development of chloroplast ultrastructure in mesophyll cells of senescing cucumber leaves in relation to plant age [d]: 16 to 20: normal, mature chloroplasts; 21 to 25: larger plastoglobules; 26 to 30: smaller grana, more plastoglobules; 31 to 35: smaller grana, thinner stroma, more plastoglobules; 36 to 40: reduced grana, thinner stroma, large plastoglobules; 41 to 45: fewer thylakoid membranes, clear stroma, large plastoglobules; 46 to 50: only the remnants of chloroplasts are present (outer chloroplast envelopes, plastoglobules).

Several authors have studied the ontogeny of chloroplast ultrastructure in mesophyll cells of primary (first after cotyledons) leaves of *Phaseolus vulgaris*. Więckowski (1967a, 1969) described the first phases of this ontogeny after leaf unfolding [d after sowing] : 4: chloroplasts often amoeboid in shape, starch and plastoglobules often present, beginning of grana formation by invagination and overlapping of primary thylakoids; 5: larger chloroplasts, ellipsoid, with grana, at most, from six thylakoids, little starch; 8: almost mature chloroplasts — formation of new grana, much starch, many plastoglobules; 12: mature chloroplasts with large grana, abundant starch, and many plastoglobules. The emergence of photosynthetic activity ($CO_2$ fixation) coincided with the grana formation (thylakoid appression, formation of thylakoid partitions). Similarly, Miller and Nobel (1972) observed an increase in number of grana thylakoids (per chloroplast section) followed (with some delay) by an increase in $CO_2$ fixation rate in developing *Pisum sativum* leaves. Barton (1966, 1970) found at the beginning of chloroplast senescence in *Phaseolus* primary leaf mesophyll cells a temporary emergence of "phytoferritinlike particles"; later, the thylakoids disintegrated.

Whatley (1971, 1974, 1977a, b, 1978, 1979, 1980) has studied in detail the ultrastructure of chloroplasts in various tissues of the primary leaf of *Phaseolus vulgaris* from its formation in embryo to its abscission (period of approximately 14 weeks). During the first 7 weeks of embryonal leaf ontogeny, the leaves contain only proplastids. After subsequent seed dormancy, during 1 week of germination, leaf (especially mesophyll) cell plastids develop into mature chloroplasts. Two weeks of chloroplast maturity are followed by 3 to 4 weeks of leaf and chloroplast senescence. Three distinct ontogenetic phases between the proplastid and mature chloroplast are described in chloroplasts of palisade parenchyma cells: a plastid surrounded by endoplasmic reticulum, an amoeboid plastid, and a young chloroplast with perforated stroma thylakoids and incipient grana. Plastid division occurs until the mature chloroplast stage. In dividing chloroplasts, the thylakoid membrane system divides first. In mature chloroplasts, the ultrastructure does not change substantially; all grana and stroma thylakoids are parallel. Chloroplast senescence starts by an increase in plastoglobule number. After one week of senescence, the parallel arrangement of thylakoids is lost (curved thylakoids), grana are often indistinct, many plastoglobules are present, chloroplasts sometimes form pseudopodia. On the basis of these observations, Whatley (1978) proposes a new general pathway of the developmental interrelationships of leaf plastids, reviving the old idea of cyclic plastid transformations of Schimper (1885) against the hypothesis of Frey-Wyssling *et al.* (1955) suggesting monotropic plastid development, *i.e.* from proplastid to leucoplast, chloroplast, and finally to chromoplast. According to Whatley's scheme, mature chloroplasts may give rise to young plastids as an alternative to senescent development. The most primitive plastid stage is called here "eoplast", the term "proplastid" being broader, involving eoplast, amyloplast, and amoeboid plastids. In addition to current plastid developmental phases between the eoplast and mature chloroplast, some facultative ones are described, *e.g.* a plastid containing phytoferritin; an etioplast with prolamellar body developing in light; a plastid with stromacentrum. In further plastid. ontogeny, chromoplast formation may be regarded as such a facultative plastid phase (or stage). Generally, the temporary accumulation of various reserve substances in plastids seems to be a consequence of a lowered rate of thylakoid membrane formation (or thylakoid decay). Irradiance and other ecological factors are responsible for these deviations from the "basic pathway" of plastid ontogeny. The thylakoid system of mature chloroplasts is in the state of dynamic equilibrium (permanent formation and disintegration of thylakoid membranes). In chloroplasts of mesophyll cells of a slowly growing and greening mutant (*"virescent"*) of *Phaseolus vulgaris* a slow chlorophyll accumulation is correlated with a slow development of the stacking of thylakoids (Dale and Heyes 1970). Maximal dimensions of the chloroplasts and maximal number of grana in primary leaves of 12 d old *Phaseolus* plants (from the beginning of germination) (see Fig. 2.3) are connected with high chlorophyll, carotenoid, and protein contents in the chloroplasts (Naito *et al.* 1981). The number of grana remains constant up to 30 d when it starts to decrease. The length of the grana is constant up to end of the ontogeny of leaves whereas the number of thylakoids per granum increases at the end of the ontogeny. The number of plastoglobules reaches a first maximum at the beginning of the ontogeny (6 d), a minimum at 22 d, and increases intensely at the end of the ontogeny, whereas the size of the plastoglobules increases during the whole ontogeny, but most intensely at the end. During senescence of the chloroplasts, shrinkage and disorganization of their membrane system are apparent.

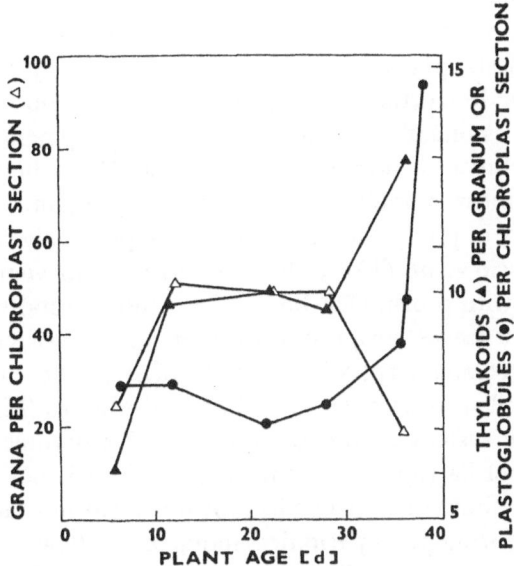

**Fig. 2.3.** Quantitative characteristics (grana and plastoglobules per chloroplast section, thylakoids per granum) of mesophyll cell chloroplasts during ontogeny of primary leaves of intact plants of *Phaseolus vulgaris* (after Naito *et al.* 1981).

In the 1970s, also the ontogeny of chloroplasts of woody plant leaves began to be studied. In developing young (March), mature (May), and yellowing or yellow (October) birch leaves (Dodge 1970), thylakoid numbers per chloroplast section or per granum increase during chloroplast development but they strongly drop after the beginning of leaf yellowing. Grana number per chloroplast decreases regularly, but more intensely during yellowing. Plastoglobules increase in size, but their number does not substantially change. Chloroplasts are the first of the cell organelles to show senescent changes, but they persist in cells of yellowing leaves for the longest period of all of them. Stearns and Wagenaar (1971) obtained similar results in mature, early and fully senescent leaves of *Acer ginnala*, *Populus tremuloides*, and *Rhus typhina*. In contrast to young (1 d after unfolding) and mature (15 d) leaves of *Populus deltoides*, in senescent (60 d) leaves the thylakoid membranes were most densely packed but the grana were hardly recognizable (Hernández-

Gil and Schaedle 1973, 1974). After achieving definitive leaf area, photophosphorylation and $CO_2$ fixation rates decreased with leaf age.

Simultaneous study of leaf anatomy, chloroplast ultrastructure (quantitatively), and accumulation of leaf pigments in maturing leaves of *Lactuca sativa* (Bourdu *et al.* 1975) showed an increase in chloroplast section area as well as in the frequency and dimensions of the grana up to the achievement of the definitive leaf blade area. Ehara and Misawa (1975) found that during ontogeny of tobacco leaves, the proportion of juvenile, mature, and senescent chloroplasts was regularly changed (seven ontogenetic phases of chloroplasts were distinguished). In senescent (degenerating) chloroplasts, plastid volume decreased probably as a consequence of thylakoid membrane destruction. Horak and Zalik (1975) observed in developing leaves of *"virescens"* mutant of barley a slower development of photosynthetic apparatus than in leaves of normal barley plants. Mean number of grana per chloroplast section, mean number of thylakoids per granum, chlorophyll content in leaves, and activity of photosystems 1 and 2 were positively correlated.

According to Rascio *et al.* (1976, 1980), proplastids in young leaves of maize give rise to plastids with prolamellar bodies (etiochloroplasts) which develop into chloroplasts. In older leaves, proplastids of intercalary meristem develop directly into chloroplasts. Here, an interplay between the plastid differentiation gradient on a leaf and that on a plant occurs. In the *Sinapis alba* leaves (Havelange 1977) the thylakoid membrane system in the mesophyll cell chloroplasts develops most rapidly in leaf primordia (up to 0.5 mm in length) and more slowly up to the achievement of the definitive length of a leaf (65 mm). No substantial changes occur in leaf chloroplasts during photoperiodical induction of flowering.

Several authors have observed changes in the thylakoid membrane system appearance (stainability) in chloroplasts during leaf ontogeny. Platt-Aloia and Thomson (1977), Casadoro *et al.* (1977), Casadoro and Rascio (1977, 1978a, b, 1979a, b), and Rascio and Casadoro (1979) described in chloroplasts of young leaves of some plants (*e.g. Atropa belladona, Helianthus annuus*) thylakoids with dark loculi and lightly stained membranes. In such thylakoids, inclusions called "membrane bound bodies" have been frequently observed. During leaf maturation, this "negative staining" is usually transformed into a "normal" one (dark membranes, light loculi). According to Casadoro and Rascio (1978b), this pathway of chloroplast development is widespread throughout the plant species. The lightly stained thylakoid membranes probably do not have a complete lipoprotein composition. However, they may be photosynthetically active. Casadoro and Rascio (1979b) have proposed three alternative models of ultrastructure development from proplastids to fully developed chloroplasts in addition to the current one. During this development, (1) lightly stained membranes may be formed, (2) prolamellar bodies may persist alongside the thylakoids, (3) the combination of (1) and (2) may occur. All these phenomena are transient ones. Similarly to Whatley (1977b), they attribute the deviations from the "basic" developmental pathway of chloroplasts to discrepancies between the rates of cell and chloroplast maturation evoked by environmental factors. Van Steveninck and Van Steveninck (1980a, b) described the ontogeny of chloroplasts in floating leaves of *Nymphoides indica*. Whereas in young leaf primordia (up to 5 mm long), there are chloroplasts with light thylakoid membranes and dark loculi in all leaf cells, in older leaves (from 25 mm long to mature ones) this type of staining is restricted to the chloroplasts of the lower epidermis and subepidermal cells. This pattern is probably a consequence of cell differentiation in the floating leaves. The electron dense substance in thylakoid loculi was proved cytochemically to be of a phenolic nature.

However, negatively stained thylakoid membranes have not been observed only in young but also in senescent chloroplasts (Młodzianowski and Młodzianowska 1973, Kutík *et al.* 1981). Negative thylakoid staining may be a consequence of an inconvenient fixation procedure (Soikkeli 1980) or an influence of tannin type substances in leaf cells (especially in conifers; Olesen 1978); it may also be due to extreme environmental conditions (low $CO_2$ concentration in the air; Tripodi 1980).

Various more or less electron-dense thylakoid inclusions, resembling "membrane bound bodies" noted above, were observed in chloroplasts of young leaves not only together with the lightly stained thylakoid membranes ("thylakoid or prethylakoid bodies", see, *e.g.*, Ames and Pivorun 1974, Hurkman and Kennedy 1977, Mares *et al.* 1979). These inclusions, frequently observed especially in neoplastic and hyperplastic cell plastids probably form sites for accumulation of proteins used in the formation of thylakoid membranes during subsequent chloroplast ontogeny. In tumor transformed cells they are more frequent in plastids probably because of the suppression of thylakoid system development in these plastids.

The course of senescence of mesophyll cell chlororoplasts may be different in isolated and attached leaves (Hurkman 1979): During natural senescence of barley leaf (10 d old leaves and green, yellow-green and yelow parts of 21 d old leaves were compared) following senescent changes were observed in chloroplasts, successively: emerging of the plastoglobules and reorientation of the thylakoid membrane system; dilatation of thylakoids, narrowing of the grana, increase in plastoglobule volume, and disappearance of the stroma; finally, disintegration of the thylakoids into isolated vesicles, rupture of the chloroplast envelope, and destruction of the chloroplast. Dunaeva (1979) recognizes five ontogenetic phases of chloroplasts in mesophyll cells of developing wheat leaves: (1) leaves having one half of maximum length, 10 to 12 d after leaf emergence: increase in partial volume of the thylakoid membrane system, many chloroplast ribosomes; dividing chloroplasts; (2) mature leaves, 1 to 25 d after achieving definitive length: well developed thylakoid system, lower chloroplast ribosome density; (3) phase preceding leaf yellowing, 30 to 40 d after leaf maturity: destruction of stroma thylakoids, growth of the grana (*e.g.* by merging); "old" chloroplasts; (4) phase of leaf yellowing, 40 to 45 d after leaf maturity: decrease in chloroplast volume, distortion and disorientation of the thylakoids, indistinct appearance of thylakoid membranes, increase in dimensions but not in number of the plastoglobules; "degraded" (senescent) chloroplasts; (5) yellow leaves without turgor pressure: no distinct chloroplasts apparent. The author agrees with the opinion that the relatively short life time of the chloroplast ribosomes determines the commencement of chloroplast senescence. Damsz (1979) studied morphometrically chloroplast maturation in successive zones (marked by the colour) of the leaves of two orchids, *Coelogyne cristata* and *Cymbidium insigne*. During a rapid increase in the number and size of the grana, a roughly constant ratio between the number of appressed and non-appressed thylakoids (4:1) is maintained.

Mustárdy and Brangeon (1978) and Brangeon and Mustárdy (1979) have proposed on the basis of their study of chloroplasts in five successive parts of a *Lolium multiflorum* leaf a new detailed model of the ontogeny of chloroplast ultrastructure. They suppose that in a proplastid, there is one growing thylakoid only; later it gives rise, by a division, to several (primary) thylakoids lying in a parallel position; these thylakoids become perforated; grana are formed by invagination and overlapping of margins of the holes; the individual thylakoids divide, branch, and merge again; thus, the complicated spiral arrangement of the thylakoids (lamellae) around the cylindrical grana is formed — the "grana-fretwork" ultrastructure of a mature chloroplast. Whereas the number of grana formation sites in thylakoids is given at an early phase of the chloroplast development (it may be genetically coded), the size of the grana (number of appressed thylakoids) and the arrangement of intergrana thylakoids are determined later in the chloroplast development. This "secondary build up" of the thylakoid system probably forms a base for the well known flexibility and dynamic nature of the chloroplast ultrastructure. Mutual recognition and response of thylakoid membranes is probably the fundamental mechanism of regulation of the ontogeny of chloroplast ultrastructure.

In older chloroplasts of leaves of *Callisia fragrans* (*Commelinaceae*) an unusually dense packing of thylakoids, and besides current plastoglobules also inclusions called plastosomes, bound by thylakoid membrane and lying thus in thylakoid lumina were observed by Wise and Harris (1980). Floyd and Noble (1980) compared quantitatively the ultrastructure of palisade parenchyma cell chloroplasts and content of chlorophylls in developing leaves of eight plant species characterictic for three vegetational floors of a deciduous forest. Shade leaves were evaluated in May, June, and July (to follow influences of "canopy closure" on chloroplasts). In woody plants, there was a substantial increase in thylakoid number per granum and decrease in starch volume per chloroplast during the whole vegetation season. In herbs, the size of grana increased between May and June only. During development, the chloroplasts of all plant species studied acquired a "shade" character (see Section 2.2.2). The commencement of chloroplast senescence was apparent already in July (large lightly stained plastoglobules, dilatation of thylakoids). According to Hudák (1981), the senescence of chloroplasts in cells of yellowing cotyledons of *Sinapis alba* is characterized by an increase in number and size of the plastoglobules. Before the cotyledons fall, vacuolization of plastids and releasing of the plastoglobules is apparent. Sometimes, tubular formations similar to the prolamellar bodies are observed in the senescent chloroplasts. In yellow cotyledons, a green border is sometimes apparent. Here, chloroplasts with "giant grana" and fewer plastoglobules are present. Valanne et al. (1981) followed early phases of the ontogeny of chloroplasts in leaves of three birch species (Betula pendula, B. pubescens, B. pubescens var. tortuosa). In leaves closed in buds (April), there are young chloroplasts rich in starch, with negatively stained thylakoid membranes and dark thylakoid loculi. During opening of the buds mid May, large light regions (with DNA) in stroma and chloroplast ribosomes attached to thylakoid membranes are apparent. After leaf unfolding, there are small prolamellar bodies here and the number of thylakoids increases. Grana are present already in very young chloroplasts (in buds) but they reach the largest dimensions in chloroplasts of fully developed leaves (beginning of June). The rate of $CO_2$ fixation and chlorophyll content in leaves increases until full anatomical and ultrastructural differentiation of the leaves is reached. Kutík et al. (1981) studied the development of mesophyll cell chloroplast ultrastructure in leaves of Tilia cordata from the opening of buds to leaf yellowing. In opening buds (in May), in the chloroplasts thylakoid system with small grana is observed. During leaf expansion, growth of the thylakoid system continues. Its maximal development is achieved after the end of leaf blade area growth (mid June). Later (in July), the largest grana stacks are present in chloroplasts, and plastoglobules increase considerably in size. In senescent chloroplasts (at the end of September), negatively stained, mostly stacked thylakoid membranes and several large, lightly stained plastoglobules per chloroplast are observed.

On the basis of recent literature on the ontogeny of chlorenchyma (mesophyll) cell chloroplast ultrastructure in developing leaves of various herbaceous and deciduous plants, some general features of this development may be outlined, considering leaf ontogeny from its unfolding to abscission under a natural day/night regime:

After leaf unfolding, chloroplasts have a system of thylakoid membranes forming small grana; "primary thylakoid" layers are usually apparent; in some cases, prolamellar bodies are temporarily present. Young chloroplast stroma is relatively dense, with numerous plastid ribosomes, some small plastoglobules, some starch inclusions, and with several, more transparent regions containing plastid DNA. Roughly in the middle of leaf ontogeny, the starch inclusions reach the largest dimensions. The most developed thylakoid membrane system is present after the end of leaf area expansion — in mature leaves. At this time, the chloroplasts may be considered to be ultrastructurally mature as well. The stroma is less dense here. Before the beginning of leaf senescence, particularly large grana are often formed in chloroplasts; at the same time, stroma thylakoids are sometimes destroyed. Chloroplast senescence (in yellowing leaves) is usually manifested by a decrease in chloroplast volume, change in chloroplast shape from lens-like to more rounded, dilatation of stroma and (later) grana thylakoids, loss of parallel arrangement of the thylakoids, increase in number and size of the plastoglobules,

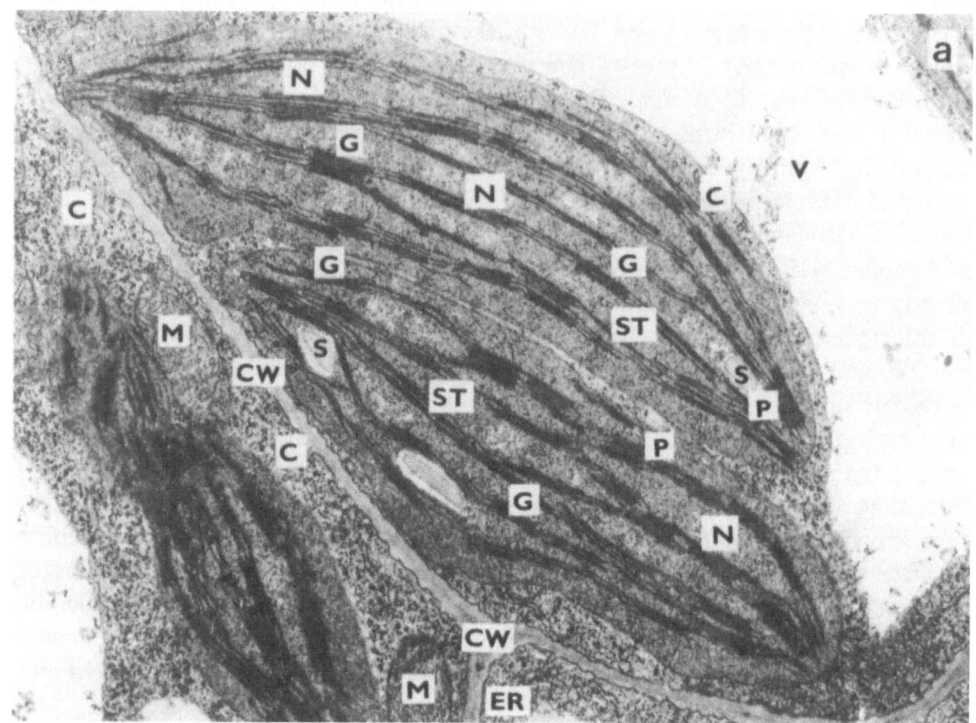

and destruction of chloroplast envelopes. Finally, the thylakoid membrane system is destroyed; the last remnants of the chloroplasts are the plastoglobules.

The described main phases of mesophyll cell chloroplast ontogeny may be illustrated by electron micrographs of developing chloroplasts in the mesophyll cells of primary leaves of *Phaseolus vulgaris* (see Fig. 2.4).

### 2.2.3.2 Leaves of Evergreen Woody Plants

Special attention has been devoted to the ontogeny of chloroplast structure in sempervirent plants, where leaves persist through several vegetation seasons. Here, the questions of chloroplast life duration (prolongation of life or additional chloroplast division) and the mode of overwintering of chloroplasts are of great importance. Moreover, in conifers, the most important group of evergreens, greening of plastids in the complete absence of light was ascertained (Laudi 1964, Walles and Hudák 1975, Öquist 1981).

During the 1970s, the ontogeny of conifer chloroplasts was investigated in the most important species, namely, in pine (Harris 1971, Campbell 1972, Walles and Hudák 1975, Martin and Öquist 1979, Lewandowska and Öquist 1980,

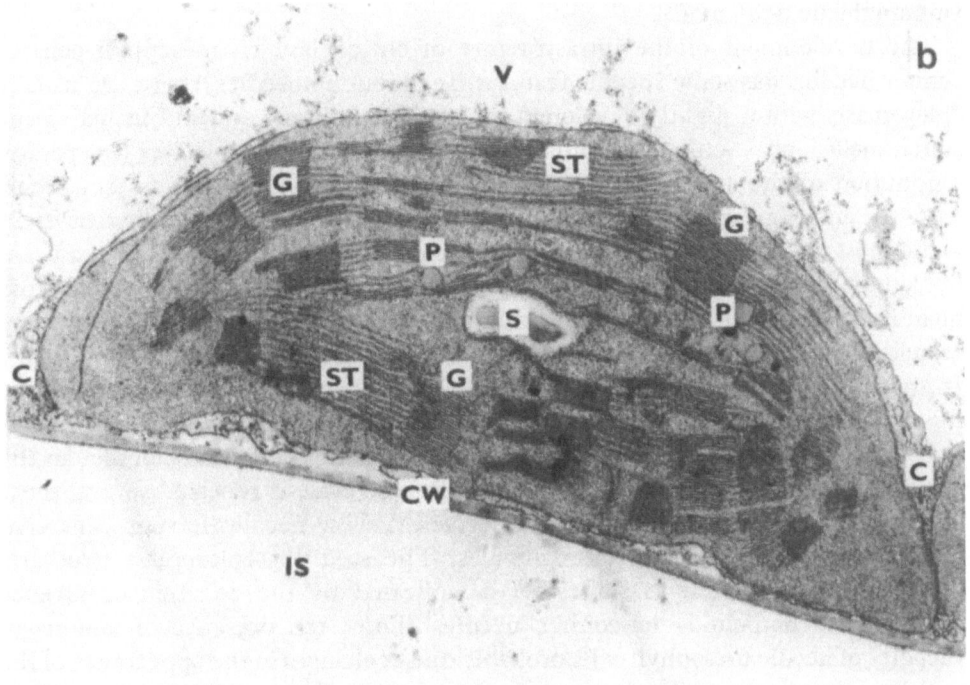

**Fig. 2.4.** The system of thylakoids is the most developed in the chloroplasts of mature leaves where large grana are visible; in senescent leaves plastoglobules become conspicuous but many thylakoid profiles are still seen. Also form and dimensions of the chloroplasts vary considerably during leaf development. Ultrathin sections (fixation: glutaraldehyde followed by osmium tetroxide, embedding in synthetic resin, staining of the sections by uranylacetate followed by lead citrate) of mesophyll cell chloroplasts from young (a — 7 d after sowing), mature (b — 19 d), and senescent (c — 29 d) primary leaves of intact plants of *Phaseolus vulgaris* L.; total magnification 35 000 ×. C — cytosol; CW — cell wall; ER — endoplasmic reticulum; G — granum; IS — intercellular space; M — mitochondrium; N — chloroplast region presumably with plastid DNA; P — plastoglobule; S — starch inclusion; ST — stroma thylakoid; V — cell vacuole. (Orig. Kutík.)

Soikkeli 1980), spruce (Senser *et al.* 1975, Walles and Hudák 1975, Medeghini-Bonatti and Bonetta Conte 1976, Soikkeli 1978, 1980, Senser and Beck 1979), larch (Walles and Hudák 1975, Medeghini-Bonatti and Bonetta Conte 1976, Tageeva *et al.* 1981), fir (Chabot and Chabot 1975), and yew (Kufner *et al.* 1978, Cunninghame *et al.* 1979).

The development of the ultrastructure of chloroplasts in mesophyll cells of conifer needles has some specific features. In sprouting needles, there are usually chloroplasts with a slightly developed thylakoid membrane system but with large starch inclusions — chloroamyloplasts. During growth of the needles (in spring), diminution of the starch inclusions and growth of the thylakoid system occur. After the end of growth of the needles (in summer), the thylakoid membrane system is the most developed. During autumn, in some cases starch is accumulated again but later it disappears; growth of plastoglobules and some degradation of thylakoid membranes are then apparent. The chloroplasts outlive winter usually in clusters around cell nuclei; their thylakoid membrane system is preserved, though reduced. In the spring of the second year of the needle's life, large starch inclusions are formed again in the chloroplasts and the thylakoid membrane system is fully restored. During the second summer, there are more numerous plastoglobules in the chloroplasts than in the first one. This ontogenetic cycle is repeated several times until yellowing and abscission of the needles. Yellow needles contain senescent chloroplasts, filled up with plastoglobules. The study of chloroplast structural development in conifers is somewhat complicated by the common occurrence of aromatic compounds in conifer needles. Thus, the presence of tannins in vacuoles of needle mesophyll cells probably causes changes in the appearance of the thylakoid membranes (negative contrast; Harris 1971, Olesen 1978).

Some phenomena observed in developing conifer chloroplasts are probably less common. Thus, Medeghini-Bonatti and Bonetta Conte (1976) described the so-called "magnograna" in young chloroplasts (from the needles in resting buds) of spruce and larch. These long, not very high grana stacks, not interconnected by non-appressed (stroma) thylakoids are transformed during sprouting of the needles into a normal thylakoid membrane system. Reversible winter yellowing of the needles of some forms of spruce, pine, and thuja, during which grana are destroyed but stroma thylakoids remain, is probably caused by some mutation leading to chlorophyll photooxidation, the autumn lowering of temperature being the mutation factor (Khodasevich *et al.* 1978). Some destruction phenomena observed in conifer chloroplasts may be due also to atmospheric pollution or to the inconvenient mode of fixation for electron microscopy (Soikkeli 1980). After excluding these influences, intact chloroplasts and other cell organelles may be observed in conifers during all the seasons of the year.

As regards non-coniferous evergreens, Lichtenthaler and Weinert (1970) found in young (still coiled) leaves of *Ficus elastica* chloroplasts containing thylakoids, prolamellar bodies and plastoglobules. After unrolling of the leaves, thylakoids are formed from prolamellar bodies and the number of plastoglobules is reduced. Later, during leaf senescence, the number and size of the plastoglobules increase with leaf age, simultaneously with the destruction of the thylakoids. Excess lipoquinones, carotenoids and other lipoid substances from destroyed thylakoids are accumulated here. Baker *et al.* (1975) studied quantitatively the development of chloroplast ultrastructure during the ontogeny of cocoa leaves (leaf chlorophyll content served as a measure of leaf age). Maximal ultrastructural development of the chloroplasts is achieved here after growth of leaf blade ends.

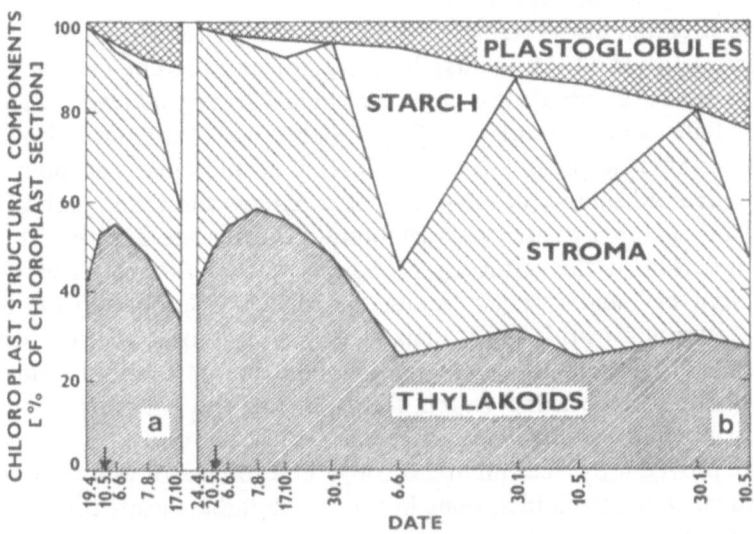

**Fig. 2.5.** Structural composition of palisade parenchyma cell chloroplasts [% of chloroplast section] during leaf development in a deciduous woody plant *Fraxinus ornus* (a) and evergreen woody plant *Phillyrea media* (b). Arrow: moment of the completion of leaf blade expansion. (From Gamaleï und Kulikov 1978.)

A quantitative comparison of chloroplast ultrastructure and their life span in palisade parenchyma cells of leaves of deciduous and evergreen woody plants from the family *Oleaceae* was done by Gamaleï (1975) and Gamaleï and Kulikov (1977, 1978). In evergreens, there is no additional chloroplast division; thylakoid turnover seems to ensure a longer life of their chloroplasts. The ontogenetic changes in the chloroplast ultrastructure (growth and diminution of the thylakoid system, starch accumulation, growth of the plastoglobules) repeat every year during the life of an evergreen leaf, but the senescent character of chloroplasts becomes more pronounced (increase in the plastoglobule volume; see Fig. 2.5). There is a good correlation between the changes in the chloroplast ultrastructure and those in the chemical composition of leaves during leaf ontogeny; the correlation between the ultrastructural and functional changes ($CO_2$ fixation rate) is not as strong. It may be summarized that in evergreens, chloroplast life span is prolonged considerably.

### 2.2.3.3 Chloroplast Ontogeny in Relation to $C_3$, $C_4$, or CAM Pathway

The study of the development of chloroplasts during leaf ontogeny is useful in solving an interesting problem of photosynthesis research — the occurrence of the different pathways of photosynthetic carbon fixation. From the structural point of view, the differences between plants with the common RuBPC pathway only ($C_3$ plants) and those plants with an additional $C_4$-dicarboxylic acid pathway ($C_4$ plants) or plants with Crassulacean acid metabolism (CAM plants) are reviewed by Laetsch (1974) and Bourdu (1976).

Very little attention has so far been paid to the chloroplast structure in relatively recently distinguished CAM plants; however, three different chloroplast types according to starch accumulation were recognized in *Bryophyllum* (Sarda *et al.* 1975). After Salema and Brandão (1978), the presence of microtubules is typical for chloroplasts of CAM plants; they even develop in chloroplasts of some $C_3$ plants forced (by enhanced salinity of environment) to follow the CAM pathway.

In $C_4$ plants, parenchymatic vascular bundle sheaths are differentiated in the leaves. Comparison of the ontogeny of "dimorphic" chloroplasts in $C_4$ plants (*i.e.*, mesophyll cell and vascular bundle sheath cell chloroplasts) has attracted the greatest attention in this field; maize has been the most studied plant here. Already Zirkle (1929) and Rhoades and Carvalho (1944) found a structural dimorphism (mainly concerning starch formation) of the chloroplasts in mature leaves of maize. Hodge *et al.* (1955), Brangeon (1973), Kirchanski (1975) and other authors have proved it by electron microscopy. The absence of well developed grana and abundance of starch inclusions in vascular bundle sheath cell chloroplasts of maize and other $C_4$ plants (*e.g.*, sugar cane — Laetsch and Price 1969; sorghum — Downton and Pyliotis 1971; $C_4$ dicotyledons *Gomphrena globosa* — Appiano *et al.* 1979, and *Haloxylon persicum* — Voznesenskaya 1981) was proved. However, the comparison of mature and young leaves of $C_4$ plants showed that chloroplasts of both tissues do not differ substantially in young leaves when both have a thylakoid membrane system with small grana. During chloroplast maturation, growth of the grana in the mesophyll cell chloroplasts and their destruction (and repression of their further formation) in the chloroplasts of vascular bundle sheath cells occur. "Agranal" chloroplasts of $C_4$ plants seem to be the functional specialization. They accumulate assimilates and form starch; for non-cyclic electron flow in photosynthesis, the presence of some thylakoid appression (grana) seems to be necessary (Downton and Pyliotis 1971).

Differences between both types of chloroplasts in $C_4$ plants as well as those among the chloroplasts of $C_3$, $C_4$, and CAM plants are rather quantitative than qualitative ones. Thus, in addition to the presence or absence of the grana, the peripheral reticulum is not restricted to agranal chloroplasts of $C_4$ plants (*e.g.* Voznesenskaya 1981) but it is present even in chloroplasts of $C_3$ plants (*e.g.* Gamaleï and Kulikov 1978). Starch is formed, under conditions favourable for photosynthesis or in isolated leaf segments, in the mesophyll cell chloroplasts of $C_4$ plants as well as in those from bundle sheath cells (Rhoades and Carvalho 1944, Orsenigo and Rascio 1976, Kutík and Beneš 1981); it is formed also in the mesophyll cell chloroplasts of young leaves (Miyake and Maeda 1978). Thus, a comparison of the ontogeny of various chloroplast types confirms the idea of plasticity of the chloroplast structure.

## 2.2.4 Chloroplast Substructure during Leaf Development

Chloroplast substructure development (the ontogeny of internal structure of thylakoid membranes) during leaf ontogeny has been studied much less than the corresponding development of chloroplast ultrastructure. Only substructural aspects of the transition of etioplasts into chloroplasts have been more studied (see reviews in Section 2.2.1). Generally, any studies on the chloroplast substructure (or supramolecular structure) are methodically more pretentious than those on the chloroplast ultrastructure. For a considerable advance in substructural studies, the development of such delicate electron microscopic methods such as freeze-fracturing and freeze-etching was necessary, along with the application of other appropriate methods (e.g. enzymatic digestion, immunocytochemistry). An analysis of the ontogenetic state of thylakoid membranes has been complicated by their high complexity and plasticity. However, some ontogenetic changes in chloroplast substructure have been recognized.

During maturation of chloroplasts of some plants, as well as in some other cases, transitions in stainability of the thylakoid membranes occur (see Section 2.2.3.1). Negative contrast of the membranes ("light membranes") may be a consequence of the incomplete assembly of chlorophyll-protein-lipid complexes of the thylakoid membranes. In senescent chloroplasts of some plants, plastoglobules become more electron transparent which is probably caused by changes in their lipid composition (Khodasevich et al. 1979). Hernández-Gil and Schaedle (1973) observed (on the ultrathin sections) some thinning of the thylakoid membranes during senescence of mesophyll cell chloroplasts of Populus deltoides. Armond et al. (1977) found that during greening of etioplasts of pea leaf cells, the density of Photosystems 1 and 2 particles did not change substantially but their size increased, particularly that of Photosystem 2 particles in appressed thylakoid regions. Similar studies on chloroplasts of naturally developing leaves are still lacking, though such data would be very useful for a closer study of the structure-function relationship in chloroplasts.

## 2.2.5 Environmental or Experimental Modifications of the Chloroplast Ontogeny

Among various environmental and experimental influences which may modify the ontogeny of the chloroplast structure during leaf development, radiation is the most important factor (Sundqvist et al. 1980). It forms a necessary condition for the development of a normal structure of mature chloroplasts. Light sensitive plastid pigment phytochrome (P) plays an important role in this process, together with protochlorophyll, see, e.g., Mohr (1977), Girnth et al. (1978), Mohr (1981). An active form of P synthesized under red or white irradiation (Pfr) is required for the formation of lipid material of plastid membranes and also for the incorporation of chlorophyll-protein complexes into it, which is necessary for normal thylakoid and grana formation. Thus, both phytochrome and (proto)chlorophyll(ide) (PChl or Chla) are required for normal chloroplast development, both pigments absorbing incident radiation (Pr $\xrightarrow{hv}$ Pfr or PChl $\xrightarrow{hv}$ Chla). An outline of the influence of a quantity of radiant energy on mature chloroplasts structure ("sun" or "shade" type chloroplasts) was given in Section 2.2.2. Continuous irradiation probably enhances senescent development of chloroplasts (formation of many plastoglobules: Osipova and Ashur 1965, Orsenigo and Rascio 1976). Also the spectral composition of radiation influences chloroplast ultrastructure: red radiation simulates shade conditions, blue radiation sun conditions (see, e.g., Silaeva 1978a, b, Sundqvist et al. 1980 for review). Differences between various plant taxonomic groups concerning radiant energy requirements for their chloroplast development should be noted here (e.g. formation of the chloroplast structure in conifers in darkness). Little is known about influences of temperature and humidity on the natural chloroplast development, leaving aside pathological effects of extreme temperatures or water stress.

Among other factors which influence plastid (chloroplast) ontogeny, natural and synthetic plant growth regulators ("growth substances") play an important role. Already in the early 1960s, kinetin was known to retard leaf senescence (Osborne 1962). In senescent chloroplasts of *Nicotiana rustica* leaves, cytokinin (6-benzylaminopurine) induces "rejuvenation" of their ultrastructure: formation of new thylakoids and later new grana, probably also formation of new ribosomes, disappearance of plastoglobules (Sveshnikova et al. 1966). Similar results were obtained by Młodzianowski and Młodzianowska (1973), Woźny and Szweykowska (1975), Borzenkova and Mokronosov (1976), Borzenkova and Bortnikova (1978), Naito et al. (1981), etc.; for review see Parthier (1979). According to Borzenkova and Nefedova (1981), in potato leaves the formation of photosynthetic apparatus (number of chloroplasts per mesophyll cell and their dimensions, chlorophyll and fraction 1 protein content, net photosynthetic rate) is completed at the time of maximal biological activity of endogenous cytokinins. The action of cytokinins on the chloroplast ontogeny seems to be light dependent. Cytokinins act on chloroplast ultrastructure indirectly, through stimulation of the accumulation of chlorophyll, plastid enzymes, and ribonucleic acids, and through stimulation of photophosphorylation. In the young intact plants where endogenous cytokinins are synthesized in roots, exogenously applied cytokinins do not influence the differentiation of chloroplasts.

As concerns other groups of growth regulators, their influence on chloroplast development is less known, comparing with cytokinins, especially in naturally developing leaves (for review see Parthier 1979, Sundqvist et al. 1980). Generally, auxins probably influence plastid ontogeny only slightly (they have an influence on starch formation), gibberellins have a retarding effect on chloroplast senescence, like the cytokinins, whereas abcisic acid and ethylene have an opposite influence (acceleration of the chloroplast senescence; Toyama 1980).

Many other substances influence the development of chloroplasts. Important are those used as herbicides (e.g. Meier et al. 1980, Pallett and Dodge 1980) but their effects on the natural chloroplast ontogeny have probably not been studied much so far. Protein synthesis inhibitors (some antibiotics) have been used as tools for the study of the involvement of chloroplast and cytoplasmic ribosomes (i.e., chloroplast and nuclear genome) in biosynthesis of chloroplast proteins during plastid ontogeny. Streptomycin, chloramphenicol, and 2-thiouracil retard and thus prolong the ontogeny of chloroplasts in leaves of potato, though their effects on chloroplast ultrastructure differ somewhat from each other (Döbel 1963). Wolf (1977) has reviewed influences of various chemical substances (antibiotics, phytohormones and their synthetic analogues, etc.) on chlorophyll synthesis and chloroplast development, but his review concerns mainly the greening of etioplasts.

Proteosynthesis, chlorophyll formation, and thylakoid membrane assembly in chloroplasts are also influenced by mineral nutrition of plants. Thus, Hudák and Herich (1976) observed injuries of developing chloroplasts in primary leaves of sunflower both during shortage or surplus of boron. The concentration of boron probably influences the activity of ribonucleases in the leaves. D'Agostino and Pennazio (1981) found in mature leaves of $C_4$ plant Gomphrena globosa (Amaranthaceae) typical symptoms of chloroplast senescence (large plastoglobules, degeneration of the system of photosynthetic membranes) under the influence of overall mineral deficiency (watering with tap water only). Ultrastructural changes, induced in the leaves by mineral deficiency, resemble those due to viral infection. Various metal cation and other mineral ion deficiencies may influence the ontogeny of chloroplast structure, especially those of nitrogen, magnesium, calcium, iron and manganese (for review see Silaeva 1978, Sundqvist et al. 1980). The study of influences of various substances on chloroplast development is at present of increasing importance in connection with problems of environmental pollution and its impact on photosynthesis.

Also some surgical treatments of plants influence the development of leaf chloroplast structure. Ljubešić (1968) followed the ultrastructure of chloroplasts during yellowing of tobacco leaves and their regreening induced by the removal of all younger parts of the plant. During leaf senescence, a substantial decrease in chloroplast volume, disappearance of stroma (non-appressed) thylakoids (grana or appressed thylakoids remain unchanged), and increase in number and size of plastoglobules occur. During regreening, full restoration of mature chloroplast structure (with several large plastoglobules persisting) occurs here. After prolonged yellowing, chloroplast senescence becomes irreversible (destruction of grana thylakoids, overfilling of the chloroplasts by plastoglobules). Huber and Newman (1976) and Tuquet and Newman (1980) observed "rejuvenation" of senescent chloroplasts in cells of yellow soybean cotyledons regreening after decapitation of the plants epicotyl. Intactness of chloroplast envelopes seemed to be a necessary condition for this process. In cotyledons of *Sinapis alba*, the rejuvenation of senescent chloroplasts may also be achieved by the decapitation of the plants (Hudák, unpublished). In mesophyll cells of primary leaves of continuously debudded *Phaseolus vulgaris* plants, prolongation of leaf and chloroplast life is observed. Senescence of these chloroplasts is prolonged and in senescent chloroplasts, conspicuous starch accumulation is observed in comparison with senescent chloroplasts of intact bean plants (Kutík *et al.* 1984). In such cases of prolonged chloroplast life span or rejuvenation, accumulation of cytokinins (transported from roots) and photosynthates in the leaves studied is probably the cause. Simola (1973) compared the ontogeny of chloroplasts in mesophyll cells of developing leaves of intact *Atropa belladona* plants with that in mesophyll cells of leaves differentiating in callus cultures derived from the stems of these plants. The chloroplasts developing *in vitro* have smaller grana and smaller starch inclusions than the naturally developing ones but the overall course of the ontogeny of both types of chloroplasts is similar.

Thus the ontogeny of chloroplast structure during leaf development is controlled by a complicated collaboration of nuclear and chloroplastic genome influenced by environmental factors (particularly by irradiance and quality of radiation), mediated mostly by hormonal activity.


## 2.3    CONCLUSIONS

Leaving aside studies on greening of etiolated leaves or various in vitro systems, a general scheme of mesophyll cell chloroplast structural ontogeny during leaf development may be outlined: In young leaves (sprouting from buds), a simple thylakoid system consisting of several layers of primary thylakoids with small grana is usually present in the chloroplasts. In some cases, prolamellar bodies may also be present here. Maximal degree of development of the thylakoid membrane system is achieved after the end of leaf area growth. Such chloroplasts may be considered as ultrastructurally mature. Soon afterwards, chloroplasts senescence begins, culminating during leaf yellowing. In senescent chloroplasts, the thylakoid system is gradually reduced simultaneously with the growth of plastoglobules where lipoid degradation products of thylakoid membranes are probably accumulated. The senescence of chloroplasts may be reverted or slowed down, *e.g.* by decapitation of a plant. In some groups of plants, deviations from this general scheme may be observed. Thus, in sempervirents, chloroplast life is extremely prolonged, or, in plants with $C_4$ photosynthesis, in a certain type of chloroplasts, grana are absent at chloroplast maturity. The whole chloroplast ontogeny described here may be considered as a certain part of the assumed pathway of plastid developmental interrelationships.

# 3 CHLOROPHYLLS AND CAROTENOIDS DURING LEAF ONTOGENY

*Z. Šesták*

## 3.1 THE IMPORTANCE OF CHLOROPHYLL AND CAROTENOIDS

Life on Earth begins and ends with chlorophyll: There is no photosynthesis, no accumulation of solar energy and its transformation into energy of organic substances without chlorophyll *a* or related pigments of photosynthetic bacteria (bacteriochlorophylls, bacteriorhodopsin). And without photosynthesis and plant production animals and men cannot live.

Photosynthetic reactions start as soon as the minimum necessary amount of chlorophyll is formed in thylakoid membranes (Smith 1949, *etc.* — see Chapter 5). Photosynthesis stops when the amount of chlorophyll in active complexes *in vivo* drops below the minimum level. Photosynthetic rate reaches its equilibrium with respiration when the so-called chlorophyll compensation point (Šesták 1964) is attained. Expression of photosynthetic activity in terms of chlorophyll amount, called assimilation number, has been the topic of many papers starting with the book by Willstätter and Stoll (1918). This magnitude has been replaced in recent papers by the expression photosynthetic unit (which characterizes the basic conglomerate of pigments, transporters of electrons and enzymes that realize the photosynthetic process) or only by $P700$ (the reaction centre chlorophyll of Photosystem 1). Thus the units for the expression of photosynthetic activity change, but the principle — pigment amount — remains, being only refined by relating to the main pigment acting in photosynthesis. And already the classics of chlorophyll research, Lyubimenko (1910, 1916) and Willstätter and Stoll (1913) mentioned that the chlorophyll metabolism and activities were strongly related to leaf ontogeny. The vast literature on chlorophyll changes connected with leaf ontogeny has been reviewed in detail (Šesták and Čatský 1967c, Šesták 1977a, 1983, 1984, Chaïka and Savchenko 1981).

The function of carotenoids in photosynthetic tissues of higher plants is still not clear (for a review see Goodwin 1980). They certainly function as antenna pigments, accepting radiant energy and transmitting it with some losses to chlorophyll *a* molecules. The importance of two other possible functions alternates in the literature with some periodicity: the photoprotection of chlorophylls against bleaching, and the direct participation of xanthophylls in the mechanisms of photosynthesis (the violaxanthin cycle participating in oxygen transport, the function of $\beta$-carotene in the C-550 absorbance change). These metabolic functions of carotenoids have been supported by recent findings of the presence of carotenoids in all native chlorophyll complexes in thylakoids, and by their abundance in the chloroplast envelope.

Ontogenetic changes in the contents of carotenoids have not been studied as often as those of chlorophylls (for review see Šesták 1978). The reason is not only their relatively lesser importance in photosynthesis, but also difficulties connected with the quantitative determination of carotenoids. The spectra of individual carotenoids are superimposed and, hence, their determination necessitates a preliminary chromatographic separation, which is tedious in large experimental series. The most often used column, paper and thin layer chromatographic methods of carotenoid separation usually yield four main zones of leaf carotenoids in amounts appropriate for spectrophotometric determination. They are $\beta$-carotene, lutein, neoxanthin, and violaxanthin. Their purity is often not satisfactory, the zones or spots may contain two pigments (lutein + zeaxanthin, cis + trans isomers, etc.) and, hence, the spectra of eluates differ from those of pure compounds described in the literature, and quantitative analysis is not accurate enough. Carotenoids are labile substances, sensitive to oxidation, irradiation, etc., and the analysis must be done with extreme care (see Davies 1976). The direct spectrophotometric measurement of the rough pigment extract at 480 or 440.5 nm yields the value for total carotenoids burdened with a large experimental error. Hence the evaluation of carotenoid contents given in the literature always requires a check of the method used for their determination.


## 3.2 BIOSYNTHETIC PROCESSES

### 3.2.1 Chlorophyll Biosynthesis

The content of chlorophyll in a leaf is the result of the balance of a steady chlorophyll synthesis and degradation. Experiments with leaves grown in darkness for a various number of days (etiolated leaves) and then irradiated, show that the amounts of chlorophyll precursors and the capacity for chlorophyll biosynthesis increase to a certain phase of leaf development and then decline. The phases of maximum accumulation of precursors and maximum capacity to chlorophyll synthesis are often different in the same plants.


#### 3.2.1.1 *Precursors of Chlorophyll*

The formation of immediate precursors of chlorophyll starts soon after the beginning of germination. Protochlorophyllide is detectable after 24 h (Rebeiz *et al.*

1970) and the rate of its synthesis in cucumber cotyledons rapidly grows during the 3rd d of germination. In primary leaves of field grown wheat the amount of protochlorophyllide per leaf increases up to 11 d and then declines (Chaïka and Savchenko 1975), while in etiolated leaves the rise is usually observed only to the age of 5 d (barley — Shlyk and Averina 1973; wheat — Klockare 1980). In dark-grown leaves of *Phaseolus vulgaris* the rapid phase of protochlorophyllide synthesis appears between 3 and 6 d (per fresh matter) or 4 and 10 d (per leaf) (Akoyunoglou and Siegelman 1968) and is followed by a plateau (Akoyunoglou and Argyroudi-Akoyunoglou 1969). Stimulation of protochlorophyllide formation by kinetin application is highest in 5 to 6 d seedlings (Shlyk and Averina 1973).

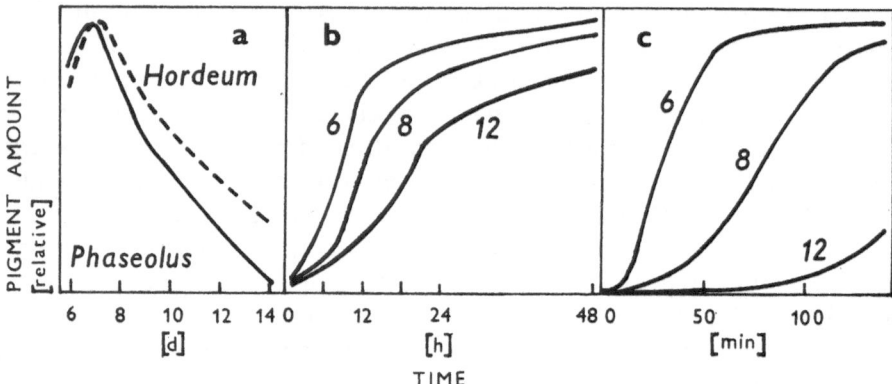

**Fig. 3.1.** Biosynthetic ability for chlorophyll production in etiolated leaves increases to *ca.* 7 d ' of age (abscissa) and then declines (a — *Phaseolus vulgaris* L., top value 500 mg chlorophyll *a* per kg fresh matter, from De Greef and Caubergs 1972; *Hordeum vulgare*, top value 300 μg chlorophyll per 100 leaf tops, according to Nadler and Granick 1970). The lag phase in chlorophyll formation prolongates and the final amount of synthetised chlorophyll declines with age (figures at curves [d]) of etiolated leaves (b — *Triticum vulgare* Vill., top value 800 mg chlorophyll (*a* + *b*) per kg, from Sundqvist *et al.* 1975). Also the rate of protochlorophyllide regeneration (abscissa: time of measuring absorbance at 650 nm) after a short irradiation declines with age of etiolated leaves (c — *Triticum aestivum* L., maximum difference of absorbance 0.2, according to Sundqvist *et al.* 1975).

Phytyl ester of protochlorophyllide, protochlorophyll, starts its slow accumulation soon after the beginning of germination (1 d in cucumber cotyledons — Rebeiz *et al.* 1970). 2 d old etiolated *Phaseolus vulgaris* leaves contain equal amounts of protochlorophyllide and protochlorophyll, while in 7 d leaves protochlorophyllide prevails (Lancer *et al.* 1975). The regeneration of protochlorophyllide and proto-chlorophyll following leaf irradiation (Fig. 3.1 c) is slower and has a longer lag period in older leaves than in the younger ones (wheat — Virgin 1961, Sundqvist *et al.* 1975; maize — Spruit and Raven 1970, Raven 1973; *Phaseolus vulgaris* — Akoyunoglou and Siegelman 1968, Akoyunoglou 1970, De Greef and Caubergs 1972, Lancer *et al.* 1975, 1976; *Pharbitis nil* — Ogawa *et al.* 1978). These calculations hold per unit of leaf biomass; per one leaf an increase and decrease may be observed with a maximum in 3 to 4 d old leaves (Virgin 1961).

The decrease in photochlorophyllide amount during ageing of an etiolated leaf may be accompanied by a decrease (barley — Hendry and Stobart 1978) or increase (wheat — Klockare 1980) in the ability to produce chlorophyll upon irradiation. This ability usually peaks in young leaves (2 d in *Pharbitis nil* — Ogawa *et al.* 1978; 4 d in *Cucumis sativus* — Dei 1978) and then declines (Fig. 3.1 a). Cultivation under long far-red radiation ($\lambda > 710$ nm) induces a higher protochlorophyllide content and higher chlorophyll synthetic capacity than cultivation in darkness (Klockare 1980). Phytylation rate grows linearly with the age of *Phaseolus* leaves from 5 to 12 d (Akoyunoglou and Michalopoulos 1971), which is in agreement with the increase in phytol content with leaf age (Liljenberg 1974). As a result, the chlorophyll amount formed after the same time of irradiation is greater in younger cotyledons (melon — Onwueme and Lawanson 1975) or leaves (barley — Nadler and Granick 1970, Valcke 1978; oats — Treffry 1975; wheat — Virgin 1966, Sundqvist *et al.* 1975; maize – H. Koch 1976; tobacco — Harris and Naylor 1968a, b, 1970; *Phaseolus vulgaris* — De Greef and Caubergs 1972, Lancer *et al.* 1976; sunflower — Ford *et al.* 1977) than in the older ones. The lower the temperature is (in the range of 35 to 20 °C) during the irradiation of etiolated leaves, the greater is the decline in rate of chlorophyll synthesis induced by leaf age (H. Koch 1976). Young leaves use various substrates (glycine, succinate) for chlorophyll synthesis more effectively than old leaves (Harris and Naylor 1968b). The initial lag phase of chlorophyll *a* synthesis (Fig. 3.1 b) is longer in older leaves (Spruit and Raven 1970, Raven 1972). It does not appear until the leaves reach some definite age, *i.e.* 5 to 6 d (Sisler and Klein 1963, Akoyunoglou and Argyroudi-Akoyunoglou 1969, Nadler and Granick 1970).

In the first hours of irradiation of etiolated leaves the synthesis of chlorophyll *a* is more rapid than that of chlorophyll *b*, later the *a/b* ratio stabilizes at 2 to 3. After the darkening of leaves the *a/b* ratio increases, while re-irradiation induces a new decline (green radish leaves — Lichtenthaler *at al.* 1981a). Under the flashing regime the *a/b* ratio is higher than in older leaves (Akoyunoglou and Argyroudi-Akoyunoglou 1969). Also the rate of chlorophyll synthesis in light (2 min) – dark (98 min) cycles (Argyroudi-Akoyunoglou and Akoyunoglou 1970) and under red and far-red radiation treatment (Akoyunoglou 1970) depends on leaf age.

The specific activities of chlorophylls in green leaves exposed to $^{14}CO_2$ show that chlorophyll regeneration is more rapid in young that older barley leaves in various phases of plant development (Gaponenka *et al.* 1977). In *Ceratophyllum demersum* $^{14}C$ is incorporated more into chlorophylls *a* and *b* in young upper than in old lower leaves (Shlyk *et al.* 1960, Shlyk and Fradkin 1961, 1962): this difference is observed for both the phytol and phorbine parts of chlorophyll molecules (Shlyk and Kukhtenko 1961). Young top parts of a *Ceratophyllum* shoot regenerate chlorophyll more rapidly than old base parts (Gaponenko *et al.* 1974).

### 3.2.1.2  Enzymes of Chlorophyll Biosynthesis

The activities of enzymes participating in chlorophyll biosynthesis and degradation also change with leaf age. The specific activity of δ-aminolevulinate dehydratase reaches its maximum sooner in younger leaves (Sundqvist et al. 1975). In the third leaf of *Capsicum frutescens* its activity declines rapidly with blade expansion to 5 cm² and then declines only very slowly (Steer 1973). In radish cotyledons the activity of this enzyme decreases after the chlorophyll content per fresh matter reaches a constant level (Shibata and Ochiai 1975, 1976).

The highest chlorophyllase activity was found in the leaves of middle insertion in tobacco (Zakaryan 1966) and *Aristolochia sipho* (Tsoneva 1969/70), while in tea (Ogura 1969), spinach and sugar beet (Vorob'eva and Krasnovskiĭ 1966), and in *Taxus baccata* and pine (Tsoneva 1969/70, 1972/3) in young leaves. Old yellowing leaves of sugar beet contain chlorophyllase mainly in its labile form, extractable with *Triton X-100*, while in young and mature leaves the enzyme is mainly in its bound form (Sud'ina et al. 1975, 1976). The sum of both chlorophyllase activities mostly increases and afterwards declines during leaf ageing (Sud'ina et al. 1976).

### 3.2.1.3  Chlorophyll Biosynthesis in Various Parts of the Leaf Blade

Differences in tissue age and structure on the area of the leaf blade and, hence, also in synthetic capacity must be taken into account in studies of chlorophyll biosynthesis (Shlyk et al. 1960, 1966, Shlyk and Averina 1969, etc.). In darkness, protochlorophyll content declines from the leaf tip to base (maize — Chaïka and Savchenko 1973; barley — Henningsen and Boynton 1974). Shlyk and Kostyuk (1972) found a more active protochlorophyllide formation in the presence of δ-aminolevulinic acid in the dark in the upper 2 cm of the first barley leaf than in the lowest 2 cm. In light grown primary wheat or barley leaves the maximum protochlorophyllide and chlorophyll contents were in the middle or second fifth of the leaf blade from the base (Shlyk and Savchenko 1967, 1970), the maximum ratio of protochlorophyllide to chlorophyll being in the leaf base. In maize leaf grown in the light the maximum protochlorophyllide content was in the second zone from the tip and the maximum chlorophyll content in the apical part. Hence the ratio of protochlorophyllide to chlorophyll *a* showed a general increase in the direction of the leaf base (Tageeva et al. 1969), but this difference declined with leaf age (Chaïka and Savchenko 1973). With prolonged irradiation, the maximum chlorophyll amount shifted from the middle zone of the leaf to zone 2 to 6 mm distant from the tip (Obendorf and Huffaker 1970). The position of the zone of maximum chlorophyll accumulation depends on the length of leaf growth in the dark. Specific activities of chlorophylls are higher in younger base than older top zones of maize leaves and in leaves of younger than older plants (Gaponenka et al. 1977, 1980,

Gaponenko *et al.* 1977). These differences are larger for chlorophyll *a* than *b* (Gaponenka *et al.* 1977).

The first hours of darkening of the youngest fully developed barley leaf induce a small decline in chlorophyll *a* content. The content of protochlorophyllide increases in the dark in the upper and lower parts of both the youngest developing and fully developed leaf (Savchenko and Chaïka 1975). These changes are reflected also in the heights of fluorescence maximum (see Section 3.6.1). In the dark, chlorophyll accumulation was *ca.* five times as rapid in the apex than in the base of *Posidonia* leaves (Adamson and Hiller 1981).

### 3.2.1.4  *Chlorophyll Biosynthesis and Chloroplast Ultrastructure*

The chlorophyll accumulation is, of course, controlled not only by the rates of processes of chlorophyll biosynthesis, but also by the formation of chloroplast ultrastructure (*cf.* Chapter 2) . Henningsen and Boynton (1974) showed that the slowing down of chlorophyll accumulation in the middle parts of primary leaves of barley with plant age from 5 to 11 d is connected with the decrease in ability to reorganize the prolamellar body membranes into primary thylakoid layers. The slowing down of chlorophyll synthetic capacity during ageing in the dark is also in relation to the decline in soluble protein content (Obendorf and Huffaker 1970). The rate of chlorophyll synthesis in 4 and 5 d etiolated barley shoots was higher than in isolated prolamellar body membranes of the same age, but the membranes reached the peak value 1 d sooner and after a longer etiolation (8 or 9 d) were more active in chlorophyll synthesis (Griffiths *et al.* 1976). The sigmoid time course of chlorophyll formation during the 45 h irradiation of primary leaves of 16 d old etiolated *Phaseolus vulgaris* plants was similar when related per chloroplast or per leaf (Gyldenholm 1968).

Chlorophyll formation is localized in special parts of the thylakoid membrane. Thus Harris and Naylor (1968a) irradiated young (upper three) and old (lower three) leaves of tobacco for 2 to 24 h and then measured the amounts of chlorophyll formed in fragments of various sizes separated from blended leaves by differential centrifugation. The decreasing order of chlorophyll content of the pellets was 5000, 50 000, 1500, 500, 100, and 105 000 $\times$ *g* for both types of leaves, but young leaves always formed more chlorophyll in every pellet type than the old leaves. Chlorophyll *a* and *b* formation continued even in isolated prothylakoid vesicles prepared from etioplasts and etiochloroplasts of oats (Wellburn and Hampp 1979). The ratio of chlorophyll *a/b* increased during the formation of chloroplast ultrastructure: a transitory increment was connected with the formation of prolamellar bodies (Valanne *et al.* 1981).

**Table 3.1** Examples of changes in chlorophyll contents during leaf ontogeny. Abbreviations and explanations:
↓ — decrease, ↑ — increase, ↑ to one (per dry matter) or two (per leaf area) decimal points. All data mean changes in chlorophyll $(a + b)$, if not stated otherwise. Only values for normal (untreated) plants are given.

| Plant (and cultivation) | Leaf age (numbered from the oldest one) | Amounts of chlorophyll $(a + b)$ in g per | | Change in chlorophyll amount and $a/b$ ratio induced by ageing | Author(s) and year of publication |
| | | kg dry matter | m² of leaf surface | | |
| (1) | (2) | (3) | (4) | (5) | (6) |
| **Herbs:** | | | | | |
| *Cucumis sativus* L. | cotyledons, 14 to 22 d | 0.1...0.4 mg per cotyledon | | ↑ to 16 d and ↓ | Lewington and Simon 1969 |
| *Cucumis sativus* L. cv. Long Green Trailing (trays with John Innes potting compost, glasshouse) | first true leaves, 11 to 33 d | 0.1...3.1 mg per leaf | | rapid ↑ to 17 d and slow ↓ | Eilam *et al.* 1971 |
| *Gossypium hirsutum* L. cv. C4727 (pots with soil + sand) | leaves of 7th–8th nodes | | 0.26...0.43 | ↑ to 10 d and ↓ | Khodzhaev *et al.* 1978 |
| *Lactuca sativa* L. (vermiculite with nutrient solution, growth room) | leaf 2, 8 to 21 d | 0...0.1 | 0.01...0.08 | per area or leaf: ↑; per dry matter: ↑ to 15 d and ↓; $a/b$ ↑ to 9 d and ↓ | Bourdu *et al.* 1975 |
| *Phaseolus vulgaris* L. cv. Jantar (pots with soil, growth chamber) | primary leaves, 7 to 30 d | | 0.14...0.65 | ↑ and ↓; $a/b$ ↑ and ↓ | Šesták *et al.* 1975, 1977; Strnadová and Šesták 1974 |
| *Phaseolus vulgaris* L. cv. Yamashiro-kurosando-saito (growth chamber, vermiculite) | primary leaves, 6 to 36 d | 0.4...1.1 mg per leaf | | ↑ to 18 d and ↓ | Naito *et al.* 1979 |

| Species | Leaf / age | Ratio | Content | Notes | Reference |
|---|---|---|---|---|---|
| *Solanum tuberosum* L. "seyanets 63-13-12", cv. Imandra, Belyi Kochubei (field) | leaf 5, 1 to 65 d | | 0...0.47 | both *a* and *b* ↑ up to 41 d, then ↓; a small trough at 20 d | Mokronosov *et al.* 1973 a, b |
| *Solanum tuberosum* L. cv. King Edward (growth chamber, pots with loam, two photoperiods) | leaves 6, 9, 12, 15, 18 and 21 | | 0.20...0.58 | ↑ and ↓; max in leaf 6 – small differences in short-day plants | Frier 1977 |
| *Spergula vernalis* Willd. | leaves 1 to 4 | 3.2...9.4 | | ↑ and ↓, max leaf 3; *a/b* ↓ | Symonides 1974 |
| *Triticum aestivum* L. (growth chamber, pots with clay-loam, sand and perlite) | primary leaf, 5 to 50 d | 2...112 µg per leaf | | ↑ to 14 d and slow ↓ | Peoples and Dalling 1978 |
| *Triticum aestivum* L. cv. 232 (dwarf) (water culture with Knop solution) | leaves 1, 3, 5 and 7 | | 0.07...0.43 | ↑ and ↓; max in later developed leaves | Chuchalin *et al.* 1977 |
| *Vicia faba* L. (soil culture, glasshouse or garden) | cotyledon | 0.3...70.0 µg per cotyledon | | ↑ to the length of 26 mm and ↓ | Millerd *et al.* 1971 |
| Evergreen plants: | | | | | |
| *Coffea arabica* L. (pots with soil, outdoors) | leaf pair 3, 58 to 125 d | 2.8...13.3 | | ↑ to 63 d and ↓ in both sun and shade leaves | Bergmann *et al.* 1970 |
| *Ficus elastica* Roxb. (glasshouse) | bud – 2 d – 20 d – 30 d – 1 year – many years – yellow – brown | | 0...0.36 | ↑ and ↓; max 1 to more years; *a/b* no difference | Lichtenthaler 1969b, 1972 |
| *Pinus nigra* Arn. ssp. *nigricans* Host. var. *austriaca* (Höss.) Novak (9 year trees) | needles 1 to 37 months old | 2.2...2.9 | | ↑ to 13 months and ↓; *a/b* ↓ | El Aouni and Mousseau 1974 |
| *Pinus radiata* D.Don (6 year tree, forest) | needles 1 to 4 years old | 2.3...4.2 | 0.14...0.37 | per dry matter or volume: ↓, per area: ↑ to 1 year and ↓; *a/b* similar to (*a* + *b*) | Wood 1974 |

### 3.2.2 Carotenoid Biosynthesis

The effects of leaf age on processes of carotenoid biosynthesis have been rarely studied. The amount of carotenoids in dark grown leaves moderately increased to *ca.* 6 d of age, but the synthetic capacity after irradiation decreased with leaf age (wheat — Virgin 1966). Both carotenoid degradation in the dark and renewed synthesis after "blue" irradiation were more rapid in young than old leaves (maize — Brandt and Tageeva 1967). The study of six categories of particles of different sizes (sedimenting at 100 to 105 000 × $g$) from young or old etiolated tobacco leaves showed that irradiation stimulated the carotenoid synthesis mainly in young leaves, which contained more carotenoids (Harris and Naylor 1968a). Also the specific activity of $^{14}$C-labelled $\beta$-carotene declined during the first two years of life of *Picea pungens* needles, but the final low level of specific activity did not change from the 3rd to the 11th year of needle life (Khodasevich *et al.* 1970).

## 3.3 PIGMENT CONTENTS DURING LEAF ONTOGENY

### 3.3.1 Chlorophyll Contents

While the material for studies of chlorophyll biosynthesis are mostly etiolated leaves, mainly young primary leaves, and the studies are often related to the activities of growth regulators, quality of irradiance, *etc.*, studies on ontogenetic changes in chlorophyll amounts deal with leaves growing in a normal environment (field, forest, greenhouse) or in growth chambers with a controlled regime. The amount of values in the literature is large, but only a part of the papers contains a sufficient number of reliable measurements for the characterization of the whole leaf ontogeny. Generally, during leaf ontogeny chlorophyll accumulates up to some maximum level and afterwards the rate of degradation processes overtakes the rate of synthetic processes (for the problems of chlorophyll degradation during leaf senescence see Kufner 1980/1).

#### 3.3.1.1 *Herbs and Deciduous Trees*

The general trend of chlorophyll ($a + b$) accumulation during the ontogeny of each leaf is an increase to a maximum followed by a decrease (Table 3.1). The rate of increase is usually more rapid than the rate of decrease, especially in leaves formed in the later phases of plant ontogeny, the whole life span of which is longer. Detectable quantities of chlorophyll are found very soon after the soaking of seeds (after 40 h in *Phaseolus*), later on (after 24 h) a rapid chlorophyll synthesis starts, often linear per leaf area, exponential per one leaf (*Phaseolus* — Więckowski 1969, Gruber *et al.* 1973). After unfolding the leaf expands in size and at the same time its thickness increases. Both these values reach their maximum at about the same time. The amount of dry matter per leaf area unit often continues to increase slowly up to the end of metabolic activity of the leaf. The rate of chlorophyll synthesis per unit leaf area or dry matter is high in the first half of the leaf expansion phase, then it slows down. Chlorophyll accumulation per unit leaf matter stops usually prior to reaching the maximum leaf area, per unit leaf area prior to, at, or after reaching the maximum leaf area (sometimes as late as 14 to 16 d afterwards — Treharne et al. 1968). Afterwards the chlorophyll amount per both relation units declines.

In leaves of a long life span (e.g. leaves of trees) usually a plateau in chlorophyll amount appears (e.g. Lichtenthaler 1971), or chlorophyll accumulates further on with the accumulation of leaf dry matter (Pieters 1974). Similar effects are observed also in leaves whose life span is prolonged artificially, e.g. in leaves of plants whose new leaves are continuously removed immediately after buds appear (Ïordanov and Popov 1967, Srivastava and Atkin 1968, Harnischfeger 1973, Šesták et al. 1977, Lane and Thompson 1978 — see also Fig. 3.2c), in rooted leaves (Yusufov and Ashurova 1975) or intact leaves treated with growth regulators (Adedipe et al. 1971).

Analogous changes as those observed per unit dry matter of leaf are found also per unit fresh matter, but this unit should not be used in plant physiological or biochemical observations.

The changes found for whole leaves may be related to ontogenetic changes in the leaf anatomy and chloroplast ultrastructure. The amount of chlorophyll per chloroplast increased with the age of spinach leaf (Bottrill and Possingham 1969). In potato leaves, the amount of chlorophyll molecules per chloroplast increased from the 6th to the 20th d and then declined to the 35th d in both palisade (0.93 to 1.86 $\times$ 10$^9$ chlorophyll molecules per chloroplast) and spongy (1.20 to 2.45 $\times$ 10$^9$ molecules) parenchyma; the highest ratio of chlorophyll molecules per palisade/spongy chloroplast was on the 10th d (0.93) and the lowest one on the 35th d (0.58) of leaf life (Mokronosov 1981). The expansion of the second leaf of lettuce (Bourdu et al. 1975) took place between the 8th and 21st d of plant life. Chlorophyll accumulation per leaf stopped on the 17th d, per leaf area and dry matter units on the 15th d. The multiplication of chloroplasts stopped on the 11th d, in the phase of rapid expansion of leaf blade, and, hence, between the 11th and 13th d a sharp decline in chloroplast amount per leaf area unit was observed. During this time the formation of new thylakoids and grana in chloroplasts continued, which resulted in a sharp increase in chlorophyll amount per chloroplast. The dimensions of chloroplasts characterized by their section area increased to the 15th d. The number of chlorophyll molecules per area unit of thylakoid membrane declined from the 8th to 11th d and then slowly increased. The ratio of chlorophyll a/b started to decline from the 11th d. In leaves 4 and 5 of spinach the leaf area, the number of chloroplasts per mesophyll and palisade cells, and the amount of chlorophyll per area unit increased between 10 and 24 d, while the cell amount per leaf area unit declined (Possingham and Saurer 1969). In *Gomphrena globosa* the formation of stabilized chloroplast ultrastructure was accompanied by a stable chlorophyll content (Appiano *et al.* 1979). In *Cyphomandra betacea* leaves the final phases of chlorophyll accumulation and the early phases of chlorophyll decline were connected with the formation of small vesicles and tubular apparatus in the inner chloroplast membrane (Harris 1978). During the development of the 5th tier leaf of potato, chlorophyll amount per chloroplast increased from 0.48 (8 d) to 0.61 (13 d) and 0.78 (26 d) pg under long-day, while only from 0.43 (8 d) to 0.48 (13 d) and 0.58 (26 d) pg under a short-day photoperiod (Mokronosov and Bagautdinova 1974). In the leaf of the 7th tier of potato, the blade of which reaches the maximum area at the age of 23 to 25 d, the volume of cells and the number of chloroplasts per cell increases to the 15th d, the chloroplast volume to the 8th d, and the amount of chlorophyll ($a + b$) per chloroplast in one year of cultivation to the 4th d, in another year much longer (Borzenkova and Nefedova 1981). During the 20 d of development of cotton leaves of the 7th and 8th node, chlorophyll amount per chloroplast changed from 0.24 (1 d) to 0.77 (10 d) and 0.58 (20 d) pg, which represented 160, 520 and 390 million chlorophyll molecules, and 540 000, 1 700 000 and 1 300 000 photosynthetic units, respectively (Khodzhaev *et al.* 1978). These time specifications are, of course, valid only for the given plant species, cultivar, and growing conditions.

Synthetic activities of nucleic acids are a necessary condition not only for the development of leaf mass, but also for the formation of chlorophyll and its *in vivo* forms. Thus the expansion of primary leaves of *Phaseolus* was preceded by a rapid increase in amounts of RNA (maximum in 11 d plants followed by a decline; a similar time course was followed by the amounts of ribosomes and polyribosomes) and DNA (maximum at 12 d followed by a steady state): maxima of leaf area and chlorophyll ($a + b$) content were reached on days 14 and 16, respectively (Makrides and Goldthwaite 1981).

Similar changes as in true leaves were observed also in cotyledons (Fig. 3.2) whose area and total chlorophyll contents increased during the 15 to 40 d of their metabolic activity, the result being an increase and decline in the chlorophyll amount per cotyledon (cucumber — Draper 1969, Lewington and Simon 1969, Ferguson and Simon 1973; *Vicia* — Millerd *et al.* 1971; mustard — Hong and Schopfer 1981; soybean — Abrahamsen and Mayer 1967; *Acer* species — Ampofo *et al.* 1976; *Fagus silvatica* — Ampofo *et al.* 1976; various plant species — Lovell and Moore 1970) or cotyledon area unit (cucumber — Harnischfeger 1973; *Acer* species — Ampofo *et al.* 1976).

The ratio of chlorophyll $a/b$ is high at the beginning of chlorophyll synthesis in young chloroplasts (chlorophyll $a$ is synthesized preferentially), the following rapid decline is replaced by a slow increase and final decline. Nevertheless, other ontogenetic courses were also observed, *e.g.* a long steady state followed by a rapid decline (primary leaves of wheat — Fedtke 1973; rape — Diepenbrock and Geisler 1978), a more or less continuous decline (*Spergula vernalis* — Symonides 1974; tobacco – Amiri 1971; wheat — Sirohi and Ghildiyal 1975; Patterson and Moss 1979; *Sinapis* — Wild *et al.* 1981c), and various courses (rice leaves of different insertions — Youn and Ota 1973).

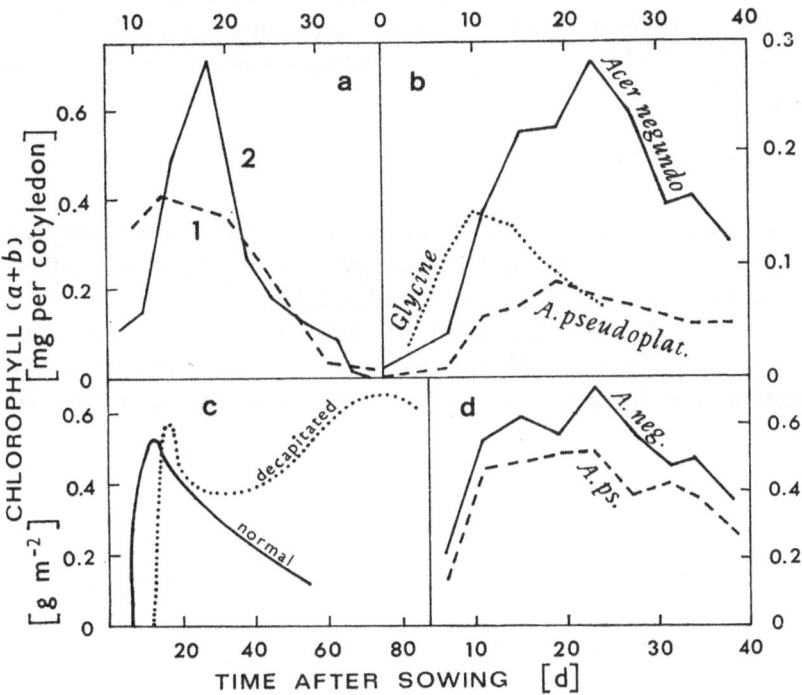

**Fig. 3.2.** Changes in chlorophyll content during ontogeny of a cotyledon. Even in the same species and cultivar grown in the same medium in the same greenhouse (a) different absolute amounts and time courses may be found due probably to different climatic conditions and time of year (*Cucumis sativus* L. cv. Long Green Trailing, John Innes potting compost, greenhouse at Queen's University, Belfast: 1 — Draper 1969, 2 — Ferguson and Simon 1973). Cotyledons of different species of the same genus (b, d: *Acer negundo* L. and *Acer pseudoplatanus* L. — values from Ampofo *et al.* 1976) show different ontogenetic courses of chlorophyll contents per cotyledon (b) than per leaf area unit (d). Cotyledons of different plant species may differ not only in chlorophyll amounts, but also in the life span of the cotyledon (b — *Glycine max* (L.) Merr. — values from Abrahamsen and Mayer 1967). Decapitation or continuous debudding of a plant prolongates the life span of the cotyledons and postpones chlorophyll degradation (c — *Cucurbita maxima*, from Harnischfeger 1973).

The ontogenetic changes in chlorophyll *a/b* in mesophyll and bundle sheath cells of the secondary maize leaf were different: in mesophyll the ratio declined from 2.6 on the 6th d to 2.0 on the 12th d and then increased to 3.0 on the 21st d, while in bundle sheath it increased from 2.2 on the 6th d to 4.6 on the 12th d and then declined (Bishop *et al.* 1971, Andersen *et al.* 1972). Similar values were observed by Bourdu and Krivitzky (1972).

The ontogenetic changes in chlorophyll amount and chlorophyll *a/b* ratio were often screened by other factors affecting the physiological state of the leaf, *e.g.* leaf hydration level (Pospíšilová *et al.* 1976), photoperiod (Aérov 1966), wavelength of radiation (Wild and Holzapfel 1980), seed spacing (Bedenko 1980), mineral nutrition (Lipskaya 1971); therefore all accounts of changes in chlorophyll amount during leaf ontogeny must be based on analyses of plants grown in an optimal environment.

### 3.3.1.2  *Evergreen Plants*

In leaves whose life span is longer than one year all ontogenetic changes proceed more slowly (Fig. 3.3). Chlorophyll amount per leaf or other unit usually does not reach its maximum in the first year of leaf life (Table 3.1).

In *Ficus elastica* the highest amounts of chlorophyll per leaf area unit were in green leaves older than 1 year (Lichtenthaler 1969b). In spruce the chlorophyll *a* and *b* amounts per dry matter increased from needles of the current year to 5

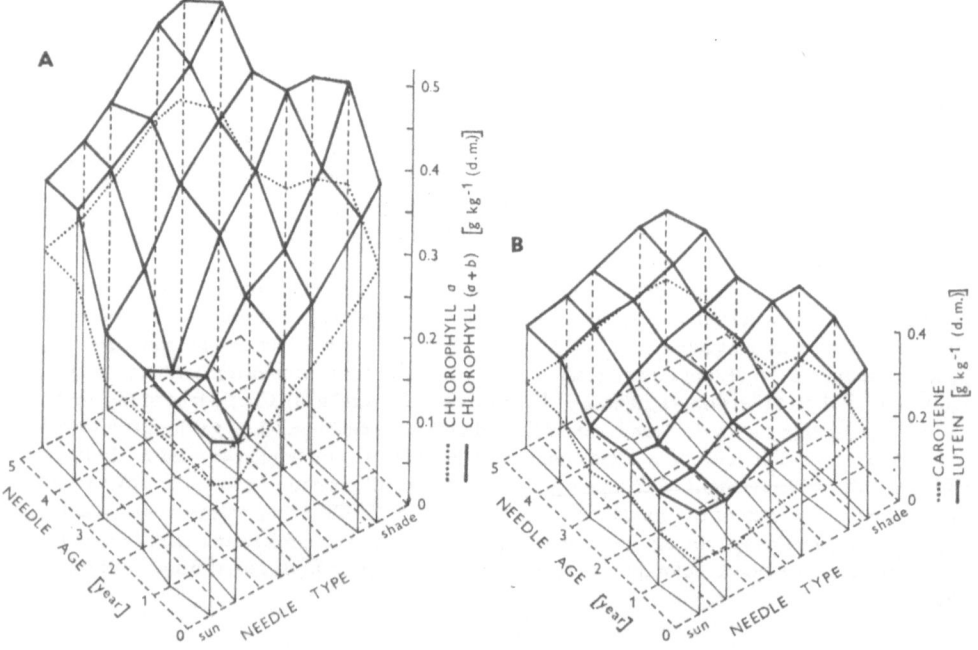

**Fig. 3.3.** The amounts of pigments in needles of evergreen plants (*Picea abies* Karst.) change rather slowly during their long life span also in dependence on the position in tree crown determining their irradiation. The differences in amounts of chlorophyll *a*, (*a* + *b*) (A) and α-carotene and lutein (B) were more or less parallel. (From W. Koch 1976.)

year old needles; more than twice as much chlorophyll was present in needles grown in the shade than in full sunlight (W. Koch 1976). On the other hand, in *Pinus radiata* the chlorophyll amount per both dry matter or leaf area units declined from 1 year to 4 year old needles (Wood 1974). The chlorophyll *a/b* ratio usually steadily declines during leaf ontogeny (*Theobroma* — Baker and Hardwick 1976; *Citrus* — Freeman *et al.* 1978).

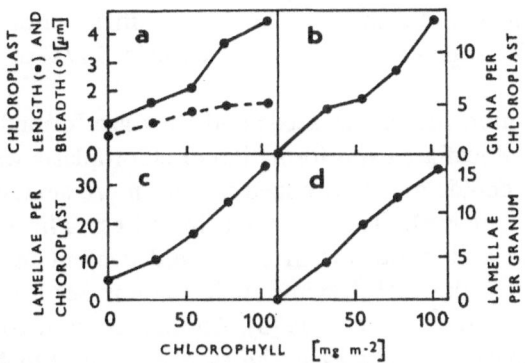

**Fig. 3.4.** Increase in chlorophyll (*a* + *b*) amount per leaf area unit (*Theobroma cacao* L.) is connected with increase in chloroplast dimensions (a) and formation of chloroplast ultrastructure, *i. e.* amount of grana (b) and thylakoids (lamellae) per chloroplast (c) or granum (d). (From Baker *et al.* 1975.)

Changes in chlorophyll content are related to the changes in leaf anatomy and chloroplast ultrastructure. In cocoa leaves, the numbers of lamellae per plastid and grana per plastid increased almost linearly with the amount of chlorophyll per area of leaf blade (Baker and Hardwick 1975a, b, Baker *et al.* 1975 — see Fig. 3.4). Also by the year-round changes of environment are important: thus the cultivation in a plastic greenhouse changes the ontogenetic course and induces an increase in chlorophyll amount in comparison to growing in the open (*Pinus silvestris, Picea abies, Picea mariana* — Linder 1980). In addition to the general climate the micro-climate inside the tree crown, especially the radiation regime, are important. These cyclic seasonal changes in chlorophyll content often superimpose the changes dependent on leaf age (Senser *et al.* 1975).

## 3.3.2 Carotenoid Contents

### 3.3.2.1 *Herbs and Deciduous Trees*

The general trend of content of the sum of carotenoids during leaf ontogeny is similar to that of chlorophylls. Nevertheless, the time course of carotenoid accumulation may be different from that of chlorophyll.

There are three types of changes in carotenoid content during leaf ontogeny, occurring in different plant species: In some species the increases and declines in chlorophyll and carotenoid contents are more or less parallel (*Quercus robur* — Goodwin 1958; sugar beet — Godnev and Shabel'skaya 1966), in some species the amounts of both chlorophylls and carotenoids remain rather high at the end of leaf life (*Prunus nigra* — Goodwin 1958), and in other species carotenoids are formed and/or persist longer than chlorophylls and induce a brown, red or yellow colouration, *e.g.* in leaves of some deciduous trees in autumn (*Acer pseudoplatanus* — Goodwin 1958; *Fagus silvatica* — Lichtenthaler 1971) or in rye (Kaval'chuk and Vechar 1973).

Carotenoids are often present in seeds and they are formed even in darkness. During the growth of leaf area and formation of chloroplasts with fully developed ultrastructure, carotenoids are synthesized usually more slowly than chlorophylls and thus the ratio chlorophylls/carotenoids increases. The calculation per dry matter even showed a slow decline in carotenoid content in the second leaf of lettuce from the 8th to the 21st d, while their amount per leaf and leaf area unif increased to a plateau (Bourdu *et al.* 1975). An increase in carotenoid amount per fresh matter of primary leaves of *Phaseolus vulgaris* from 0 to 48 h of leaf growth was followed by a decline up to 192 h (Więckowski 1961). An almost stoichiometric decrease in carotenoid concentration with chlorophyll accumulation was observed in young tobacco leaves (Harris and Naylor 1968b). During irradiation of etiolated leaves of maize seedlings the amount of the sum of carotenoids per leaf area unit increased up to 25 h and then decreased, while the amount of chlorophyll continuously increased (Tageeva *et al.* 1971). A decline in the ratio chlorophyll/carotenoids may be calculated for the interval of 4 to 19 d of life of the first leaf of rye cv. Belta (12.9 → 4.0), Druzhba (8.5 · → 4.8), and Benyakonskae (11.6 → 3.8) (Kaval'chuk and Vechar 1973).

Similar changes as in true leaves were observed in cotyledons (radish — Ochiai *et al.* 1971a, b; mustard — Hong and Schopfer 1981). In leaves of a long life span (*e.g.* leaves of deciduous trees) usually a plateau in the amount of carotenoids appeared (*e.g. Fagus silvatica* — Lichtenthaler 1971).

The accumulation of all major carotenoids proceeds most often in parallel, but with various velocities. The amounts of individual carotenoids declined usually in the sequence lutein — carotenes — neoxanthin — violaxanthin as evident from the results of Więckowski (1961) with primary leaves of *Phaseolus vulgaris*, but after leaf unfolding all carotenoids accumulated only till 48 h of continuous irradiation, and then their amounts per leaf fresh matter declined, while per one leaf the increase continued up to 144 h and even then declined.

The character of ontogenetic changes in carotenoid content depends not only on plant species, but also on plant cultivar (potato — Kislyakova *et al.* 1967; wheat — Parshina *et al.* 1972; rye — Kaval'chuk and Vechar 1973; *etc.*) and environmental conditions.

### 3.3.2.2 Evergreen Plants

In leaves of life span longer than one year, the carotenoid amount per leaf usually does not reach its maximum during the first vegetation season (Fig. 3.3 B) and often not even after several years (*Picea pungens*, *P. excelsa* — Godnev *et al.* 1969; *P. abies* — W. Koch 1976). The amounts of carotenoids change usually during the vegetation season, sometimes reaching the highest amount in autumn and winter months (fir, pine — Sirotkin and Anufrieva 1973) or summer months (*Pinus silvestris*, *Picea abies* — Linder 1972). In addition to temperature also irradiance of leaves varying inside the crown controls the carotenoid content (W. Koch 1976).

The ratio of chlorophyll/carotenoids mostly slowly increases with leaf age (Linder 1972, W. Koch 1976), in some plants it remains almost unchanged for several years and then rapidly declines prior to leaf abscission (*Ficus* — Lichtenthaler 1969b, Lichtenthaler and Weinert 1970). In *Cereus peruvianus* the ratio fell with leaf age (Lichtenthaler 1969a). An inverse relaation of chlorophyll and carotenoid amounts in fir and pine needles was observed in the course of a year (Sirotkin and Anufrieva 1973). The amounts of individual carotenoids mostly change in parallel (*Taxus baccata* — Kufner *et al.* 1978). In cocoa flush leaf after a lag phase the carotenoid concentration as well as chloroplast dimensions and the number of grana and thylakoids per one chloroplast increased parallel to the chlorophyll amount from 50 to 150 mg m$^{-2}$. The leaves with 150 mg m$^{-2}$ chlorophyll contained the following amounts of carotenoids [mg m$^{-2}$]: lutein 13.0, neoxanthin 4.5, $\alpha$-carotene 1.9, $\beta$-carotene and violaxanthin 1.3 (Baker *et al.* 1975).

## 3.4   LEAF INSERTION AND PIGMENT CONTENTS

### 3.4.1   Chlorophyll and Leaf Insertion

In the course of the vegetation season leaves are formed successively. This means that each leaf is formed in another phase of plant ontogeny, when the plant has different amounts of substrates for disposal, and under various climatic conditions. Also when the plants are grown under controlled conditions, the heterogeneity of leaves is preserved. Primary leaves often differ in shape from secondary leaves. The first leaves formed reach smaller maximum dimensions than the leaves of middle insertion, and the last formed leaves are usually smaller than the middle leaves.

Similar differences are typical for other leaf parameters, including chlorophyll amounts (Fig. 3.5). Changes analogous to those during the ontogeny of one leaf, *i.e.* increase and decrease, occur also along the insertion gradient, *i.e.* when all leaves on one plant are analysed simultaneously (Table 3.2). Thus the highest maximum chlorophyll contents are mostly reached by the leaves of medium insertion. Nevertheless, the position of leaf with the maximum chlorophyll amount per area or dry matter unit shifts during the vegetation season from the lowest leaf insertion to the highest insertion (see, *e.g.*, *Brassica napus* — Skośkiewicz 1973; *Nicotiana rustica* — Wróblewska 1973; *Zea mays* — Kupka and Truong Quang Tan 1975). In plants of tobacco cultivars after floral maturity chlorophyll amounts

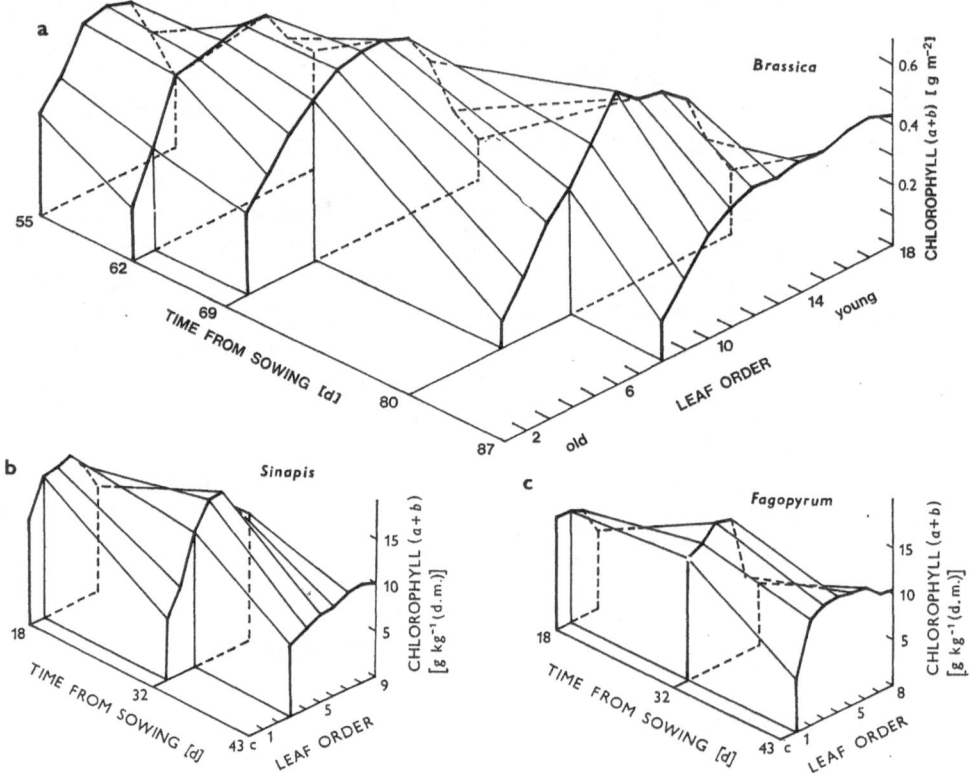

**Fig. 3.5.** The leaf insertion gradient of chlorophyll ($a + b$) content changes during the ontogeny of plants, *e.g. Brassica oleracea* L. convar. *acephala* (DC.) Alef. var. *medullosa* Thell. (a — per unit leaf area; Šesták and Čatský 1967a), *Sinapis alba* L. and *Fagopyrum esculentum* Much. (b, c — per unit dry matter; values from Gej 1966). Each individual leaf as well as all leaves on a plant exert an increase in chlorophyll amount to a maximum value followed by a decrease.

per leaf area declined from young to old leaves, while the amount of pheophytin was lowest in the leaves of middle insertion (De Jong and Woodlief 1978). Even here different results are often reached per different relation unit: thus in wheat plants maximum chlorophyll amounts per fresh matter were reached in the flag leaf, while per leaf area in the 7th leaf (Tschakalova and Hoffmann 1976).

The chlorophyll insertion gradients found in a plant naturally depend on leaf anatomy. Thus in 21 d spinach plants having nine leaves the largest leaf area was in leaves 3 + 4, the highest number of chloroplasts per palisade cell in the oldest leaves 1 + 2, and the highest chlorophyll amounts per leaf area in leaves 5 + 6 (Possingham and Saurer 1969). Their values allowed a relative calculation of chlorophyll amount per chloroplast, which was the highest in leaves 5 to 8 and declined in the oldest leaves.

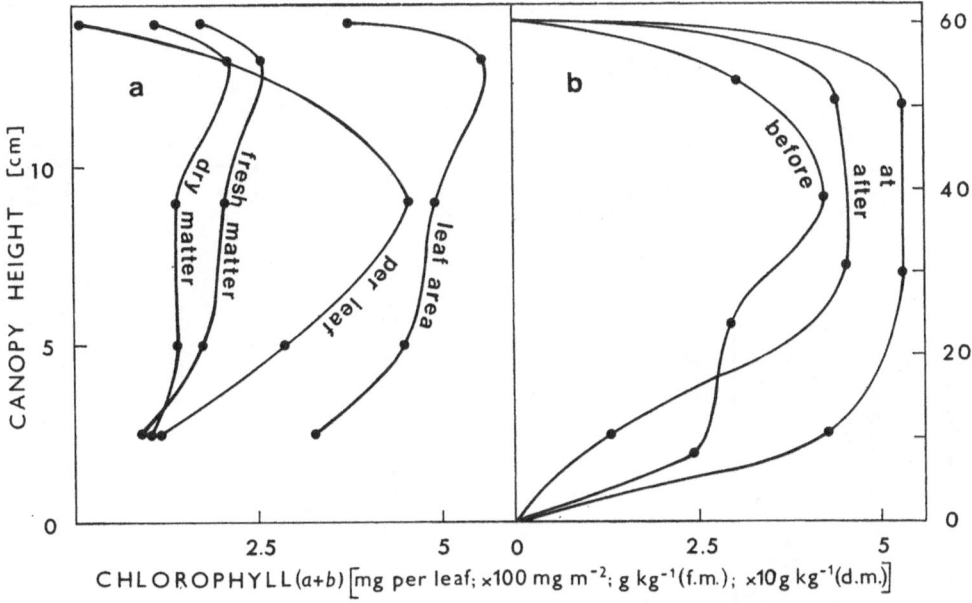

**Fig. 3.6.** Chlorophyll distribution in leaves of different ages situated in individual height layers of a canopy depends also on the relation unit, *i.e.* per one leaf, leaf area, fresh or dry matter units (a — *Phaseolus vulgaris* L. var. *nanus* cv. Sabo, 49 d-plants, values from Bolhár-Nordenkampf 1975). During the development of a canopy chlorophyll distribution changes according to leaf area index (LAI), *i.e.* before, at, or after reaching optimum LAI (b — *Medicago sativa* L., 150 g sown at 5 × 5 cm spacing and thinned to 5 × 10 cm, from Okubo *et al.* 1975b).

Also the leaf life span and duration of chlorophyll accumulation in leaves of various insertion levels may be very different, thus affecting the chlorophyll insertion gradients per various relation units (Fig. 3.6 a). In wheat the highest chlorophyll amounts per leaf area or leaf dry matter were in leaf 3 whose average life span was 50 d, while the maximum chlorophyll amount per one leaf was found in leaf 9 having the longest life span of 82 d (Gautam 1963).

In plant canopies, tree crowns, *etc.* the relations are more complicated: they are affected by microclimatic gradients, namely by the irradiance gradient. A model example for this situation is a cabbage-head: chlorophyll amount decreases in a logarithmic gradient from outer to inner leaf layers, more rapidly in red than in green cabbage cultivars (Hoffmann *et al.* 1977). The induced changes in the rate

**Table 3.2** Examples of differences in chlorophyll content in leaves of different insertion. For abbreviations see Table 3.1.

| Plant (and cultivation)  (1) | Leaf insertion (numbered from the oldest leaf)  (2) | Amounts of chlorophyll $(a + b)$ in g per kg dry matter  (3) | m² of leaf surface  (4) | Change in chlorophyll amount and $a/b$ ratio induced by ageing  (5) | Author(s) and year of publication  (6) |
|---|---|---|---|---|---|
| *Beta vulgaris* L. var. *saccharifera* cv. Dobrovická A (field) | leaves 1 to 25 at 7 phases of vegetation season (27 June to 4 Oct.) | | 0.20...0.62 | ↑ and ↓ ; max on 25 July in leaf 2 from the top, on 10 Aug. leaf 14, on 4 Oct. leaf 2 | Hodáňová 1970 |
| *Brassica oleracea* L. var. *medullosa* Thell. (pots with soil, air-conditioned chamber) | leaves 3 to 8 from the top in plants 17, 24 and 32 d old | | 0.32...0.72 | ↑ and ↓ , max in leaf 5 (17 d), 4 (24 d) and 5 (32 d) from the top | Šesták and Čatský 1967b |
| *Cucumis sativus* L. | cotyledons and leaves 1 to 5 | 5.0...17.0 | 0.17...0.26 | ↑ to leaf 4 and ↓ | Frydrych 1976 |
| *Fagopyrum esculentum* Mönch. | cotyledons, leaves 1 to 9 at 16, 25 or 42 d of plant life | 1.0...13.2 | 0.04...0.34 | ↑ to the 2ⁿᵈ or 3ʳᵈ leaf from the top and ↓ ; $a/b$ ↑ and ↓ | Gej 1971 |
| *Glycine max* (L.) Merrill cv. Harosoy, Wasekogane, Shinanomejiro, Shiromeyutaka, Norin NO1, YL-2-1 (soil pots) | 3ʳᵈ to 8ᵗʰ trifoliate leaf | | 0.08...0.56 | ↑ and ↓ | Watanabe 1973 |
| *Helianthus annuus* L. cv. Borowski IHAR (pots with sand) | insertion in 18 and 31 d plants | 6.3...15.9 | 0.08...0.16 | ↑ (18 d) or ↑ to 3ʳᵈ true leaf and ↓ ; $a/b$ ↑ and ↓ | Gej 1966 |
| *Lactuca sativa* L. cv. Golden State (pots with sand) | leaves 1 (old) to 11 | 1.4...11.1 | 0.10...0.25 | ↑ to leaf 3 and ↓ ; $a/b$ ↑ to leaf 6 and ↓ | Gej 1976 |

| Species | Leaves | | | Trend | Reference |
|---|---|---|---|---|---|
| *Nicotiana rustica* L. (sand pot cultures in glasshouse and climate chamber) | leaves 3 to 14, plants 20 to 122 d old | 0.10...0.42 | | ↑ to max and ↓ ; max shifts with plant age (leaf 3 at 20 d, 4 at 34 d, 5 + 6 at 55 d, 9 at 71 d, 8 at 80 d, 11 at 104 d, 12 + 13 at 122 d) | Wróblewska 1973 |
| *Nicotiana sanderae* hort. (pots with soil, air-conditioned chamber) | leaves 2 to 9 from the top | 0.25...0.41 | | ↑ to leaf 3 (from the top) and ↓ | Šesták and Čatský 1962 |
| *Phragmites communis* Trin. | leaves 11 (bottom) to 1 (top) | 0.31...0.58 | 4.7...7.7 | ↑ to leaf 9 and ↓ | Rejmánková 1973 |
| *Zea mays* L. | leaves 1 to 11 | | 1.9...3.2 | ↑ to leaf 7 and ↓ | Lozova 1963 |
| *Zea mays* L. cv. Stupická raná (pots with soil, glasshouse) | leaves 4 to 14 in plants 38, 52, 66 and 94 d old | 0.07...0.35 | | ↑ and ↓ | Šesták and Václavík 1965 |

of growth of individual leaves affect also the insertion gradient of chlorophyll amounts (Fig. 3.6 b). Leaves of different ages are also differently sensitive to various factors, which induce chlorophyll degradation (*e. g.* industrial air pollution — Tanaka and Sugahara 1980 a, b). For examples of vertical distribution of chlorophyll in canopies of wheat and sugar beet see Aliev and Safarov (1978) and Hodáňová (1973), respectively. In tree canopies, the life span of individual leaves often determines the chlorophyll gradient with leaf insertion. Generally, the highest amount of chlorophyll per leaf area unit is in the top (*Ulmus scabra*) or middle (*Quercus robur*, *Tilia cordata*, *Acer platanoides*) part of the tree crown (Goryshina *et al.* 1979). In 11 to 12 m tall trees of *Picea sitchensis* the seasonal changes in chlorophyll and carotenoid contents and in chlorophyll *a/b* ratio were more or less parallel in three canopy levels (nodes 1 to 4, 5 + 6, and 7 + 8 from the top — Lewandowska and Jarvis 1977).

As concerns the degradation processes, chlorophyll in the dark was more rapidly degraded in young apical than old basal leaves of *Tradescantia albiflora* (Adamson and Hiller 1981).

In some cases the differences induced by various environmental regimes change the rate of growth of individual leaves and some of their parameters, but the insertion gradient of chlorophyll remains very similar. Thus, the three cultivars of *Dactylis glomerata* formed their leaves in day/night temperatures [°C] of 21/13 or 29/21 at different rate (age differences for individual leaves up to 6 d between cultivars and up to 3 d between temperature regimes), but nevertheless in the phase of 12 leaves the highest chlorophyll amount per leaf area unit (0.60 to 0.71 mg m$^{-2}$) was always in leaf 8 (12 to 16 d old) (Treharne et al. 1968). The trend of increase in chlorophyll amount from the youngest to the oldest leaf of 9 d maize plants was not influenced by the temperature of roots, but the adverse effect of low (8 to 10 °C) temperature was highest in the oldest leaf (Andreenko and Titova 1957).

Similarly as the ontogenetic changes found in one leaf are not satisfactory for characterizing the ontogeny of all other leaves, the insertion profile measured in one phase of plant ontogeny is not satisfactory for predicting all insertion profiles occurring during plant life: this holds true not only for chlorophyll, but the majority of leaf characteristics.

### 3.4.2 Carotenoids and Leaf Insertion

Each leaf undergoes its own ontogenetic changes in the contents of carotenoids, and a similar course to a maximum followed by a decline is typical also for the insertion gradient of carotenoids. Simultaneous analysis of all leaves on a plant gives a distribution typical for the given plant age. The insertion gradient changes in the course of plant life, the maximum amount of carotenoids shifting from lower to upper leaves. Complete insertion gradients are rare in the literature (see Tarnowska 1964 and Landi and Antongiovanni 1973 for carotene).

The ratio of chlorophylls/carotenoids from young to old leaves increases (*Vicia faba* — Novitskaya et al. 1977) or declines (2.3 → 1.5 in wheat — Popov and Bak''rdzhieva 1970; 4.1 → 2.6 in tobacco — Fleck-Gerndt 1971a; 1.2 → 0.6 in *Phyllanthus* — Salema and Abreu 1972; 8.5 → 3.4 in maize — K''drev and Georgieva 1975) or changes only a little (4.2 …4.7 in potato — Kislyakova et al. 1967; 4.3…4.7 in *Tradescantia* — Keresztes and Faludi-Dániel 1973). In relation to chlorophyll content, the amount of carotenes is a little lower and xanthophylls a little higher in old than young leaves of *Eucharis grandiflora* and *Billbergia forgetiana* (Lichtenthaler 1968a).

As concerns the individual carotenoids, the changes in their ratio depend not only on leaf age, but often mainly on environmental factors. In wheat the amount of carotenoids declined in the sequence carotene — lutein — violaxanthin — neoxanthin, and for all these pigments from young to old leaves (Parshina et al. 1972), while in tobacco cultivars the sequence was lutein — β-carotene — neoxanthin — violaxanthin (De Jong and Woodlief 1978). In *Ficus elastica* and *Cereus peruvianus* the relative concentrations decreased in the sequence lutein — β-carotene — violaxanthin — neoxanthin, and their ratio did not change much during leaf life with regard to the sum of carotenoids, but increased by ca. 50% with regard to chlorophylls (Lichtenthaler 1969a, b). In *Vicia faba*, the ratio of amounts of four measured carotenoids to chlorophyll amount [mg carotenoids per kg chlorophyll] was always higher in leaves of the $2^{nd} + 3^{rd}$ tier of young (10 d) than old (45 d) plants, but the difference was highest in violaxanthin (44.4: 19.2) and decreased through carotene (89.0: 45.0) and lutein (63.0: 51.4) to neoxanthin (33.4: 32.0) (Novitskaya *et al.* 1977).

In potato the ratio lutein/violaxanthin increased from upper to lower leaves in all phases of vegetation, while the absolute amounts of carotene, lutein and violaxanthin per leaf area and dry matter mostly declined from upper to lower leaves (Kislyakova *et al.* 1967). In *Tradescantia* the old leaves contained 80% of all carotenoids of young leaves, and the ratio carotene: xanthophyll declined in a similar way (from 0.70 to 0.55) (Keresztes and Faludi-Dániel 1973). In apple, pear and apricot trees the amounts of carotene, lutein and violaxanthin per leaf area or dry matter were higher in older, mature leaves than in the young ones (Shishkanu 1970).

A complicated situation is in plant canopies or tree crowns, where the microclimatic factors often change the insertion gradient. More carotene, lutein and violaxanthin was always in the upper layer of the maize canopy than in its lower layer (Bazhanova *et al.* 1967). The carotenoid gradient along the branch in a tree crown often oscillates during the vegetation season and with regard to cardinal points, but the highest amounts of carotenoids are mostly in upper or middle leaves (apple tree — Shishkanu 1970, Shishkanu *et al.* 1970, Shatkovskiĭ 1972). These results seem contradictory to the observations of the same authors, that mature and old leaves contain more carotenoids than young leaves: hence the effect of irradiance of leaves in the crown probably outweighs the effect of leaf age.

## 3.5    PIGMENT DISTRIBUTION IN LEAF BLADE

### 3.5.1    Chlorophylls

The leaf blade is far from being homogeneous. Its parts differ in thickness, amounts and types of cells and chloroplasts, and, as a result, in chlorophyll content. The tissues of the leaf tip are usually older than the tissues of the leaf base. Leaf centre is mostly thicker than the leaf margin.

In leaves growing perpendicularly to the sun, more chlorophyll per leaf area unit is in the leaf centre than in its margin. This difference increases with leaf age (*Parthenocissus* — Moore 1975) and changes with the vegetation season (*Phaseolus vulgaris* and *Sinapis alba* — Dontschev and Lossner 1976). The differences between the leaf tip and base depend mainly on the relation unit. Thus there was no difference in chlorophyll amount between the tip and base of a 2 cm long spinach leaf per fresh matter, but per cell more chlorophyll was in the tip and per chloroplast more chlorophyll in the base (Detchon and Possingham 1972). Comparison of tips of leaves 2, 4 and 7 cm long showed the highest amounts of chlorophyll per cell and chloroplast in the largest, *i.e.* oldest leaf. In tobacco higher chlorophyll content per unit area was in leaf tips than bases, but per fresh or dry matter in leaf bases (Andrews and Svec 1975).

In vertically growing leaves of monocotyledons environmentally (by irradiance) induced chlorophyll distribution prevails. In these leaves the chlorophyll amount per leaf area or dry matter unit declines from the apical to the basal zone of the leaf blade (wheat — Hoffmann and Michaelis 1976; *Typha angustifolia, T. latifolia, T. laxmanii, Glyceria aquatica* — Molyaka *et al.* 1973, Szczepański 1974; barley — Berner 1971, Lipskaya 1980; maize — Signol 1965, Tageeva *et al.* 1969, Baszyński 1971, Leech *et al.* 1972, 1973, Dezhi 1974, Hawke *et al.* 1974, Leech 1974, Baker and Leech 1977, Leese and Leech 1977, Gaponenka *et al.* 1978; *Lolium* — Prioul *et al.* 1980 a, b; onion — Genchev 1970). Sometimes a small increase preceding the decrease is observed (barley — Henningsen and Boynton 1974; *Dactylis glomerata* — Schäfer and Tirtapradja 1970; maize — Hopkins *et al.* 1975; sugar cane — Alexander and Kumar 1974). In a young leaf, the gradient per dry matter unit may be different, as the leaf tip is thin (maize — Hawke *et al.* 1974). The expressiveness of the gradient declines with leaf age (maize — Hopkins and Walden 1977).

The differences in chlorophyll *a/b* ratio on the area of one leaf blade are usually small (maize — Baker and Leech 1977).

The largest chloroplasts are in the apex of leaves, and their volume decreases in the direction of the leaf base. The ratio chlorophyll *a*: chloroplast volume was similar in the two upper quarters of the leaf blade, and maximal in the third quarter (maize — Tageeva *et al.* 1969).

Some authors observed differences between the left and right halves of leaves (wheat — Parshina *et al.* 1974a, b), or between the left and right leaflets (secondary *Phaseolus vulgaris* leaves — Urmantsev 1970). In compound leaves the situation is more complicated: In the first (young) and ninth leaf of 6 or 9 year old oil palm plants the highest chlorophyll amounts per dry matter were in leaflets formed in the rachis part nearest the base (Bolle-Jones 1968).

### 3.5.2    Carotenoids

The distribution of carotenoids in the leaf blade has been described only for a few plant species. In apple tree leaves the highest amounts of carotene, lutein and violaxanthin per leaf area unit were always in the middle part of the blade, in calculation per dry matter the leaf tip contained the highest amounts of carotene and violaxanthin (Shishkanu 1970, Shishkanu *et al.* 1970). The ratio chlorophylls/carotenoids declined from leaf base to leaf tip (Shishkanu 1970). In the second leaf of 14 d seedlings of maize the content of $\beta$-carotene in dry matter declined from apex to base (Baszyński 1971). An increase to a maximum in the upper third of leaf blade followed by a decrease to the base was observed in $\beta$-carotene, lutein-epoxide and violaxanthin contents of 7 d etiolated wheat leaves (Virgin 1966). Some differences were observed even between the contents of carotene, lutein and violaxanthin in left and right leaflets of the secondary leaves of *Phaseolus vulgaris*; more carotenoids per dry matter of leaf area were in the left leaflets (Urmantsev 1970).

## 3.6  STATE OF PIGMENTS *in Vivo*

### 3.6.1  Chlorophylls *in Vivo*

Chlorophyll is *in vivo* bound in lipoprotein complexes, differing in pigment aggregation and type of protein and hence in their optical properties, molecular mass, sedimentation patterns, *etc.*\* As they are supposed to function in different reactions of photosynthesis, the changes in their content during leaf ontogeny have often been studied. The formation of complexes of chlorophyll with protein moiety is a necessary condition for running chlorophyll biosynthesis, as shown in experiments with protein inhibitors (for reviews see Kirk 1974, Kirk and Tilney-Bassett 1978, Treffry 1978).

The oldest methods used for the characterization of chlorophyll *in vivo* forms are those of differential extraction. Leaves are ground (sometimes after thermal degradation, sometimes with the addition of $CaCO_3$) and then left for some time in darkness with a solvent mixture less appropriate for chlorophyll extraction (*e.g.* 60% acetone, petroleum ether, 5% acetone in petroleum ether). The fraction thus obtained ("easily extractable", "labile" or "free" chlorophyll) is compared with the residual fraction extractable only by the normal procedure, *i.e.* with 85% acetone or 95% ethanol ("bound" chlorophyll). By step extraction more fractions may be obtained. In some cases the extractions are done under continuous grinding. The results reached by these methods depend on the conditions of extraction, *e.g.* on the size and water content of extracted particles. In addition to this, the solvents used for extraction act mostly in a contrasting manner, *e.g.* petroleum ether extracts mainly newly formed ("young") chlorophyll molecules, while acetone "old" chlorophyll molecules (Shlyk and Mikhaïlova 1967, Shlyk *et al.* 1970).

The differences in quantity of individual fractions separated by the differential extraction method from leaves of various age are not uniform (see Table 4 in Šesták 1977a). Most of the chlorophyll was either in the "labile" chlorophyll fraction and

---

\* To distinguish these properties of chlorophyll *in vivo*, the terms forms, types, states, species, complexes, *etc.* are used having mostly a similar meaning. For simplicity, the classic term forms *in vivo* (French 1958) is mainly used here.

declined with leaf age (*Pinus silvestris, Fraxinus excelsior, Helianthus annuus* — Osipova 1947; *Zea mays* — Lozova 1963, Smirnova 1967; *etc.*) or in the "bound" fraction with increasing amount of the "labile" chlorophyll during leaf ontogeny (tobacco — Steffens 1960; wheat — Popov and Bak"rdzhieva 1970; *Taxus baccata* — Popov and Tsoneva 1966/7). In *Prunus silvestris* needles the amount of "labile" chlorophyll was medium in summer, low in winter and high in spring of the 2nd year (Khodasevich *et al.* 1973). In barley and maize the amount of "labile" chlorophyll diminished from tip to base of the leaf blade (Gaponenko *et al.* 1970, Gaponenko 1976). Nevertheless, due to very variable results the method is rarely used nowadays.

More promising were the studies of Kahn and Chang (1965) who found always more chlorophyll-protein complex extractable with *Triton X-100* in mature than in expanding or senescent leaves of spinach, lettuce and soybean.

The spectroscopic studies, often completed with derivation or computer analysis of absorbance curves (French 1967), yield satisfactory information on chlorophyll *in vivo* forms, when a spectrophotometer of top quality is used, and when the extremely thin leaves, monolayers of chloroplasts (*e.g.* spread on ultrafiltration membranes) or homogenates are used. In thicker leaves the absorption peaks and shoulders are broadened and shifted as a result of selective scattering. Low (liquid nitrogen) temperatures are preferable for the measurement (for discussion of possible limitations see Meister 1977). The optical properties are affected also by the changes in chloroplast dimensions and ultrastructure.

As concerns protochlorophyllide, etiolated 2 d cotyledons of flax or etiolated 2 d leaves of barley contain mainly protochlorophyllide of the 635 nm form, while from the 3rd to 12th d the formation of long-wave (650–652 nm), directly phototransformable protochlorophyllide form is prevalent (Akulovich *et al.* 1970, 1973, Akulovich and Raskin 1971). The maximum content of both protochlorophyllide forms was found in leaves of the same age in cucumber (11 to 12 d) and barley (8 to 9 d) (Akulovich *et al.* 1974). In older leaves (13 to 17 d) the amount of the 635 nm form increased anew and hence the ratio of both protochlorophyllide forms approached 1. This was reflected also in the formation of chlorophyll forms after irradiance with two flashes: in 2 d cotyledons the 673 nm form prevailed, while in the older cotyledons the long-wave (682–684 nm) form of chlorophyll was formed (Akulovich *et al.* 1970, Akulovich and Raskin 1971). In 3–4 d dark grown leaves of barley the form of protochlorophyllide absorbing at 650 nm prevailed, while with increasing age the 636 nm peak became increasingly dominant (Virgin 1975). Cultivation under long wavelength far-red radiation ($> 700$ nm) stimulated the formation of the 635 nm form (maximum in 5 d wheat leaves — Klockare 1980). The higher content of the short-wave form in 5 d than 7 d old dark or far-red grown wheat leaves was confirmed also by fluorescence emission spectra (Klockare 1980). The ratio of fluorescence emission at 655 and 630 nm in etiolated *Phaseolus vulgaris* seedlings increased to a value of 2.7 at 15 d and then decreased (Boardman *et al.* 1971). In far-red grown *Quercus robur* leaves the amount of protochlorophyllide per leaf declined with leaf age, while the ratio of fluorescence at 632 and 657 nm changed in the sequence 4.5 — 3.6 — 3.9 — 4.9 (Axelsson *et al.* 1981).

Concerning the chlorophyll forms in the course of greening, the duration of the Shibata shift (absorption shift from 682–684 to 672–673 nm) increases with the age of etiolated leaves (*Phaseolus* — Akoyunoglou and Michalopoulos 1971; wheat — Axelsson 1977, Klockare 1980) or etiolated cotyledons (*Pharbitis nil* — Ogawa *et al.* 1978) and is related to the duration of the lag period of protochlorophyllide regeneration (see Section 3.2.1.1). The relative photostability of both 673 and 684 nm forms is lower in older (12 d) than young (5 d) wheat leaves (Axelsson 1977). The fluorescence lifetime increases during the Shibata shift from 3.1 to 4.1 ns in 8 d leaves and from 1.9 to 3.6 ns in 21 d leaves; the structure of lifetime curves becomes more complex during leaf ageing from 3 to 21 d (*Phaseolus* — Goedheer and van der Cammen 1981). In very young leaves (3 to 4 d) the Shibata shift is absent.

The green leaves or isolated chloroplasts show one main absorption peak at *ca.* 678 nm, the analysis of which gives at least three distinct chlorophyll *in vivo* forms (at *ca.* 670, 680 and 705 nm); their formation is more rapid in younger leaves (Butler 1965).

The position of the main absorption peak of chlorophyll *a* in the red spectral region shifts with the age of maize leaves to longer wavelengths (Šesták 1972). This *ca.* 1 nm shift observed in leaves and isolated chloroplasts at 77 K corresponds to an increase in the amount of chlorophyll forms absorbing at 684–686 nm. A similar 1.5 nm shift from 678.5 to 680 nm was observed in the earliest phases of development of the primary leaf of *Phaseolus* (Więckowski 1967b, 1969). Also the absorbance at 695 and 710 nm was increased (Šesták 1972), similarly as in the experiments of Goedheer (1967) with spinach leaves. Old leaves of winter rye contained a lower amount of the monomeric chlorophyll form absorbing at 670 nm (Baranov *et al.* 1974), while in *Phaseolus* the second derivative spectra showed an increase in the short-wavelength chlorophyll forms (*ca.* 674 nm) with leaf age (ageing of a primary leaf — Ĭordanov *et al.* 1973; young and old yellowing leaves — Saakov *et al.* 1978). In rice leaves of the same insertion from the top (2nd and 3rd leaves), the absorption maximum shifted to a shorter wavelength and the absorbance in longer wavelengths declined with plant age from 68 to 142 d (Inada 1977). Second leaves from the top in wheat exert, during the flower formation, a shift of the main red chlorophyll peak from 678–680 nm to 675 nm and back to 679–680 nm (Kozhushko and Chernysheva 1976); in one phase the main chlorophyll peak splits into two maxima (674–675 and 679–682 nm). Leaf insertion induces spectral changes in yellowing leaves of *aurea* forms of jasmine and tobacco (Kaler and Akulovich 1965, Aerov and Manuil'skiĭ 1967).

The changes in the quantitative ratio of chlorophyll *in vivo* forms during leaf ontogeny are confirmed by fluorescence spectra; similarly to absorption spectra their quality declines with the increasing thickness of leaves.

The fluorescence excitation spectra of chloroplasts isolated from *Cucurbita maxima* cotyledons in the expansion phase and in advanced age did not differ in shape (Harnischfeger 1973).

The fluorescence emission spectra (usually excited with the wavelength 435 nm) have usually two peaks: $I_1$, at 686 to 690 nm (sometimes split into a double band 686–694 nm — Khodasevich and Lis 1980, Jenkins *et al.* 1981b), and $I_2$, at 734 to 740 nm. Their ratio varies in the literature from 0.4 to 2.0. Mostly an increase in the $I_1/I_2$ ratio with leaf age was observed: from 1.40 to 1.92 in *Sinapis alba* leaf grown 7 to 35 d under "blue light" (Horváth *et al.* 1973), from 0.79 to 1.06 in 15 to 34 d old primary leaf of *Phaseolus vulgaris* (Raafat *et al.* 1969), from 0.53 to 0.88 in 24 and 42 d old primary leaf of *Phaseolus vulgaris* (Jenkins *et al.* 1981b), from 1.6 to 2.0 in the third leaf of 22 to 50 d old barley plants (Mader *et al.* 1981), and from 0.42 to 0.61 in 1 to 4 years old needles of *Picea pungens* (Khodasevich and Lis 1980). Other authors observed a decline in this ratio with leaf age: from 0.51 to 0.46 in a leaf insertion gradient of pea (Paromenskaya *et al.* 1975). Growing under "blue" radiation induced a higher $I_1/I_2$ ratio (2.62 and 1.42 for 1 and 3 d) than growing under "red" radiation (2.23 and 1.13 for 1 and 3 d) (primary leaves of barley — Lichtenthaler *et al.* 1980).

Also the character of fluorescence induction curves, *i.e.* the time course of fluorescence intensity at wavelength of fluorescence peak (*e.g.* 685 nm at 293 K and 695 nm at 77 K), changed in the course of leaf senescence (Fig. 3.7): the first and second peaks were smoothed out, the initial maximum fluorescence level, the peak induction intensity and the steady level declined, the time to reach the first peak was shorter, but the ratio of variable to maximum steady fluorescence (which estimated the maximum yield of primary photochemistry of Photosystem 2) did not change (Jenkins *et al.* 1981a). (This ratio increases very rapidly during the first ten hours of leaf greening — Baker and Butler 1976.) The ratio of fluorescence at 770 and 685 nm (representing emission from Photosystems 1 and 2, respectively) increased in the course of leaf senescence (Jenkins *et al.* 1981a). Unfortunately, these experiments were not done with leaves prior to their maturity. Variable differences in ontogenetic changes of parameters of fluorescence induction curves of the third and the seventh barley leaf grown in the field were observed by Mader *et al.* (1978). In 6 d old etiolated leaves of *Phaseolus vulgaris* only one peak (10 s) in the fluorescence induction curves was observed, while in 10 to 14 d leaves a second flat peak appeared after ca. 30 s (Jouy 1982). In the desiccation tolerant plant *Borya nitida* rehydration of air dried plants lead to regreening in young and mature leaves reflected in the restoration of fluorescence induction curves started after 47 h of water supply, while in old leaves even after 64 h the fluorescence activity was not restored (Hetherington and Smillie 1982).

The analysis of induction curves of fluorescence in wheat leaves showed a similar course (increase to a maximum followed by a decrease) of the ratios of maximum to steady fluorescence from the first to the youngest (sixth) leaf. The ratios were between 2.33 and 4.47; also the shapes of induction curves found in various parts of the leaf blade were different (Nesterenko and Sid'ko 1980). The relative level of DCMU-induced fluorescent changes declined from 190 to 0 in the young to senescent leaves of *Ipomoea pentaphylla* (Kulandaivelu and Daniell 1980).

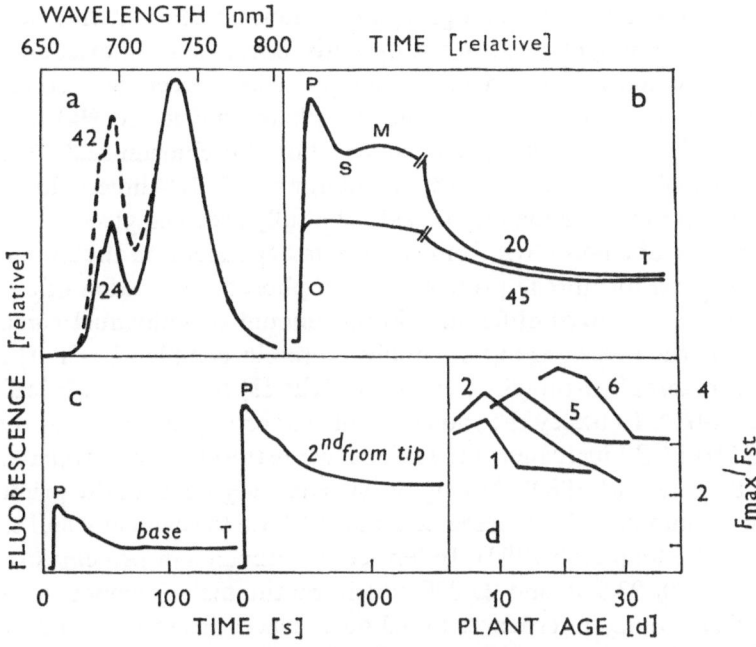

**Fig. 3.7.** The ratio of peaks (at 686, 694 and 734 nm) in the fluorescence emission spectrum changes in the course of leaf senescence (a — 24 and 42 d old primary leaf of *Phaseolus vulgaris* L., fluorescence measured at 77 K; from Jenkins *et al.* 1981b), which also changes the kinetics of fluorescence emission (b — 20 and 45 d old primary leaf of *Phaseolus vulgaris* L., fluorescence emission at 685 nm after a 632.8 nm excitation, measured at 273 K; O, P, S, M, T — phases of the kinetics curve; from Jenkins *et al.* 1981a). The shape and size of the induction curve of fluorescence are different also in various parts of the leaf blade (c — third leaf of 20 d old *Triticum aestivum* L. plant divided into six segments, curves for the segments with lowest and highest intensity of fluorescence, fluorescence emission at 685 nm after a 350–550 nm excitation measured at 273 K; from Nesterenko and Sid'ko 1980). The ratio of maximum (at P = $F_{max}$) to stationary (at T = $F_{st}$) fluorescence in leaves of various insertions (1, 2, 5, 6 from the oldest one) changes during their development (d — the same plant, conditions, and reference as in c).

On the area of a 7 d old secondary leaf of maize the ratio of variable to maximum fluorescence ($F_v/F_m$) at 695 nm slowly increased from base to leaf middle and then declined, while the ratio of $F_v/F_m$ at 735 and 695 nm and the ratio of variable to minimum fluorescence peaked sharply in a ca. 3 cm distance from the leaf base (Baker and Miranda 1981); the ratio of fluorescence emission at 685 to 735 nm declined with the distance from leaf base.

The increase in delayed light emission from chloroplasts, especially of its induction phase, with leaf age (5 to 10 d wheat plants) was connected with a slower induction and a lower steady state of photosynthetic $CO_2$ uptake by leaves (Klimov et al. 1978). The intensity of delayed light emission at all excitation irradiances increased from upper to lower leaves of sunflower (Volodarskiï and Bystrykh 1975).

While the fluorescence spectra give information on the radiative de-excitation processes of chlorophylls in vivo, the recently developed photoacoustic spectroscopy detects the non-radiative de-excitation processes. Greening of etiolated radish cotyledons lead to an increment of the red photoacoustic (20 Hz) maximum at 680 (after 6 h) and later (after 5 d) at 688 nm (Buschmann and Prehn 1981); the spectrum and its reaction to 3-(3,4-dichlorophenyl)-1,1-dimethylurea was more expressed on the lower than upper side of a fully green cotyledon.

The presence of chlorophyll forms in vivo is dependent on the amount of individual chloroplast membrane proteins. Electrophoretic separation of chlorophyll-protein complexes showed differences in the amount of individual complexes. In Pisum sativum the percentage of chlorophyll-protein complex I (CP I) decreased and that of chlorophyll-protein complex II (CP II) increased with leaf age (Valanne et al. 1979). In barley the ratio of light-harvesting chlorophyll-protein complex (LHCP) to CP I increased with leaf age, while the chlorophyll a/b ratio slightly declined (Prenzel et al. 1980). Among the six equal segments of the primary leaf of barley the amount of LHCP increased from the base to segment 4 and then declined (Viro and Kloppstech 1980). In lettuce the amounts of proteins of molecular masses of 25 000, 27 500, and 23 000 which are the major components of LHCP increased from young inner leaves to old outer leaves (Henriques and Park 1976). The amounts of two chloroplast protein fractions increased with ageing of primary leaves of Vicia faba (Novitskaya et al. 1977). Also the decline in protein content in senescing rice leaves was simultaneous to the decline in chlorophyll content (Kar and Mishra 1976).

Analysis of the circular dichroism signal of intact chloroplasts from developing primary leaves of Phaseolus vulgaris gives evidence for a decline in the amount of LHCP and chlorophyll-protein complex of Photosystem 2 with leaf ageing (Šesták and Demeter 1976).

As concerns the amount of pigment of the reaction centre of Photosystem 1, P700, during 7 d of greening of the primary leaf of Phaseolus vulgaris a small increase to a steady amount was observed between the days 1 and 2 (by means of the electron paramagnetic resonance signal 1 – Więckowski 1975).

Concurrently with the quantitative changes in chlorophyll in vivo forms the size of the photosynthetic unit [characterized as amount of chlorophyll $(a + b)$ molecules per 1 molecule of P700] changes with leaf age, e.g. from 290 in young maize leaves to 863 in old leaves (Keresztes and Faludi-Dániel 1973). The size of the photosynthetic unit increased and then declined in the course of ontogeny of cotton (Khodzhaev et al. 1978) or Phaseolus vulgaris (Šesták and Demeter 1976) leaves.

### 3.6.2    Carotenoids *in Vivo*

There are no special theories explaining the differences in functions of carotenoids in plant organisms by their different binding with protein or aggregation as in the case of chlorophylls. Nevertheless, the method of differential extraction (see Section 3.6.1) utilized for distinguishing chlorophyll forms *in vivo* was used several times to obtain carotenoid fractions differing in their extractability and, thus, in the strength of their binding in the thylakoid membrane.

In apple tree leaves the strength of the binding of carotenoids was always greater in lower than upper leaves on the branch (Aėrov and Likholat 1967). In wheat the percent of "bound" carotenoids declined from young to old leaves (Popov and Bakardshijeva 1967, Popov and Bak"rdzhieva 1970). In radish the differences were rather small and they were affected to various degrees by kinetin, quercetin or N-allyl-N'-3-hydroxy-4-carboxyphenyl-thiourea treatment (Karanov *et al.* 1970). Variable amounts of three fractions of carotenoids differing in extractability were found in *Pinus silvestris* and *Taxus baccata* by Popov and Tsoneva (1966/7).

The above mentioned results were not complemented with spectral or functional studies and hence their value is questionable.

### 3.7    LEAF AGE EFFECTS OBSERVED IN DETACHED LEAVES

The experiments on stimulation or retardation of chlorophyll degradation by means of growth regulators and other substances are mostly done with detached leaves or leaf segments. Chlorophyll degradation is usually more rapid in old than young detached leaves (*Ginkgo biloba* — Specht-Jürgensen 1967; tobacco — Wu 1971), nevertheless these experiments do not represent the course of chlorophyll degradation in attached leaves. Tetley and Thimann (1974) did not find important differences in the chlorophyll degradation pattern in the first leaves detached from 5 to 14 d old oat plants. Benzyladenine application to oat roots slowed down the chlorophyll degradation more in the lower than the upper halves of leaves put in the dark (Thimann *et al.* 1974). Benzyladenine, kinetin and kinetin riboside slowed down the chlorophyll degradation in leaf discs floating on the solution surface in old leaves of Brussels sprouts, but stimulated it in discs of young leaves (Dennis *et al.* 1967). In some plant species, *e.g. Phaseolus vulgaris*, kinetin stimulated chlorophyll degradation in detached leaves in the light: the degradation was more rapid in 15 than 9 d old primary leaves (Wachowius and Wachowius 1969 in Hoffmann 1970). In *Brassica oleracea* var. *italica* the response of chlorophyll to surface application of kinetin was as great in young as in old leaves, but the reduction of the kinetin effect by naphthaleneacetic acid was greater in young leaves (Abrams and Pratt 1966). The synthesis of chlorophyll was more inhibited by chloramphenicol in the 3rd than in the 2nd leaf pair (Osipova *et al.* 1967). Gibberellin application stimulated chlorophyll degradation in the light in discs from primary leaves of 26 d plants of *Phaseolus vulgaris*, but slowed it in leaf discs from 15 d plants (Artamonov and Kuramagomedov 1973).

Darkness induced labilization of aggregated chlorophyll forms in both mature and senescing leaves (Shabel'skaya and Gvardiyan 1978). The ratio of fluorescence emission peaks at 685 and 735 nm declined in detached primary leaves of barley kept in darkness from 3.3 (0 d) to 2.5 (7 d) (Biswal *et al.* 1979).


## 3.8    CONCLUSION

Even if many differences among plant species exist (sometimes genetically fixed, sometimes induced by environmental conditions, sometimes caused by the researcher's negligence), the general trend — rapid increase to a maximum and slower decline is characteristic for changes in chlorophyll and carotenoid amounts during ontogeny of a leaf in an insertion gradient, in the composition of pigment forms *in vivo*, etc. These changes reflect the changing structure and composition of the thylakoid membranes (see Chapter 2) and are closely related not only to changes in leaf optical properties (see Chapter 4), but also to the changes in the composition and activities of the electron transport chain and photophosphorylation (see Chapter 5), and in the carbon fixation activities (see Chapter 6). All these components must be analysed together for the biochemical basis of photosynthetic activity to be understood.

# 4 LEAF OPTICAL PROPERTIES

*Danuše Hodáňová*

## 4.1    INTRODUCTION

A leaf has to be exposed to radiant energy in order to photosynthesize, grow and develop during its ontogeny and keep the thermodynamic balance in relation to the environment. The radiation incident upon the leaf surface and passing through the leaf may be reflected (R), transmitted (T), scattered (S) and absorbed (A) by leaf tissues, cells and their organelles. Each of these characteristics is an important indicator of leaf responsibility to radiation as it results from plant phyllogeny, adaptations to different ecological niches and the instantaneous physiological state as well.

Conventionally, radiation which is absorbed ($I_A$), transmitted ($I_T$) and reflected ($I_R$) by a leaf, respectively, has been expressed as a fraction of the incident radiation ($I$) and referred to either a single wavelength ($\lambda$), the wavelengths constituting a particular spectrum, or it may be averaged over a specific waveband. Then, leaf absorptance, A ($I_A/I$), reflectance, R ($I_R/I$) and transmittance, T ($I_T/I$) are determined as follows:
A(R, T)$_\lambda$ — leaf absorptivity (reflectivity or transmissivity); A(R, T)$_{\lambda \text{ to } \lambda}$ — leaf absorptance (reflectance or transmittance) spectrum; A(R,T) coef$_{\lambda \text{ to } \lambda}$ — leaf absorption (reflection or transmission) coefficient.

If T is determined as the transmittance of the medium disregarding boundary effects ($T$), the quantity absorbance ($A$) is used as an expression of log $1/T$.

The symbols A, R and T, respectively, are used here as designations of optical phenomena without a distinction between particular quantities such as, *e.g.*, the absorptance, the absorbance, the absorptivity, *etc.* In tables and figures, each quantity is characterized by the corresponding symbol and unit.

Generally, relations between A, R, T and S (specified by quantitative patterns and spectral parameters — see Fig. 4.1) determine leaf optical properties which change — similarly as other physiological attributes of the leaf do — with leaf age and insertion level on a plant.

The responses of leaves to the ultraviolet (250 to 400 nm, UV), the photosynthetically active (400 to 700 nm, PAR), and the infrared (beyond 700 nm, IR) radiation, respectively, are evidently of an adaptive even though contradictious character which reflects a compromise between two requirements: to carry on metabolic processes by utilizing necessary quanta of energy and to avoid those effects of radiation which may destroy the leaf's vital functions.

Thus, the photobiologically effective but mostly injurious UV is almost completely absorbed by cuticular waxes and leaf epidermis so that it hardly enters the deeper layers of the leaf where metabolic processes take place. The leaf epidermis is thick enough (the T of UV through leaf epidermis is generally less than 10% — Robberecht and Caldwell 1978) to protect the mesophyll tissue from, *e.g.*, the disintegration of deoxyribonucleic acid (McLaren and Luse 1961, Deering 1962),

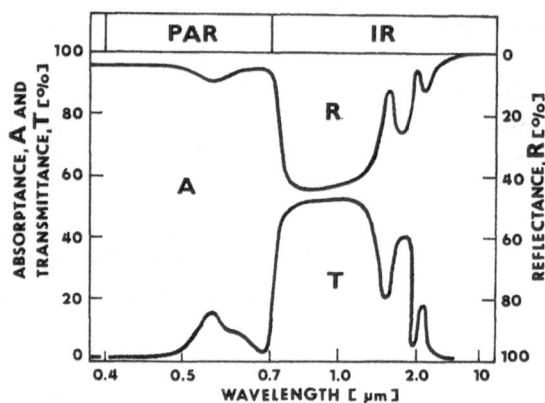

**Fig. 4.1.** Idealized relation between leaf absorptance, reflectance and transmittance as a function of wavelengths of incident radiation.

the inhibition of enzymatic activity (Mantai *et al.* 1970), the inhibition of chlorophyll synthesis, irregularities in cell growth and development (the reduction of cell division and enlargement, the dry matter accumulation — Basiouny *et al.* 1978), *etc.* Such negative effects would be expected if the protective shield of the stratospheric ozone had been removed and UV had increased above 7% of the total solar radiation reaching the Earth's surface at present. For the variety of UV-elicited phenomena in plants in the region of wavelengths 250 to 280 nm (UV-C), 280 to 320 nm (UV-B) and 320 to 400 nm (UV-A), respectively, see, *e.g.*, Allen *et al.* (1975) and Caldwell (1981).

The leaves of higher plants are well disposed structurally and functionally to absorb about 50 to 95% of PAR and the fraction of the near IR up to the 800 nm wavelength (Brandt and Tageeva 1967, Shul'gin 1967, 1973) which satisfies the energy requirements of single biophysical and biochemical reactions (most chemical bonds require energies ranging between 1.25 and $5.0 \times 10^5$ J mol$^{-1}$ which are equivalent to the energy contents of a quantum of radiation of wavelengths between 950 and 237 nm). Radiation of these wavelengths is converted through photosynthesis into chemical energy, used as a stimulus of the photoinductive processes and a regulator of the photomorphogenetic processes (Morgan and Smith 1981) or can bring about leaf photonastic and phototropic orientations (Dennison 1979) and chloroplast structural transformations (Zurzycki and Gabryś 1977) and phototactic movements (Britz 1979) before being dissipated as fluorescence or heat. Regardless of high A, the green leaf can convert the absorbed PAR with a maximum 27% efficiency only. As PAR is only a part of the entire solar spectrum and represents about 45% of the energy available, leaf photosynthesis can at best convert 12% of the total solar radiation into organic material. In this way, the primary conversion photoact in the reaction centres of photosynthesis is adaptively provided by a several times higher absorption capacity of the leaf.

In order "to prevent" the overheating of leaf tissue by excess radiation (and because of a different transparency of leaf absorbing materials to radiation of particular wavelengths), the mature leaf is capable to reflect and transmit, respectively, about 10 to 20% of the incident PAR and 40 to 50% of the near IR (700 to 1350 nm).

The far IR is not sufficiently energizing to build chemical bonds and organize groupings of complex organic molecules. Nevertheless, the leaf absorbs it to a great extent. Most of this energy goes into heat (latent and sensible heat). This is important in maintaining the leaf at a proper thermodynamic level (so that the chemical reactions can proceed at a suitable rate) and is used to evaporate water from the leaf and help to drive water transport in the xylem. Radiant energy in the form of heat is also reradiated by a leaf to the environment.

As follows from the "wave-like" and "particle-like" behaviour of radiation, the interactions between the leaf and radiation must obey not only the relationships of geometrical and physical optics but also the laws of quantum physics. Hence, the amounts of radiant energy which is absorbed, emitted and transferred by convection and partly by conduction, and which is put into or taken out of storage in leaf tissues, are important components which determine the energy and mass budget of the leaf consistently with particular environmental conditions and the leaf ontogenetic stage (Monteith 1973, Gates 1980, Campbell 1981).

Prior to estimating optical properties with respect to the leaf developmental scale, a detailed analysis of responses of the mature leaf to radiation is useful here at least for two reasons. When maturing, (1) the functional activity of a leaf is mostly the highest so that it can serve as a good criterion for determining the range of ontogenetic deviations in leaf optical properties, (2) the leaf is usually fully differentiated as to its internal structure so that the mechanisms of leaf A, R, T and S, respectively, in their comprehensiveness and relative to the highest capacity of photoreceptors may be determined.

## 4.2    PHOTORECEPTORS AND MECHANISMS

Various structures and mechanisms operating in the leaf enable it to respond selectively to radiation of different wavelengths and energy content and the totals of radiant energy striking the leaf surface per unit area and in unit time (for the physical nature of radiation see, *e.g.*, Monteith 1973, Shul'gin 1973, Gates 1980). The mechanisms of leaf A, R and T are different and still specified as regards the units of the organizational levels of the leaf (electron/proton — atom/molecule— organelle/cell — cell/tissue). Each unit functions as it on its own (as a bearer of intrinsic optical properties) as in the interactions with the other units (as a constituent of a hierarchical system of the leaf organizational levels).

### 4.2.1    Absorption

As the physical phenomenon, the A of radiation is a function of changes in the spin and angular momentum of electrons, transitions between orbital states of electrons in particular atoms (A of UV and PAR) and vibrational-rotational modes within the polyatomic molecules (A of IR). Energies characterizing these changes are different (electronic energy states are approximately one hundred times greater than vibrational energies and these are $10^4$ to $10^5$ times greater than rotational energies) and invoke specific biophysical and photochemical responses in the leaf: *e.g.*, owing to the electronic energies the leaf can build and/or interrupt single chemical bonds; the vibrational and rotational energies allow the leaf to lose water through transpiration, *etc.* (for further details see, *e.g.*, Clayton 1980 and Gates 1980).

In PAR, most radiation is absorbed by the chlorophylls (which amount to 65 to 75% of the total pigments in the mature leaf — *cf.* Chapter 3) and carotenoids as accessory pigments in chloroplasts of leaf epidermal (guard cells) and mesophyll cells to be used in photosynthesis (the blue and red A peaks of chlorophyll *a* and chlorophyll *b* in the ranges of 400 to 500 nm and 600 to 700 nm, respectively, and the action band of carotenoids between 350 and 500 nm). Further photoreceptors such as the phytochrome system $P_r$ and $P_{fr}$ (with the A maximum at 660 and 730 nm, respectively) and the blue-light receptors of flavoprotein type (the action maximum at 370 and 450 nm) are beyond the scope of this chapter.

The trapping of incident quanta of PAR by reactive pigments' molecules is the first step in converting radiant energy into chemical energy. Owing to the electronic transitions, A of photons by antennae chlorophylls and carotenoids causes the pigments to pass from their lower energy states to higher ones so that at least 8 photons entering reaction centres of the photosystem (PS) 1 and PS 2 might drive the photochemical reactions of photosynthesis (for review see, *e.g.*, Campbell and Black 1978; *cf.* Chapter 5). Besides the energy which migrates to the reaction centre, energy not used must dissipate quickly or irreversible photooxidation of chlorophylls occurs: much energy is thus lost as heat and little is lost by fluorescence as the electron falls back to the ground state. Apparently, there is approximately a tenfold optical disproportion in the capacity of the photosynthetic unit (PSU): under saturating irradiance, chlorophyll molecules in PSU collect about 2000 excitations per second per trap while dark reactions of photosynthesis can transfer 100 to 200 electrons per second only. The ratio between the electron input and utilization seems to have a strongly adaptive character which is rooted in the multiple reinforcement of the carboxylation function to the "detriment" of the electron transport function of photosynthesis. At the same time, the energy transfer probability in the appropriate centres seems to remain the same regardless of variations in the size of PSU (*e.g.* 630 to 940 and 220 to 540 chlorophylls per reaction centre in leaves of sun and shade plants, respectively — Malkin and Fork 1981).

Photosynthesis as the major synthetic process in the leaf is ultimately dependent on the quanta of energy captured by chlorophylls and carotenoids in between 400 to 700 nm. At the same time, it is also affected by radiation of other wavelengths which is absorbed by other chemical compounds.

In UV, cuticular waxes, flavonoids and phenolic compounds (including the anthocyanin group) in cell walls and vacuoles of leaf epidermis, abscisic acid, indole-3-acetic acid, $P_r$ and $P_{fr}$, proteins, nuclear deoxyribonucleic acid and ribonucleic acid are supposed to be the common potential receptors which mediate UV-induced changes at the electron/proton and atom/molecule levels in the leaf (Caldwell 1981) and determine leaf A curve as being high and flat and only continuously declining throughout the biologically most significant UV-B and UV-A spectra (Gausman *et al.* 1975). Optical activities of some of these photoreceptors protect the leaf from ultraviolet damage to the photosynthetic system, *e.g.* the disruption of lamellar membranes of chloroplasts, the destruction of plastoquinone and the inhibition of electron transport associated with PS 2 and the inactivation of PS 1 (Mantai *et al.* 1970) and, consequently, the rate of net photosynthesis (Van *et al.* 1976, Teramura *et al.* 1980). But photosynthesis can be also stimulated to a small degree by UV-A (McCree and Keener 1974) since the A spectra of chlorophylls and accessory pigments partly transgress into this region. Also the activation of the ribulose-1,5-bisphosphate carboxylase in response to UV-A may by suggested (Daley *et al.* 1978).

In IR, the liquid water in cell walls and cytosol is responsible for high leaf A because the vibrational absorptance bands of triatomic water molecules are at 10 to 15 times higher frequencies than are the electronic absorptance bands of the other compounds and pigments mentioned. The long-wave A spectrum of the leaf has three peaks at 1450, 1950 and beyond 3000 nm (Allen *et al.* 1970, Gates 1980). Above the 3000 nm, the leaf absorbs radiation almost completely (95 to 97%) and behaves like a black body. Owing to the rotational transitions in water molecules, some secondary leaf A peaks in the near IR also occur. The "visible window" between 400 and 700 nm is relatively free of A by water in leaf tissues but in the UV spectrum of less than 300 nm, A by water molecules becomes again important. Even though IR radiation, absorbed by water molecules, is not a driving force in leaf photosynthesis, it strongly affects this process by influencing enzymatic steps to $CO_2$ fixation due to changes in leaf temperature and the efficiency of water use in photosynhesis.

## 4.2.2 Transmission

Radiation is transmitted through the leaf because of the selective susceptibility and transparency of single photoreceptors mentioned and due to the gaps in and between absorbing and reflecting surfaces, respectively, which a radiant beam can pass through (the "sieve effect" — Rabinowitch 1951, Fukshansky 1978). Leaf T (Fig. 4.2, *top left*), similarly as leaf R, is then dependent on the size of particles

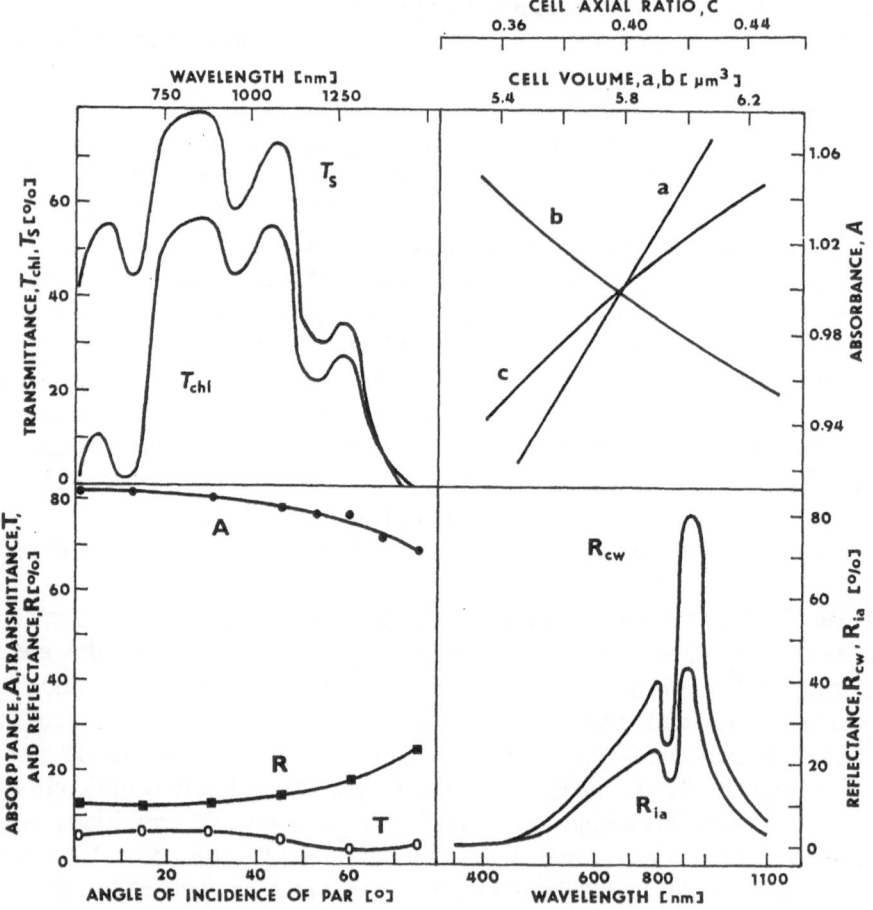

interacting with radiation (Rayleigh and Lorenz-Mie scattering phenomena rooted in a comparability of leaf particle sizes with a particular wavelength of radiation) and on the size and number of gaps over individual planes inside the leaf. The latter two characteristics are functions of spatial dispersion and overlapping (density) of particles constituting the leaf and interacting with radiation.

Generally, spectral T curve of the leaf is opposite to leaf A curve. Its highest values are in the green region (550 nm) and, particularly, in the near IR (700 to 1350 nm). Changes in leaf T are parallel to those of leaf R: T is great/small at the wavelengths at which R is also great/small (*cf.* Fig. 4.1).


### 4.2.3 Reflection

The reflection of radiation is the expression of refractive index discontinuity at the solid/liquid surface and angle of incidence of a radiant beam on an exposed plane (cosine law, mirror-like effects). Following the laws of optics, radiation must pass from a material with a high index of refraction to a material with a low one on the condition that the angle of incidence is sufficiently large. When irradiating the leaf, its R is almost constant between 0° and 40° but it sharply increases with the increasing angle of incidence of radiation over 45° (Brandt and Tageeva 1967, Yates 1981 — Fig. 4.2, *bottom left*).

The leaf reflects radiation because of refractive index discontinuities at the air-cuticle and the intercellular air spaces-hydrated cell walls interfaces (Willstätter and Stoll 1918, Sinclair *et al.* 1973). Thus in soybean (Woolley 1975), the indexes of refraction amount to 1.415 (air-hydrated walls of mesophyll cells) and 1.48 (air-external surface of cuticular hairs) in comparison to the index of refraction for the air-pure water interface which equals 1.33. Because of refractive index differences, cytosol, chloroplasts, nucleus, *etc.* can also cause radiation to reflect and thus diffuse inside the leaf (Gausman 1973 a, b, 1977; *cf.* Fig. 4.2, *bottom right*). In addition, cell walls are also supposed to act autonomously (Sinclair *et al.* 1973) owing to their radiation conducting microfibrilar structure.

---

←

Fig. 4.2. Transmittance spectrum of resuspended chloroplasts ($T_{chl}$) from homogenized *Spinacia oleracea* L. leaves is lower than that of subcellular particles remaining in the supernatant of chloroplasts preparation ($T_s$) (*top left*, after Gausman 1973b). A theoretical approximation by Latimer (1979) — *top right* — shows absorbance of radiation (*A*) by homogeneous nonphotosynthesising cell suspensions as a function of cell size (volume) during growth without division (a), when cell dry matter and shape remain constant (b), and as a function of changing cell axial ratio (c). Reflectance (R) of *Sorghum almum* L. leaf increases with the increased angle of incidence of PAR from 40° to 75° by about 11 % and absorptance (A) decreases by about the same rate, while leaf transmittance (T) remains nearly constant (*bottom left*, after Yates 1981). Hydrated walls of mesophyll cells of *Agave americana* L. reflect more radiation ($R_{cw}$) than internal cell areas ($R_{ia}$) (*bottom right*, after Gausman 1977).

Most leaves exhibit a combination of two reflectance types: the specular (external) and the diffuse (internal) R. According to Woolley (1971), more than half of the total leaf R is specular. Conformably to this, Shul'gin (1973) found leaf diffuse R as contributing to the total R of holly and French bean leaves by only 16 and 33%, respectively.

In contradistinction to the specular R, diffuse R caused by heterogeneous cell particles may have a more selective character because of their discriminating A and T. Differences between single cell components in their diffusive properties seem to be larger in the near IR than in PAR (*cf.* Fig. 4.2, *bottom right*).

## 4.3 LEAF AS AN OPTICAL SYSTEM

### 4.3.1 Scattering and Optical Path of Radiation Inside a Leaf

With regard to varietal structures and mechanisms contributing to leaf A, R and T, an entire leaf acts as a highly heterogeneous medium, the optical properties of which are affected by multiple scattering (S) of radiation inside the leaf (Butler 1964, Bryant *et al.* 1969). This is apparently a complex function of refractive index discontinuities between leaf structural constituents and of differences between refractive indexes of suspended particles and the medium (see Section 4.2.3). Considering spatial relationships, the S of radiation inside the leaf is dependent on (a) direct R from the exposed surface of the leaf which decreases irradiance of the first layer of chloroplasts, (b) reverse R from various deeper layers of the leaf which adds radiation to that coming from the leaf surface, and (c) the angular distribution of radiation which is changed by diffuse R from an initial direction to various shallower angles. Leaf S determines the mean effective optical path of radiation inside the leaf the length of which exceeds the leaf thickness ("detour effect" — Rabinowitch and Govindjee 1969) and thus exerts an additional effect on leaf A and T (an increased probability that much radiation would be intercepted by pigments and less of it would leave the leaf). Thus, reducing the scattering area at the intercellular air space-cell wall interfaces by, *e.g.*, the infiltration of the leaf by water decreases leaf R and partly A and increases leaf T (Brandt and Tageeva 1967).

The radiation penetrating into and through the leaf is scattered to the extent to which leaf material is diffuse, *i.e.* heterogeneous and dispersed within the leaf. Leaf A, R and T spectra, respectively, may be thus broadened and intensified, or narrowed and suppressed, relatively to the path of radiation through the leaf which may be lengthened, or shortened effectively, due to variations in the amounts of chlorophylls (Kasanaga and Monsi 1954, Hiroi and Monsi 1966, Kirita and Hozumi 1973, Phan *et al.* 1979; *cf.* Fig. 4.3, *top left*), the ratio of their *in vivo* forms (Inada 1977, Eller *et al.* 1981), the size and density of chloroplasts and other organelles in cells and leaf tissues (Gausman 1973b), the number, size and shape of cells (Fig. 4.2, *top right*) and intercellular air spaces (Thomas *et al.* 1967, Gausman *et al.* 1969a, b, 1970, Allen *et al.* 1971), the occurrence and compactness of cuticular waxes (Clark and Lister 1975), density of hairs, thorns, *etc.* (Ehleringer and Björkman 1978, Ehleringer and Mooney 1978, Eller 1979), the thickness of

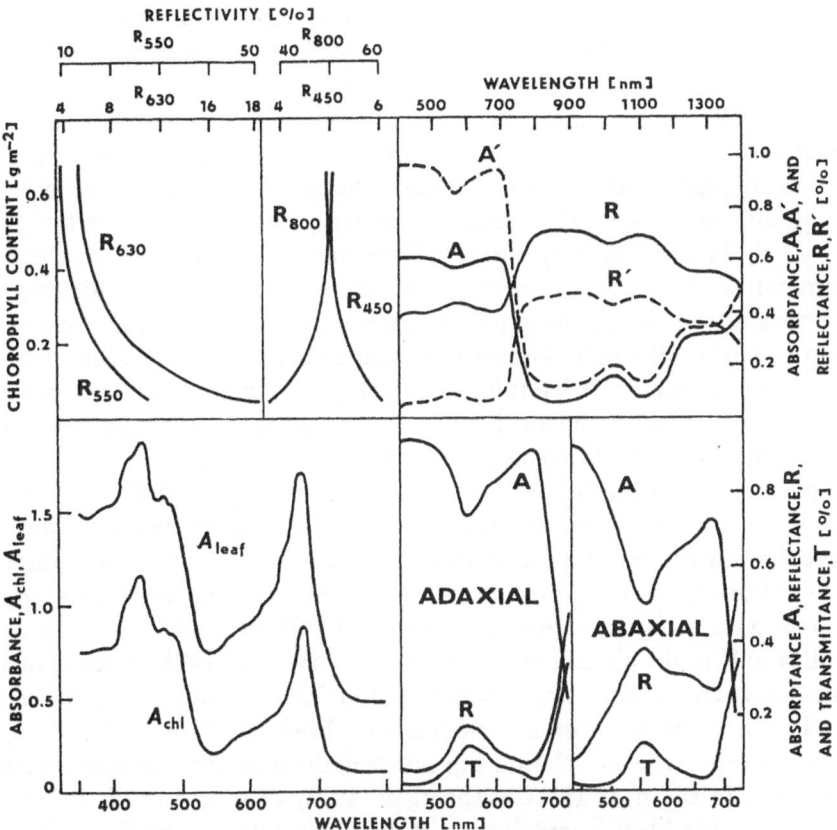

**Fig. 4.3.** At 550 and 630 nm, reflectivity (R) of *Cinnamomum camphora* Sieb. leaf increases almost exponentially with a decrease in chlorophyll content, while at 450 and 800 nm, it is almost constant except at low chlorophyll content (*top left*, after Aoki *et al.* 1980). After removing waxes from the adaxial surface, absorptance (A') of *Kalanchoe pumila* Baker leaf increases and reflectance (R') decreases in comparison with those (A, R) of the untreated leaf. Absorptances obtained for the leaves with and without waxes come close together in the near IR where chlorophylls do not absorb, while the differences between reflectances are obvious over the entire PAR and the near IR spectrum (*top right*, after Eller and Willi 1977). The absorbance of *Sinapis alba* L. leaf ($A_{leaf}$) over the PAR region is about 1.7 to 3.6 times higher than that of chloroplast suspension with equal chlorophyll content ($A_{chl}$). The greatest differences occur between 500 to 600 nm where chlorophyll absorption is minimum (*bottom left*, after Rühle and Wild 1979). Absorptance (A) spectrum of a *Rhododendron ferrugineum* L. leaf irradiated from the adaxial surface is higher and the reflectance (R) spectrum is smaller than those of the leaf irradiated from the abaxial surface. Transmittance (T) curves remain almost the same in both cases (*bottom right*, after Eller 1975).

the leaf and relative water content in leaf tissue (Carlson *et al.* 1971, Gausman *et al.* 1973), chloroplast reorientations (Lechowski 1974), nitrogen content in the leaf (Takano and Tsunoda 1970), *etc.*

A close relationship between the scattering phenomena, the pathlength of radiation beam inside the leaf and structural heterogeneity of leaf laminae is well evidenced by the A, R and T characteristics of the entire leaf which are generally higher in comparison to those of leaf homogenates, pigment extracts and suspensions of subcellular particles (Fig. 4.3, *bottom left*).

Single structural components and the leaf physiological state affect optical properties of a mature leaf to a different extent. Thus leaf A in PAR does not change proportionally, *e.g.*, with changes in leaf chlorophyll content: bean and potato leaves respond to 63 and 36% decrease in chlorophyll concentration by lowering leaf A by only 17 and 7%, respectively. Differencess in A relative to chlorophyll concentration are less apparent in blue and red spectral regions in which the maximum absorbance bands of chlorophylls occur but are more pronounced in the green region in which leaf R and T dominate (Brandt and Tageeva 1967, Shul'gin 1973).

Leaves differ in their relative thickness. High/low values of leaf area/dry matter ratio correspond to low/high leaf T while R appears to be relatively insensitive to changes of this parameter (Carlson *et al.* 1971).

Changes in leaf water state cause the internal surface area of the leaf to change by changing cell dimensions and leaf air volume: *e.g.* in maize and soybean, the 30% decrease in the relative water content of the leaf results in a 35 and 47% decrease in leaf air volume (Woolley 1973). In this way, leaf water content affects the shape and volume of intercellular air spaces (Cutler *et al.* 1977) and, consquently, the diffuse R at the air-cell wall interfaces (Allen *et al.* 1971). Leaf R increases with the decreasing relative water content below 80%; at greater values no or few changes in R have been reported (Thomas *et al.* 1966, Shul'gin 1973).

As may be expected, the relationship between the relative water content and leaf R is strongest in the near IR (Carlson *et al.* 1971, Gausman *et al.* 1971b). The infrared R of a dried leaf is largely that of diffuse cellulose R, while the fresh leaf infrared R curve depends on a combination of diffuse R with water absorption bands (Woolley 1971). Similarly, the influence of freezing and thawing of a leaf causes leaf R to change to a different extent.

### 4.3.2  Dorsiventrality of a Leaf

Under natural conditions, the leaf is differently irradiated from both sides. Hence, making allowance for unequal irradiance from the upper (adaxial) and lower (abaxial) sides of the leaf, actual differences in leaf A, R and T are expected. However, even under the same irradiance from either the adaxial or abaxial surface, the response of the leaf may be different. While in monocotyledonous plants (unifacial leaf), A is the same for both sides (Sinclair *et al.* 1973), in dicotyledonous plants (dorsiventral leaf), A of the leaf irradiated from the adaxial surface (normal position) is higher than if the leaf is irradiated at an inverse position from the abaxial surface (McCree 1971/1972, Sinclair *et al.* 1973) partly because of its smaller R (Sheehy 1975, Meyer and Walker 1981 — see also Fig. 4.3, *bottom right*).

Differences in A are maintained over the entire PAR and IR regions while R curves for the leaf in normal and inverse position, respectively, cross over between the 800 and 1300 nm (Sinclair *et al.* 1973). At the wavelengths longer than 1300 nm, R from the lower surface again exceeds that from the upper one (Gupta and Woolley 1971).

In PAR, a dorsiventral leaf transmits radiation almost equally through both surfaces (Sălăgeanu 1965) but in IR, T of the leaf irradiated from the upper side is smaller in comparison to the leaf exposed to radiation from the lower side (Gupta and Woolley 1971).

Different optical responses of the leaf in the normal and inverse position, respectively, are mainly due to differences in roughness and composition of layers on adaxial and abaxial epidermes (pubescence, epicuticular waxes, *etc.*), and thickness of both epidermes and adjacent mesophyll tissues. The palisade parenchyma can absorb more PAR because of a higher thickness (volume) of cell layers (Dornhoff and Shibles 1976) with a higher chlorophyll ($a + b$) content (Moss 1964) and higher chlorophyll $a$ to $b$ radio (Malkina 1976a), higher number of chloroplasts per cell, higher mean volume of chloroplasts in cells (Wild and Wolf 1980) and smaller thickness of chloroplast grana (Skene 1974). In comparison to spongy parenchyma, the palisade mesophyll can also absorb more of the far IR because of a larger cellular volume (Parkhurst 1982) related with a higher % of water content per unit volume of the tissue. Because of the larger surface area of cell wall exposed (Nobel *et al.* 1975, Outlaw *et al.* 1976), the palisade tissue may also scatter more in the near IR in which chlorophylls are translucent (total leaf R increases because of the increased diffuse R).

The parts that the palisade and spongy parenchyma play in the scattering of radiation inside the leaf still remain an object of possible theoretical considerations to the present time. According to Willstätter and Stoll (1918), R at the air-cell wall interfaces of spongy parenchyma is critical for R and T of an entire leaf. Knipling (1970) supposes that air spaces within the palisade mesophyll may be more important in scattering radiation than air spaces in the spongy parenchyma layers. Sinclair *et al.* (1973) assume that the spongy mesophyll can act as a barrier preventing some of the radiation reflected from the palisade tissue from emerging on the adaxial side as leaf R.

### 4.3.3 Attenuation of Radiation Inside a Leaf

It follows from the diffusive pathways of radiation within the leaf (see Section 4.3.1) that at least in thin leaves the internal R may be sufficient to cause uniform distribution of radiation and make difficult an exponential attenuation of radiation across leaf laminae, *i.e.* from the usually more irradiated adaxial towards the less or non-irradiated abaxial side of the leaf (Brandt and Tageeva 1967). On the other hand, in thick leaves, an exponential gradient of irradiance of mesophyll layers underlying one another may be reasonable according to the Lambert-Beer law (Rabinowitch 1951, Laïsk 1969, Oya and Laïsk 1976) since larger amounts of leaf material would be expected to attenuate more of the incident radiation.

It was verified by means of mathematical approximations that the internal geometry of the leaf brings about a more uniform distribution of radiation within the leaf than would be expected from the Lambert-Beer law (Oya and Laïsk 1976). However, the leaf scatters too little radiation to be able to behave as an ideal diffuser or specular reflector (Howard 1966, Woolley 1971). In a thick dorsiventral leaf, the departure of the radiation distribution pattern from the Lambert-Beer law is probably closely related to the optical properties of palisade and spongy parenchyma, respectively. If irradiated from the adaxial leaf surface, the former absorbs more radiation immediately before it enters the less absorbing spongy parenchyma. However, a reverse pattern, *i.e.*, more radiation transmitted and scattered by the palisade parenchyma, and, consequently, more of it absorbed by the spongy one, may be also supposed if chloroplast movements under strong irradiance are considered (Starzecki 1962).

The unequal contribution of the palisade and spongy parenchyma to the attenuation profile of radiation across leaf laminae can be of an adaptive character which seems to be related largely to differences in photosynthetic activity between both parenchyma tissues (Outlaw *et al.* 1976, Mokronosov 1981 — *cf*. Section 7.2.5.2).

The efforts to quantify complex scattering phenomena and determine the attenuation profile of radiation in the leaf with regard to details in spatial and functional patterns of leaf absorbing, reflecting and transmitting constituents, have advanced due to the use of mathematical models. Allen and Richardson (1968) tried to descri-

---

→

**Fig. 4.4.** Approximately 40 % of plant species absorb about 80 % of the incident PAR which represents about 40 % from the total radiation (300 to 4000 nm) (*top left*, after Shul'gin 1973). The absorption coefficient decreases from succulents to xerophytes and mesophytes and reaches the smallest value in hydrophytes; the transmission coefficient changes inversely while the differences in reflection between single plant groups are rather small (*top right*, data from Shul'gin 1973). The differences in the absorption spectra between succulents, xerophytes, mesophytes and hydrophytes in the PAR region are mostly pronounced at 500 to 600 nm (*bottom left*, after Shul'gin 1973). The ranges of leaf absorption coefficients over 400 to 700 nm are different between single groups of desert plants being the highest in shrubs (29 to 89 %) and the smallest in trees (73 to 87 %) (*bottom right*, after Ehleringer 1981).

be leaf R and T in the near IR by applying the K-M theory (Kubelka and Munk 1931) of the attenuation of radiation in a diffusing medium to a single leaf. The flat plate model specified by two optical constants (an effective index of refraction and an effective coefficient of A) was used to explain diffuse R and T of a compact leaf without intercellular air spaces (Allen *et al.* 1969), and a lacunose leaf containing intercellular air spaces (Allen *et al.* 1970). Tucker and Garratt (1977) related the radiation absorbed, reflected and transmitted by a leaf to scattering properties of palisade and spongy parenchyma, and to the specular R from leaf cuticle; they simulated optical function of an entire leaf based on a random process making up a Markov chain with discrete states (a stochastic model). The model of Fukshansky (1978, 1981) took into account the multiple scattering of radiation inside a leaf and the occurrence of the sieve effect together with the A and R spectra of leaf pigments to characterize irradiance gradients in leaf tissues.

Nevertheless, up to now, no exhausting and explicit conception which would quantify distributive patterns of radiation in the leaf has been advanced.

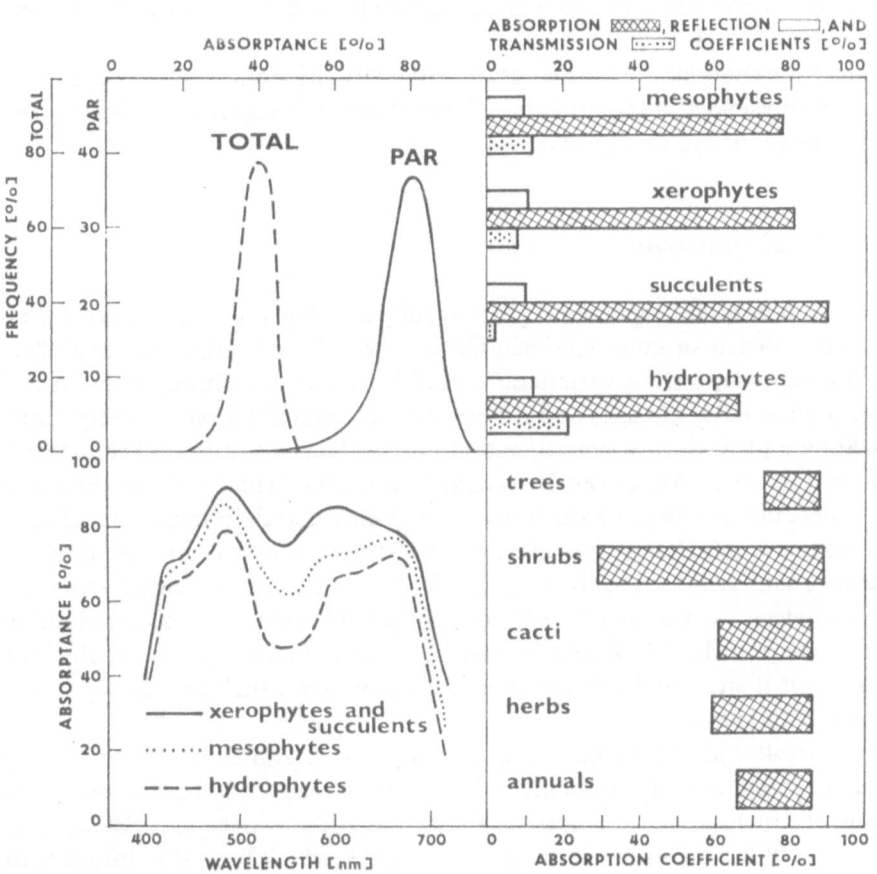

## 4.4  VARIABILITY IN LEAF OPTICAL PROPERTIES

Differences in leaf internal structure in subdividing leaf tissues, cells and their organelles cause optical properties of leaves to vary among the plants. Leaf A, T and R characteristics differ in herbs, grasses, evergreen and deciduous shrubs and trees, in mesophytes, xerophytes, succulents and aquatic plants, and in plants of different taxonomic ranges as well (Fig. 4.4, *top right, bottom left* and *right* — Kasanaga and Monsi 1954, Inada 1976, Gates 1980, Gausman *et al.* 1981). Leaf A, R and T change with the direction of the incident radiation (Sun—leaf geometry) and atmospheric conditions (Brandt and Tageeva 1967, Shul'gin 1973) and vary with plant growth as being related to, *e.g.*, latitudinal, longitudinal and altitudinal gradients (Kuznetsova *et al.* 1974). Similarly, the plants adapted to sun and shade habitats (Rühle and Wild 1979, Eller *et al.* 1981), to field, growth chamber and greenhouse conditions (McCree 1971/1972) and plants subjected to short- and/or long-termed ecological stresses such as nutrient deficiency (Thomas and Oerther 1972, Terry 1980), water (Thomas *et al.* 1971) and temperature (Cameron 1970) stresses, salinity and osmotic stress (Gausman and Cardenas 1968) also show distinct optical properties.

Hence, a genotypic and phenotypic diversity in optical properties — as in other leaf physiological parameters — has not to be disregarded if changes during leaf development are considered.

### 4.4.1  Leaf Ontogeny

In most studies on leaf optical properties, fully developed mature leaves have been used as the objects of investigations. Until now, relatively little has been done in estimating age-dependent variations in leaf A, R and T as brought about by leaf ontogeny (chronological age) and differences between the leaves subsequently developed on a plant during the vegetation period (leaves of different sequential age and insertion level). Moreover, if attainable, the data (Table 4.1) are inconsistent with respect to leaf optical characteristics described and determinations of leaf age and insertion level. Hence, it is difficult to distinguish between non-identical characteristics and avoid undesirable confusions to make some general conclusions.

Nevertheless, regardless of a lack of conformable information, a general ontogenetic pattern of leaf A, R and T may be forecast indirectly, *i.e.* on the basis of changes in leaf structural and physiological attributes which are closely related to leaf optical properties.

Theoretically, in the young compact leaf, low R and A and high T may be expected. Immature protoplasmic cells are too small and contain only a small amount of chlorophylls to intercept much radiation. There are also only a few small intercellular air spaces in the leaf mesophyll which limit the diffuse R inside the leaf to reach higher values. As the leaf matures, differentiation into palisade and spongy parenchyma becomes more pronounced (see Sections 1.4.1 and 1.4.4) and the number and volume of intercellular air spaces increase (see Section 1.4.3) similarly as the leaf cells increase in size and quantity (see Sections 1.2.1 and 1.2.3). The effective pathlength of radiation through the leaf is lengthened and leaf S increases. Consequently, total leaf R tends to be higher due to the increased diffusive component. Leaf A progressively increases till the leaf attains its matu-

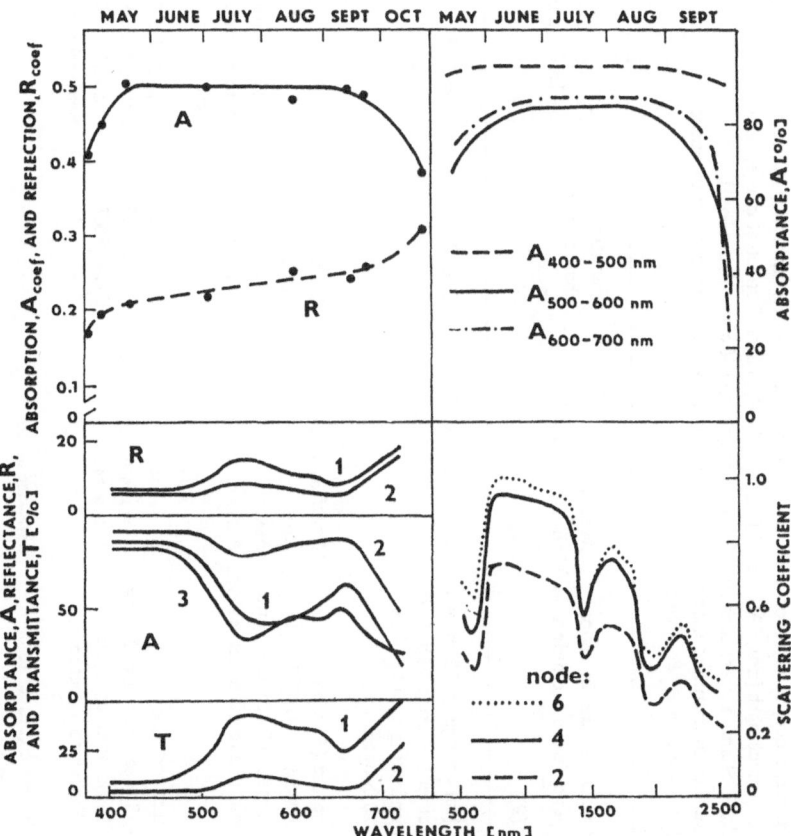

**Fig. 4.5.** The absorption (A) and reflection (R) coefficients (in UV, PAR and IR) of *Populus deltoides* L. leaves change during the vegetation period (*top left*, data from Gates 1980). The absorptance (A) changes during the ontogeny of *Tilia cordata* L. leaves are more expressed in the red and green than blue waveband (*top right*, data from Brandt and Tageeva 1967). Comparison of absorption (A), reflection (R) and transmission (T) spectra throughout the PAR region shows differences with ageing of *Betula verrucosa* L. leaves: 1, 2, 3 — 0, 116 and 146 d after leaf unfolding, respectively (*bottom left*, after Brandt and Tageeva 1967). Also the scattering coefficients for different wavelengths of the incident radiation differ in leaves of *Gossypium hirsutum* L. of different insertion level: leaves representing nodes 2, 4 and 6 basipetally have after emergence ages 20, 29 and 35 d, respectively (*bottom right*, after Gausman et al. 1971 a).

**Table 4.1** Changes in optical properties with leaf ontogeny and insertion level.
Abbreviations and explanations:
Plant and cultivation: where cultivation is not given, plants are taken from natural habitats.
Measured characteristics: A — absorptance, R — reflectance, S — scattering, T — transmittance [% of the incident radiation]: $A(R, S, T)_\lambda$ — A, R, S and T, respectively, at a given wavelength [nm]; $A(R, S, T)_{\lambda \text{ to } \lambda}$ — A, R, S and T spectral curve, respectively, in a given range of wavelengths [nm]; $A(R, S, T) \text{coef}_{\lambda \text{ to } \lambda}$ — A, R, S and T, respectively, averaged over a specific waveband [nm].
Variations in characteristics measured: minimum to maximum values if numerical results are given by authors.
Changes induced by leaf ageing: ↑ — increase, ↓ — decrease.

| Plant (and cultivation) (1) | Leaf age and insertion level (from the oldest one) (2) | Measured characteristics (3) | Variations in characteristics measured (4) | Changes induced by leaf ageing (5) | Author(s) and year of publication (6) |
|---|---|---|---|---|---|
| *Acer carpinifolium* S. et Z. | leaf ontogeny, 12 measurements during 218 d after unfolding (16 April to 20 November) | $T\text{coef}_{400 \text{ to } 700}$ | 5.0...21.0 | ↓ to 198 d and ↑ | Kasanaga and Monsi 1954 |
| *Acer platanoides* L. | young (4 to 6 d), intermediate (10 to 14 d), mature (20 to 25 d) and old (70 to 80 d) leaves | $A\text{coef}_{400 \text{ to } 700}$ $A_{441}$ $A_{545}$ $A_{685}$ | 58.0...84.0 88.0...96.0 44.0...71.0 54.0...81.0 | ↑ ↑ ↑ to mature leaf ↑ to mature leaf and ↓ | Malkina 1976b |
| *Betula verrucosa* L. | leaf ontogeny, 13 measurements during 146 d after unfolding (7 May to 1 October) | $A_{400 \text{ to } 500}$ $A_{500 \text{ to } 600}$ $A_{600 \text{ to } 700}$ | 9.00...96.8 60.8...87.5 51.9...88.7 | ↑ to 104 d (19 August) and ↓ ↑ to 116 d (1 September) and ↓ | Brandt and Tageeva 1967 |
| *Brassica oleracea* L. (field) | leaves 1 (outer) to 6 (inner) | $T\text{coef}_{400 \text{ to } 700}$ | 4.8...44.4 | ↓ | Kasanaga and Monsi 1954 |
| *Euonymus bungeanus* Max. | leaf ontogeny, 9 measurements during 262 d after unfolding (26 March to 13 December) | $A\text{coef}_{300 \text{ to } 2500}$ $R\text{coef}_{300 \text{ to } 2500}$ | 31.5...54.2 22.1...35.6 | ↑ to 192 d (4 October) and ↓ ↑ | Gates 1980 |

| Species | Measurements | Parameter | Value | Changes | Reference |
|---|---|---|---|---|---|
| *Ginkgo biloba* L. | leaf ontogeny, 12 measurements during 222 d after unfolding (17 April to 25 November) | $Tcoef_{400 \text{ to } 700}$ | 2.0...24.5 | ↓ to 38 d and ↑ from 197 d | Kasanaga and Monsi 1954 |
| *Gossypium hirsutum* L. (field) | leaves of 20, 25, 29, 33 and 35 representing nodes 2, 3, 4, 5 and 6 (basipetally) | $R_{500 \text{ to } 2500}$<br>$S_{500 \text{ to } 2500}$<br>$A_{500 \text{ to } 2500}$ | | ↑<br>↑<br>↑ at 500 to 750 nm and at 1350 to 2500 nm; ↓ at 750 to 1350 nm | Gausman *et al.* 1971a |
| *Liriodendron tulipifera* L. | leaf ontogeny, 7 measurements during 200 d after unfolding (4 April to 1 November) | $Acoef_{300 \text{ to } 2500}$<br>$Rcoef_{300 \text{ to } 2500}$ | 28.2...52.8<br>14.3...37.7 | ↑ to 53 d (8 June) and irregular changes<br>↑ to 186 d (18 October) and ↓ | Gates 1980 |
| *Oryza sativa* L. | leaves 2 and 3 (basipetally) on plants 68, 83 and 142 d old | $A_{350 \text{ to } 800}$ | | ↓ ; largest differences at 550 nm | Inada 1977 |
| *Populus deltoides* Marshal. | leaf ontogeny, 8 measurements during 181 d after unfolding (27 April to 25 October) | $Acoef_{300 \text{ to } 2500}$<br>$Rcoef_{300 \text{ to } 2500}$ | 27.8...50.8<br>16.9...36.2 | ↑ to 64 d (1 July) and irregular ↓<br>irregular ↑ | Gates 1980 |
| *Tilia cordata* L. | leaf ontogeny, 16 measurements during 145 d after unfolding (9 May to 1 October) | $A_{400 \text{ to } 500}$<br>$A_{500 \text{ to } 600}$<br>$A_{600 \text{ to } 700}$ | 84.0...96.3<br>49.7...86.6<br>41.3...88.5 | ↑ to 102 d (19 August) and ↓ | Brandt and Tageeva 1967 |
| *Triticum vulgare* Vill. (field) | upper and lower leaves on 19 May and 2 June | $Acoef_{400 \text{ to } 700}$<br>$Rcoef_{400 \text{ to } 700}$<br>$Tcoef_{400 \text{ to } 700}$ | 74.8...80.5<br>10.1...13.0<br>9.0...12.2 | ↑ ; ↓<br>↓ ; ↑<br>↓ ; ↑ | Mitrofanov *et al.* 1969 |

Table 4.1 (continued)

| Plant (and cultivation) | Leaf age and insertion level (from the oldest one) | Measured characteristics | Variations in characteristics measured | Changes induced by leaf ageing | Author(s) and year of publication |
|---|---|---|---|---|---|
| (1) | (2) | (3) | (4) | (5) | (6) |
| *Vicia faba* L. var. *equina* (field) | leaves 5, 7, 10 and 14 | $A_{400\ to\ 800}$ | | ↓ to leaf 7, ↑ to leaf 10 and ↓ to leaf 14 at 550, 700 and 800 nm | Brandt and Tageeva 1967 |
| *Vitis vinifera* L. (field) | leaves 1 to 6 (basipetally) | $T\mathrm{coef}_{250\ to\ 3300}$ | 10.0...35.0 | ↓ | Kriedemann 1968 |

rity. This increase is related to the increased number of chloroplasts per cell, chloroplast size (see Section 1.4.5), chlorophyll content (*cf.* Chapter 3) and higher hydration level of leaf tissues (water-filled vacuoles in cells develop): leaf A and R seem to be positively linked with leaf thickness and water content determined on a dry-matter basis (Gausman *et al.* 1971a, b, 1973) which usually increase with leaf age (see Section 1.2.1). As the leaf becomes mature, T decreases (due to the increased amounts of the absorbing materials per unit leaf area) being negatively correlated with leaf thickness and water state. With leaf senescing, A decreases namely because of structural degradation of chloroplasts which causes the blue and red A bands of chlorophylls to weaken (in the yellowing and dying off leaf, A is usually smaller than that in the young leaf). Owing to the reduction in the amount and activity of the absorbing materials in the senescing leaf, R continues to increase and T also increases.

Changes in A and R coefficient obtained with *Populus* (Fig. 4.5, *top left*), *Liriodendron* and *Euonymus* leaves (Gates 1980) confirm that such an ontogenetic pattern of leaf optical properties may indeed be expected if the leaf responses to radiation are integrated over the entire short- and long-wave spectra. However, the pattern of ontogenetic changes may be different with respect to single wavelengths or wavebands of radiation with which the leaf interacts. For example in PAR, leaf A attains the maximum with leaf maturation but leaf R after the initial increase at the beginning of leaf expansion decreases in parallel with leaf T (*Hedera, Cinnamomum* — Kasanaga and Monsi 1954; *Betula, Tilia* — Brandt and Tageeva 1967). Some increase in T is usually observed in yellowing and dying off leaves (*Ginkgo, Acer* — Kasanaga and Monsi 1954).

The extent of variations in optical properties with leaf ontogeny is rather narrow, especially if the average A, R and T coefficients throughout the UV, PAR and the near IR are estimated (*cf.* Fig. 4.5, *top left*), but it is more obvious if leaf spectral responses to single wavelengths of radiation are compared. To give an example, in PAR, the ontogenetic changes in leaf A are more pronounced in the green and red regions than in the blue one (Fig. 4.5, *top right*). Different ontogenetic capabilities in reflecting the wavelengths of the near IR were also observed (oak — Gates 1980).

The shape of leaf A, R and T ontogenetic curves, respectively, is, *e.g.*, dependent on leaf pubescence which is usually denser in the young than in the old leaf (*Verbascum* — Gates 1980) and the progressive development of leaf xeromorphic structure (*Rhododendron* — Eller 1975).

Leaf ontogeny as a consequence of developmental stages from leaf emergence to the dying off of the leaf is closely linked to leaf morphogeny: The leaf blade consists of developmentally heterogeneous tissues in which some cells divide (leaf base with the youngest tissue), elongate (the middle part of leaf laminae) and are fully differentiated (leaf apex with the oldest tissue). Conformably to this, optical characteristics change along the leaf blade (maize — Brandt and Tageeva 1967, Hopkins *et al.* 1975). In PAR, A of the maize leaf increases from the base to the apex in a strong correlation with chlorophyll content and chloroplast dimensions and ultrastructure. The proportions in T between single leaf blade regions are of a reverse character. Similarly, R of the leaf base is the smallest while the highest R occurs in the region immediately beyond the leaf apex.

### 4.4.2 Insertion Level and Sequential Age of Leaves

If leaf age is considered independently, one would expect that individual leaves on a plant have the same optical properties at comparable stages of ontogeny. However, endogenous developmental processes in plants not only cause the earlier-formed leaves to become senescent as the new leaves are produced, but they also cause the latter to be positioned above and develop under different microclimatic conditions. Hence, leaf A, R and T are influenced by both the sequential age and insertion level of leaves.

Changes in leaf optical properties induced by the effects of the above mentioned factors are of a diverse character in single plant species (*cf.* Table 4.1). Both the decline (cabbage — Kasanaga and Monsi 1954; grape – Kriedemann 1968; *Eucalyptus* — Cameron 1970; wheat — Intykbaeva and Sokolova 1974; rice —

Inada 1977) and the increase (cotton — Gausman *et al.* 1971a; ivy — Kirk and Goodchild 1972; magnolia — Gates 1980) in A, R and T, respectively, from the young (upper) to the old (bottom) leaves and the attainment of the maximum (*Eucalyptus* — Cameron 1970; apple tree – Shishkanu *et al.* 1970; maize — Al Abbas *et al.* 1974; magnolia — Gates 1980) or minimum (sunflower — Gates 1980) values in the adult (intermediate) leaves similarly as irregular changes (cotton, maize — Gausman *et al.* 1971b, 1973; wheat — Mitrofanov *et al.* 1969) are reported in the literature. No comfortable patterns of age-dependent and insertion changes may be described even with respect to leaf spectral responses (horse bean, cucumber — Brandt and Tageeva 1967; cotton — Gausman *et al.* 1971a; soybean — Gupta and Woolley 1971; Norway maple — Malkina 1976b; *cf.* also Fig. 4.5, *bottom left* and *right*).

Obviously, the above data do not permit to answer the question as to what extent changes in leaf optical properties may be derived from internal changes caused by leaf ageing or comprise a response to unequal environmental conditions which the sequentially developing leaves meet with.

## 4.5    CONCLUSIONS

Leaf optical properties may give information about the structure of particles constituting the leaf; differences in spectral characteristics may reveal the functioning of single photoreceptors as likewise the dynamics of energy transitions may help to detect the order and kinetic steps in complex biochemical reactions. Leaf optical properties may also indicate indirectly parameters of the physiological state of the leaves and plants. The determination of leaf optical properties at particular wavelengths of radiation used in non-destructive remote sensing methods (*e.g.*, aerial infrared colour photography) aids the crop management practice to estimate more effectively deviations from the optimum course of physiological, growth and developmental processes in plants, and to detect and avoid conditions producing various stresses (*e.g.*, the moisture stress, nutrient defficiency) during the vegetation period.

The general features of interactions of leaves with radiation are known. However, with regard to the leaf as an entire optical system, much remains to be learnt about mechanisms involved at each organizational level of the leaf. The determination of the optical function of the leaf with respect to scattering and attenuation of radiation by leaf tissues is dependent on a further improvement of experimental methods and a more intense use of mathematical models. These should help to explain leaf optical properties as changing to the spatial (leaf internal structure) and chronological (leaf age) scale as well. Information is needed to distinguish changes in leaf A, R and T resulting from endogenous processes and from sequence of events set in motion by external stimuli.

Because of a great importance of PAR for the most vital functions of plants, much remains to be done in the analysis of leaf optical properties relative to ontogenetic changes in leaf photobiological activity in this part of the solar spectrum.

As the leaf grows old in natural conditions, its irradiance may decrease almost to 10 to 15% of that of the young expanding leaf, but the range of ontogenetic changes in leaf A, R and T, respectively, is usually much narrower. The ontogenetic changes in leaf optical properties relative to changes in the spectral composition of radiation are more pronounced: with senescence, lower position and corresponding decrease in irradiance of the leaf with the depth in the canopy, leaf A in the green, blue and red regions, respectively, changes to a different extent being closely linked with changing contributions of single wavebands to the total PAR. As the leaf A spectrum and action spectra of photosynthesis and chlorophyll synthesis always exert two peaks, the activity of carotenoids is maximum in the blue region, *etc.*, the effectivity of each waveband with respect to the amount of quanta incident and absorbed by a developing leaf has to be taken into consideration: the amount of radiation at which the effectivities of single wavebands are the same or completely different has to be determined with more preciseness because the pattern of ontogenetic changes in single photobiological processes is closely dependent on combinations between quantitative-to-qualitative parameters of the available radiation.

# 5 CHANGES IN ELECTRON TRANSPORT CHAIN COMPOSITION, AND ACTIVITIES OF PHOTOSYSTEMS AND PHOTOPHOSPHORYLATION DURING LEAF ONTOGENY

*Z. Šesták*

## 5.1 PRINCIPLES OF PHOTOSYNTHETIC ELECTRON TRANSPORT CHAIN AND PHOTOPHOSPHORYLATION

Photosynthesis *sensu stricto* is a series of biophysical and biochemical processes driven by radiant energy in the range of *ca.* 400–700 nm resulting in the formation of $NADPH_2$ and ATP. Both these compounds are necessary for the fixation of $CO_2$ to the appropriate acceptor and the redution of the primary product to the primary saccharide, triose phosphate. The formation of $NADPH_2$ and ATP takes place in thylakoid membranes situated in the chloroplast (see Chapter 2). Absorption of radiant energy by chlorophylls and other photosynthetic pigments located in the thylakoid membrane produces a separation of positive and negative charges in special pigments of the reaction centres of Photosystems 1 and 2 (PS 1 and PS 2), *i.e.* $P700$ and $P680$. The functionning of PS 2 is bound with the splitting of water: oxygen is evolved, and the electrons obtained pass through a series of acceptors, located in various parts of the thylakoid membrane, to NADP. The redox components involved form the photosynthetic electron transport chain. Its formulation with two photoreactions in series suggested by Hill and Bendall (1960), the so-called Z-scheme, is generally accepted today. Nevertheless, alternative hypotheses (of oxygenic and anoxygenic photosystems — see Arnon *et al.* 1981) still exist.

The final effect of the electron transport is the lowering of its redox potential, resulting in a better electron donating capacity (more negative redox potential). The transport of an electron from $P680$ in PS 2 (one photoreaction) passes probably through pheophytin *a*, acceptor $Q$ (presumably a special plastoquinone), a plastoquinone pool, cytochrome *f*, and plastocyanin to PS 1. Here it replaces the electron ejected from $P700$ in the second photoreaction. The electron passes then through non-haem iron-sulphur proteins (probably two) to ferredoxin, and the final reduction of $NADP^+$ is mediated by the ferredoxin-NADP-reductase.

The electron transport chain may function in the described manner (see Fig. 5.1) as a non-cyclic electron flow, or in the short cyclic form including only PS 1 ($P700$ — non-haem iron-sulphur acceptor — cytochrome $b_{563}$ — plastoquinone — cytochrome *f* — plastocyanin — $P700$). Both these forms of electron transport are coupled with the formation of ATP from ADP and inorganic phosphate. The energy for these photophosphorylations is provided by the flow of electrons to more positive redox potentials (*e.g.* from $Q$ to $P700$). The coupling mechanism is explained either by a generation of a very reactive chemical intermediate which drives ATP synthesis (chemical hypothesis) or by the formation of a proton gradient on both sides of the thylakoid membrane, which sets up a membrane potential plus a pH difference, which drive the phosphorylation (chemi-osmotic hypothesis).

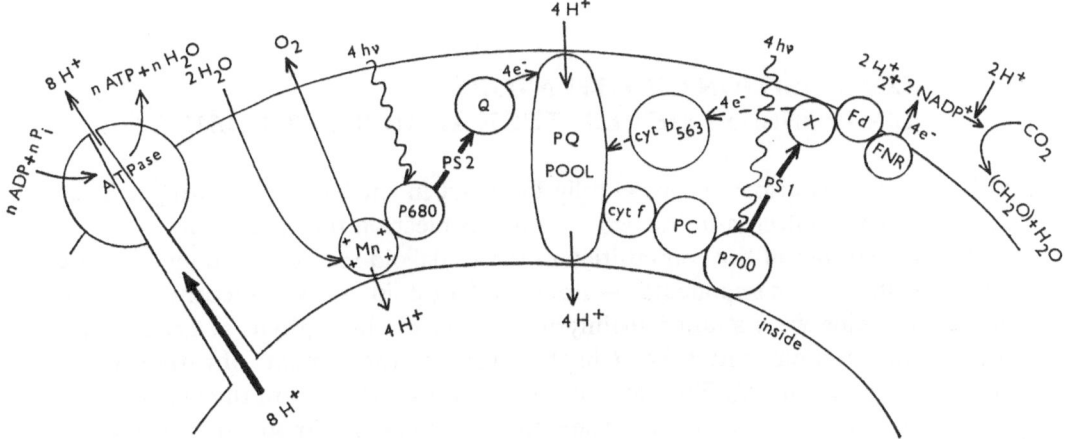

**Fig. 5.1.** Scheme of location of the electron transport chain and photophosphorylation mechanism in the thylakoid membrane. Radiant energy absorption and use in the Photosystem 2 (PS 2) reaction centre ($P$ 680) is bound with splitting of water in the Mn-catalyzed system and with electron transport through primary acceptor $Q$, the plastoquinone (PQ) pool, cytochrome $f$ (cyt $f$), and plastocyanin (PC). The electrons replace those ejected from the reaction centre ($P$ 700) of Photosystem 1 (PS 1) and accepted by non-haem iron-sulphur proteins (X), and then transported through ferredoxin (Fd) and ferredoxin-NADP-reductase (FNR) to NADP. Reduced NADP is then used for reduction of the primary product of $CO_2$ fixation to the primary photosynthate ($CH_2O$). Cytochrome $b_{563}$ (cyt $b_{563}$) functions in the shortened cyclic electron transport. Electron transport and proton ($H^+$) translocation are coupled with phosphorylation (using ADP and inorganic phosphate, $P_i$). The scheme calculates with the transport of four electrons, splitting of two molecules of water, and production of one molecule of oxygen, with the minimum possible eight-quantum requirement.

The thylakoid membrane is a very dynamic system, consisting of a lipid bilayer into which various proteins are embedded. Some proteins bear photosynthetic pigments (see Chapter 3), some function as enzymes or bear coenzymes. The lipids ensure fluidity of the membrane, the proteins decrease their mobility. The functionning of the electron transport chain and photophosphorylations is ensured by a localization of all redox components, enzymes and pigment complexes in given parts of the thylakoid membrane, which prevents the undesirable back-flow of quanta or electrons.

The structure and composition of the thylakoid membrane changes in the course

*129*

of chloroplast development (see Chapter 2) and leaf ontogeny. These changes are evident in the size of the photosynthetic unit (Chapter 3) and in the activites of both photosystems and of the non-cyclic and cyclic photophosphorylation, as well as in the amounts of individual components of the electron transport chain and enzymes. The information in this field is determined mainly by the availability of appropriate methods — while numerous data exist on the ontogenetic differences in the activity of the Hill reaction, data on some components either do not exist or have been obtained only recently.

## 5.2    COMPOSITION OF THYLAKOIDS:
## COMPONENTS OF ELECTRON TRANSPORT CHAIN

As the electron transport chain is fully localized in the thylakoid, changes in its composition are reflected in the activities of the electron transport chain.

From the studies of the composition of the thylakoid membrane, only a few are related to its basic components — lipids and proteins — as independent compounds. Proteins were studied mainly in relation to chlorophylls *in vivo*, photosystems and reaction centres (see Chapter 3) and to the contents of carboxylation enzymes (see Chapter 6). The ratio of seven main proteins from thylakoids of tobacco chloroplasts changed with leaf age: these changes corresponded to an increase in proteins belonging to PS 2 and decline in proteins of PS 1 (Wolińska 1976).

Lipid studies dealt mainly with whole chloroplasts and only rarely with isolated thylakoids (Öquist and Liljenberg 1981). In chloroplasts of primary leaves of three cultivars of wheat, the amounts of monogalactosyl diglycerides, digalactosyl diglycerides and sulpholipids increased between 6 and 15 d of plant life, while the content of phosphatidylglycerides declined (Bocharov *et al.* 1977). In cucumber cotyledons the amounts of monogalactosyl diglycerides, digalactosyl diglycerides and sulpholipides increased to the age of 18 d and then declined (Ferguson and Simon 1973); other authors found the increase only in digalactosyl diglycerides (to 14 d), while they determined a steady decline in monogalactosyl diglycerides content per cotyledon (Draper 1969). The changes in the levels of mono- and digalactosyl diglycerides in chloroplasts were in correlation with the activity of the enzyme UDP-galactose-diglyceride galactosyl transferase in chloroplast envelopes: in soybean it increased to 11 d of cotyledon age and then declined (Dalgarn *et al.* 1979). In tobacco, the ratio of monogalactosyl diglycerides to digalactosyl diglycerides declined with leaf age (Wolińska 1976). In chloroplasts from rye leaves the ontogenetic increase per dry matter was observed in all lipid groups (Kaval'chuk and Vechar 1973). The amounts of individual lipid groups per chlorophyll unit changed irregularly (rye — Vecher *et al.* 1978). The ratio of unsaturated to saturated fatty acids of lipids in thylakoids increased from young to mature leaves of *Betula pendula* and then declined in all above mentioned groups of lipids; the quantitative composition and ratios were significantly changed by the level of irradiance (Öquist and Liljenberg 1981).

The amounts of components of the photosynthetic electron transport chain are sometimes used to characterize its capacity. Mainly the amounts of cytochromes and plastoquinone are determined (mostly in relation to the amount of chlorophyll *a*) to represent the capacity of electron transport between the electron donor of PS 2 and acceptor of PS 1. The drawback to these characteristics is that the amount of the component of the electron transport chain which is minimum (hence probably other than the determined component) decides the resulting rate of electron transport, and that the total amount of the determined component may be different from the amount of its active molecules.

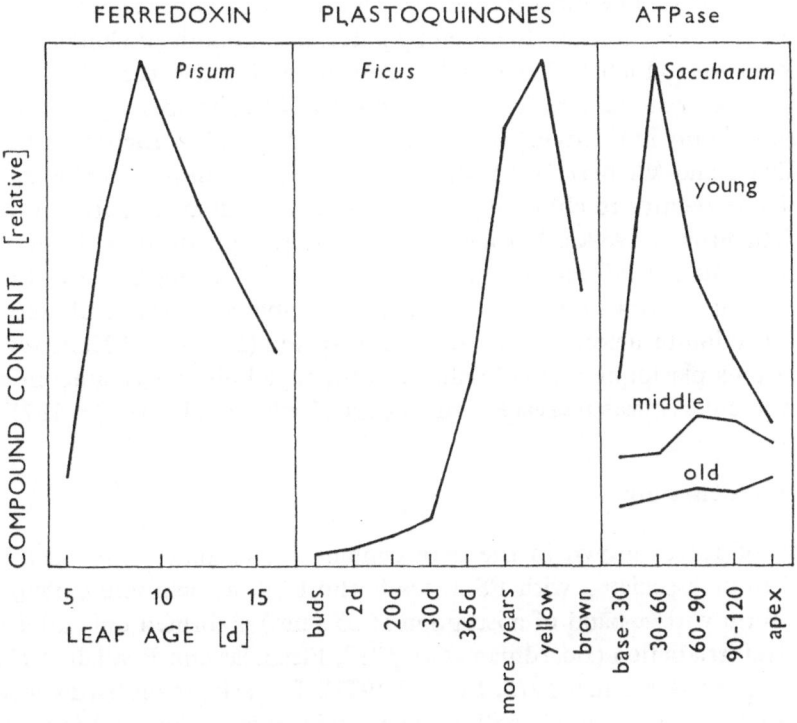

**Fig. 5.2.** Contents of components of the electron transport chain or enzymes of photophosphorylation mostly increase to a maximum and then decline. Increase to the peak is usually more rapid than decline (*left*: ferredoxin content per fresh matter during ageing of pea leaf, which reached its maximum area at the age of 12 d — from Smillie 1962). The contents of plastoquinones increase up to the end of leaf life (*middle*: sempervirent leaves of *Ficus elastica*, contents per leaf area unit — from Lichtenthaler 1969b) and decline only during leaf browning. The content of ATPase is different in various parts of leaf blade and declines from young to old leaves (*right*: sugar cane, contents per protein unit — from Alexander and Kumar 1974).

### 5.2.1 Plastoquinones

During greening (0 to 96 h light) of etiolated maize, oat, pea and broad bean leaves the amounts of plastoquinones *A* and *C* increase, while plastoquinone *B* disappears (Barr and Crane 1970, Lichtenthaler 1977).

The amount of total plastoquinones in dry matter increased with age (4 to 19 d) of the primary leaf in three rye cultivars (Kaval'chuk and Vechar 1973). The amount of plastoquinone-9 per unit chlorophyll increased with age of the primary wheat leaf (8 to 27 d), while per leaf area unit the rapid increase to 12 d was replaced by a slow decline (Bardat *et al.* 1981). In two cultivars of potato an increase in plastoquinone *A* content per both chlorophyll and dry matter units was observed from 2 to 45 d (Vecher *et al.* 1978). The amounts of plastoquinones *A* and *C* increased in the insertion gradient from young to senescent tobacco leaves (Barr and Arntzen 1969). The ratio of chlorophyll *a*/plastoquinones showed during the vegetation season an increase followed by a decrease in both sun and shade leaves of *Fagus silvatica* (Lichtenthaler 1971). In sempervirent leaves of *Ficus elastica* the amount of plastoquinone increased from leaf buds to many years old leaves and decreased only in the dying brown leaves (see Fig. 5.2 — Lichtenthaler 1969a, Lichtenthaler and Weinert 1970). In the cactus *Cereus peruvianus* the amount of plastoquinone relative to chlorophyll *a* increased with tissue age from 1–2 to 5–6 years (Lichtenthaler 1969c). The amount of plastoquinone declined from the apex to the base of the second leaf of maize (Baszyński 1971). Young leaves of *Vicia faba* of the same age grown in winter contained negligible amounts of plastoquinone, while its maximum amounts were formed in spring (Bucke and Hallaway 1966).

The surplus plastoquinone is localized in plastoglobuli, whose amount and dimensions in chloroplasts increase with leaf age (Lichtenthaler 1969b, 1971).

### 5.2.2 Cytochromes

Etioplasts of leaves grown in the dark contain cytochromes $f$, $b_6$ and $b$-559$_{LP}$ supposed to be associated with PS 1, while the PS 2 cytochrome $b$-559$_{HP}$ (high potential form with α-band of absorption at 559 nm) is formed only after several hours of leaf irradiation (Boardman *et al.* 1972, Plesničar and Bendall 1972, 1973, Henningsen and Boardman 1973, Bendall 1977). In dark grown barley leaves the levels of cytochromes $f$ and $b$-559$_{HP}$ per protein unit are low and increase only after irradiation, while the content of cytochrome $b$-559$_{LP}$ increases to 5 h only and then declines (Nasrulhaq-Boyce and Jones 1981). The contents of cytochromes $b_{559}$ and $C$-550 increase during the greening of 7 d etiolated primary leaves of *Phaseolus vulgaris* simultaneously with the contents of $P700$ and reach the maximum value per leaf area after *ca.* 30 h of irradiation (Baker and Butler 1976). The ability to photooxidise cytochrome $f$ by PS 1 and photoreduce it by PS 2 develops rapidly after 2—3 h of greening (Boardman *et al.* 1972).

The cytochrome contents and reactivities have only rarely been studied in the course of ontogeny of unfolded leaves. The chloroplasts in cotyledons of decapitated pumpkin plants have a steady chlorophyll/cytochrome ratio of about 500 from 10 to 38 d, but at the end of the expansion phase of the cotyledon this ratio increases to *ca.* 1300 (52 d) and then drops again (Harnischfeger 1973). The experiments on the ontogeny of primary leaf of mustard show that the ratio chlorophyll/cytochrome *f* increases from 700 to 1000 between the 13th and 22nd d (Wild *et al.* 1981c). The ratio depends on growing conditions of the plants: when plants are grown under "white" or "blue" radiation it rises from the 6th to the 34th d only very slowly, while under equienergetic "red" radiation an increase from 1100 to 1600 between the days 14 and 24 has been shown (Wild and Holzapfel 1980). The amount of cytochrome *f* per unit area of these leaves slowly increases from the 15th to the 42nd d (Zerbe and Wild 1981). The highest amounts of cytochromes *f* and $b_6$ in dry matter are in the middle leaf tier of potato and the oldest lower leaves contain only traces of cytochrome, in barley leaves the amounts of cytochromes *f* and $b_6$ increase from 3–5 d to 8–10 d and decline again (Vecher and Predkel' 1974). Changes of absorbance at 554 nm after irradiation, which show the rapid oxidation of cytochrome *f*, are lowest in the youngest leaves and highest in the leaves of middle insertion (sunflower – Volodarskiĭ and Bystrykh 1975).

### 5.2.3    Ferredoxin

The amount of ferredoxin (formerly called photosynthetic pyridine nucleotide reductase) increases in pea leaves to the 9th d after germination and then slowly declines (Fig. 5.2); this change is parallel to the change in photosynthetic rate of the leaf (Smillie 1962). Ferredoxin II, which migrates slightly more slowly during polyacrylamide gel electrophoresis, is synthesized in etiolated leaves later than ferredoxin I; the metallic ions are incorporated after leaf irradiation (3rd leaves of pea — Khristin and Akulova 1975). The maximum amount of both ferredoxins I and II per unit fresh matter is in 12–14 d seedlings of pea, but the amounts and ratios of both forms are dependent on the length of day (12 or 15 h day — Dutton *et al.* 1980).

### 5.3    ACTIVITIES OF PHOTOSYSTEMS

The activities of photosystems and photophosphorylation are mostly determined in chloroplast preparations, from which arises one basic problem: To what degree are the values obtained in this way representative of the photoreactions in an attached leaf? It is always necessary to choose the most delicate and appropriate method of chloroplast isolation, to select the optimum conditions for the reaction, and to be careful in interpreting the results. The above mentioned prerequisites are not always fulfilled, and thus we often find in the literature contradictory data, whose interpretation is difficult as the description of methods, state of plant material, *etc.* is not complete. The verification of values obtained with isolated whole or fragmented chloroplasts by measurements in whole leaf tissues is usually difficult and therefore very rare, and its results are generally not statisfactory. Hence the usual confirmation of values is indirect, from comparison with gas exchange or energy accumulation measurements. Nevertheless, the measurements on isolated chloroplasts are at present mostly the only way to explain the importance of individual photoreactions for the ontogenetic changes of the overall photosynthetic process.

## 5.3.1 Photosystem 2

The activity of PS 2 in isolated chloroplasts has been determined in laboratories since the discovery by Hill (1937). Hence the Hill activity has been the photochemical reaction of photosynthesis measured most consistently. This is reflected in the number of papers in which the changes in activity of Hill reaction in the course of leaf ontogeny have been described (see Table 1 in Šesták 1977b). It is supposed nowadays that the Hill reaction, especially when measured spectrophotometrically with ferricyanide as electron acceptor (system $H_2O \rightarrow K_3[Fe(CN)_6]$), does not characterize the activity of PS 2 only, but the joint activities of PS 2 and PS 1. Nevertheless, the determination of Hill reaction using various electron donors and acceptors will probably be taken for an index of the PS 2 rate till a better characteristic is found.

During greening of etiolated leaves, detectable PS 2 activity (all the data below are related to unit chlorophyll amount) was observed after 2 to 6 h irradiation of leaves, depending on plant species (for reviews see Treffry 1978, Bradbeer 1981). In the following 5 h the activity sharply increased. In the early phases of chloroplast development the activity depended on the type of PS 2 reaction: activity measured as oxygen production with ferricyanide as electron acceptor peaked after 2 h of greening, then declined and finally slowly increased, while the 2,3',6-trichlorophenol indophenol (TPIP) reduction measured spectrophotometrically increased during the 6 h of greening, and the 2,6-dichlorophenol indophenol (DPIP) reduction with diphenyl carbazide as electron donor increased for 4 h only (Henningsen and Boardman 1973). The increase to a peak value after 14 h greening, followed by a decline and a slow increase was found also by Raïtsina *et al.* (1968) in the system $H_2O \rightarrow K_3[Fe(CN)_6]$. The slow formation of chlorophyll during the initial lag phase does not limit the development of PS 2 (Nadler *et al.* 1972). The rate of PS 2 saturated with radiant energy is higher and is sooner reached in barley chloroplasts after a longer (70 h) greening than after a 2 h greening (Henningsen and Boardman 1973).

In the course of normal leaf development, mostly only a few phases of leaf ontogeny or leaves roughly grouped according to insertion (young — mature — old) were compared (see Table 1 in Šesták 1977b). These studies give evidence of a simple ontogenetic course of PS 2 activity: an increase to maximum rates and a a slow decline. However, detailed studies show that the situation is probably more complicated.

Using the system $H_2O \rightarrow K_3[Fe(CN)_6]$ Vecher and Lebedeva (1967) found in the first + second leaves of pea a decline in rates from 5 to 12 d (in other experiments to 33 d — Vecher et al. 1967) leaves and a slow increase afterwards, and in the system $H_2O \rightarrow DPIP$ a decline to the $32^{nd}$ d and a small increase after 39 d. In winter-grown pea leaves a small peak was observed in 10 d old leaves and a final decline after 15 d (Vecher et al. 1967). In *Phaseolus vulgaris* leaves, the system $H_2O \rightarrow DPIP$ showed a trough in PS 2 activity at the age of 15 d and then a steady increase with a peak value in 42 d old leaves; with $H_2O \rightarrow K_3[Fe(CN)_6]$ a small decrease was between 6 and 21 d and the maximum appeared again in the oldest leaves analysed (Vecher and Lebedeva 1967). Oscillations in PS 2 activity with three to four peaks of declining maximum rate were observed also during ontogeny of primary leaves of *Sinapis alba* (Fig. 5.3 — Wild et al. 1981a). Ontogenetic course of PS 2 activity with two peaks was observed also in the flag leaf of two cultivars of wheat grown in the field (from ear emergence to milk ripeness), but the position and heights of peaks differed when the $H_2O \rightarrow K_3[Fe(CN)_6]$ or $H_2O \rightarrow DPIP$ system was used (Fig. 5.3 — Volodarskiĭ et al. 1978). Other studies showed three maxima of PS 2 activity during the ontogeny of one leaf: the first maximum after leaf unfolding, the second peak after attaining the maximum leaf area and chlorophyll content, and a third peak preceding the end of metabolic activity of the leaf. This ontogenetic course was observed in primary leaves of *Phaseolus vulgaris* (Fig. 5.3 — Strnadová and Šesták 1974, Šesták et al. 1975, 1977, 1978a, b) and in pumpkin cotyledons (Harnischfeger 1974). In contrast to the above mentioned papers, the general ontogenetic course was not changed by the method of determination of PS 2 activity (spectrophotometry with $K_3[Fe(CN)_6]$ or DPIP, potentiometry with $K_3[Fe(CN)_6]$ — Strnadová and Šesták 1974). (Nevertheless, the method of PS 2 activity determination, e.g. method of chloroplast isolation, the buffer used, presence of ADP, uncouplers, etc., determines not only the rates obtained, but also the magnitude of ontogenetic differences — cf. Bezuglov and Chernysheva 1974, Plesničar and Bogdanovič 1976, Šesták et al. 1978b, Volodarskiĭ et al. 1978.) Harnischfeger (1974) tried to explain this course by the ontogenetic changes in the response of PS 2 to pH, but his observations were not confirmed in experiments using isolation and/or reaction media of different pH (Šesták et al. 1978b). According to Fenchuk (1972) chloroplasts from young pea leaves reduced DPIP at the highest rate at pH 8.0 to 8.5 under high irradiance (300 W $m^{-2}$), while under non-saturating irradiances the pH of ca. 7.3 was the optimum. He supposed the action of two enzyme systems of Hill reaction: one labile, working in basic media and under high irradiance, and active in chloroplasts from young leaves, and the other active in acid media, saturated with low amount of photons, and functioning mainly in chloroplasts from old leaves. In his further study (Fenchuk 1980) pH near 7.8 was optimum, but this effect was more apparent in chloroplasts from young (2 to 15 d) than old (22 to 30 d) pea leaves.

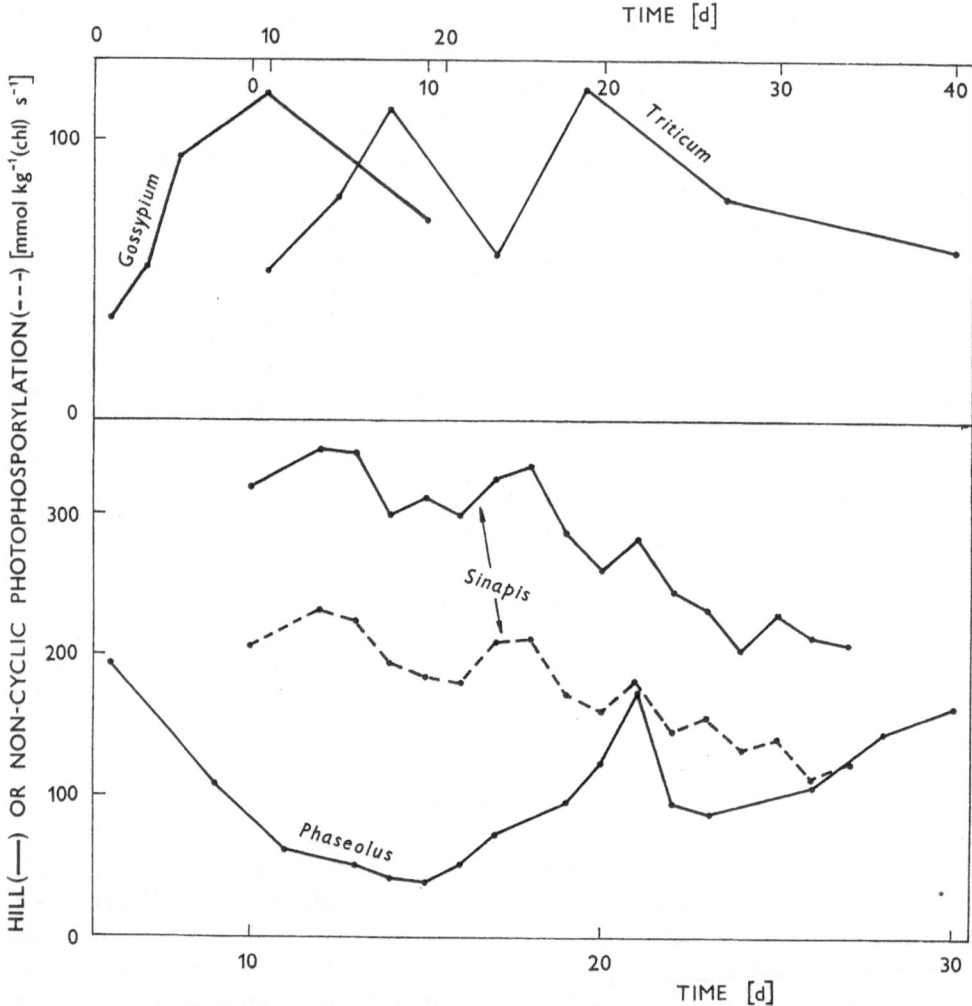

**Fig. 5.3.** The activity of Hill reaction ($H_2O \rightarrow K_3[Fe(CN)_6]$) in the course of leaf ontogeny either rapidly increases to a peak and slowly declines (*top left*: leaves of the 7th + 8th nodes of *Gossypium hirsutum* — from Khodzhaev *et al.* 1978) or undergoes a time course with two or three peaks (*top right*: flag leaf of winter wheat — from Volodarskiï *et al.* 1978; *bottom lowest curve*: primary leaf of *Phaseolus vulgaris* — from Šesták *et al.* 1977) or the oscillations seem to be irregular with a declining tendency (*bottom highest curve*: primary leaf of *Sinapis alba* — from Wild *et al.* 1981a). The ontogenetic course of Hill reaction activity is usually parallel to that of non-cyclic photophosphorylation (*bottom middle curve*: primary leaf of *Sinapis alba* — from Wild *et al.* 1981a).

The life span of the leaf determines the time of appearance of the maxima of PS 2: *e.g.*, the second peak in primary leaves of *Phaseolus vulgaris* appears *ca.* 2 d later, when the buds of secondary leaves are continuously removed (Šesták *et al.* 1977). Chloroplasts from evergreen leaves (needles of the first and second year from *Picea abies*) exert a seasonal course of PS 2 activity with peak values in the summer months, the maximum activity being in young needles (Senser and Beck 1978). In *Picea sitchensis* the PS 2 activity per needle area declines and per dry matter unit increases from top to the bottom of the canopy; the differences per chlorophyll amount are irregular and almost not significant (Lewandowska *et al.* 1977). The activity of PS 2 increases with cotyledon age of evergreens (*Pinus nigra* — Plesničar and Bogdanovič 1976), while in annual species the activity increases and declines or even continuously declines (radish — Buschmann and Lichtenthaler 1977) in the course of cotyledon life span.

The development of the plant is naturally also reflected in the ontogenetic changes in PS 2 activity during the development of individual leaves. Thus the leaves which appeared later reached a higher maximum rate of PS 2 than the older leaves (3rd and 7th leaf of field grown barley — Mader *et al.* 1978). Maximum activities of Hill reaction were observed either in the phase of flowering (wheat — Aliev and Azizov 1975, Zelenskiĭ *et al.* 1978; cotton — Krasichkova *et al.* 1977) or during the vegetative stage (pea — Bezuglov and Chernysheva 1974). The ontogenetic course in PS 2 activity and the maximum activities reached are also influenced by the previous history of plants, *e.g.* the photoperiod (Miller 1960), growth irradiance (Buschmann and Lichtenthaler 1977, Wild *et al.* 1981a), season of the year (Vecher *et al.* 1967), water relations (Todd and Basler 1965, Keck and Boyer 1974, Pospíšilová *et al.* 1976, Volodarskiĭ and Bystrykh 1976), mineral nutrition (Zaĭtseva 1970), *etc.* Chloroplasts from leaves of different ages exert a different sensitivity of PS 2 to growth substances (Karanov *et al.* 1970), or substances stimulating or inhibiting the Hill reaction such as EDTA, methylamine hydrochloride, diphenyl carbazide, trifluralin, or TRIS (Więckowski 1972, Hieke 1974, Jenkins and Woolhouse 1981b). PS 2 ($H_2O \rightarrow$ DPIP) in mature tomato leaves was less sensitive to storing the leaves in the dark at 0–4 °C for 3 d than that from expanded and young leaves (Kaniuga *et al.* 1978).

On a leaf blade, the activity of PS 2 of chloroplasts declines from leaf base to the older cells of leaf tip; this difference is more expressed at low irradiances (maize — Baker and Leech 1977). In greening etiolated leaves this distribution may be converse (maize — Klyuchareva 1974; wheat — Golod 1964).

PS 2 activity may be characterized also by the amount of the active primary electron acceptor $Q$ calculated from the kinetics of chlorophyll fluorescence *in vivo*. According to Harnischfeger (1973) the ratio of chlorophyll/$Q$ was steady during the life span (22 to 110 d) of cotyledons of decapitated pumpkin plants. The ratio of the variable fluorescence to maximal constant level of fluorescence estimates the maximum yield of primary photochemistry of PS 2: no difference in this parameter was found in the course of senescence (18 to 48 d) of primary leaves of *Phaseolus vulgaris* (Jenkins *et al.* 1981a).

The changes in Hill activity are probably connected with the changes in chloroplast and thylakoid ultrastructure (see Downton and Pyliotis 1971, Andersen *et al.* 1972, Horak and Zalik 1975), but no exact ontogenetic studies have been made until now. The rapid increase in activity of PS 2 coincides with the rapid formation of grana per plastid and lamellae per granum (Boardman *et al.* 1972). One very important factor is certainly the consistency of the chloroplast envelope and chloroplast fragility, which determine the penetration of electron acceptor into the thylakoid membrane. Thus in tobacco the chloroplast stability increases with leaf age, and at the same time the activity of PS 2 declines (De Jong and Woodlief 1974). In secondary leaves of maize the maximum PS 2 activity appears 7 d after sowing in mesophyll chloroplasts, while in bundle sheath chloroplasts two peaks — on the 7th and 12th d — are observed (Andersen *et al.* 1972).

According to Raafat *et al.* (1970a) the changes in PS 2 activity with leaf ontogeny (decline and a small increase at the end of leaf life) are in relation to the changes in the oxidation-reduction potential of chloroplast suspensions, measured with the system Pt electrode — calomel electrode.

During ageing of spinach leaves the saturation rates decline, but not the zero intensity quantum requirements of Hill reaction (Drury and Park 1968). Hence the decline in photosynthetic activity with leaf age may be connected with diminishing activities of dark reactions. The photon fluence rate curves of PS 2 for chloroplasts from primary leaves of continuously debudded 7–9 and 11–16 d old plants of *Phaseolus vulgaris* exert a similar character; both leaf groups differ only in absolute values, including the saturation level of PS 2. Chloroplasts from older primary leaves (from 19 to 26 d plants) show relatively high activities of PS 2 at low photon fluence rates and relatively low saturation plateau, *i.e.* they acquire a shade character (Šesták *et al.* 1977). Similarly chloroplasts from bottom nodes of canopy of *Picea sitchensis* show a 50% saturation of PS 2 activity at much lower photon fluence rate than chloroplasts from higher nodes (Lewandowska *et al.* 1977). Also the shape of curves expressing the dependence of PS 2 activity on irradiance is the same for chloroplasts from 18–21 and 35–41 d primary leaves of *Phaseolus vulgaris*, even if the absolute values are different (Jenkins *et al.* 1981b).

## 5.3.2 Photosystem 1

The number of studies on the ontogenetic changes of PS 1 is rather small, and they were measured by various methods.

During leaf greening, the active PS 1 is found sooner than PS 2, just after the beginning of irradiation of barley or *Phaseolus vulgaris* leaves and peaks already after 2 h (for reviews see Treffry 1978, Bradbeer 1981).

Normal ontogeny of leaves is characterized by changes of PS 1 similar to those of PS 2. In primary leaves of *Phaseolus vulgaris* three maxima of PS 1 activity (Mehler reaction) are found, the appearance of the second peak coinciding with that of PS 2 (Šesták *et al.* 1977, 1978b). The expressiveness of this peak depends on the pH of the isolation and reaction media (Šesták *et al.* 1978b). A smaller decline in the first phases of leaf ontogeny followed by a secondary peak of PS 1 (DPIP red. → NADP) was observed also in mesophyll chloroplasts from secondary maize leaves (Andersen *et al.* 1972). In tobacco a rapid decline in PS 1 activity from young to mature leaves is followed by relatively high rates per chlorophyll amount in old yellowing leaves (Wolińska 1976). In field grown barley higher maximum activity of PS 1 is reached in the older (3rd) than younger (7th) leaf (Mader *et al.* 1978).

Comparison of chloroplasts isolated from different parts of the leaf blade shows a rapid decline (to 12%) of PS 1 activity from the physiologically young base to leaf tip; the difference is more expressed under low irradiances (Baker and Leech 1977).

During a rapid development of radish cotyledons the activity of PS 1 declines (Buschmann and Lichtenthaler 1977).

The activity of PS 1 is also dependent on the phase of plant ontogeny: thus maximum activities of PS 1 of chloroplasts are observed at the beginning of flowering of cotton (Krasichkova *et al.* 1977).

The measurements of dependence of PS 1 activity on photon fluence rate show no difference in quantum yield extrapolated to zero photon fluence between the phases of the two maxima and the first trough of activity during ontogeny of primary leaves of decapitated *Phaseolus vulgaris* plants. The highest quantum yields are typical for the first maximum of PS 1 activity (Šesták *et al.* 1977). The shape of curves expressing the dependence of PS 1 activity on photon fluence rate is not significantly different for chloroplasts from needles of different insertion in the *Picea sitchensis* canopy (Lewandowska *et al.* 1977). The saturated PS 1 activity was higher in needles from the top than bottom of the canopy (Lewandowska and Jarvis 1978). The shape of curves expressing the dependence of PS 1 activity on irradiance is the same for chloroplasts from 18–21 and 35–41 d primary leaves of *Phaseolus vulgaris* (Jenkins *et al.* 1981b).

Another characteristic of PS 1 — the amount of $P700$ (see also Section 3.6.1) — rapidly increases during leaf greening (Boardman *et al.* 1972, Plesničar and Bendall 1972, 1973, Baker and Butler 1976). The ratio of chlorophyll/$P700$ declines in pea leaves between 3 and 8 h of irradiation from 3900 to 1200. Evaluation of the electron paramagnetic resonance signal 1 amplitude did not show any change in the relative content of $P700$ between the 2nd and 7th d of development of the primary leaf of *Phaseolus vulgaris* (Więckowski 1975). During the further phases of leaf ontogeny the relative amount of $P700$ declined and increased anew (Šesták and Demeter 1976). In mature leaves of three herb layer species from a deciduous forest community the relative amount of $P700$ declined during the vegetation season (Harvey 1980).

With the electron flow from reduced PS 1 the formation of an oxygen-free-radical called superoxide, $O_2^-$, may be connected. Under a limited supply of $NADP^+$, electron from the reduced ferredoxin may be passed to $O_2$ instead of to the Calvin cycle; another way of formation of $O_2^-$ is directly connected with PS 1. In higher plants, the toxic superoxide is rapidly removed by a copper-zinc enzyme super-oxide dismutase, most of which is located in the chloroplast stroma. The amount of superoxide dismutase per unit protein rapidly declines from young upper (11th) to middle (6th) leaves and then remains almost stable (maize — Asada et al. 1977): this decline is more rapid than the ontogenetic decline in PS 1 activity and thus the resistance to light-oxygen toxicity in old leaves is rather low.

### 5.3.3   Ratio of Activities of Photosystems 1 and 2

The ratio of activities of both photosystems (PS1/PS2) declines during the ontogeny of leaf (*Phaseolus vulgaris* — Šesták et al. 1977, 1978b). In radish cotyledons an increase in the PS 1 /PS 2 ratio was observed between the days 3 and 8 (Buschmann and Lichtenthaler 1977), while during senescence of soybean cotyledons (8 to 21 d) the PS1/PS2 ratio declined (Bricker and Newman 1982). The ontogenetic trend was not changed by adverse conditions of plant cultivation, *e.g.* water deficit (Po-spíšilová et al. 1976), but it was reversed by using isolation and reaction media of inappropriate pH (Šesták et al. 1978b). The decline of the ratio PS 1/PS 2 was in agreement with the changes in the ratio of subchloroplast particles enriched with PS 1 and PS 2, which were isolated from chloroplasts of young, mature and old leaves (spinach, radish — Šesták 1969, 1970; see Fig. 5.4). Raafat et al. (1970b, c) connected these changes with the ratio of "bound" (representing the pigment system of PS 1) and "labile" (pigment system of PS 2) chlorophyll forms *in vivo* (see Section 3.5.1). Also during the early phases of greening of etiolated leaves the initial higher amount of particles enriched with chlorophyll forms of PS 1 is replaced by a higher amount of chlorophylls of PS 2 (Więckowski 1971).

The combined activity of both photosystems may be characterized by NADP photoreduction ($H_2O \rightarrow NADP$). This activity declined from young to old leaves in one cultivar of tobacco and increased in another (Haraguchi and Shimizu 1970). In secondary maize leaves the NADP photoreduction rate increased to the 12th d and then declined in mesophyll chloroplasts, while it continuously declined in bundle sheath chloroplasts (Bishop et al. 1971, Andersen et al. 1972). The combined activity of both photosystems measured by the coupled system $H_2O \rightarrow$ methyl viologen declined in the course of senescing of primary leaves of *Phaseolus vulgaris* (15 to 41 d — Jenkins and Woolhouse 1981a); it reached higher peak values in the 3rd than 7th leaf of field grown barley (Mader et al. 1978).

On the area of maize leaf blade, the ratio of PS 1/PS 2 declined to *ca.* 1/4 from leaf base to tip; this decline was accompanied by an increase in chloroplast dimorphism and in average number of grana per section area unit (Baker and Leech 1977).

The activity of the whole photosynthetic electron transport chain was studied in leaves of various tiers also by physical methods — delayed light emission, electron paramagnetic resonance, and induced absorbance changes at 554 and 520 nm

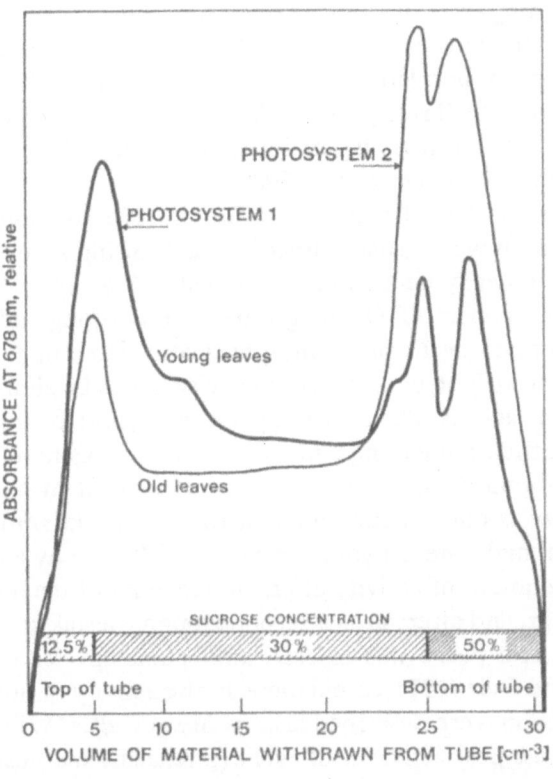

**Fig. 5.4.** The ratio of activities of Photosystems 1 and 2 declines in the coures of leaf ontogeny. This is reflected also in the ratio of particles enriched with individual photosystems which may be isolated from leaves of different age: young leaves contain more smaller particles enriched with Photosystem 1 (results of sucrose gradient separation of French press particles from spinach leaves — from Šesták 1970).

(Bystrykh and Matorin 1975). These methods confirm the slowing down of photosynthetic electron transport and efficiency of utilization of primary photoproducts in photosynthetic reactions with the ageing of leaves.

Quantum requirements for both PS 1 and PS 2 were lower in needles from the top of *Picea sitchensis* canopy (nodes 1–4 from the top of the 2–year-old seedlings)

than from the bottom (nodes 7 + 8); this difference was much higher for PS 1 than PS 2 (Lewandowska and Jarvis 1978). The difference was not only induced by growing at high and low irradiance in different parts of the canopy.


## 5.4    PHOTOPHOSPHORYLATION

Photosynthetic phosphorylations as energy storing processes start soon after the beginning of greening of etiolated leaves. Cyclic photophosphorylation, bound to PS 1, is active usually at the same time as PS 1, *i.e.* immediately after irradiation. Non-cyclic photophosphorylation starts 2 to 5 h later, at the same time as Hill activity (for reviews see Treffry 1978, Bradbeer 1981). The coupling of photophosphorylation to electron transport increases during chloroplast greening and grana stacking (wheat — Duysen *et al.* 1980).

In further days of leaf development the activities of non-cyclic and cyclic photophosphorylation were usually described as showing an increase and a following decrease or a steady decline only (see Table 2 in Šesták 1977b). The only exception is in the papers of De Jong (1974) or De Jong and Woodlief (1974), describing a rise of non-cyclic photophosphorylation from upper (young) to lower (old) leaves in tobacco. The ontogenetic course seems to be simple when a mixture of leaves of various ages is taken during plant ontogeny: In this way Smillie and Krotkov (1959) in their pioneering paper describing ontogenetic changes in photophosphorylation found a steady decline in rates with plant age from 9 to 28 d. Another possibility is ontogenetic course with three peaks similar to that of PS 2 activity: the initial high rates in young leaves are followed by a decline, alternated by the second maximum of activity after the reaching of maximum leaf area and chlorophyll content, and afterwards a second trough (Šesták *et al.* 1975). A similar time course with two peaks was in the same plant species (*Phaseolus vulgaris*) found by Heyes and Dale (1971), oscillations in the activity of non-cyclic and cyclic photophosphorylation were described also in *Sinapis alba* (Wild *et al.* 1981a). In needles of *Picea abies* peak activities of cyclic (phenazine methosulphate mediated) photophosphorylation are in winter months, while those of non-cyclic phosphorylation in spring months (Senser and Beck 1978).

The dependence of photophosphorylation on pH and temperature of the incubation medium may be different in chloroplasts from different parts of the leaf blade: thus in primary leaves of wheat the maximum activity of chloroplasts from the leaf tip was observed at higher pH and temperature than in chloroplasts from leaf base (Osakovskiĭ and Solomonova 1980).

The ontogenetic courses of photophosphorylation and photosystems activities are similar (Fig 5.3; Šesták *et al.* 1975; *cf.* also Tables 1 and 2 in Šesták 1977b). Sometimes some discrepancies are observed, *e.g.* the maximum activity of cyclic photophosphorylation may appear 2 d later than the maximum activity of Hill reaction (Raïtsina et al. 1968). The ratio of activities of non-cyclic photophospho-rylation and PS 2 activity (the P/O ratio) increases from young upper to old lower leaves (tobacco — De Jong 1974). The increased photophosphorylation rate of old leaves may be caused by a high percentage of intact chloroplasts in these prepara-tions. A similar ratio P/2 e expressing the coupling of photophosphorylation to electron transport either declines in the course of senescing of primary leaves of *Phaseolus vulgaris* ($H_2O \rightarrow$ ferricyanide, or $H_2O \rightarrow$ oxidized $p$-phenylenediamine — Jenkins and Woolhouse 1981b) or does not change in the same plant material ($H_2O \rightarrow$ methyl viologen — Jenkins and Woolhouse 1981a), depending on the part of the electron transport chain involved. In *Sinapis alba* the ratio declines in plants cultivated under low irradiance, while at high irradiance only oscillations in the ratio are observed (Wild et al. 1981a). Quantum requirements of ferredoxin--catalyzed cyclic photophosphorylation as well as those of NADP reduction and concurrent sum of photophosphorylations were higher in chloroplasts from old (42 d) than young (27 d) spinach plants (Chain and Arnon 1977).

The basic condition is the concentration of substrate — ADP — and the enzyme ATPase. The contents of ADP and ATP per unit fresh matter declined from the flag (1st) to the oldest (4th) leaf of wheat — the molar ratio for leaves 1/4 was 2.72 for ADP and 2.19 for ATP (Ching et al. 1975). Maximum contents of ADP and ATP per one leaf were in leaf 2. These amounts were affected by fertilization with $NH_4NO_3$. The amount of macroergic phosphorus and the rate of formation of macroergic compounds declined from leaves of upper to lower insertion, the absolute amounts depending on levels of N, P and K (Khlyastikov 1977). The findings on ATPase were of a similar character: in the two lowest leaves of kale steady ontogenetic decline in ATPase activity per chlorophyll unit was observed, while in *Phaseolus vulgaris* the decline was alternated by a peak in the period of flowering (Vecher and Raïtsina 1967). In the primary leaf of wheat grown in light an increase in ATPase activity per fresh matter unit to 7 d was found in distilled water and to 6 d when grown in Knop's nutrient solution (generally lower activi-ties!), and then a decrease (Ponomareva 1958). ATPase activity decreased from the youngest to the oldest leaves of sugar cane (Fig. 5.2, *right* — Alexander and Kumar 1974), cotton (Imamaliev and Zikiryaev 1972), and eight monocotyledonous and eight dicotyledonous plant species growing in India (Patra and Mishra 1981). The zone of maximum ATPase activity per protein unit was situated in the youngest cotton leaves 30 to 60 cm from the leaf base. It shifted during leaf ageing in the di-rection of the leaf apex (Alexander and Kumar 1974). During the vegetation season maximum ATPase activity per protein unit shifted from the leaves of lowest tiers of potato to medium (in the period of bud formation) and upper (after flower-ing) tiers (Kislyakova et al. 1967, Bogacheva et al. 1970).

The ontogenetic course of photophosphorylation is affected by various factors. The genetic basis is important — per chlorophyll unit the tetraploids are more active than diploids (Vechar et al. 1970), virescent mutants more active than normal plants (Heyes and Dale 1970), *etc*. Plant decapitation shifts the apearance of peaks and changes the rates (Ïordanov and Popov 1967, Yordanov 1970, Šesták *et al*. 1977, Patra and Mishra 1981). The changes in climatic factors during the vegetation season act in plants grown outdoors: Maslow (1964) found a sharp increase in photophosphorylation rates during October. Uncouplers affected cyclic photophosphorylation to a lesser degree in chloroplasts from young lettuce leaves, containing more stroma lamellae than chloroplasts from old leaves (Arntzen *et al*. 1971). Cyclic photophosphorylation in younger (6 weeks) Swiss chard leaves was more stimulated by incubation of chloroplasts in solutions with physiological concentrations of indole-3-acetic acid than in old leaves (Tamàs *et al*. 1972).

The question remains how representative are the measurements on isolated chloroplasts for photophosphorylation in the whole tissue. Yordanov (1971) compared non-cyclic photophosphorylation in chloroplasts isolated from 10 and 20 d primary leaves of intact or decapitated *Phaseolus vulgaris* plants with photophosphorylation in discs from the same leaves floating on Knop's solution with $KH_2{}^{32}PO_4$, but did not obtain corresponding values.

## 5.5 GENERAL COURSE OF PHOTOCHEMICAL ACTIVITIES

The supposed course of activities of PS 1 and PS 2 and non-cyclic and cyclic photophosphorylation during leaf ontogeny and in a leaf insertion gradient is an increase to a maximum followed by a decrease. Nevertheless, the ontogenetic course of all these photoreactions may be even more complicated, with two to three peaks or oscillations. The first peak may be caused by a rapid formation of the electron transport chain, preceding the accumulation of antenna chlorophyll. The following trough may be connected with the increase in size of the photosynthetic unit. The second maximum coincides with the phase of optimally formed chloroplast ultrastructure. The following decline in photochemical activities is a senescence phenomenon, while the eventual final activity peak reflects the rapid degradation of antenna chlorophylls. The ratio of activities of Photosystems 1 and 2 declines during leaf ontogeny. In leaves of a long life span the ontogenetic changes are slow and reflect also microclimatic conditions of leaf (plant) environment. Leaves formed in *ca*. one third of plant ontogeny reach usually maximum activities of photochemical reactions of photosynthesis. The changes are partly due to changes in chloroplast ultrastructure, rigidity of its envelope, sensitivity to the conditions of assay, *etc*. The amounts of components of the electron transport chain, *i.e.* plastoquinones, cytochromes, and ferredoxin, are probably not a limiting factor.

# 6 CARBON FIXATION PATHWAYS, THEIR ENZYMES AND PRODUCTS DURING LEAF ONTOGENY

*J. Zima and Z. Šesták*

## 6.1 CARBON FIXATION PATHWAYS

While the processes of photosynthesis connected more or less directly with utilization of radiant energy were treated in Chapter 5, this chapter deals with biochemical reactions of fixation and reduction of carbon dioxide, and photosynthates formation. In contrast to the photochemical processes, where quanta are directly used and ATP and NADPH are formed, in the biochemical processes these compounds are utilized for conversion of carbon dioxide to primary photosynthates. As this conversion can occur without the direct participation of radiant energy, these processes are often called "dark reactions" of photosynthesis. But, even here, radiant energy plays an important role in control and coordination of photosynthetic reactions and their interaction with other metabolic processes by modification of enzyme activities (*cf*. Buchanan 1980).

Carbon fixation processes are not the most typical reactions of photosynthesis: they are not even peculiar only to plants, but they occur also in heterotrophic cells. The unique peculiarity of plant photosynthesis is the utilization of photon energy for its own production of NADPH and ATP used for fixation and reduction of $CO_2$ and transformation of primary photosynthates.

The basic pathway of photosynthetic carbon fixation is the well known Calvin (or Calvin-Benson) cycle. It is called also the $C_3$ cycle, as its primary product is 3-phosphoglycerate (a $C_3$ compound). The enzyme catalysing its basic reaction — fixation of $CO_2$ to ribulose-1,5-bisphosphate (RuBP) — is ribulose-1,5-bisphosphate carboxylase (RuBPC — E.C. 4.1.1.39) which however catalyses also oxygenation of RuBP. Plants growing in extreme environments developed in the course of their phylogeny additional cycles (for reviews see Huber and Sankhla 1976, Raghavendra 1980) serving the accumulation of $CO_2$ to levels necessary for the reaction catalysed by RuBPC (this enzyme has usually a high $K_m$ for $CO_2$, and $CO_2$ competes for the reaction site with $O_2$). These cycles are called the $C_4$ cycle or Hatch-Slack cycle, and CAM (Crassulacean Acid Metabolism) cycle. In both of them, $CO_2$ is fixed in a reaction catalysed by phoshoenolpyruvate carboxylase (PEPC — E.C. 4.1.1.31). In further reactions organic ($C_4$) acids are formed (mainly aspartate or malate) which are decarboxylated producing $CO_2$. The $CO_2$ is then refixed in the Calvin cycle. The decarboxylation takes place after the transport of the organic acids into other cells (in $C_4$ plants) or during another period of the 24 h-cycle (in CAM plants). $C_4$ plants are above all tropical grasses and species of monocotyledons and dicotyledons from habitats of high irradiance and temperature, while plants of the family *Crassulaceae* and some similar taxons of a succulent character, with nocturnal $CO_2$ assimilation and nocturnal acidification, belong to CAM plants. $C_3$, $C_4$ and CAM plants differ not only in the first carboxylation enzyme and products, but also in leaf anatomy and other photosynthetic characteristics (see Chapters 1, 2, 7 and 8).

Unfortunately, even the consideration of the metabolic types of plants has not simplified the evaluation of the literature data on the ontogenetic differences in the "dark" reactions of photosynthesis, their enzymes and products. Not only various methods of determination and various relation units were used, but often the characteristics of carbon fixation activity were measured only in a few phases of leaf ontogeny. All these differences and uncertainties make the comparison rather complicated.

## 6.2    ENZYME ACTIVITIES DURING LEAF ONTOGENY

### 6.2.1    RuBPC and PEPC in $C_3$ Plants

The most abundant protein in higher plant leaves was originally called fraction 1 protein (F1P). Further research showed, that the activity of RuBPC was more or less linearly correlated with the amount of that protein (Steer 1973, Blenkinsop and Dale 1974). Generally, the content  of F1P in leaves increases during leaf expansion and then it declines (Fig. 6.1). The maximum appears about at the time of reaching maximal leaf area. Such a time course was found, *e.g.*, by Callow (1974) in the third leaf of cucumber, and Gordon *et al.* (1978) in the first  leaf of pea. When the amount of F1P is related to the whole leaf, as in the above two papers, the final shape of the ontogenetic changes can be determined mostly by the change of leaf area. When the amount of F1P was related to the leaf area unit, a similar result was found in *Perilla frutescens* leaves (Kannangara and Woolhouse 1968), but more often a decrease in the amount of F1P per leaf area unit or total protein content was described (*cf.* Zima and Šesták 1979). Different time-courses are typical for experiments not including the whole leaf life. During the development of the second leaf of wheat, the incorporation of amino acids into the large sub-unit of F1P peaked on the 11th d, *i.e.* prior to reaching the maximum plastid polyribosome content on the 13th d, which again preceded the maximum content of this sub-unit per leaf blade (Brady and Scott 1977a, b).

When the activity of RuBPC was directly measured, the increase to maximum (mostly prior to or at the end of leaf blade expansion) followed by a decrease was also the most usual result (Fig. 6.1), and this ontogenetic course held also for the recalculation per amount of protein or F1P, for one leaf blade, leaf area unit, or leaf fresh matter of properly watered plants (*e.g. Atriplex hastata* — Downton and Slatyer 1971; barley — Lloyd 1976; *Capsicum* — Steer 1971, 1973; *Phaseolus*

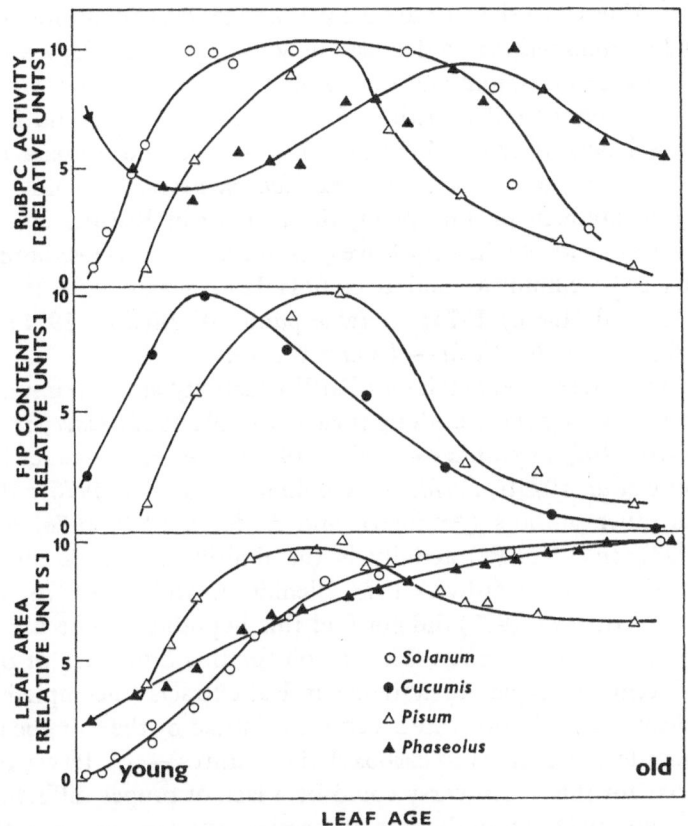

**Fig. 6.1.** Ribulose-1,5-bisphosphate carboxylase (RuBPC) activity (*top*) and also the fraction 1 protein (F1P) content (*middle*) increase during the ontogeny (8 to 20 d) of the first leaf of *Pisum sativum* L. to the maximum on the 16th d of 0.24 nmol s$^{-1}$ per leaf and 0.47 mg per leaf, respectively, attained at the time of reaching the maximum leaf area (0.181 × 10$^{-3}$ m$^2$) (data after Gordon *et al.* 1978). Similar ontogenetic (5 to 48 d) change of F1P content was found in the third leaf of *Cucumis sativus* L. (*middle* — maximum 26.4 mg per leaf in the 14 d old leaves) (after Callow 1974). A broad maximum of RuBPC activity (*top*) reaching 13.8 mmol s$^{-1}$ kg$^{-1}$(prot.) was found in *Solanum tuberosum* L. leaves 8 to 17 d old prior to the end of leaf area expansion (*bottom*, max. 6.1 × 10$^{-3}$ m$^2$) (after Mokronosov and Nekrasova 1977). In the primary leaves of decapitated plants of *Phaseolus vulgaris* L. the more complicated course of RuBPC activity was found (*top*): a decline in young leaves (11 to 18 d) followed by an increase in the middle part of leaf life to the maximum (18 μmol s$^{-1}$ m$^{-2}$ on the 39th d), and again a decline in old leaves (41 to 49 d). The increase of RuBPC activity took place when the rate of leaf area expansion slowed down (*bottom*, max. 4.7 × 10$^{-3}$ m$^2$ ) (after Zima *et al.* 1981).

— O'Toole *et al.* 1977; onion — Steer and Darbyshire 1979; pea — Smillie 1962, Gordon *et al.* 1978; *Perilla* — Batt and Woolhouse 1975; potato — Mokronosov *et al.* 1973b, Mokronosov and Nekrasova 1977, Nekrasova 1978, Borzenkova and Nefedova 1981, Mokronosov 1981; wheat — Brady *et al.* 1971, Peoples and Dalling 1978; *cf.* also Table 1 in Zima and Šesták 1979). When the activity of RuBPC was measured three times a week in 10 to 50 d primary leaves of continuously decapitated plants of *Phaseolus vulgaris,* a decrease of activity in young leaves was followed by an increase to the maximum and then a final decrease (Zima *et al.* 1981).

Ontogenetic changes similar to those in true leaves were found in cotyledons (castor bean — Dockerty *et al.* 1977; radish — Ochiai and Shibata 1970, Ochiai *et al.* 1971a), and in protonemata of *Ceratodon purpureus* (Valanne *et al.* 1978). In the insertion gradient on one plant, the activity of RuBPC either decreases from young (upper) to old (lower) leaves, or increases to a maximum and then declines (tobacco — Shumway and Kleinhofs 1973, Daley *et al.* 1978; *Antirrhinum* — Hedley and Harvey 1975; *Atriplex patula* — Medina 1971; *Dactylis* — Treharne and Eagles 1970; *Vicia* — Li and Wu 1982).

A correlation between the maximum RuBPC activity and maximum photosynthetic rate was found by some authors (pea — Smillie 1962; *Dactylis* — Treharne and Eagles 1970; *Atriplex patula* — Medina 1971; *Capsicum* — Steer 1971; potato — Mokronosov *et al.* 1973b; *Perilla* — Woolhouse and Batt 1976): they believe, together with other authors (Andreeva and Avdeeva 1970, 1976, Avdeeva and Andreeva 1973), that RuBPC activity is the limiting factor of photosynthesis. Other authors (Baker and Hardwick 1973, Blenkinsop and Dale 1974, Thomas and Thorne 1975, Wirth *et al.* 1977) did not find this hypothesis to be correct, at least for some phases of leaf ontogeny. The reason for this difference of opinion may be due to different carboxylating activities in leaf extracts (incomplete extraction) and in intact leaves (the reaction rate is certainly limited by the complicated transfer of $CO_2$ through leaf structures to carboxylation centres — *cf.* Jarvis 1971). Other possible reasons for this controversy could be a lack of proper differentiation between enzyme activity (equal to the rate of reaction catalysed under optimal conditions *in vitro* — as far as such conditions are known) and reaction (process) rate (affected also by interactions of various processes). Some experiments also ascertained that the change in RuPBC activity induced by some factor (*e.g.* leaf shading) was greater than that of the photosynthetic rate (barley — Blenkinsop and Dale 1974). Yellowing of leaves, accompanied always by a decline in photosynthetic rate, is marked also by a final decline in total RuBPC activity (*Perilla* — Batt and Woolhouse 1975), or RuBPC activity related to RuBPC protein determined immunologically (wheat — Hall *et al.* 1978).

PEPC activity in $C_3$ plants is usually lower by one order than the RuBPC activity. Eventual high PEPC activity always pointed to a connection with the $C_4$ metabolism ($F_1$ hybrid of *Atriplex patula* $\times$ *A. rosea* — Pearcy and Björkman 1971).

The usual course of PEPC activity during leaf ontogeny is similar to that of

RuBPC, *i.e.* an increase followed by a decline, in relation to any unit used (tobacco — Wada 1971; *Antirrhinum* — Hedley and Rowland 1975). Only an increase or only a decline are characteristic for experiments of a short duration. In potato, Mokronosov *et al.* (1973b) found in juvenile leaves a high PEPC activity, which was replaced by a high RuBPC activity in older leaves. Complicated ontogenetic courses described by Marco *et al.* (1976, 1979) may be explained by changing climatic conditions, as they analysed field-grown wheat plants. In the leaf insertion gradient, the activity of PEPC per protein amount increased from young to old leaves in wheat, while in barley it showed a minimum in leaf 3 (Wirth *et al.* 1977).

Changes of the activities of RuBPC and PEPC measured during leaf growth in darkness depend on the particular conditions of the experiments. Thus in etiolated primary leaves of *Phaseolus*, Bradbeer (1971) found a decline in RuBPC activity and an increase in PEPC activity per leaf between days 12 and 19. Expansion of leaf age ranging from 6 to 27 d in further experiments (Bradbeer *et al.* 1974) showed an increase to a maximum followed by a decline: the peak activity of RuBPC (14 d) preceded that of the PEPC (21 d). Both these maxima appeared after reaching the maximum amount of cells per leaf (12 d); maximum amount of plastids per cell was reached in 20 d old leaves. An increase in both RuBPC and PEPC activities (calculated per unit amount of protein or fresh matter) with the increase in tissue age of etiolated French bean leaves from 4 to 14 d was found by Akoyunoglou and Argyroudi-Akoyunoglou (1972). The relative increase in RuBPC activity after continuous irradiation of etiolated bean leaves was the highest in 5 d leaves, while only a decline in activity was observed in 11 and 15 d leaves; under intermittent irradiation (1 ms light + 15 min darkness) the increase in activity was the highest in 6 d leaves, lower in 8 d, and the lowest in 4 d leaves (Dassiou and Akoyunoglou 1969). RuPBC activity per fresh matter increased in dark grown primary leaves of barley to the 6th, 7th, 8th or 9th d (various leaf parts from tip to base of the leaf), while after 24 h under 10.76 klx the highest activity was in all parts of the 7 d old leaf (Obendorf and Huffaker 1970). A steady increase in RuBPC activity was in the same leaf of the same species found by Kannangara (1969) and in the primary leaf and coleoptile of barley by Feierabend (1969).

## 6.2.2 RuBPC and PEPC in C$_4$ Plants

In C$_4$ plants PEPC catalyses the primary fixation of CO$_2$ from the air, while RuBPC catalyses the secondary fixation of CO$_2$ which is evolved by decarboxylation of C$_4$-dicarbonic acids produced with the participation of PEPC. The important role of PEPC in these plants is underlined by its mostly higher activity.

The changes in activities of both these enzymes during leaf ontogeny or in the leaf insertion gradient of C$_4$ plants have been studied by a few authors only. The most general course of these changes in PEPC and also RuBPC activities are a continuous increase in activity or increase followed by a decrease (Fig. 6.2) (maize — Möller *et al.* 1977, Stamp 1978, 1979, Crespo *et al.* 1979, Bassi and Passera 1982; *Setaria* — Wu *et al.* 1982; *Atriplex spongiosa* — Downton and Slatyer 1971).

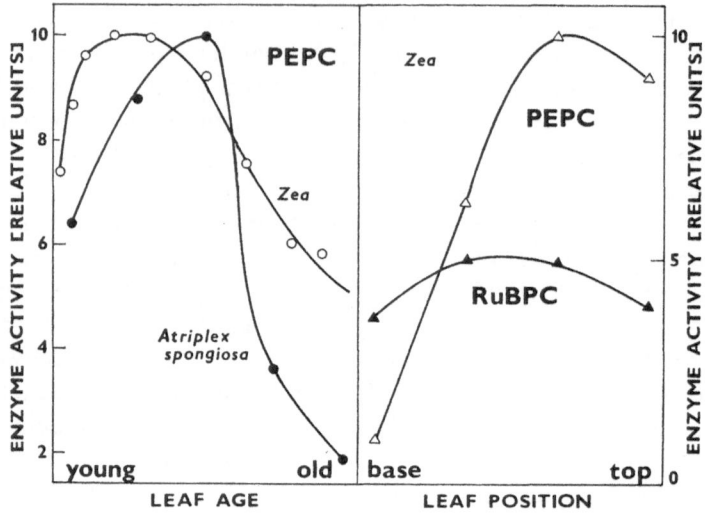

**Fig. 6.2.** The changes in phosphoenopyruvate carboxylase (PEPC) activity during the $C_4$ plant leaves ontogeny were similar to those in ribulose-1,5-bisphosphate carboxylase (RuBPC) activity in $C_3$ plants, *i.e.* an increase in activity in young leaves followed by a decrease. In young leaves of *Atriplex spongiosa* F. v. M. followed for 4 weeks (*left*), the maximum of PEPC activity – 21.8 mmol $s^{-1}$ $kg^{-1}$(prot.) — was observed after two weeks of development (after Downton and Slatyer 1971). During the development of the first leaf of *Zea mays* L. (5 to 53 d), the maximum — 465 μmol $s^{-1}$ $kg^{-1}$(f.m.) appeared in 14 d old leaves (after Möller *et al.* 1977). In the leaf insertion gradient (*right*) of *Zea mays* L. leaves 3 to 6, the maximal activity of PEPC – 44.4 μmol $s^{-1}$ $kg^{-1}$(f.m.) was found in the 5th leaf, while the maximum of RuBPC activity — 22.2 μmol $s^{-1}$ $kg^{-1}$(f. m.) — was in the 4th leaf (after Crespo *et al.* 1979).

## 6.2.3  RuBPC and PEPC in CAM Plants

CAM plants show similar changes in PEPC activity in the leaf insertion gradient and/or during leaf ontogeny as $C_4$ plants, *i.e.* continuous increase in activities or increase to the maximum followed by a decrease with the age of the tissue (Fig. 6.3) (*Mesembryanthemum crystallinum* — Willert *et al.* 1976, 1977, Greenway *et al.* 1978, Winter and Greenway 1978; *Prenia* and *Sceletium* — Willert and Willert 1979; *Bryophyllum* — Nishida 1978, Jones *et al.* 1981; Kalanchoë — Buchanan-Bollig *et al.* 1980).

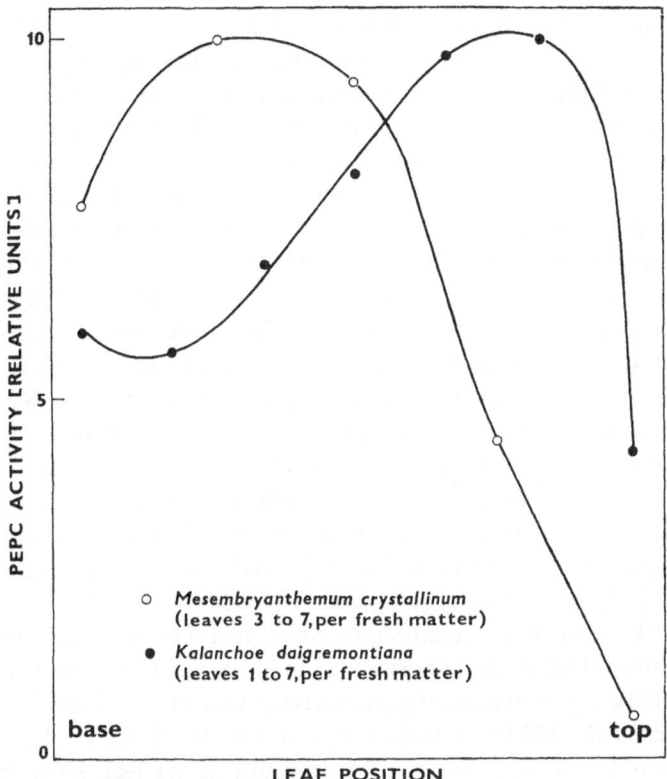

**Fig. 6.3.** Phosphoenolpyruvate carboxylase (PEPC) activity increases to the maximum (in the 4th leaf) in the leaf insertion gradient (leaves 3 to 7) of a CAM plant *Mesembryanthemum crystallinum* L. and then declines (after Willert *et al.* 1976). In the leaf insertion gradient (leaves 1 to 7) of *Kalanchoë daigremontiana*, a slight decrease in PEPC activity was followed by an increase to the maximum of 106 $\mu$mol $s^{-1}$ $kg^{-1}$(f.m.) in the 6th leaf (after Buchanan-Bollig *et al.* 1980).

### 6.2.4   Other Enzymes of Carbon Metabolism in C₃ Plants

Several enzymes of the Calvin cycle were also determined during leaf ontogeny. From the point of view of the time of reaching maximal activities, Batt and Woolhouse (1975) analysed the third pair of leaves of *Perilla frutescens* and distinguished two groups of enzymes: enzymes which reach their maximum activity prior to the completion of leaf expansion and then it rapidly declines [RuBPC as stated above, and phosphoribulokinase (E.C. 2.7.1.19) and NADPH-glyceraldehyde-3-phosphate dehydrogenase (E.C. 1.2.1.13 — NADPH-GPD)], and those which maintain high activity till a late stage of senescence [phosphoglycerate kinase (E.C. 2.7.2.3), alkaline fructose-1,6-bisphosphatase (E.C. 3.1.3.11), and ribose-5-phosphate isomerase (E.C. 5.3.1.6)]. Enzymes of the first group reach maximum activity related to both leaf area and protein content between the 23rd and 28th d of leaf life, while the rapid decline in activity of alkaline fructose-1,6-bisphosphatase and phosphoglycerate kinase (enzymes of the second group) occurs after the 42nd and 35th d of leaf age, respectively. The expression of ribose-5-phosphate isomerase activity on the leaf area basis produces a decline in the course of leaf development, while the expression on the protein basis shows an increase. On the other hand, in the first leaf of pea also an increase in activity of this enzyme to a maximum (still preceding the maximum of RuBPC) was followed by a decrease, when expressed per one leaf blade (Gordon et al. 1978). Similarly, a maximum of glyceraldehyde-3-phosphate dehydrogenase activity per one leaf preceding the full expansion of the first barley leaf was found by Lloyd (1976): the activity of phosphoglycerate kinase reached a maximum 3 d after full expansion of this leaf. NADPH-GPD activity on the protein basis increased with leaf length between 20 and 80 mm in the primary leaf of *Phaseolus vulgaris* growing under continuous illuminance (Schwarz 1971). In contrast to this, a decrease in activity of this enzyme related to fresh matter in broad bean leaves between the 10th and 20th d was found by Avdeeva and Andreeva (1973). An increase in NADPH-GPD activity with age of rye seedlings from 2 to 7 d was found by Feierabend (1969).

NADPH-GPD activity was maximal in wheat leaf 5 (numbered from the oldest one) on the chlorophyll content basis, and it only declined with leaf age on the protein content basis; in oat minimum activity was in leaf 3 on both the bases used. As concerns the NADPH-malate dehydrogenase (E.C. 1.1.1.82), minimum activity was found in wheat leaf 3 and maximum in oat leaf 3 per both relation units used (Wirth et al. 1977). In contrast to that, maxima of NADPH-GPD and NADPH-malate dehydrogenase activities related to protein content unit were found in the eighth leaf of tobacco (De Jong and Woodlief 1979). Bradbeer et al. (1974) determined also the activities per one leaf of pentosephosphate isomerase, phosphoribulokinase, NADP-triosephosphate dehydrogenase, triosephosphate isomerase, fructose-bisphosphate aldolase and transketolase in extracts from etiolated primary French bean leaves of different ages. Excepting the first enzyme activity, the others reached a maximum between 14 and 18 d. The activity of pentosephosphate isomerase increased to 14 d, remained constant between the 14th and 18th d, and then increased again. The rate of increase in enzyme activities at the beginning of ontogeny was different for various enzymes. An increase followed by a decrease of phosphoribose isomerase activity in various sections of 5 to 10 d old first barley leaves growing in the dark was found by Obendorf and Huffaker (1970). The position of maximum varied with the leaf section. They also found a slight increase in phosphoribulokinase activity in the same time interval.

Activity of transketolase, the enzyme involved in both photosynthesis and respiration, attained a maximum per fresh matter unit similarly to RuBPC in pea leaves 9 d old, prior to full expansion of the leaf. Its activity in young leaves (5 d) was relatively higher than that of the photosynthetic enzyme RuBPC, which was in agreement with the activities of other respiratory enzymes measured (Smillie 1962).

The activity of sucrose synthetase (E.C. 2.4.1.14) believed to be the principal enzyme of sucrose synthesis in plants, attained a maximum in *Lolium temulentum* leaves on the 11th d after emergence of the leaf tip (Pollock 1976).


### 6.2.5    Other Enzymes of Carbon Metabolism in C₄ and CAM Plants

Only a few data are available on ontogeny of leaves of $C_4$ and CAM plants in relation to these enzymes. An age induced decrease in NADPH-GPD activity in 10, 20, and 30 d maize leaves was found (Avdeeva and Andreeva 1973). The activity of malate dehydrogenase (E.C. 1.1.1.37) was maximal in an extract from mature leaves of *Portulaca oleracea* (Kennedy 1976). The activity of pyruvate orthophosphate dikinase in *Kalanchoë pinnata* and *K. blossfeldiana* peaked in mature leaves (Sanada and Nishida 1982).


## 6.3    CHANGES IN CARBON FIXATION PATHWAYS AND PRIMARY PHOTOSYNTHATES

A leaf age induced transition from one prevailing pathway of carbon fixation to another has only exceptionally been observed in plant species included in the $C_3$, $C_4$, or CAM groups.

In juvenile (2 to 3 d) leaves of the $C_3$ plant potato a high absolute rate of PEP carboxylation (70 to 80% of total $CO_2$ fixation) was observed, but the $C_4$ acids formed were used in the non-photosynthetic metabolism of amino acids and organic acids (Mokronosov 1973, 1981, Mokronosov *et al.* 1973b). In 25 d old and older potato leaves 80 to 90% of the $CO_2$ was fixed by means of RuBPC. Maximum amount of primary photosynthates of the $C_3$ cycle was in the 20 to 24 d leaves, prior to the end of leaf expansion. Contrary to this, in the $C_3$ plants wheat and oat the ratio of $C_3$ and $C_4$ products of carbon fixation declined from the flag leaf to the oldest leaf: this trend was similar to, but more expressed than the ratio of activities of RuBPC and PEPC (Wirth *et al.* 1977). In the course of development of the leaf of the third tier of cucumber the relative incorporation of $^{14}C$, after 1 min of photosynthesis, into amino acids declined, into sugars and organic acids it increased and then rapidly declined, while the incorporation into other groups of substances was more or less irregular (Karpilova *et al.* 1982). Products of short-time $^{14}CO_2$ photosynthesis in mature mulberry leaves were mainly alanine, glyceric acid, sugars, glycine and serine, while in young leaves 3-phosphoglyceric acid was the most heavily labelled compound (Yamashita *et al.* 1975).

In the natural $C_3$-$C_4$ intermediate *Mollugo nudicaulis* the photosynthetic characteristics continuously changed from the $C_3$ to the $C_4$ ones between the youngest and the oldest leaf: while 3-phosphoglycerate and sugar phosphates were the primary photosynthates in young leaves, old leaves (with Kranz anatomy, high photosynthetic and low photorespiratory rates) formed mostly $C_4$ acids malate and aspartate (Raghavendra *et al.* 1978).

During the 2 s exposure to $^{14}CO_2$ of the $C_4$ plant *Portulaca oleracea* young leaves incorporated relatively more $^{14}C$ into alanine and malate and relatively less $^{14}C$ into aspartate than mature leaves (Kennedy and Laetsch 1973). After a 10 s photosynthesis the ratio of alanine and aspartate accumulation in these leaves was reversed; senescent leaves formed lesser amounts of these photosynthates and relatively more malate. The largest difference after the 10 s exposure to $^{14}CO_2$ was in the relative amount of phosphoglycerate that increased from young to senescent leaves. In the experiments of Hatch (1975), young leaves of the same species incorporated more $^{14}C$ into alanine than mature leaves after 4, 21 and even 60 s of photosynthesis.

In maize the relative amount of $^{14}C$ after the 10 s photosynthesis increased with leaf age in malate and decreased in aspartate, while the amount of labelled phosphoglycerate was the same. The relative amount of $C_4$ acids was 75% in young, 72% in mature, and 66% in senescent leaf tissues (Williams and Kennedy 1977). In 16 d old maize plants the ratio of $^{14}C$ incorporation into malate and aspartate was higher in the 2nd than the 1st or 3rd leaf (Morot-Gaudry *et al.* 1979). When only four early photosynthates were studied, 86.7% of all radioactivity in 4 week-old maize leaves was found in malate and aspartate and none in saccharose and alanine, while the percentage for 10 week-old leaves was 40.3 and 53.0%, respectively; a similar difference in distribution was observed after 30 s photosynthesis (Reynolds *et al.* 1974).

In CAM plants young leaves may fix $CO_2$ by the $C_3$ pathway, while older leaves acquire the CAM pathway (*Kalanchoë pinnata* — Thompson *et al.* 1977). The CAM cycle develops not only with the acquisition of the succulent character of leaves: in *Echeveria columbiana* the succulence index increased from the youngest to the 9th leaf and then hardly changed at all, while the night $CO_2$ fixation capacity increased with leaf age up to the 17th and 18th leaf and only then, in senescent leaves, declined (Medina and Delgado 1976). The observation of $\delta^{13}C$ values and free acidity in leaf pairs 1 (young) to 5 (old) of *Bryophyllum calycinum* showed an increase in relative activity of the CAM fixation pathway with leaf age (Lerman *et al.* 1974); old leaves contained less $^{13}C$ in the soluble fraction (malate and other organic acids, amino acids, sugars). In the same plant species, mature leaves incorporated more $^{14}C$ into malate and aspartate and less into sugar phosphates than young leaves (Nishida 1978). In *Mesembryanthemum crystallinum* the rate of malate formation in darkness depended on the NaCl availability, PEPC activity and leaf age: the highest amount of malate was formed in older leaves (Willert *et al.* 1977).

## 6.4    SECONDARY PHOTOSYNTHATES

The variance in products of long term $^{14}CO_2$ metabolism in the light with leaf age has been observed in various plant species, nevertheless, the generalization is rather difficult as various fractions of substances were analysed in experiments of different schemes.

In the course of the fifth potato leaf ontogeny starch formation (per leaf dry matter unit) prevailed in the first half of the leaf blade expansion, peaking in ca. 15 d leaves, while later more saccharose was formed (peak in 27 d leaves). The amount of $^{14}C$ used for protein formation declined with leaf age (Mokronosov et al. 1973b). Similar results (relative increase in ethanol-soluble photosynthates and a decline in $^{14}C$ in starch and proteins) were observed after 1 h photosynthesis of 40 to 67 d leaves of *Perilla frutescens* (Hardwick et al. 1968). Study of a 12 h $^{14}C$ fixation per 9 to 35 d old primary leaf of *Phaseolus vulgaris* showed the highest saccharose, citrate and malate amounts in 14 d leaves, while glucose, fructose, ribose, phosphorylated sugars, and most organic acids (glucuronic, gluconic, tartaric, $\alpha$-ketoglutaric, succinic, and fumaric) peaked in 21 d leaves, and amino acids peaked either at 14 or 21 d (Raafat and Höfner 1971). Total amounts of $^{14}C$ sugars and organic acids were highest in 21 d primary leaves of *Phaseolus*, while those of $^{14}C$ amino acids were highest in 14 d leaves (Raafat et al. 1971).

The changes per area unit of leaf blade during leaf expansion (17 to 100% of final size) were not very dramatic: the amount of total carbon metabolites increased by about 30% and the highest increment was in the soluble carbon fraction (7th leaf of tomato – Ho 1977).

The amount of non-structural saccharides per unit fresh mater rapidly declines from the basal, meristematic section of a 10 cm long onion leaf to the middle and apical parts of the blade. The decline was most abrupt in the content of trisaccharides and most gradual as regards glucose (Steer and Darbyshire 1979).

The comparison of secondary photosynthates formation in leaves of various insertion levels in tobacco showed an increase in $^{14}C$ accumulation per dry matter unit from young to old leaves as regards sugars, organic acids, phosphate compounds and starch, while both soluble and insoluble proteins were formed mainly in young leaves, and soluble amino acids were synthesized most rapidly in leaves of medium insertions (Ongun and Stocking 1965). A lesser amount of soluble sugars was always found in upper tobacco leaves than in lower leaves, while the relative amount of starch per dry matter unit was on the 13th and 22nd July higher in lower leaves and later (31st July, 12th August) in upper leaves (Kakie 1972).

After a 20 min exposure with $^{14}CO_2$ the younger (2nd) leaf of soybean accumulated relatively more amino acids and starch than the 5th leaf, which was more active in the formation of sugars, organic acids and proteins (Chub 1975).

Primary leaves of barley contained more ethanol-insoluble products than secondary leaves; the relative amount of ethanol-insoluble photosynthates declined with leaf age from 6 to 10 d (Felippe and Dale 1972). In half-expanded and youngest mature leaves of two cultivars of *Lolium perenne* no important difference in the proportion of $^{14}$C labelled water soluble photosynthates (alanine, glycine, serine, glutamine, glutamic acid, aspartic acid, malic acid) and storage products (sucrose and other saccharides) was found (Taylor *et al.* 1971). During June and July the upper leaves of spring wheat exerted a higher relative radioactivity in ethanol-soluble organic acids and amino acids after a 24 h $^{14}$CO$_2$-assimilation than lower leaves (Höfner and Orlovius 1977). The decline in amount of total saccharides and starch with leaf age was more pronounced in halophytic (*Triglochin mariti-mum, Scorzonera parviflora*) than non-halophytic plant species (*T. palustre, S. austriaca*) (Ladenburger and Albert 1981). In *Brassica pekinensis* the amounts of total saccharides and reducing sugars declined to older leaves, while the content of crude starch increased (Yukimoto *et al.* 1978).

There was no large difference in the ratio of alkaline-soluble substances and insoluble residue found in the shoot apex of *Pisum arvense* after $^{14}$CO$_2$ photosynthe-sis of single leaf 4, 6, 7 or 8 (Patel 1966).

Senescent sugar cane leaves contained after 1 h photosynthesis more $^{14}$C in total amino acids and organic acids than mature leaves, in which sugars prevailed (Nimbalkar and Joshi 1975). Upper leaves of maize contained per dry matter unit more non-reducing sugar and starch and less reducing sugars than lower leaves (Allison and Weinmann 1970); this study similarly to some above mentioned studies measured the total amounts of substances in leaves, not taking into account the changes induced by translocation, transformation and utilization of photo-synthates.


## 6.5    CONCLUSION

The changes in amounts and activities of the carboxylation enzymes RuBPC and PEPC in the course of leaf ontogeny are usually similar to those in the activi-ties of photochemical reactions of photosynthesis (*cf.* Chapter 5) and net CO$_2$ uptake (*cf.* Chapter 7). The rather rapid increase in contents and activities of these and other enzymes of the carbon fixation pathways is mostly followed by a slower decline. The relative amounts and activities of RuBPC, PEPC and other enzymes depend on the type (C$_3$, C$_4$, or CAM) of carbon fixation. The respective pathway determines also the presence and quantities of primary and secondary photosynthates. A transition from the C$_3$ to the C$_4$ pathway or *vice versa* is some-times observed during leaf ontogeny: these changes are characteristic for the natural C$_3$-C$_4$ intermediate species.

# 7 GAS EXCHANGE AND DRY MATTER ACCUMULATION DURING LEAF DEVELOPMENT

*Ingrid Tichá, J. Čatský, Danuše Hodáňová,*
*Jana Pospíšilová, M. Kaše and Z. Šesták*

## 7.1 PHOTOSYNTHETIC GAS EXCHANGE

The primary processes of photosynthesis (see Chapter 5) supply NADPH and ATP for the fixation and reduction of carbon dioxide (see Chapter 6) which is transported into the leaf through leaf anatomical structures (see Chapter 1) characterized by different conductances for $CO_2$ transfer (see Chapter 8). Reduction of the fixed $CO_2$ requires a hydrogen ion which is produced by photosynthetic water splitting; in this way stoichiometric amounts of oxygen are produced. Photosynthesis is thus connected with gas exchange, *i.e.* $CO_2$ uptake and $O_2$ efflux.

Simultaneously, respiratory processes in the leaf are operating, during which time oxygen is consumed and $CO_2$ released: first, there is the process of photorespiration which is tightly coupled with photosynthetic carboxylation (*cf.* Chapter 9) and serves to remove the product of the oxygenation reaction of the key enzyme of photosynthesis, ribulose-1,5-bisphosphate carboxylase/oxygenase (RuBPCO).

The other respiratory process in the photosynthesizing leaf is "dark" respiration (tricarboxylic acid cycle and the cytochrome electron transfer system). This process is partly inhibited by light in photosynthesizing cells but still some 25% of the dark rate seems to be preserved (*cf.* Section 10.8). Also nonchlorophyllous cells, *e.g.* epidermal cells, contribute to the respiration rate in light.

At every moment the net $CO_2$ influx into a leaf is therefore a result of the $CO_2$ uptake by photosynthetic carboxylation minus the $CO_2$ release by photorespiration, $R_L$, and partial "dark" respiration, $R'_D$. Moreover, $CO_2$ produced by photorespiration may be immediately completely ($C_4$ plants) or partly ($C_3$ plants) reassimilated by photosynthesis. Thus, the rate of true photosynthesis (gross photosynthesis, $P_G$), *i.e.* the total $CO_2$ uptake at the fixation site of photosynthesis, cannot be assessed accurately because the portion of $CO_2$ released in photorespiration and reassimilated can hardly be measured. Indirect arguments suggest that the reassimilation may reach in $C_3$ species as much as 30% to 40% of the photorespiration (D'Aoust and Canvin 1974). Net photosynthetic rate or apparent photosynthetic rate, $P_N$, is the rate of the net exchange of $CO_2$ between the leaf and the atmosphere:

$$P_N = P_G - R_L - R'_D .$$

**Table 7.1** Comparison of the $C_3$ and $C_4$ syndromes. Abbreviations: $E$ — transpiration rate; PEPC — phosphoenolpyruvate carboxylase; RuBPC — ribulose-1,5-bisphosphate carboxylase; WUE — water use efficiency.

| Parameter (1) | $C_3$ (2) | $C_4$ (3) |
|---|---|---|
| Leaf structure | dorsiventral: mesophyll divided into layers of palisade parenchyma and spongy parenchyma | wreath-like or „Kranz" leaf anatomy: vascular bundles are surrounded by concentric layers of bundle sheath cells and mesophyll cells |
| Tissue density per unit leaf volume | smaller than in $C_4$ | larger than in $C_3$ |
| Interveinal distance | larger than in $C_4$ | short |
| Stomata density | smaller than in $C_4$ | 1.5 times higher than in $C_3$ |
| Chloroplasts | one type with grana and starch grains | mostly dimorphic; in bundle sheath cells large chloroplasts without grana but with starch grains; in mesophyll cells chloroplasts with grana |
| Primary carboxylation by | RuBPC (substrate: $CO_2$) | PEPC (substrate: $HCO_3^-$) |
| First product of $CO_2$ fixation | $C_3$ substance: phosphoglycerate | $C_4$ substance: (oxalacetate), malate, aspartate |
| Net photosynthetic rate ($P_N$) | $\leq 0.8$ mg $CO_2$ m$^{-2}$ s$^{-1}$ | 1.5 to 2.5 mg $CO_2$ m$^{-2}$ s$^{-1}$ |
| Temperature optimum of $P_N$ | 10 to 25 °C | 30 to 45 °C |
| Saturating irradiance for $P_N$ | $\leq 950$ μmol m$^{-2}$ s$^{-1}$ | no saturation of $P_N$ between 1900 to 2800 μmol m$^{-2}$ s$^{-1}$ |
| $CO_2$ saturation of $P_N$ in contemporary atmosphere | not saturated | near saturation |
| Photorespiration ($R_L$) | yes (20 to 50% of $P_N$) | yes?, but either very low rates or $R_L$ inhibited by the high $CO_2$ concentration in the chloroplasts of bundle sheath cells at the site of RuBPC |

| | | |
|---|---|---|
| Approximate internal $CO_2$ concentration on surface of mesophyll cells before RuBPC carboxylation | 480 mg m$^{-3}$<br>150 mg m$^{-3}$ | 200 mg m$^{-3}$<br>400 mg m$^{-3}$ |
| $CO_2$ compensation concentration ($\Gamma$) | 70 to 130 mg m$^{-3}$ (at 21% $O_2$, 25 °C) | 0 to 18 mg m$^{-3}$ (independent of $O_2$ and temperature) |
| Sensitivity to $O_2$ (Warburg effect) | yes | no |
| $\delta^{13}C\star$ | −24 to −30‰ | −10 to −18‰ |
| Usual values of stomatal conductance to $CO_2$ transfer ($g_s$) | 7 mm s$^{-1}$ | 5 mm s$^{-1}$ |
| Usual values of intracellular conductance to $CO_2$ transfer ($g_M$) | 1.5 mm s$^{-1}$ | 10 mm s$^{-1}$ |
| Transpiration ratio $E/P_N$ | 190 mol H$_2$O mol$^{-1}$ $CO_2$ | 80 mol H$_2$O mol$^{-1}$ $CO_2$ |
| WUE | lower | higher |

$\star\ \delta^{13}C\,[‰] = \dfrac{^{13}C/^{12}C\ \text{sample} - {}^{13}C/^{12}C\ \text{standard}}{^{13}C/^{12}C\ \text{standard}} \times 1000$

The value $P_N$ is commonly used for characterizing photosynthetic activity of the leaf or plant.

In this chapter, only $P_N$ of leaves of higher plants during leaf development is treated. Photosynthesis in whole plants or canopies is not dealt with. From the literature, only those papers are referred to here where the whole ontogeny of a leaf is followed. The amount of information on ontogenetic and sequential changes in $P_N$ and related characteristics reported in the literature is large (cf. reviews by Šesták and Čatský 1967c, Šesták 1979, 1981, Mokronosov 1981, and Tichá and Čatský 1981) but only a part of it contains a sufficient number of reliable measurements to characterize the whole leaf ontogeny.


## 7.1.1    CO₂ Fixation Types and their Response in CO₂ Exchange

According to the pathway of carbon dioxide fixation, $C_3$, $C_4$ and CAM plants may be distinguished (for their anatomical characteristics see Section 1.4.1) which differ also in $CO_2$ and water vapour exchange parameters.

$C_3$ plants: In $C_3$ plants (Table 7.1) the first stable product of $CO_2$ fixation by ribulose-1,5-bisphosphate carboxylase (RuBPC) is a three-carbon compound — 1,5-disphosphoglycerate. In these plants the reductive pentose phosphate cycle (Calvin-Benson cycle) is working which is the basic biochemical pathway whereby carbon dioxide is fixed and converted to sugar phosphates (cf. Chapter 6). This pathway is apparently ubiquitous in all photoautotrophic green plants. The plants with the $C_3$-syndrome have medium net photosynthetic rates under natural $CO_2$ concentration, a considerable rate of photorespiration, and higher stomatal and lower intracellular conductances for $CO_2$ transfer than $C_4$ plants, etc. (Table 7.1).

$C_4$ plants: In plants with the $C_4$-syndrome (Table 7.1) there is an additional pathway via which carbon dioxide is first incorporated into four-carbon acids or amino acids and is later released to be refixed via the reductive pentose phosphate cycle. The primary carboxylation by means of phosphoenolpyruvate carboxylase (PEPC) in chloroplasts of the mesophyll cells (the "$CO_2$ pump" or $CO_2$ concentrating mechanism) supplies the secondary carboxylation by RuBPC in chloroplasts of the bundle sheath cells with high $CO_2$ concentrations. Thus, carboxylation of ribulose-1,5-bisphosphate (RuBP) is stimulated and oxygenation of RuBP is competitively inhibited by $CO_2$. No photorespiration rate in the form of $CO_2$ output can be measured because all $CO_2$ leaking from the bundle sheath cells is immediately trapped and refixed by PEPC in the mesophyll cells. Low $CO_2$ compensation concentrations are found in these plants. $C_4$ plants are photosynthetically very efficient and have high $P_N$ even under natural $CO_2$ concentration. Their intracellular conductance for $CO_2$ transfer is higher than in $C_3$ plants but the "carboxylation conductance" is similar (Table 7.1, cf. Chapter 8). The $C_4$

plants are classified into three groups, based on the type of major decarboxylating enzyme present, which may be either NADP-malic enzyme, NAD-malic enzyme, or phosphoenolpyruvate carboxykinase (Hatch *et al.* 1975; Chapter 6); the groups, however, do not differ greatly in $CO_2$ exchange parameters.

Natural $C_3$-$C_4$ intermediate plants: Some plants such as *Mollugo verticillata, M. nudicaulis, Panicum* species of the Laxa group, *Moricandia arvensis, M. sinaica, Flaveria pubescens, F. anomala, etc.* possess anatomical, biochemical or physiological features that are intermediate between $C_3$ and $C_4$ plants (for reviews see, *e.g.*, Huber and Sankhla 1976, Apel and Peisker 1979, Rathnam and Chollet 1980, Rathnam-Chaguturu 1981). Also such $CO_2$ exchange parameter as $CO_2$ compensation concentration is intermediate.

CAM (Crassulacean Acid Metabolism) plants: As their stomata are open during the night, the CAM plants show dark $CO_2$ uptake from the atmosphere. $CO_2$ is primarily fixed *via* PEPC and stored, *e.g.*, as malic acid, while during the day the net $CO_2$ uptake is depressed or interrupted, the stomata are closed, the accumulated malic acid is decarboxylated, and the $CO_2$ released is used in the reductive pentose phosphate cycle (for reviews see, *e.g.*, Kluge and Ting 1978, Kluge 1979). The CAM plants represent an adaptation to arid environment because the closed stomata during the day protect the plant from water loss by transpiration.

Shifts between carboxylation types: Recently, there has been some evidence that the type of $CO_2$ fixation within a plant species is not as constant as was previously assumed. The effects of growth conditions, nutrition, water supply, growth regulators, and mainly leaf or plant ontogeny may result in shifts from one to another pathway of $CO_2$ fixation (*cf.* Section 6.3; for reviews see Huber and Sankhla 1976, Raghavendra 1980). According to others authors, there is no evidence that the carbon assimilation pathway is essentially changed (Osmond *et al.* 1982).

## 7.2 NET PHOTOSYNTHETIC RATE

### 7.2.1 Net Photosynthetic Rate during Leaf Ontogeny

In general, $P_N$ increases rapidly during leaf development reaching a maximum ($P_{Nmax}$) before full leaf area expansion is finished, then decreases during late developmental stages or senescence. Some examples of ontogenetic patterns of $P_N$ in various plant species are compared in Table 7.2 and Figs. 7.1, 7.2, 7.3, 7.5, 8.2, 8.3, 8.4, 8.5, 10.1, 10.2, 10.3, 10.4. The general ontogenetic pattern of $P_N$ was found in different plant types, *e.g.*, herbs (*Fagopyrum* and *Phaseolus* — Saeki 1959; *Cucumis* — Hopkinson 1964; *Sorghum, Gossypium, Helianthus* — Elmore *et al.*

**Table 7.2** Net photosynthetic rate as affected by leaf age. A choice of data from the literature. Abbreviations and explanations: IRGA — infrared gas analyser; max — maximum; $P_G$ — gross photosynthetic rate; $P_N$ — net photosynthetic rate; RH — relative humidity; VPD — vapour pressure deficit; ↑ — increase; ↓ — decrease

| Plant (and culti-vation) | Age of leaf (leaves numbe-red from the oldest one) | Method of $P_N$ mea-surement, experi-mental conditions (radiation, tempe-rature, $CO_2$ con-centration, RH or VPD) | Net photosynthetic rate [mg m$^{-2}$ s$^{-1}$] if not indicated otherwise (from young to old leaves) | Changes in $P_N$ induced by ageing | Author(s) and year of publication |
|---|---|---|---|---|---|
| (1) | (2) | (3) | (4) | (5) | (6) |
| *Atriplex spongiosa* F.v.M. (C$_4$); *Atriplex hastata* L. (C$_3$) (aerated water culture, glasshouse) | "young" and "ageing" leaf of three weeks old plants with eight leaves, 7 to 23 d | open system, IRGA, 460 W m$^{-2}$ (400––700 nm), 23 °C | C$_4$: "young": 2.16...1.22 "ageing": 2.15...0.97 C$_3$: "young": 1.19...0.82...1.25 "ageing": 1.18...0.73...1.11 | ↓ ↓ ↓, ↑ ↓, ↑ | Slatyer 1970 |
| *Capsicum annuum* L. cv. Market Giant (1:1:1 soil–sand–peat moss mixture, glasshouse) | leaf inserted one internode above the first flower, 0 to 84 d after anthesis | open system, IRGA, 125 W m$^{-2}$ (400––700 nm), 25 °C, 1.85 kPa | 0.39...0.28...0.41...0.30 | ↓, ↑, ↓ | Hall and Brady 1977 |
| *Cucurbita pepo* L. (glasshouse) | first leaf, 2 to 26 d | open system, IRGA | 0.44...0.61...0.21 | ↑, ↓ (max d 4) | Sisson 1981 |
| *Glycine javanica* L. (controlled envi-ronment) | youngest full expanded leaf, 0 to 42 d after unfolding | open system, IRGA, 100 klx, 30 °C, 545 mg $CO_2$ m$^{-3}$, VPD 17 mm | 0.36...0.83...0.28 | ↑, ↓ (max d 6) | Wilson and Ludlow 1970 |

| Species (conditions) | Leaves / age | Method | $P_{N\,max}$ | Trend | Reference |
|---|---|---|---|---|---|
| *Glycine max* (L.) Merrill (controlled environment) | leaves 1, 3, 5, 7, 9, 0 to 93 d from plant emergence | open system, IRGA | 1: 0.48...0.51...0.05<br>3: 0.22...0.63...0<br>5: 0.22...0.73...0<br>7: 0.10...0.78...0<br>9: 0.21...0.79...0 | ↑, ↓<br>↑, ↓<br>↑, ↓<br>↑, ↓<br>↑, ↓ | Woodward 1976 |
| *Gossypium hirsutum* L. cv. Deltapine 16 (glasshouse) | leaves 5, 7, 9, 0 to 68 d | open system, IRGA, 2000 µmol m$^{-2}$ s$^{-1}$, 27.8 °C, $CO_2$ m$^{-3}$, 600 mg VDP 2.08 kPa | 0.18...1.08...0.40 | ↑, ↓ | Constable and Rawson 1980 |
| *Helianthus annuus* L. cv. Suncross 51 (glasshouse) | mean of leaves 7, 11, 15, 21, 0 to 60 d | open system, IRGA | 0.60...1.43...0.60 | ↑, ↓ | Rawson and Constable 1980 |
| *Hevea brasiliensis* Muell. Arg. (controlled environment) | Leaf Blade Class 2 or 5 to 9 + 50 d (*i.e.* 17 to 84 d) | open system, IRGA | 0.06...0.55...0.40<br>0.025...0.26...0.14 | ↑, ↓<br>↑, ↓ | Samsuddin and Impens 1979a, b |
| *Lycopersicon esculentum* Mill. cv. Potentate (pots with soil, glasshouse) | leaves 3 and 5, 0 to 46 d | open system, IRGA | $P_{N\,max}$ (light saturated:)<br>leaf 3: 0.76...1.05...0.04<br>leaf 5: 0.75...0.82...0.11 | ↑, ↓<br>↑, ↓ | Peat 1970 |
| *Nicotiana tabacum* L. cv. Mammoth 17L (pots, controlled environment) | leaves 2 to 10, 20 to 90 d from sowing: leaf 12, 0 to 45 d | open system, IRGA | leaf 2: 0.44...0.11<br>3: 0.47...0.53...0.08<br>4: 0.28...0.56...–0.10<br>5: 0.56...0.07<br>6: 0.33...0.56...0.14<br>7: 0.26...0.67...0.06<br>8: 0.44...0.64...0.27<br>9: 0.08...0.64...0.14<br>10: 0.28...0.63...0.28<br>12: 0.20...0.90...0.05 | ↓<br>↑, ↓<br>↑, ↓<br>↓<br>↑, ↓<br>↑, ↓<br>↑, ↓<br>↑, ↓<br>↑, ↓<br>↑, ↓ | Rawson and Hackett 1974, Rawson and Woodward 1976 |

**Table 7.2** (continued)

| Plant (and cultivation) | Age of leaf (leaves numbered from the oldest one) | Method of $P_N$ measurement, experimental conditions (radiation, temperature, $CO_2$ concentration, RH or VPD) | Net photosynthetic rate [mg m$^{-2}$ s$^{-1}$] if not indicated otherwise (from young to old leaves) | Changes in $P_N$ induced by ageing | Author(s) and year of publication |
|---|---|---|---|---|---|
| (1) | (2) | (3) | (4) | (5) | (6) |
| *Phaseolus vulgaris* L. cv. Pencil Pod Black Wax (sandy loam–peat moss–turface mixture, growth chamber) | primary leaf, 2 to 22 d; first trifoliate leaf, 2 to 19 d; second, 2 to 14 d; third, 2 to 8 d | radiometric method, closed system, IRGA | primary leaf: 0.31...0.49...0.21...0.35 leaf 1: 0...0.38...0.24 2: 0.01...0.42...0.33 3: 0.08...0.30...0.23 | ↑,↓,↓,↑ ↑,↓ ↑,↓ ↑,↓ | Fraser and Bidwell 1974 |
| *Phaseolus vulgaris* L. cv. Jantar (growth chamber) | primary leaf, 2 to 20 d from sowing | open and closed systems, IRGA, 1250 μmol m$^{-2}$ s$^{-1}$ (400–700 nm), 28 °C, 600 mg $CO_2$ m$^{-3}$ | $P_{N\,max}$ (light saturated): 0.27...1.05...0.25 | ↑,↓ | Čatský *et al.* 1976, Čatský and Tichá 1980 |
|  | 11 to 22 d |  | 0.54...0.64...0.27 | ↑,↓ | Čatský and Tichá 1979 |
| *Pisum sativum* L. cv. Alaska (growth chamber) | leaves 3, 6, 9, 12, 20 to 70 d | gas-exchange system, 1500 μmol m$^{-2}$ s$^{-1}$, 21°C | leaf 3: 0.25...1.36...0 6: 0.69...0.76...0 9: 0.29...0.65...0 12: 0.50...0.78...0 | ↑,↓ ↑,↓ ↑,↓ ↑,↓ | Bethlenfalvay and Phillips 1977 |
| *Populus* x *euramericana* cv. Wisconsin–5 .4, 8, 19, 36 (growth chamber) | leaf 8 at LPIs | closed system, IRGA, 500 μmol m$^{-2}$ s$^{-1}$ (400–700 nm), 25 °C | 0.12...0.27...0.18 | ↑,↓ | Dickmann and Gordon 1975 |
| *Rumex patientia* L. (glasshouse) | third leaf, 6 to 50 d | open system, IRGA | 0.14...0.33...0.10 | ↑,↓ | Sisson and Caldwell 1977 |

| Species (conditions) | Material | Method | Value | Trend | Reference |
|---|---|---|---|---|---|
| *Sinapis alba* L. (phytotron) | leaves 7 and 13, 3 to 46 d | IRGA, 135 W m⁻², 25 °C, 545 mg CO₂ m⁻³ | 1.63...0.13 | ↓ | Cornic *et al.* 1970 |
| *Sorghum almum* Parodi (controlled environment) | youngest full expanded leaf, 0 to 40 d after unfolding | open system, IRGA, 100 klx, 30 °C, 600 mg CO₂ m⁻³, VPD 17 mm | 2.10...2.42...0.51 | ↑, ↓ | Wilson and Ludlow 1970 |
| *Triticum aestivum* L. cv. Kalyansona (pots, perlite, nutrient solution, phytotron and growth chamber) | flag and penultimate leaves, 0 to 40 d from anthesis | open system, IRGA, 1200 µmol m⁻² s⁻¹ (400–700 nm). 24.5 °C, 535 mg CO₂ m⁻³, 46% RH | flag leaf: 0.82...0.52 penultimate leaf: 0.78...0.40 | ↓ ↓ | Rawson *et al.* 1976 |
| *Triticum aestivum* L. cv. Kleiber and Maris Huntsman (pots, glasshouse) | intact flag leaf, 0 to 30 d from anthesis | radiometric method ($^{14}CO_2$ and $^{12}CO_2$ uptake), 1319 µmol m⁻² s⁻¹, 26.8 °C | $P_N$: 0.59...0.08 $P_G$: 0.74...0.16 | ↓ ↓ | Thomas *et al.* 1978 |
| *Triticum aestivum* L. cv. Arthur (field) | flag leaf, 0 to 32 d after anthesis | $^{14}CO_2$-pulsing apparatus | 0.46...0.51...0.06 | ↑, ↓ (max d 12) | Wittenbach 1979 |
| *Vigna unguiculata* (field) | leaves 5, 7, 9, 0 to 40 d | mobile gas analysis open system, IRGA | 0.90...1.40...0.20 | ↑, ↓ (max d 20) | Littleton *et al.* 1981 |

1967; *Perilla* — Woolhouse 1967/8; *Nicotiana* — Wada *et al.* 1967; *Plectranthus* — Tichá 1968, 1976b; *Lycopersicon* — Peat 1970; *Calendula* — Groen 1973; *Glycine* — Dornhoff and Shibles 1974; *Nicotiana* — Rawson and Hackett 1974, Rawson and Woodward 1976; *Phaseolus* — Fraser and Bidwell 1974, Srivastava *et al.* 1975, Čatský *et al.* 1976, O'Toole *et al.* 1977, Čatský and Tichá 1979, 1980; *Manihot* — Aslam *et al.* 1977; *Pisum* — Bethlenfalvay and Phillips 1977; *Rumex* — Sisson and Caldwell 1977; *Gossypium* — Constable and Rawson 1980; *Helianthus* — Rawson and Constable 1980; *Cucurbita* — Sisson 1981; *Carica* — Lin and Ehleringer 1982), in grasses (*e.g.*, *Festuca* — Jewiss and Woledge 1967, Woledge 1971; *Triticum* — Osman and Milthorpe 1971, Feller and Erismann 1978, Hall *et al.* 1978, Wittenbach 1979, Austin *et al.* 1982; *Zea* — Thiagarajah *et al.* 1981), and trees (*e.g.*, *Populus* — Dickmann and Gordon 1975, Tsel'niker *et al.* 1981; *Hevea* — Samsuddin and Impens 1979a, b; *Malus* — Kennedy and Johnson 1981; *Pinus* — Coyne and Bingham 1982; *Prunus* — Sams and Flore 1982). The ontogenetic trends and even the maximum $P_N$ reached do not differ much between species of the same genus (*cf.* *Glycine javanica* and *G. max* in Table 7.2, *Agropyron intermedium* and *A. desertorum* — Frank 1981) or between different cultivars (*cf.* *Triticum aestivum* cultivars in Table 7.2 and *Festuca* — Wilhelm and Nelson 1978) or hybrids of the same species (*Solanum* — Zrůst and Smolíková 1977); on the other hand, growth environment and conditions of determination may affect the absolute $P_N$ values (*cf.* Table 7.2 and Sections 7.2.5 to 7.10).

The ontogenetic course of $P_N$ is associated with numerous events which appear in the developing leaf. A rapid increase in leaf area, leaf thickness, and surface and volume of mesophyll cells, leaf internal surface, number of chloroplasts per cell and chloroplast dimensions (see also Chapter 1) are the main structural events. The very low $P_N$ in newly unfolded leaves is associated with high rates of dark respiration (*Populus* — Dickmann 1971; *Betula* — Valanne *et al.* 1981; Fig. 7.2; Section 7.3), and low stomatal and mainly intracellular conductances for $CO_2$ transfer (Fig. 10.3; Chapter 8). Both conductances increase during

---

→

**Fig. 7.1.** Net photosynthetic rate ($P_N$) increases rapidly during leaf ontogeny reaching the maximum usually before full leaf area expansion (labelled by arrows) is finished, and then decreases slowly during the period of photosynthetic senescence. This general pattern of $P_N$ was found in C₃ (*Nicotiana, Raphanus, Triticum, Vitis*) and C₄ (*Sorghum*) plants, in cotyledons (*Raphanus*), leaves of herbs (*Nicotiana*), grasses (*Triticum*), or shrubs (*Vitis*). In some plants two or more maxima of $P_N$ are found during leaf ontogeny (*Glycine*): the secondary maxima in $P_N$ are ascribed to various events in the plant such as unfolding of further leaves, flowering, pod filling, *etc.* (a — *Glycine max* (L.) Merrill, after Woodward and Rawson 1976; *Nicotiana tabacum* L., after Wada 1968; *Raphanus sativus* L., after Suzuki 1976; *Sorghum almum* Parodi, after Ludlow and Wilson 1971; *Triticum aestivum* L., after Marshall and Biscoe 1980b; *Vitis vinifera* L., after Kriedemann *et al.* 1970). The ontogenetic changes in $P_N$ are similar when expressed per leaf area unit or dry matter unit (b — *Festuca arundinacea* Schreb., after Jewiss and Woledge 1967), and per leaf area unit or unit mass of chlorophyll (c — *Triticum aestivum* L., after Austin *et al.* 1982).

further leaf development, while dark respiration rate declines. In the period of leaf expansion the pigment contents increase (see Chapter 3), photosynthetic enzymes are formed and their activities increase (Chapter 6) together with the capacity of light harvesting processes, activities of electron transport chain and photophosphorylation (Chapter 5). The period of formation of the photosynthetic apparatus coincides with the period of the highest rise in cytokinin activity; the content of free auxins is changed only later (*Solanum* — Borzenkova and Nefedova 1981). The photosynthetic apparatus of the leaf is fully developed and

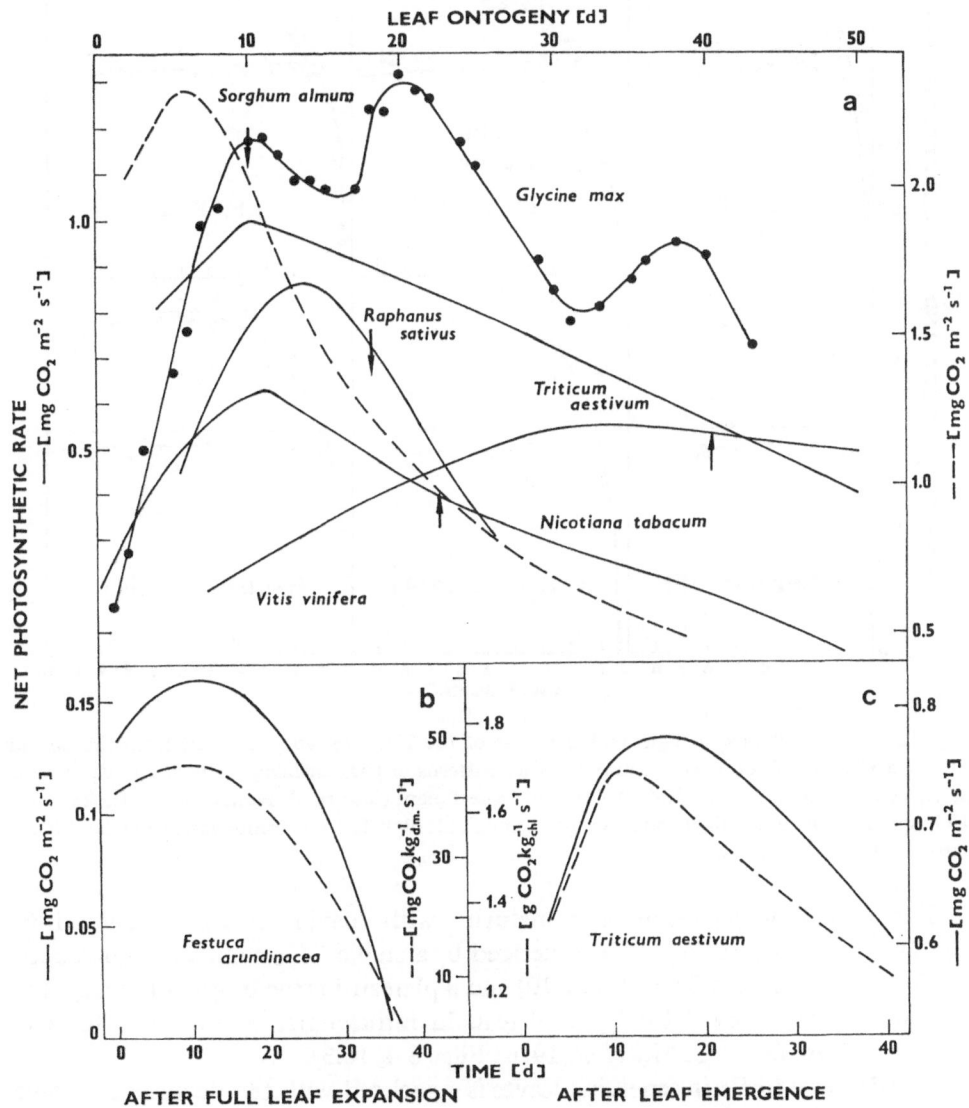

maximum $P_N$ attained at reaching 35 to 100% of the maximum leaf area (35 to 55% in *Solanum* — Borzenkova and Nefedova 1981; 37% in *Nicotiana* — Rawson and Hackett 1974; 50 to 80% in *Helianthus* — Rawson and Constable 1980; 70% in *Phaseolus* — Čatský et al. 1976; 75 to 90% in *Gossypium* — Constable and Rawson 1980; 100% or later in *Festuca arundinacea* — Jewiss and Woledge 1967).

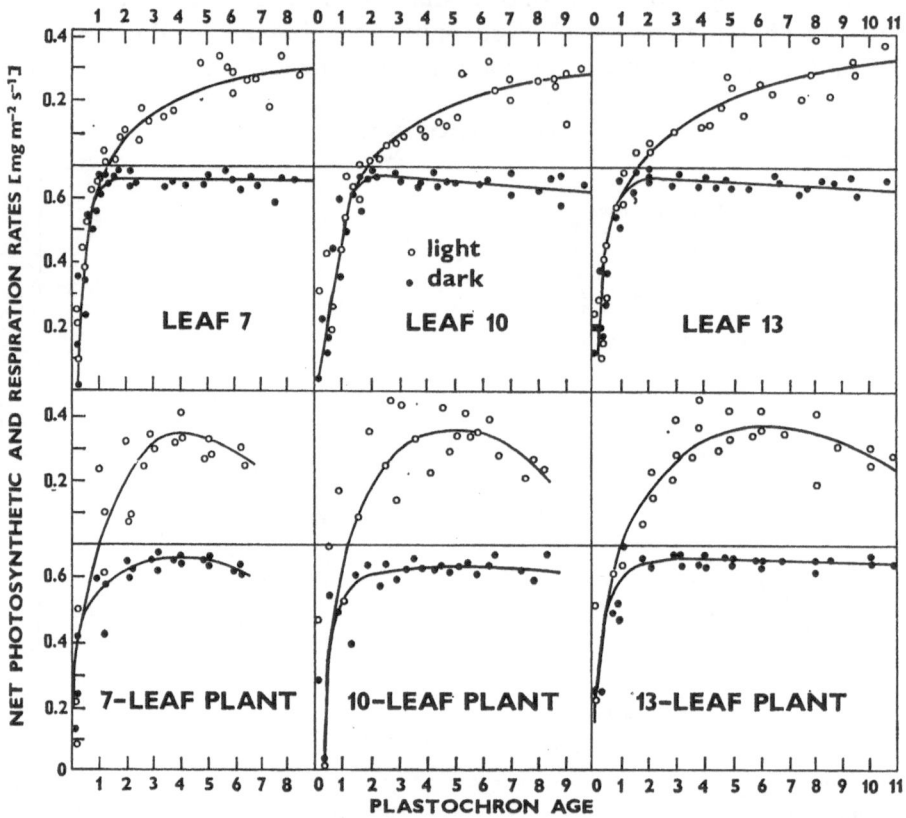

**Fig. 7.2.** $CO_2$ exchange in light and darkness of the 7th, 10th and 13th leaf from the bottom during a 17 d period (ontogeny) compared with patterns of $CO_2$ exchange in light and darkness of all successive leaves from a 7-leaf, 10-leaf and 13-leaf plant (insertion). Newly formed leaves evolve $CO_2$ at a high rate in the light and show also comparable high dark respiration rates (*Populus deltoides* Bartr. — Dickmann 1971).

The period of photosynthetic maturity with maximum $P_N$ (Tanaka 1961, Šesták and Čatský 1967c) is characterized by a stable $CO_2$ compensation concentration (*cf.* Section 7.6.2 and Fig. 7.10) and a plateau in true or gross photosynthetic rate which is paralleled by a plateau in intracellular conductance for $CO_2$ transfer (*Phaseolus* — Čatský et al. 1976; Figs. 8.4, 10.3).

The decrease in $P_N$ in senescing leaves is associated with the decrease in stomatal and intracellular conductances, chlorophyll content, enzyme activity, *etc.* Nevertheless, no single one of these factors is solely responsible for the whole ontogenetic course of $P_N$.

Leaves of different final size, from different nodes and from plants grown at different times of the year have similar patterns of light saturated $P_N$ in relation to their age (*Phaseolus* — Davis and McCree 1978; *Gossypium* — Constable and Rawson 1980).

No great differences in the ontogenetic course in $P_N$ expressed per unit leaf area or mass were reported (*e.g.*, in *Medicago* — Pearce and Lee 1969; *Festuca* – Jewiss and Woledge 1967; *Glycine* — Silvius *et al.* 1978; *Senecio* — Morozov 1980; see Fig. 7.1b). Only in the uppermost leaves of tobacco, $P_N$ per unit leaf area was considerably higher than $P_N$ per unit leaf dry mass (Fleck-Gerndt 1971a). $P_N$ per unit leaf area increased to a maximum during leaf ontogeny, and therefore declined, while $P_N$ expressed per chlorophyll content was more constant or declined over this period of time (*Dactylis* — Treharne *et al.* 1968; *Triticum* — Austin *et al.* 1982; Fig. 7.1 c), in *Phaseolus vulgaris* and *Theobroma* leaves the same pattern of development of both $P_N$ per leaf area and $P_N$ per chlorophyll content was reported (Więckowski 1966, Baker and Hardwick 1973). Comparable results were obtained for $P_N$ during leaf development on both dry matter and chlorophyll basis (*Betula* — Valanne *et al.* 1981).

Leaves with a long leaf duration (one year or longer) usually have lower $P_N$ with less pronounced maxima, *i.e.* the development of photosynthesis is extended over a longer time scale (*Lepechinia* — Field and Mooney 1983; *cf.* Fig. 10.1).

An exception from the common ontogenetic pattern of $P_N$ was found in cotyledons: $P_N$ was high in 1-day-old cotyledons and decreased, first rapidly and then slowly, during the 42 d of their life span (*Gossypium* — Lane and Hesketh 1977). Lasley and Garber (1978) found a biphasic ontogenetic pattern in $P_N$ of cucumber cotyledons: a period of relatively high $P_N$ up to 17 to 18 d from planting, followed by a decline in $P_N$ thereafter.

In some plant species two or more maxima of $P_N$ are found during leaf ontogeny: the "additional" maxima are ascribed to various events in the plant such as unfolding of further leaves, flowering, fruit development, *etc.* (*cf. Phaseolus* — Fraser and Bidwell 1974, Šesták *et al.* 1975; *Pisum* — Flinn 1974; *Glycine* — Woodward and Rawson 1976; *Capsicum* — Hall and Brady 1977 — Fig. 7.1 a). For example, $P_N$ in the flag leaf of wheat fell from anthesis, then rose to a maximum 15 to 16 d after anthesis, before falling again, at first slowly and then rapidly as the leaves senesced (25 d after anthesis). The rise in $P_N$ on the 15th day coincided with the increase in rate of grain growth which suggests that $P_N$ in flag leaves varied in response to the demand for photosynthates (Evans and Rawson 1970).

Each of the successively formed leaves on a plant passes through a course of ontogenetic changes in photosynthetic characteristics. The shapes of the courses are more or less similar, but they differ in extent of values (*e.g.*, *Glycine* — Kumu-

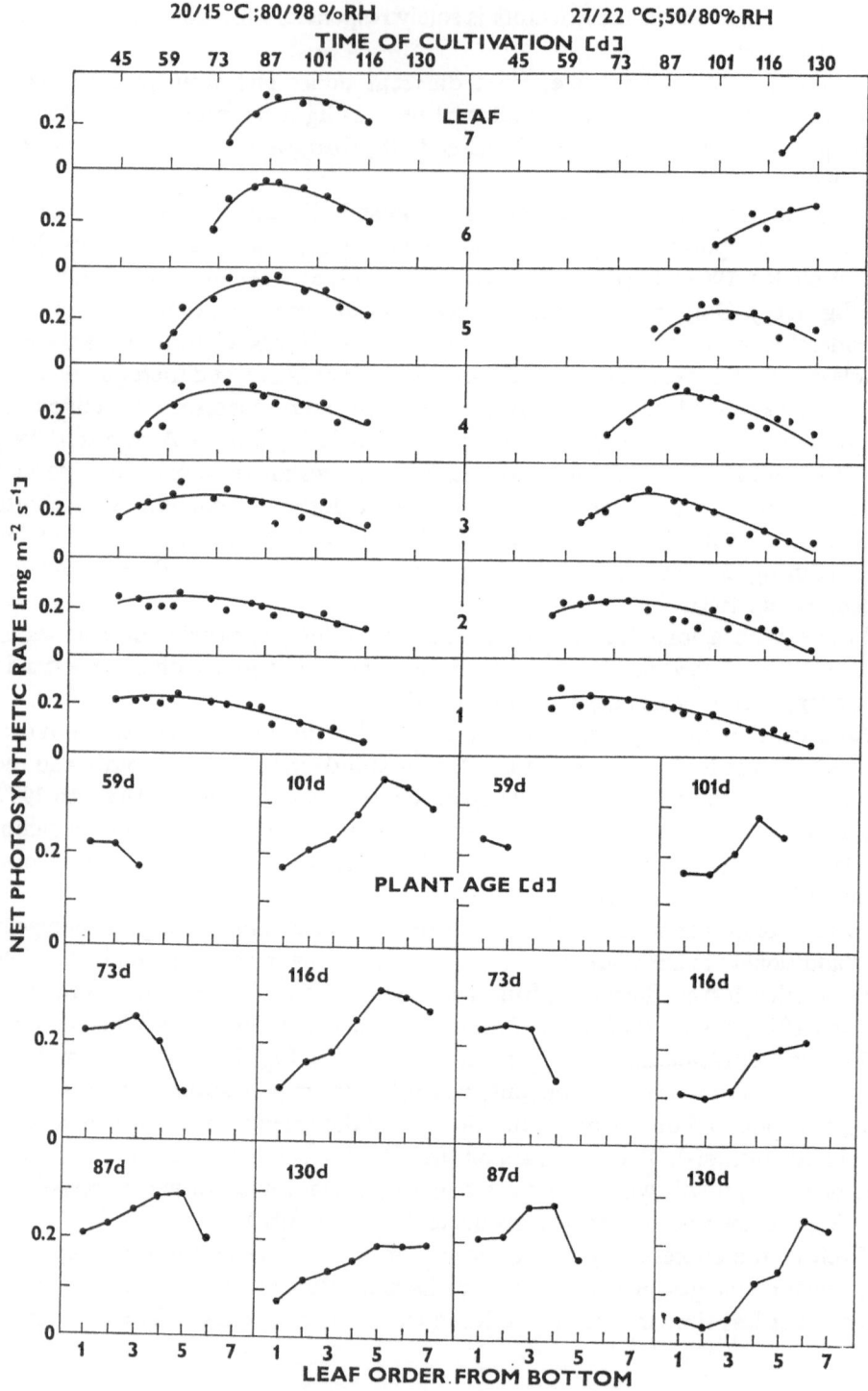

GROWTH CONDITIONS

20/15°C ;80/98 %RH          27/22 °C;50/80%RH

TIME OF CULTIVATION [d]

ra and Naniwa 1965; *Plectranthus* — Tichá 1968, 1976b; *Pennisetum* — McPherson and Slatyer 1973; *Nicotiana* — Rawson and Hackett 1974; *Phaseolus* — Tanaka and Kikuchi 1976, Tanaka and Fujita 1979; *Solanum* — Moll and Henniger 1978; *Beta* — Hodáňová 1981; *Saccharum* — Varlet-Grancher et al. 1981; see Figs. 7.2, 7.3 *top*, 7.5 *top*). The age effected decline of $P_N$ was more marked at upper positions in maize leaves (Vietor et al. 1977).

Consequently, also the relative contribution of individual leaves to whole-plant photosynthesis is different in various phases of plant development (*e.g.*, *Sorghum* — Naylor et al. 1975; *Plectranthus* — Tichá 1976a; *Picea* — Schulze et al. 1977a, b; *Phaseolus* — Ïordanov 1979; see Fig. 7.4).

In CAM plants, young leaves had similar $^{14}$C-labelling patterns to those of $C_3$ plants, mature leaves showed high incorporation of $^{14}$C into $C_4$ acids, especially at night due to the simultaneous fixation of $CO_2$ by both enzymes RuBPC and PEPC (*Bryophyllum* — Nishida 1978). Similarly in *Mesembryanthemaceae*, the establishment of the CAM syndrome seemed to be highly controlled by leaf age (Willert 1979).

Photosynthetic activity of a single, rooted leaf increased during leaf development and then maintained a fairly constant high level whereas leaf area and root volume continued to increase steadily (*Phaseolus* — Sawada et al. 1982).

The ontogenetic changes in leaf $P_N$ expressed as adaptations to changing irradiance, $CO_2$ concentration, temperature and availability of the main mineral elements were studied using a mathematical model (Kaler and Fridlyand 1978). The effect of leaf ontogeny on $P_N$ was treated in the model of Marshall and Biscoe (1980a, b) based on the analysis of the $P_N$-irradiance curves in wheat flag leaves (*cf.* also Tenhunen et al. 1980).

The ontogenetic changes in $P_N$ of individual leaves seem to be more plant-induced and inherently fixed than environment-induced. The extent of values is, of course, varied by environmental factors (*cf.* Sections 7.4 to 7.10), but the described trends are maintained. As a result of the ontogenetic changes in $P_N$ in successive leaves, the insertion profiles in $P_N$ on the plant appear (*cf.* Section 7.2.2).

---

←

**Fig. 7.3.** Comparison of the ontogenetic changes in net photosynthetic rate ($P_N$) of individual leaves on the plant (*upper part*) and the insertion gradients in $P_N$ during plant development (plant age indicated in days — *lower part*). The plants were cultivated in controlled environments which simulated a drier and hotter "summer" climate (day/night 27/22 °C, 50/80% relative air humidity), and a more humid and colder "spring" climate (day/night 20/15 °C, 80/98% relative air humidity). The trends of ontogenetic changes in $P_N$ of the successively formed leaves and in the two types of controlled environments are more or less similar but differ in the extent of values, and the same holds for the insertion gradients in $P_N$ (*Plectranthus fructicosus* L'Hérit — Tichá 1968, 1970a,b, 1976b).

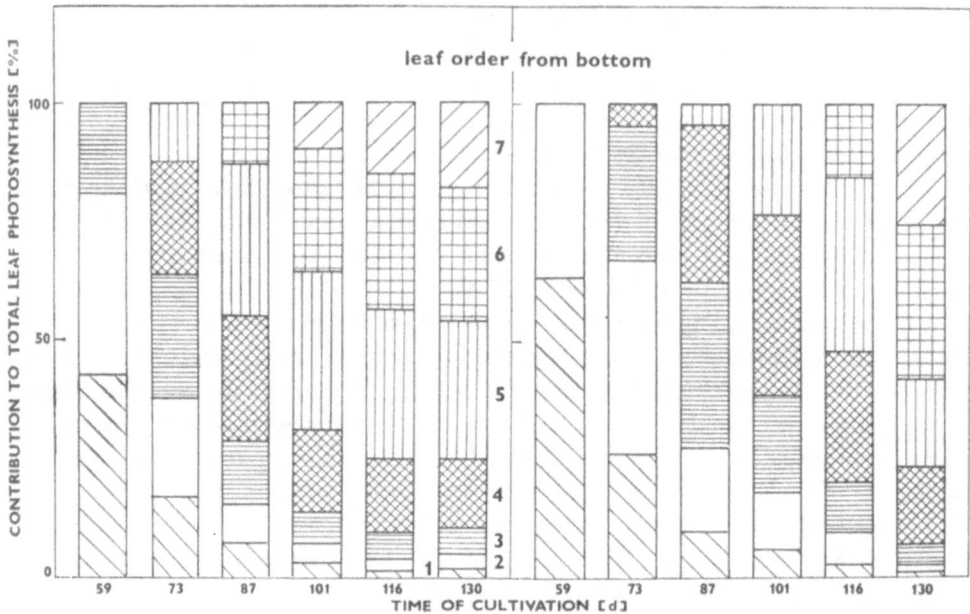

**Fig. 7.4.** Contribution of individual leaves to total leaf photosynthesis [%] during plant ontogeny (given in days of cultivation) in "spring" (20/15 °C; 80/98% relative air humidity) and "summer" (27/22 °C; 50/80 relative humidity of air) controlled environments. The contribution of individual leaves was similar in both environments, its maximum shifting from bottom leaves to leaves of middle insertion levels during plant ontogeny. Integrating the values of leaf photosynthesis for the whole vegetation period, the 5th leaf from bottom in "spring" conditions and the 4th leaf in "summer" conditions showed the highest contribution to plant photosynthesis (29% and 25% of the total net photosynthesis of the leaf apparatus, respectively: *Plectranthus fructicosus* L'Hérit — Tichá 1976a, b).

## 7.2.2   Leaf Insertion and Net Photosynthetic Rate

Even if the ontogenetic courses of $P_N$ of subsequently formed leaves are similar in shape (*cf.* Section 7.2.1; Figs. 7.3 *top* and 7.5 *top*), the rate of increase in $P_N$, the maximum rates reached and the time of reaching the maximum mostly differ (Fig. 7.5 *top*); successive leaves develop from different primordia at different phases of plant ontogeny under various microenvironmental conditions, and are differently supplied with nutrients and water. Hence, the insertion profiles or gradients in $P_N$ change during plant ontogeny, too: in the young plant they are basipetal (Fig. 7.5, *bottom* A), in middle-aged plants an increase and decrease in $P_N$ is found (Fig. 7.5 B, C; 7.6 b), and in old plants acropetal insertion gradients with maximum $P_N$ in the uppermost leaf (Fig. 7.5 D) are reported (*Plectranthus* —

Tichá 1970a, b, 1976b, 1984; Fig 7.3, *bottom*). Thus, studying insertion gradients which are measured more easily, is not a good substitution for leaf ontogeny (*cf.* also *Nicotiana* — Václavík 1975). The described changes in insertion profiles with plant age explain also why various types of distribution of $P_N$ along the plant may appear when plant age has not been carefully considered. In the literature,

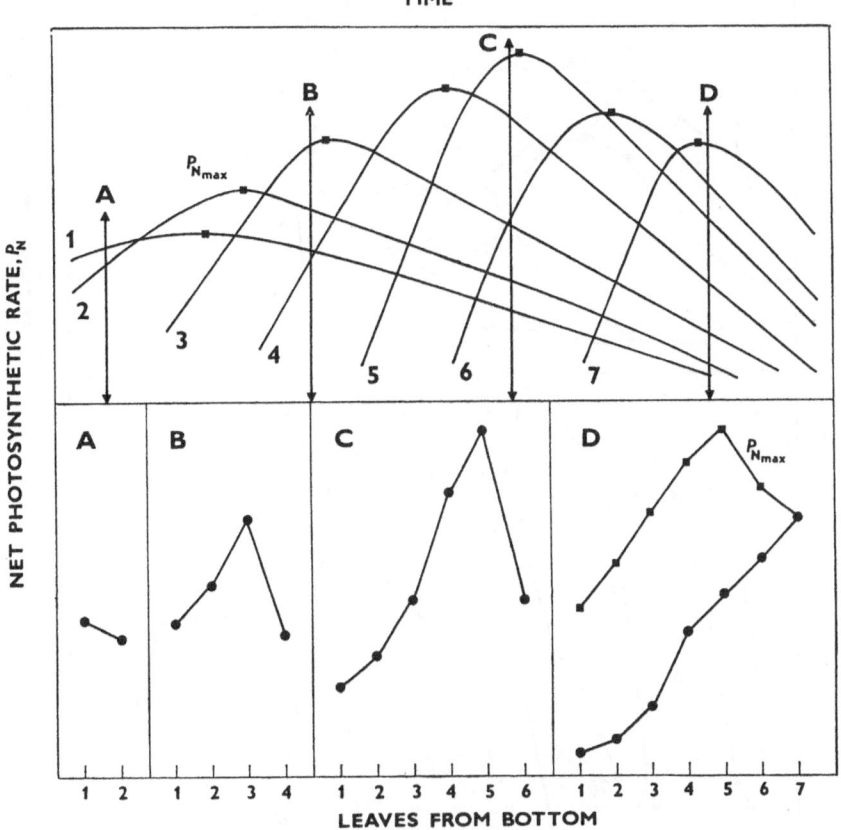

**Fig. 7.5.** Subsequently formed leaves (1 to 7) on the plant are characterized by similar ontogenetic courses in net photosynthetic rate, $P_N$ (*cf.* Section 7.2.1) but they differ in the rate of increase, the maximum net photosynthetic rate reached ($P_{N max}$ ■), and the time when this maximum is achieved (*top*). At four moments during plant development (A, B, C, D), insertion profiles in $P_N$ are demonstrated showing (*bottom*) a basipetal character (with highest $P_N$ in the lowest leaf) in young plants (A), an increase from lower leaves to a maximum in the middle leaves and a decline in the upper ones in middle-aged plants (B, C), and an acropetal insertion gradient (highest $P_N$ in the uppermost leaf) in old plants (D). Thus, (1) the ontogenetic course of leaf $P_N$ is not comparable with insertion gradients in $P_N$, namely in very young or very old plants, for insertion gradients in $P_N$ vary with plant age. Leaf insertion or position is not a good substitution for leaf ontogeny, and (2) if studying insertion profiles, plant age should be carefully indicated. The insertion gradient of $P_{N max}$ resembles in shape the ontogenetic course of $P_N$ of the individual leaf (see the curve in the upper part of D). (Relative units — after Tichá 1984).

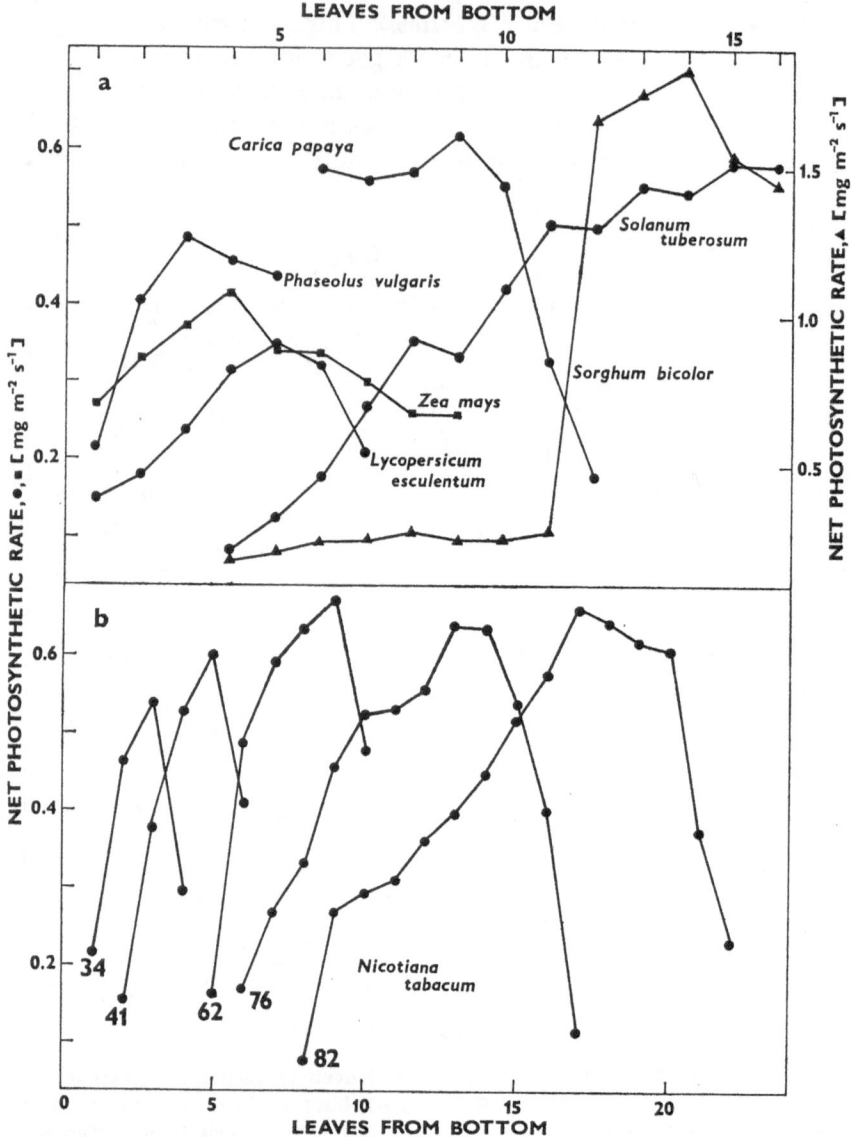

**Fig. 7.6.** Examples of insertion profiles (or gradients) in net photosynthetic rate ($P_N$) published for C$_3$ (*Carica, Lycopersicum, Phaseolus, Solanum*) and C$_4$ (*Sorghum, Zea*) plants (a — *Carica papaya* L., 120 to 150 d old plants, after Imai *et al.* 1982; *Lycopersicum esculentum* Mill., 6 to 8-leaf plants, after Ito 1970; *Phaseolus vulgaris* L., 20 d old plants, from Ïordanov 1979; *Solanum tuberosum* L., 16-leaf plants, after Tsuno 1977; *Sorghum bicolor* (L.) Moench, 60 d old plants from a canopy where lower leaves were shaded, after Naylor *et al.* 1975; *Zea mays* L., flowering plants with 10 leaves, after Meinl and Bellmann 1965). Insertion profiles of $P_N$ on middle-aged plants (34 to 82 d after sowing) are similar (b — *Nicotiana tabacum* L., after Rawson and Hackett 1974).

174

changes in $P_N$ with leaf age are frequently investigated without distinguishing between leaf ontogeny and leaf insertion level (for discussion see Tichá 1984); this misunderstanding is stimulated by the fact that the insertion profiles in $P_N$ are mostly studied on middle-aged plants, the profiles of which resemble in shape the common ontogenetic pattern in $P_N$ (profiles of type B, C on Fig. 7.5, *bottom*; for further examples see, *e.g.*, *Nicotiana* — Fleck-Gerndt 1971b, *Ipomoea* — Kato *et al.* 1979; *Ophiopogon* — Tanaka and Mizuno 1981; *Brassica* — Sasahara 1982; *Vitis* — Pandey and Farmahan 1977, Alleweldt *et al.* 1982; *Populus* — Bassman and Dickmann 1982; Table 7.3 and Figs. 7.3, *bottom*, 7.6).

The insertion profile of the maximum $P_N$ ($P_{N\,max}$) reached by the leaf during its ontogeny, resembles in shape the ontogenetic course of $P_N$ for the individual leaf (Fig. 7.5, *bottom* D — the upper curve) (*Plectranthus* — Tichá 1970a, 1984; *Pennisetum* — McPherson and Slatyer 1973; *Glycine* — Gordon *et al.* 1982). The insertion profiles of maximum $P_N$ may be more extensive in plants with numerous leaves, or more condensed in plants with a rosette or a head.

The extent of values of $P_N$ is, of course, varied by various environmental factors (see Sections 7.2.4 to 7.2.10), the described trends in insertion gradients being maintained (*cf.* Fig. 7.3 *bottom*, *right* and *left*). Only, *e.g.*, under severe water stress (see Section 7.2.8), $CO_2$ enrichment (see Section 7.2.6.3) or photoperiod treatment (*Hordeum* — Ma and Hunt 1983) acropetal gradients in $P_N$ were reported, for the uppermost leaves are preferentially supplied with water or are more sensitive to $CO_2$ enrichment or prolonged photoperiod: thus, the highest $P_N$ is found in the uppermost leaves.

In $C_4$ plants the changes in $P_N$ in leaves of different insertion levels which correspond to the above general scheme are mostly accompanied by differences in the activities of carboxylases (*cf.* Chapter 6). $CO_2$ compensation concentration is often higher in the lower leaves and very low in the others. Therefore, shifts in the type of $CO_2$ fixation were suggested (*Zea* — Crespo *et al.* 1979).

In grasses, the uppermost leaves reach a maximum $P_N$ near anthesis; leaf sheaths show a pattern of $P_N$ similar to that of the leaves, although the rates are considerably less (*Triticum* — Patterson and Moss 1979).

Complicated insertion and age gradients were found in conifers where the insertion profiles in $P_N$ of the needles of one branch are combined with the insertion profiles of $P_N$ on different branches; furthermore, the situation is usually strongly modified by environmental factors, namely irradiance (*e.g.*, *Pseudotsuga* — Woodman 1971; *Picea* — Watts *et al.* 1976, Fuchs *et al.* 1977, Schulze *et al.* 1977a, b, Leverenz and Jarvis 1980; *Pinus* — Rook and Corson 1978). Thus, four-year-old and older needles of spruce contribute more than 1/3 towards the total annual $CO_2$ uptake; the productivity of current year needles is only about 15% of the total $CO_2$ exchange (Schulze *et al.* 1977a). Similarly, carbon dioxide exchange in the crown of oak was analysed by Aubuchon *et al.* 1978, Malkina (1978a, b) and Yakshina and Malkina (1979).

**Table 7.3** Net photosynthetic rate in leaves of different insertion levels. A choice of data from the literature.
Abbreviations and explanations: chl – chlorophyll, d.m. – dry mass, IRGA – infrared gas analyser, max – maximum, N – nitrogen, $P_N$ – net photosynthetic rate, $P_G$ – gross photosynthetic rate, PI – plastochron index, RH – relative humidity, ↑ – increase, ↓ – decrease. Where leaves are numbered from the top of the plant and the overall number of leaves on the plants is not indicated, the symbols leaf $n$, leaf $(n + 1)$, to leaf $(n + x)$ are used.

| Plant species (and cultivation) | Leaf insertion (leaves numbered from the lowest one), plant age | Method of $P_N$ measurement; experimental conditions (radiation, temperature, $CO_2$ concentration, RH) | Net photosynthetic rate $P_N$ [mg m⁻² s⁻¹] if not indicated otherwise (from upper to lower leaves) | Changes in $P_N$ due to insertion level | Author(s) and year of publication |
|---|---|---|---|---|---|
| (1) | (2) | (3) | (4) | (5) | (6) |
| *Astrebla lappacea* (Lindl.) Domin (glasshouse and growth chamber) | leaves $n$ to $(n + 4)$ on shoots with six to nine fully expanded leaves | open system, IRGA | $P_N$ of the uppermost leaf = 100%: glasshouse: 100...54 growth chamber: 100...51 | → → | Doley and Yates 1976 |
| *Beta vulgaris* L. | leaves 1 to 30 | | 0.42...0.86...0.20 | ↑, ↓ | Ito 1965 |
| *Beta vulgaris* L. var. *saccharifera* Lange cv. Dobrovická A (field) | leaves 1 to 47 (64 d old plants: 1 to 15; 121 d: 12 to 36; 162 d: 27 to 47) | dry mass increment in a ventilated chamber, 200 W m⁻² (400–700 nm), 25 °C, 580 mg $CO_2$ m⁻³ | [mg (d. m.) m⁻² s⁻¹]: 64 d old plant: 0.64...0.71...0.31 121 d old plant: 0.25...0.50...0.04 162 d old plant: 0.15...0.25...0.03 | ↑, ↓ ↑, ↓ ↑, ↓ | Hodáňová 1975, 1981 |
| *Chenopodium album* L. (natural locality) | leaves 4 to 15 on plants in early flower bud | portable steady-state gas exchange system, IRGA | 0.58...1.02...1.00 | ↑, ↓ | Mooney *et al.* 1981 |
| *Cucumis sativus* L. (pots with gravel, | leaves 4 to 24 on plants with | closed system, IRGA, 420 W m⁻², | 545 mg $CO_2$ m⁻³: 0.62...0.92...0.60 | ↑, ↓ | Aoki and Yabuki 1977 |

| Species and growth conditions | Leaves investigated | Method | Values | Trend | Reference |
|---|---|---|---|---|---|
| water cultures, growth chamber) | 25 leaves | 30 °C, 90% RH | $2175$ mg $CO_2$ $m^{-3}$: 1.10...1.28...0.54<br>$4345$ mg $CO_2$ $m^{-3}$: 0.83...0.38...0.57<br>$9960$ mg $CO_2$ $m^{-3}$: 0.97...0.29 | ↑, ↓ | Schäfer and Tirtapradja 1970 |
| Dactylis glomerata L. cv. Holstenkamp (field) | leaves 1 to 7 on ca. 80 d old plants | IRGA | 0.06...0.42...0.11 | ↓, ↑ | Hofstra and Hesketh 1975 |
| Glycine max (L.) Merr. cv. Biloxi (phytotron) | middle leaflets of leaves 1 to 5 on ca. 25 d old plants | open system, IRGA, 25 klx, 30 °C | $545$ mg $CO_2$ $m^{-3}$: 0.42...0.49...0.38<br>$1810$ mg $CO_2$ $m^{-3}$: 0.43...0.13 | ↑, ↓ (max leaf 3) ↓ | |
| Gossypium hirsutum L. cv. Deltapine Smoothleaf (controlled environment) | leaves 1 to 9 on 4 to 5 weeks old plants | oxygen electrode, leaf slices | [mmol $kg^{-1}$(chl) $s^{-1}$]: 22.0...29.2...12.8 | ↑, ↓ | Jones and Osmond 1973 |
| Helianthus annuus L. (experimental field) | leaves 2 to 35; 20, 40, 60 d after germination | $P_G$ values calculated by means of a model | [mg $leaf^{-1}$ $d^{-1}$]:<br>20 d after germination: 10...240...100<br>40 d after germination: 10...550...200<br>60 d after germination: 10...480...400 | ↑, ↓<br>↑, ↓<br>↑, ↓ | Kobayashi 1975 |
| Helianthus annuus L. cv. Suncross 52 (glasshouse) | leaves 7, 15, 20 | open system, IRGA | $P_{N\,max}$: 1.95...1.55 | → | Rawson 1979 |
| Hordeum vulgare L. cv. HOR 2361 and HOR 1788 (growth chamber) | leaves 2 to 7 (after full expansion of each of the leaf) | open system, IRGA, 30 klx, 21 °C, 80–95% RH | HOR 2361: 0.58...0.59...0.58<br>HOR 1788: 0.48...0.50...0.49 | ↑, ↓<br>↑, ↓ | Apel and Lehmann 1969 |

**Table 7.3** (continued)

| Plant species (and cultivation) | Leaf insertion (leaves numbered from the lowest one), plant age | Method of $P_N$ measurement; experimental conditions (radiation, temperature, $CO_2$ concentration, RH) | Net photosynthetic rate $P_N$ [mg m$^{-2}$ s$^{-1}$] if not indicated otherwise (from upper to lower leaves) | Changes in $P_N$ due to insertion level | Author(s) and year of publication |
|---|---|---|---|---|---|
| (1) | (2) | (3) | (4) | (5) | (6) |
| *Lycopersicon esculentum* L. (field and plastic-house) | leaves 1 to 7 or 2 to 14 | open system, IRGA, 130 W m$^{-2}$, 25.8 °C | 0.21...0.35...0.15 545 mg $CO_2$ m$^{-3}$: 0.43...0.52...0.40 1810 mg $CO_2$ m$^{-3}$: 0.70...0.77...0.64 | ↑,↓ ↑,↓ ↑,↓ | Ito 1973 |
| *Medicago sativa* L. cv. Saranac CC 120 (growth chamber) | leaves 1 to 6 on plants with 9 leaves (regrowth after cutting) | open system, IRGA, 1116 µmol m$^{-2}$ s$^{-1}$ | 1.03...1.22...1.06 | ↑,↓ (max leaf 5) | Ku and Hunt 1973 |
| *Medicago sativa* L. cv. Hunter River (glasshouse) | leaves 2 to 14 on 6 month old plants | open system, IRGA, 280 W m$^{-2}$ (300–3000 nm), 25 °C | 1.07...1.10...0.30 | ↑,↓ (max leaf 12) | Hodgkinson 1974 |
| *Morus alba* L. cv. Ichinose | leaves 1 to 50 in early August (1) and late October (2) | IRGA, 25 °C, 545 mg $CO_2$ m$^{-3}$ | (1): 0.28...0.86...0.20 (2): 0.22...0.47...0.11 | ↑,↓ ↑,↓ | Murakami 1978 |
| *Nicotiana tabacum* L. cv. Virgin Gold (1) and Wisconsin 38 (2) (glasshouse) | leaves 4 to 13 on 150 to 180 d old plants | radiometric method ($^{14}CO_2$) | [mg kg$^{-1}$ s$^{-1}$]: (1): 2.52...2.95...0.58 [% of the uppermost leaf]: (2): 100...10 | ↑,↓ ↓ | Fleck-Gerndt 1971a |

| Species | Leaves | Method | $P_{N}$ values | Response | Reference |
|---|---|---|---|---|---|
| *Nicotiana tabacum* L. cv. Wisconsin 38 (growth chamber) | leaves 6 to 18 on a plant with 19 leaves longer than 10 cm (PI = 19) | open system, IRGA, 300 W m$^{-2}$ (400–700 nm), 25 °C, 635 mg $CO_2$ m$^{-3}$, 60% RH | abaxial leaf surface: 0.35...0.42...0.11 adaxial leaf surface: 0.20...0.21...0 | ↑, ↓ ↑, ↓ | Václavík 1975 |
| *Phaseolus vulgaris* L. cv. Pencil Pod Black Wax (growth chamber) | primary leaf and 1 to 3 secondary leaves | radiometric method ($^{12}CO_2 + {}^{14}CO_2$) | $P_{N\,max}$: 0.30...0.42...0.38...0.49 | ↑, ↓, ↑ | Fraser and Bidwell 1974 |
| *Phragmites communis* Trin. (littoral stand) | leaves 1 to 10 on plants with 12 leaves, leaves $n$ to $(n + 15)$ on plants with 16 leaves | dry mass increment open system, IRGA | [mg (d. m.) m$^{-2}$ s$^{-1}$]: 0.08...0.20...0.06 0.52...0.61...0.44 | ↑, ↓ ↑, ↓ | Rychnovská 1967 Gloser 1977 |
| *Phragmites communis* Trin. var. *berlandieri* (Fourn.) Fern. (stand) | leaves 5 to 16 on plants with 16 leaves | radiometric method ($^{14}CO_2$), 100 klx, 30 °C | 0.13...0.17...0.01 | ↑, ↓ | Walker and Waygood 1968 |
| *Pinus pinaster* Aiton (stand) | needles 1 to 5 years old on 15 years old trees | radiometric method, 40 W m$^{-2}$, 20 °C | [mg kg$^{-1}$ (d. m.) s$^{-1}$]: 0.94...0.30 | → | Keay *et al.* 1968 |
| *Pinus silvestris* L. (stand) | current shoots and 1 to 3 years old shoots on 20 years old trees | open system, IRGA, 800 µmol m$^{-2}$ s$^{-1}$, 15 to 20 °C | 0.59...0.64...0.28 | ↑, ↓ | Linder and Troeng 1980 |
| *Picea abies* (L.) Carst. (stand) | needles 0 to 2 years old, sun side, on 89 years old trees | mobile field laboratory, IRGA | $P_{N\,max}$ [mg kg$^{-1}$ (d. m.) s$^{-1}$]: 1.30...0.83 | → | Schulze *et al.* 1977b |
| *Populus deltoides* Bartr. (growth room) | leaves 8 to 16 | radiometric method ($^{14}CO_2$) | −0.23...0.38...0.31 | ↑, ↓ (max leaf 12) | Larson *et al.* 1972 |

**Table 7.3** (continued)

| Plant species (and cultivation) | Leaf insertion (leaves numbered from the lowest one), plant age | Method of $P_N$ measurement; experimental conditions (radiation, temperature, $CO_2$ concentration, RH) | Net photosynthetic rate $P_N$ [mg m⁻² s⁻¹] if not indicated otherwise (from upper to lower leaves) | Changes in $P_N$ due to insertion level | Author(s) and year of publication |
|---|---|---|---|---|---|
| (1) | (2) | (3) | (4) | (5) | (6) |
| *Populus euramericana* Hardtwalden (glasshouse) | leaves $n$ to ($n + 13$) on shoots with 17 leaves | open system, IRGA, 30 klx, 25 °C | 0.13...0.43...0.03 | ↑,↓ (max leaf $n + 9$) | Furukawa 1973a |
| *Solanum tuberosum* L. cv. Nōrin No. 1 (pots) | leaves 1 to 13 on plants with 14 leaves (29th June) | | 0.40...0.43...0.17 | ↑,↓ | Tsuno 1977 |
| *Trifolium pratense* L. cv. Temara (glasshouse) | leaves 1 to 4 on 55 d old plants | open system, IRGA, 230 W m⁻², 20 °C | without N: 0.52...0.30; with 35 g m⁻³ of nitrate N: 0.66...0.48 | ↓ ↓ | Maag and Nösberger 1980 |
| *Triticum aestivum* L. (field) | leaves 1 to 9 | open system, IRGA | [g m⁻² year⁻¹]: $P_N$: 584...7...25 $P_G$: 634...8...30 | ↓,↑ | Shatilov and Sharov 1978 |
| *Vitis vinifera* L. × *V. labruscana* Bailey (field) | leaves 1 to 25 on 4 years old plants | IRGA | −0.25...0.20...0.14...0.22...0.08 | ↑,↓,↑,↓ | Tezuka *et al.* 1980 |
| *Zea mays* L. cv. TA 48/65 (field) | leaves 5 to 12 on plants with 12 leaves (10th July) | dry mass increment | [mg (d. m.) m⁻² s⁻¹]: 0.46...0.65...0.62 | ↑,↓ | Sýkorová 1976 |

## 7.2.3 Net Photosynthetic Rate on Leaf Blade Area

Differences in $P_N$ are not only found during leaf ageing and in leaves of different insertion levels but also on one leaf blade which is heterogeneous also in other characteristics and parameters (*cf.* Chapters 1 to 4).

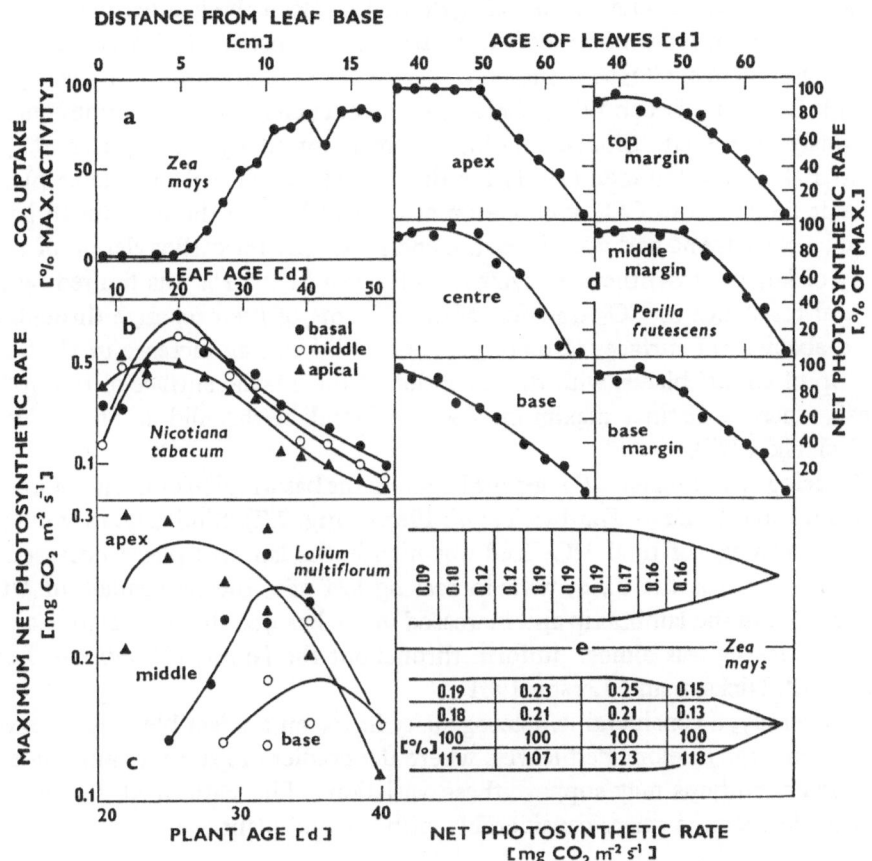

Crespo *et al.* 1979

0.60...0.64...0.38

closed system, IRGA, 1700 μmol m⁻² s⁻¹, 28 °C

leaves 1 to 5 on three weeks old plants

*Zea mays* L. cv. Kalahari Early Pearl (phytotron)

**Fig. 7.7.** Net photosynthetic rate ($P_N$) along the leaf lamina increases from leaf base towards leaf tip; at the tip of the leaf $P_N$ may again decrease somewhat (a — *Zea mays* L., second leaf, after Miranda *et al.* 1981b). On the leaf periphery $P_N$ is higher than in the leaf centre (e — *Zea mays* L., 2 cm long segments and segments of the peripheral and the central parts of the leaf blade, and relative comparison [%] of the differences, after Repka and Jureková 1981). $P_N$ of the apical, middle and basal portions of the leaf varies markedly during leaf ontogeny, the maximum $P_N$ being reached first in the apical, then in the middle and last in the basal leaf part (b — *Nicotiana tabacum* L., after Wada and Kuroda 1968; c — third leaf of *Lolium multiflorum* Lam., after Prioul *et al.* 1980a). The gradual decrease in $P_N$ in the whole leaf during leaf senescence is in fact an average of different rates of change in different parts of the lamina (d — *Perilla frutescens* (L.) Britt, after Hardwick *et al.* 1968).

$P_N$ on the leaf lamina of *Ficus* was higher at the more apical and near-margin parts of the leaf than at the basal and near-midrib parts (Takaoki 1969); in elongated leaves of grasses the highest $P_N$ was found in the middle of the leaf (*Zea* — Meinl and Bellmann 1965; *Dactylis* — Schäfer and Tirtapradja 1970). In tobacco, $P_N$ per leaf area unit and per leaf dry mass was lower in the apical part of the blade than in the basal part (17 and 32% — Slavík 1963); the differences in $P_N$ were correlated with differences in leaf hydration levels of the leaf lamina.

On a 10 cm long maize leaf only tissues more distant than 4 cm from the leaf base assimilated $CO_2$; $P_N$ increased nearly fivefold towards the leaf tip (Leech 1974, Leese and Leech 1977). Chlorophyll content and plastid volume contemporaneously increased. Similar findings were reported by Miranda *et al.* (1981b), and Baker and Miranda (1981) for the second leaf of 7 d-old maize plants: the major limitation to $CO_2$ assimilation at the leaf base in the leaf sheath lay within the chlorenchyma and was either the energy supply for carboxylation or the capacity of key photosynthetic enzymes; on the leaf blade, $P_N$ was limited by the stomatal resistance to $CO_2$ transfer. Measurements of gas exchange through adaxial and abaxial leaf surfaces of maize leaves showed also an increase in $P_N$ from base to tip of all leaf blades with the exception of the adaxial surface of the uppermost leaf, where a distinct maximum $P_N$ was found in the middle of the leaf blade (Václavík 1977).

During leaf ontogeny, $P_N$ decline begins in the basal region of the lamina adjoining the midrib (*Perilla* — Hardwick *et al.* 1968 — Fig. 7.7). Similarly, in young leaves the percentage of total $^{14}C$ fixed was high in the lamina tip and decrased almost linearly towards the base; with increasing leaf age, the percentage of $^{14}C$ fixed decreased in the lamina tip and increased in the base, so that the relative activity in mature leaves was almost uniform throughout the lamina (*Populus* — Larson *et al.* 1972, Dickson and Larson 1981).

Generally, a considerable heterogeneity in $P_N$ on the leaf blade is observed, namely on large or elongated leaves, where the gradient in the surrounding environmental conditions may support these variations. The pattern of $P_N$ distribution on the leaf lamina varies considerably with leaf and plant age.


### 7.2.4   Modifications of Developmental Pattern in Net Photosynthetic Rate

The ontogenetic course of $P_N$ and its differences in leaves of different insertion levels are inherent to any plant species or cultivar (Hedley and Harvey 1975), but the extent of values may be modified by many internal and environmental factors. Substantial seasonal variations (*Morus* — Murakami 1978; *Glycine* — Gordon *et al.* 1982), changes due to phases of plant ontogeny, *e.g.* unfolding of new leaves, flowering and fruiting, and such artificial treatments as defruiting, decapitation and defoliation by grazing, cutting, pruning, *etc.* were found (*e.g.*, Gej 1970, Satoh and Hazama 1971, Fraser and Bidwell 1974, Kriedemann *et al.* 1976, Hall and Brady 1977, Hodgkinson 1974, Bassman and Dickmann 1982).

The developmental course of $P_N$ can be varied by pathogene infection (Raggi 1978a, b, Ellis *et al.* 1981), growth regulators, *e.g.*, gibberellic acid, kinetin (Hall *et al.* 1978), CCC (Tezuka *et al.* 1980) or benzyladenine (Adedipe *et al.* 1971), ozone stress (Coyne and Bingham 1982), in mutant plants (Heyes and Dale 1971), and may change in different cultivation environments (field and growth chamber — Pearce *et al.* 1968, field and glasshouse — Rawson and Constable 1980).

Of great importance are modifications induced by environmental factors such as radiation, $CO_2$ concentration, temperature, water supply and mineral nutrition (see Sections 7.2.5 to 7.2.10; for reviews see, *e.g.*, Marcelle 1975, and Landsberg and Cutting 1977).

### 7.2.5 Irradiance

The relationship between $P_N$ and irradiance ($I$) (400 to 700 nm) of the leaf has been most often characterized by the so-called $P_N$-$I$ response curve, the action spectrum of $P_N$ and the relevant conversion indices (quantum yield, quantum requirement, photosynthetic efficiency). Owing to these characteristics, changes in $P_N$ can be specified with respect to both quanta of radiant energy incident and/or absorbed (more recommended, but rarely used) per unit leaf area in unit time, spectral composition of $I$, and leaf capability to convert $I$ into chemical energy and utilize the latter for $CO_2$ fixation.

#### 7.2.5.1 *The Dependence of Net Photosynthetic Rate on Leaf Irradiance*

The leaf $P_N$-$I$ response curve has a non-linear character given by coordinates of the following parameters: (1) the compensation irradiance, $I_{comp}$ (*i.e.* the smallest $I$ at which no net gain in $CO_2$ uptake by a leaf is achieved), (2) the more or less extensive linear region of the initial part of the curve ($P_N$ increases at a constant rate with an increase in $I$), the slope of which characterizes the maximum photosynthetic efficiency, (3) the more or less pronounced intermediate section of curvature ($P_N$ is no more proportional to the increase in $I$), and (4) the region of a plateau response of $P_N$ to $I$, $P_{N\,max}$ (no or a small increase in $P_N$ implying the saturation of the photosynthetic system by $I$, $I_{sat}$). Each of these parameters varies with leaf age and insertion level.

When maturing, the leaf has the highest $P_{N\,max}$ which is attained under the high $I_{sat}$ while the senescing leaf has the smallest $P_{N\,max}$ saturated at much lower $I$; the young expanding leaf is usually of an intermediate character (*Helianthus* —

Hiroi and Monsi 1966, Horie and Udagawa 1971; *Glycine* — Kumura 1969; *Nicotiana* — Haraguchi and Shimizu 1970; tropical grasses — Ludlow and Wilson 1971; *Cucumis* — Iwakiri and Inayama 1975; *Gossypium* — Nagarajah 1975b, Constable and Rawson 1980; *Arachis* — Trachtenberg and McCloud 1975; *Acer* — Malkina 1976b; *Vigna* — Bonhomme *et al.* 1977; *Colocasia* — Sato *et al.* 1978; *Coffea* — Yamaguchi and Friend 1979; *Phaseolus* — Čatský and Tichá 1980; *Triticum* — Marshall and Biscoe 1980a, Winzeler and Nösberger 1980) (*cf.* Fig. 9.5). In sunflower, Hiroi and Monsi (1966) found higher $I_{sat}$ in the young and old leaf than in the mature one.

With leaf ageing, the approach to the saturation level of $P_N$ is usually faster so that the convexity of the $P_N$-$I$ curve is more evident (the intermediate part of the $P_N$-$I$ response curve is shorter) in the mature and senescing leaf than in the unfolding one (*Gossypium* — Constable and Rawson 1980; *Helianthus* — Rawson and Constable 1980). However, a sharp transition between the linear part of the curve and the saturating plateau of $P_N$ in the young tobacco (Haraguchi and Shimizu 1970) as well as no marked differences in the convexity of the curve during leaf ontogeny (*Helianthus* — Hiroi and Monsi 1966) may be shown.

$P_N$ of most leaves is fully saturated with $I_{sat}$ below 2000 $\mu$mol $m^{-2}$ $s^{-1}$, but leaves of some plants require a higher $I_{sat}$ (*e.g.* mature or old leaves of *Nicotiana* — Haraguchi and Shimizu 1970, *Colocasia* — Sato *et al.* 1978, and *Triticum* — Winzeler and Nösberger 1980).

Because of small differences, no clear and explicit pattern of age-dependent changes in $I_{comp}$ may be derived from graphical presentation of $P_N$-$I$ relationship: $I_{comp}$ either does not change (*Helianthus* — Hiroi and Monsi 1966; *Triticum* — Marshall and Biscoe 1980a, Winzeler and Nösberger 1980) or changes irregularly during leaf ontogeny (*Colocasia* — Sato *et al.* 1978), or declines with age up to leaf maturity (*Gossypium* — Constable and Rawson 1980) or full leaf senescence (*Nicotiana* — Haraguchi and Shimizu 1970). In *Calopogonium, Glycine, Pennisetum* and *Sorghum* (Ludlow and Wilson 1971), *Acer* (Malkina 1976b) and *Phaseolus* (Čatský and Tichá 1980), a strong decline of $I_{comp}$ from leaf unfolding to maturation followed by a subsequent increase (up to nearly the initial value, *e.g.* in *Phaseolus*) with further senescing of the leaf was found (Fig. 7.8, *top left*).

In an insertion gradient, the highest $P_{N\ max}$ and $I_{sat}$ is usually found in the young leaves situated at the top of a shoot or tree branch and the lowest ones in the old bottom leaves, *e. g.* in orange (Kriedemann 1971), mulberry (Murakami and Takeda 1973), rice (Sato and Kim 1980a), and barley (Biscoe *et al.* 1977). In mulberry (Murakami 1982), also $I_{comp}$ is higher with younger leaves. This pattern may be substantially modified with respect to the sequential emergence and unfolding of leaves during the vegetation period (*Populus* – Shiba 1978; *Oryza* — Sato and Kim 1980a; *Helianthus* — Rawson and Constable 1980). No clear differences in $P_N$—$I$ curves with leaf insertion levels were found in sorghum and tobacco by Turner and Incoll (1971). In soybean (Kumura 1969), sugar beet (Hodáňová

1979) and ivy (Bauer and Bauer 1980), the $P_N$—$I$ response changes with leaf insertion level are similar to those during leaf ontogeny (*e.g.* $P_{N\,max}$ is higher in leaves of intermediate insertion position than in the upper and bottom leaves; a higher $I_{sat}$ is also needed to saturate $P_N$ of the young and mature leaves than the old ones — Fig. 7.8, *top right*).

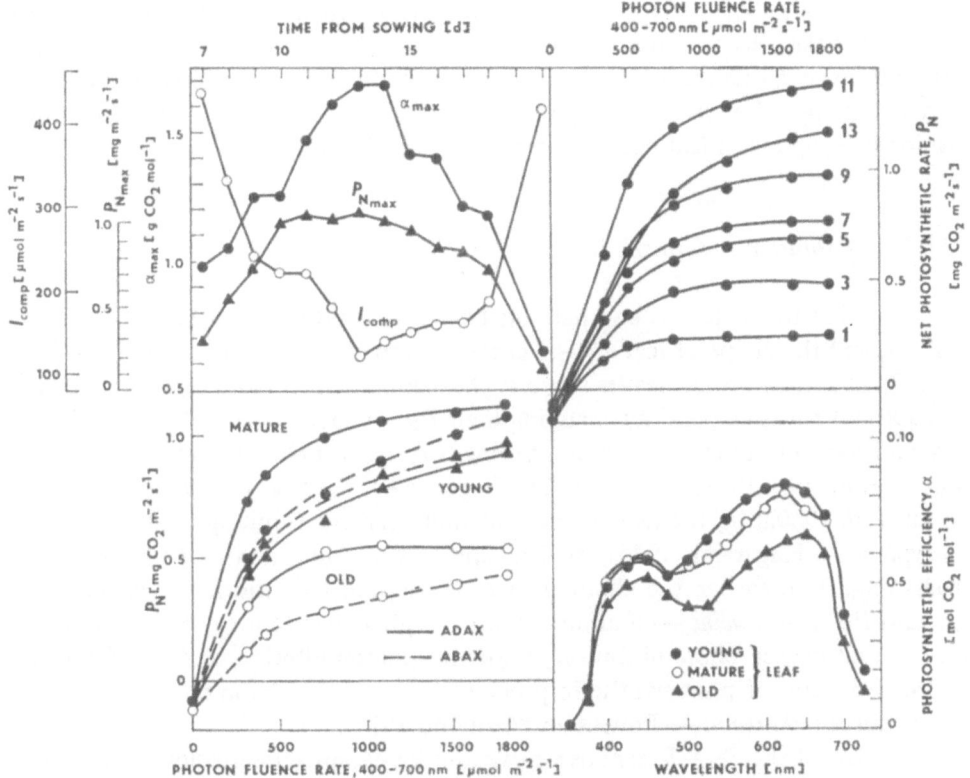

**Fig. 7.8.** The compensation irradiance ($I_{comp}$), the maximum photochemical efficiency ($\alpha_{max}$), and the maximum net photosynthetic rate ($P_{N\,max}$) under the saturating irradiance ($I_{sat}$) change during the development of primary leaves of *Phaseolus vulgaris* L. (*top left* — after Čatský and Tichá 1980). The dependence of $P_N$ on irradiance changes with leaf age (leaves numbered from the outer old to the inner young one) in *Beta vulgaris* L. (*top right* — after Hodáňová 1979). This dependence is not the same if leaves are irradiated from either the adaxial (adax) or abaxial (abax) leaf surface (*bottom left* — after Hodáňová, unpublished). The spectral dependence of photochemical efficiency, $\alpha$, is more expressed in young and mature than in old leaves of maize (*bottom right* — McCree 1971/1972).

In the literature, various mathematical approximations of the shape of $P_N$—$I$ curve are given (Michaelis-Menten equation, Blackman type curve, rectangular and non-rectangular hyperbola or asymptotic exponential) and compared (*e.g.*

Thornley 1976, Prioul and Chartier 1977, Ceulemans *et al.* 1980) and a possibility of their mutual transformations is estimated (Goudriaan 1979). Particular mathematical derivations fit experimental data to a different extent. Thus throughout wheat leaf development from maximum elongation until its complete senescence, the shape of the $P_N$—$I$ response curve remains constant and close to a Blackman type response rather than to a rectangular hyperbola (Marshall and Biscoe 1980b). Murakami and Takeda (1973) and Murakami (1982) conclude that it is difficult to express $P_N$ relatively to $I$ for mulberry leaves of different age and insertion level with one curve. Conformably to these findings, both the proportional and disproportional changes between single parameters of $P_N$—$I$ response curve which determine its shape with leaf age and insertion level may be assumed.

### 7.2.5.2 *Adaptation to Growth Irradiance*

An adaptation to shade and sun habitats brings about differences in $P_{N\ max}$, $I_{comp}$ and $I_{sat}$, and the shape of leaf $P_N$—$I$ curve as well (for detailed comparisons between $P_N$ of sun and shade leaves see, *e.g.*, Boardman 1977c, Wild 1979, and Björkman 1981). Leaves respond by changing their $P_N$—$I$ curves relatively to age, insertion, and transfer of plants to different growth $I$, *e.g.* from low/high to high/low regimes (*Glycine* — Bunce *et al.* 1977; *Fragaria* — Jurik *et al.* 1979; *Lolium* — Prioul *et al.* 1980a, b), the use of supplemental $I$ (*cf.* the thinning of the foliage in *Gossypium* — Nagarajah 1976), or the time of beginning and ceasing (duration) of shading. The earlier the shading of a developing leaf, the lower is its $I_{comp}$, $I_{sat}$, and $P_{N\ max}$ (*Glycine* — Kumura 1968). $I$ higher than that to which the plants are adapted causes photoinhibition; both the photoinhibition and the following reacclimatization of photosynthetic processes change the shape of the $P_N$—$I$ response curve (*Phaseolus* — Powles and Osmond 1979).

The shape of the $P_N$—$I$ response curve and its parameters are closely related to the anatomical structure of leaf blades. A bifacial mature and amphistomatous leaf irradiated from the adaxial surface usually attains the higher $P_{N\ max}$ under the lower $I_{sat}$ than the same leaf irradiated from the bottom (Fig. 7.8, *bottom left*). It also has a longer linear part of the $P_N$—$I$ curve with a more sharp transition between this part and the saturation plateau of $P_N$, *i.e.* a faster approach to reaching $P_{N\ max}$ (*Nicotiana* — Moss 1964; *Acer, Betula* — Malkina 1976a; *Syringa* — Oya and Laïsk 1976; *Picea* — Leverenz and Jarvis 1980). At the same time, the irradiated adaxial surface may exert the same, the increased or the decreased $I_{comp}$ in comparison with the abaxial surface. The direction of incident $I$ induces more obvious differences in $P_N$—$I$ curves in the mature and old leaves than in the young ones (Fig. 7.8, *bottom left*): this is linked with the rate of differentiation of leaf tissues across the leaf lamina. Differences in the structure, optical, diffusive and metabolic properties of palisade and spongy parenchyma are probably a reason for the non-identical leaf $P_N$—$I$ curves if either the palisade or spongy parenchyma acts as a shading tissue (*cf.* Sections 4.3.2 and 4.3.3).

Simultaneous irradiation from both sides with equal $I$ results in $P_{N\,max}$ higher than in the leaf irradiated unilaterally, but $I$ which is needed to saturate $P_N$ decreases (*Syringa* — Oya and Laïsk 1976; *Picea* — Leverenz and Jarvis 1980). In rice, equal irradiation of the adaxial and abaxial surfaces with the high $I_{sat}$ decreases leaf $P_{N\,max}$ (the photoinhibitory effect of excess $I$ — Tanaka and Matsushima 1970).

In comparison to a dorsiventral leaf, no differences in the $P_N$—$I$ response curve of the mature unifacial leaf irradiated from either the upper or lower surface were observed (*Zea* — Moss 1964). However, if $CO_2$ uptake by the adaxial and abaxial side of the maize leaf is measured separately, differences in their $P_N$—$I$ curves may be found (Bertsch and Domes 1969). While in maize the contribution of the adaxial leaf surface is lower than that of the abaxial one, in rice, the higher $P_N$—$I$ curve for the adaxial side of the mature leaf than for the abaxial one was reported (Tanaka 1972).

The longitudinal differences in morphologial and anatomical characteristics of an expanding leaf blade (internal gradient of differentiation of tissues along the leaf lamina) also cause the $P_N$—$I$ curve to vary: the leaf apex attains the highest $P_{N\,max}$ under the higher $I_{sat}$ than does the leaf base, in accordance with the progressive age of leaf tissues from the leaf base to the apex. This relation is maintained regardless of a different $I$ during plant growth but the proportions between basal, medial and apical zones change with leaf ageing (*Lolium* — Prioul *et al.* 1980a,b).

### 7.2.5.3 *Energy Conversion in Photosynthesis*

The relationship between the driving force and the reactant in photosynthesis, *i.e.* the number of quanta (400 to 700 nm) required to provide sufficient thermodynamic energy and the number of $CO_2$ molecules reduced (or $O_2$ molecules evolved) is characterized conventionally as the quantum requirement, or reciprocally, as the quantum yield. According to the photochemistry of photosynthesis, 8 to 10 absorbed quanta are required for the reduction of 1 molecule of $CO_2$ or, reversely, the primary acts enable the reduction of $0.1 - 0.125$ molecules of $CO_2$ per quantum absorbed. Similar characteristics calculated on the basis of the net $CO_2$ uptake rate by a leaf represent integrated parameters which are a result of balancing all energy requiring and supplying processes involved in and coupled to the photosynthetic $CO_2$ fixation process. These characteristics are called photosynthetic requirement ($q$) and photosynthetic efficiency ($\alpha$). In both $C_3$ and $C_4$ plants, 0.11 to 0.04 molecules $CO_2$ are fixed per quantum absorbed, *i.e.* about 9 to 23.8 quanta are required for the net uptake of 1 molecule of $CO_2$ by a leaf (Campbell and Black 1978, Björkman 1981) under the atmospheric $CO_2$ and $O_2$ pressure and optimum leaf temperature of 25 to 30 °C.

As follows from the graphical expression of leaf $P_N$—$I$ relationship, $\alpha$ is determined by the slope of the $P_N$—$I$ response curve. At low $I$, $P_N$ linearly depends on the capacity of leaf chloroplasts to absorb radiant energy and thus $\alpha$ is the highest ($\alpha_{max}$).

A decrease in $\alpha$ with the increasing $I$ up to the saturation value is usually attributed to the dominating effects of other factors than $I$ ($CO_2$, $O_2$, temperature) and other processes than those responsible for the trapping of $I$ ($CO_2$ diffusion inside the leaf, enzyme activity, etc.).

Both first-hand and derived data from the slopes of $P_N$—$I$ response curves indicate that the values of $\alpha_{max}$ may increase with leaf ageing to a maximum which occurs at different times of leaf ontogeny for different species and then decline (*Helianthus* — Hiroi and Monsi 1966; *Calopogonium, Glycine, Pennisetum, Sorghum* — Ludlow and Wilson 1971; *Acer* — Malkina 1976b; *Malus* — Watson and Landsberg 1979; *Coffea* — Yamaguchi and Friend 1979; *Phaseolus* — Tsuji et al. 1978a, b, Čatský and Tichá 1980) or remain relatively constant after the leaf elongation is completed (*Pennisetum* — McPherson and Slatyer 1973; cereals — Takeda and Udagawa 1976; *Gossypium* — Constable and Rawson 1980; *Triticum* — Marshall and Biscoe 1980b). In coffea plants grown in a growth chamber, higher $\alpha_{max}$ were reported in the young and also in senescing leaves (Yamaguchi and Friend 1979).

In conifers, photosynthetic quantum requirements increased with needle age: in pine from 8.8 to 23.0 (El Aouni and Mousseau 1974), in 3 year-old fir trees from 15 to 17 and in 2 year-old spruce trees from 13 to 17 (Auclair and Gaudillère 1975).

Changes in $\alpha_{max}$ in leaves of different insertion levels are often similar to its changes during leaf ontogeny (*Beta* – Hodáňová 1979). In rice, $\alpha_{max}$ gradually increases from the lower to the upper leaves (Sato and Kim 1980a) while in birch it decreases with higher leaf position (Öquist et al. 1982). No differences between $\alpha_{max}$ for leaves of different ages were found in leaves of *Colocasia* (Sato et al. 1978) and soybean leaves, *Calopogonium* runners, and sorghum and millet tillers (Ludlow and Wilson 1971).

Lower $\alpha_{max}$ in both the juvenile and the senescing leaf is partly linked to a lower absorptance of PhAR by the leaf at the beginning and end of its life span. Irregularities and shifts in $\alpha_{max}$ with leaf insertion level arise — to a great extent — from the acclimation to different irradiances to which the subsequent leaves are exposed (*Glycine* — Kumura 1969).

**7.2.5.4** *Photosynthesis, Energy Conversion and Spectral Irradiance*

As follows from the spectrum of leaf absorptance and physical nature of radiation, neither the action spectrum of $P_N$ nor the spectral $\alpha$ expressed both per quanta of the absorbed $I$ or per unit energy may be flat over the entire PhAR region.

The action spectrum of $P_N$ expressed per quanta of radiation is usually characterized by two peaks: a relatively broad one in the red (at about 620 nm) and a sharp one in the blue (440 nm) region. The height of the blue maximum may be the same as that of the red one (radish – Bulley *et al.* 1969; herbaceous plants – Inada 1976) or it may be lower by almost 30% (*Phaseolus* – Balegh and Biddulph 1970; field crops — McCree 1971/1972; trees – Inada 1976).

When calculated per quantum of $I$, the spectral curve of α has a similar shape as the action spectrum of $P_N$ (Inada 1976). The absolute values of the spectral α vary from 0.054 to 0.076 between single wavelengths of PhAR and with different plant species (McCree 1971/2): the red region of PhAR is the most effective to $P_N$ while the efficiency in the blue region is approximately 30% lower. Spectral α values are generally higher in young than old maize leaves, the greatest differences with leaf age occurring between 500 to 650 nm (Fig. 7.8, *bottom right*), but the relative spectral α (normalized at 620 nm) is similar for both leaves (McCree 1971/1972). In mandarin, Inada (1977) did not find any appreciable difference between the relative spectral α of the young and old leaves, but in tea slightly higher values for the old leaf were observed throughout the entire PhAR region. Also the sequential ageing of leaves subsequently produced during plant development leads to variance of the spectral α (*Oryza* — Inada 1977).

The effects of leaf adaptation to growth $I$ are very expressed in plant native habitats. Thus the irradiance of the upper leaves from sun and deep shade plants may differ more than 300 times. In addition, the lower leaves may recieve only 5 to 10% of $I$ incident on the leaves of the uppermost insertion levels. While penetrating through the canopy, also the proportion between direct and diffuse components of radiation changes (frequency in alternating sunflecks, penumbra and shadows) as well as the spectral composition of $I$. Hence, the photosynthetic capacity of all leaves on a plant has to be sufficiently large to allow an adequate response to the variability of leaf irradiance: differences in $P_N$—$I$ response curves, conversion characteristics and action spectra of $P_N$, relatively to leaf age and insertion level, are then a useful tool how to utilize most efficiently the radiant energy available along the microclimate profile.

## 7.2.6    Carbon Dioxide Concentration

As was already stated by Gaastra (1959), "under saturating photon flux densities $CO_2$ availability becomes the primary limiting factor to photosynthetic $CO_2$ assimilation". Under atmospheric $CO_2$ concentrations, photosynthesis is mostly not saturated with $CO_2$ as is indicated by analysing the $CO_2$ dependence of $P_N$, and $CO_2$ is, therefore, an important limitation to plant photosynthetic production. In principle, short term and long term effects of $CO_2$ on photosynthesis, and also during leaf development, have to be distinguished. Thus under short term effects the influence of actual $CO_2$ concentration on $P_N$, and under long term effects the influence of growing plants at different $CO_2$ concentrations on $P_N$ are understood.

### 7.2.6.1 *Short Term $CO_2$ Effects on Net Photosynthetic Rate*

The $CO_2$ dependence of net photosynthetic rate ($P_N$—$CO_2$ curve) is a more or less sigmoidal curve with a long linear part (Tichá *et al.* 1980). The initial linear relationship between $CO_2$ concentration and $CO_2$ uptake is followed by a part where $P_N$ is no more proportional to $CO_2$ concentration, and finally, $CO_2$ uptake is nearly independent of $CO_2$ concentration. The linear part of the plot of $P_N$ (ordinate) to ambient $CO_2$ concentration $c_a$ (abscissa) is characterized by two of these three parameters: (1) the $CO_2$ compensation concentration, $\Gamma$, *i.e.* the $CO_2$ concentration at which $P_N = 0$, determined as the intersection of the $CO_2$ response curve with the abscissa (see Section 7.2.6.2), (2) the $CO_2$ evolution rate of the leaf in the light including photorespiration rate and rate of "dark" respiration not inhibited by light, obtained as the rate at zero $CO_2$ concentration (intersection with the ordinate), and (3) the overall conductance for $CO_2$ transfer given by the slope of the $CO_2$ response curve ($P_N/c_a$).

If the conductance for $CO_2$ transfer in the gaseous phase is known (*cf.* Chapter 8) the intercellular $CO_2$ concentration can be calculated and the $P_N$ dependence on intercellular $CO_2$ concentration ($c_i$) analysed; then the effects of stomatal and boundary layer conductances are eliminated. The $P_N$—$c_i$ response curve leads to the same value of $CO_2$ compensation concentration as the $P_N$—$c_a$ curve, but the intersection of the curve with the ordinate indicates the rate of $CO_2$ evolution in the light from mesophyll cells into intercellular spaces at zero intercellular $CO_2$ concentration. The slope of the linear relationship between $P_N$ and $c_i$ is a measure of the "carboxylation efficiency" corresponding to the mesophyll or intracellular conductance determined as residual conductance (*cf.* Chapter 8).

$\rightarrow$

**Fig. 7.9.** The $CO_2$ dependence of net photosynthetic rate ($P_N$) during leaf ontogeny (upper part) in primary leaves of *Phaseolus vulgaris* L. The linear parts of the relationship $P_N$-ambient $CO_2$ concentration between 50 and 1100 mg m$^{-3}$ were characterized by the intersection at $P_N$ equalling zero called the $CO_2$ compensation concentration ($\Gamma$), the extrapolated rate at zero $CO_2$ evolution from the leaf in light ($R'_L$), and the slope of the $CO_2$ dependence of $P_N$ representing the overall conductance for $CO_2$ transfer ($g$). During leaf ontogeny $\Gamma$ decreased rapidly in the expanding leaf, held for a relatively long period a constant value, and increased in the senescing leaf (*lower part* – a); $R'_L$ decreased to a level which is maintained also in the senescing leaf (*lower part* – b); $g$ increased to a maximum which coincided with maximum $P_N$, followed by a slower decline (*lower part* – c). Patterns of ontogenetic changes in $P_N$ at different ambient $CO_2$ concentrations (500, 600, 800 and 1000 mg m$^{-3}$) were analogous, with a maximum $P_N$ on about the 11th day after sowing (*lower part* — d). (After Tichá *et al.* 1980.)

190

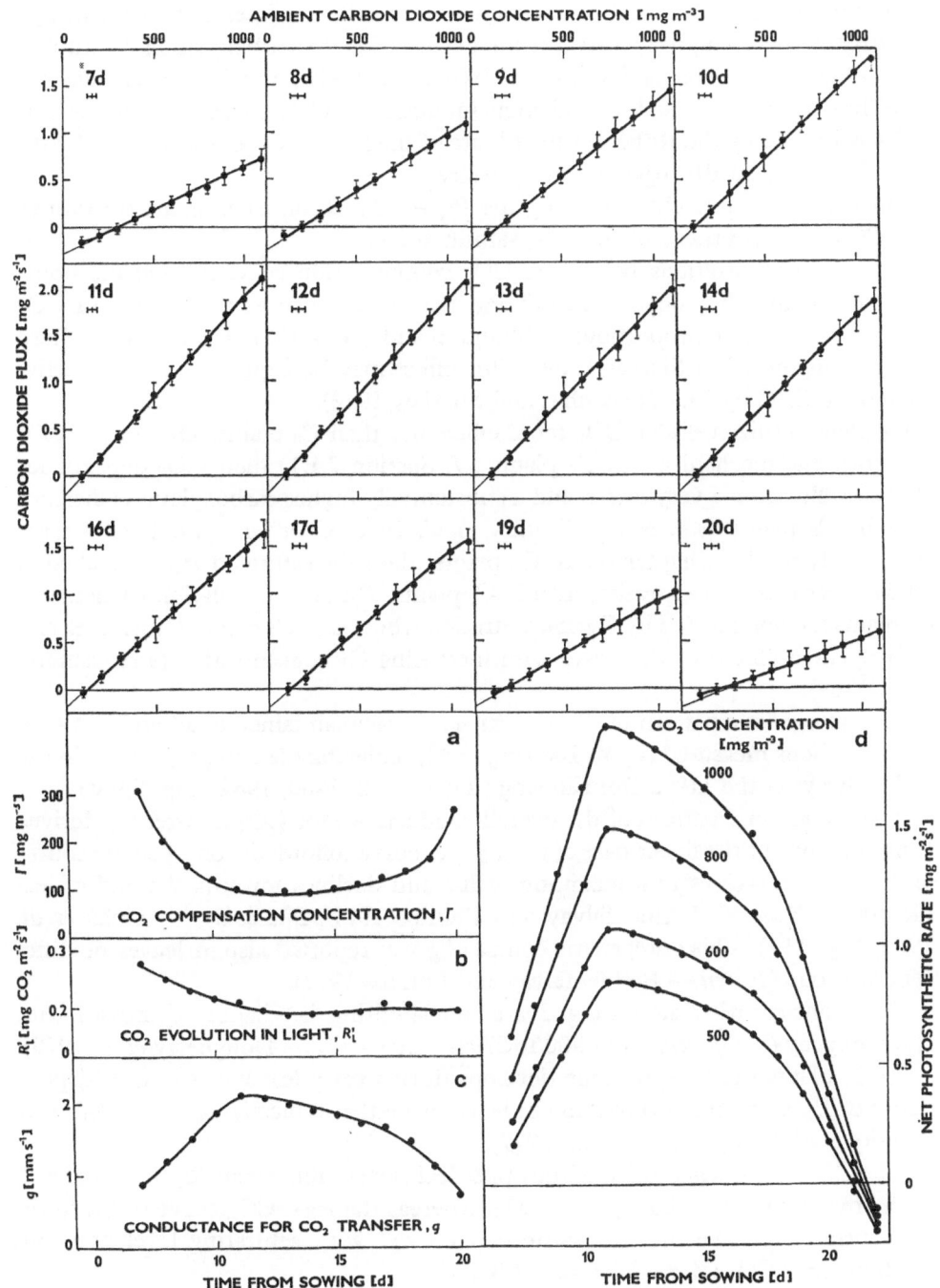

AMBIENT CARBON DIOXIDE CONCENTRATION [mg m⁻³]

CO₂ COMPENSATION CONCENTRATION, Γ

CO₂ EVOLUTION IN LIGHT, R'ₗ

CONDUCTANCE FOR CO₂ TRANSFER, g

CO₂ CONCENTRATION [mg m⁻³]

TIME FROM SOWING [d]

191

The carboxylation efficiency in $C_3$ plants is proposed to be proportional to the activity of RuBPC in the leaf. At higher $c_i$, the capacity to regenerate RuBP becomes limiting. However, net photosynthetic rate still increases somewhat with $c_i$ as RuBP is increasingly diverted from oxygenation to carboxylation (Farquhar et al. 1980, Caemmerer and Farquhar 1981). Thus, $CO_2$ uptake rate is governed by different processes at low and high intercellular $CO_2$ concentrations: at low $c_i$, $P_N$ is limited by the RuBP saturated rate of the RuBPCO, and at high $c_i$ by the rate allowed by RuBP regeneration capacity.

Further parameters characterizing the $P_N$—$CO_2$ response curves are the saturating $CO_2$ concentration and the $CO_2$ saturated $P_N$.

At $CO_2$ concentrations below the $CO_2$ compensation concentration the linear character of the $CO_2$ dependence of $P_N$ may be changed as the decline in $CO_2$ evolution is no more proportional (Holmgren and Jarvis 1967, Heath and Orchard 1968, Troughton and Slatyer 1969). This effect may be explained, e.g., by activation of RuBPC by $CO_2$ (Farquhar and Sharkey 1982).

$C_4$ plants utilize cellular $CO_2$ more efficiently than $C_3$ plants. Due to the $CO_2$ concentrating mechanisms in $C_4$ plants (cf. Section 7.1.1) their $CO_2$ dependence of $P_N$ results from a very low $\Gamma$ and a substantially higher carboxylation efficiency than in $C_3$ plants (e.g. Morot-Gaudry et al. 1976, Chartier et al. 1977, Pearcy et al. 1981, cf. also Chapter 6). In $C_4$ plants, the $CO_2$ saturated $P_N$ is reached at much lower $CO_2$ concentration than in $C_3$ plants. $\Gamma$ and carboxylation efficiency in $C_4$ plants are not affected by $O_2$ concentration whereas in $C_3$ plants $\Gamma$ increases and carboxylation efficiency decreases with increasing $O_2$ concentration (e.g. Osmond et al. 1969).

The ontogenetic pattern of $P_N$ in Phaseolus was maintained at all ambient $CO_2$ concentrations measured (up to $1000 \text{ mg m}^{-3}$), achieving the ontogenetic maximum in $P_N$ always at the 11th d from sowing (Tichá et al. 1980, 1984; Fig. 7.9 d).

The ontogenetic pattern of the overall conductance for $CO_2$ transfer ($g$) derived from the slope of the linear part of the $P_N$—$c_a$ curve follows the ontogenetic course of $P_N$, i.e. $g$ increases to a maximum value and declines towards the end of leaf life span (Pisum — Bethlenfalvay and Phillips 1977; Phaseolus — Tichá et al. 1980; Fig. 7.9c). This ontogenetic course of $g$ was reported also in leaves of different insertion (Pisum — Bethlenfalvay and Phillips 1977).

$CO_2$ concentration saturating $P_N$ at non-limiting irradiances decreases with leaf ontogeny (in Quercus from 900 to $670 \text{ mg m}^{-3} CO_2$ — Dougherty et al. 1979).

In $C_4$ plants the $C_4$ syndrome develops during early leaf life: the $CO_2$ dependence of $P_N$ in young (8-d old) maize leaves is neither typically $C_4$ nor $C_3$ (Morot-Gaudry et al. 1976, Chartier et al. 1977).

The $CO_2$ dependence of $P_N$ of the upper leaf surface follows an "optimum curve" with a maximum near $600 \text{ mg m}^{-3} CO_2$, whereas the $P_N$—$CO_2$ curve of the lower leaf surface approximates a "saturation curve" with saturating level at about $5400 \text{ mg m}^{-3} CO_2$ (Zea — Domes 1971).

### 7.2.6.2  Carbon Dioxide Compensation Concentration

The carbon dioxide compensation concentration (formerly the $CO_2$ compensation point), $\Gamma$, is reached in an illuminated leaf when $P_N = 0$. It represents an equilibrium between the $CO_2$ consuming processes (photosynthesis) and the $CO_2$ producing processes (photorespiration, "dark" respiration) (for reviews see Smith *et al.* 1976, Sirohi and Shrivastava 1978, and Canvin 1979). At given environmental conditions, $\Gamma$ is remarkably constant within and between most plant species (Moss 1971, Krenzer *et al.* 1975). Therefore, $\Gamma$ is often used for a quick classification of the $CO_2$ fixation type: usual values of $\Gamma$ at high irradiances, 25 °C, and 21% $O_2$ are for $C_3$ plants between 70 and 180 mg m$^{-3}$ (*e.g.* Black 1973, Krenzer *et al.* 1975, Zelitch 1975), for $C_4$ plants between 0 and 40 mg m$^{-3}$ (*e.g.* Black 1973, Zelitch 1975, Canvin 1979). $\Gamma$ of the so-called $C_3$–$C_4$ intermediate plants is between the values for $C_3$ and $C_4$ plants (25 to 70 mg m$^{-3}$; *e.g.* Apel and Peisker 1979, Raghavendra 1980; *cf.* Fig. 7.10 c). $\Gamma$ is highly sensitive to temperature, rising sharply above 25 °C (Fig. 7.10 b). The sharp temperature dependence is the main reason why $C_3$ plants lose a considerable part of the sugar phosphates, formed in photosynthesis, mainly on warm days with bright sunlight when growing conditions should be very favourable (*cf.* Bassham 1977, Woolhouse 1978).

$\Gamma$ is also used as an indicator of photorespiratory activity in plants since higher $\Gamma$ may indicate a higher photorespiration rate (*e.g.* Zelitch 1975, Lurie *et al.* 1979; *cf.* Chapter 9).

Leaf ontogeny: During leaf expansion in $C_3$ plants, $\Gamma$ falls from very high values to a relatively constant lower value (for review see Tichá and Čatský 1981; Table 7.4 and Fig. 7.10 a). This trend is consistent with the increasing ratio of photosynthetic to respiratory processes during the expansion of the leaf: $P_N$ increases to a maximum, while $R_D$ declines from a rather high initial rate to a minimum (*e.g. Phaseolus* — Čatský *et al.* 1976). Low conductances for $CO_2$ transfer in very young leaves (*cf.* Chapter 8) decrease their gas exchange. The resulting high internal $CO_2$ concentration may reduce photorespiratory activity (*Nicotiana* — Salin and Homann 1973). During further leaf development the conductances for $CO_2$ transfer and photorespiration rate increase (Fig. 10.3).

Hence, $\Gamma$ reaches a relatively constant minimum already before gaining full leaf expansion. The minimum lasts usually over a considerable part of leaf ontogeny (*e.g., Hordeum* — Fair *et al.* 1972, 1973; *Populus* — Dickmann and Gjerstad 1973; *Phaseolus* — Čatský *et al.* 1976, Smith *et al.* 1976; *Rosa* — Bozarth *et al.* 1982).

Thereafter, during the senescence of leaf, $\Gamma$ starts to rise again (*Populus* — Dickmann and Gjerstad 1973; *Phaseolus* — Čatský *et al.* 1976, Smith *et al.* 1976, O'Toole *et al.* 1977, Čatský and Tichá 1979, Tichá *et al.* 1980; *Pisum* — Bethlenfalvay and Phillips 1977; *Hevea* — Samsuddin and Impens 1979a; *Lolium* —

Azcón-Bieto et al. 1981; *Malus* — Kennedy and Johnson 1981). In this period, the main contribution to the rise in $\Gamma$ is an abrupt decline in $P_N$ and slow decrease in photorespiration rate, but an increase in "dark" respiration rate in light; the intracellular conductance for $CO_2$ transfer also declines (*e.g. Phaseolus* — Čatský et al. 1976, Peisker et al. 1981; *Lolium* — Azcón-Bieto et al. 1981).

The above course of ontogenetic changes in $\Gamma$ was found in all leaves of different insertion levels in pea plants (Bethlenfalvay and Phillips 1977).

Frequently only a part of leaf ontogeny was studied, and therefore a decrease in $\Gamma$ was reported when differences between young and mature leaves were determined (*e.g.*, *Hordeum* — Fair et al. 1974; *Asparagus* — Downton and Törökfalvy 1975; *Quercus* — Dougherty et al. 1979; *Bougainvillea* — Even-Chen and Sachs 1980), or no significant differences in $\Gamma$ were detected in middle-aged leaves (*e.g.*, *Pisum* — Poskuta et al. 1975; *Phaseolus* — Wort 1976; *Capsicum* — Hall and Brady 1977; *Glycine* — Sambo et al. 1977; *Sinapis* — Wild and Höhler 1978). In the period before and during senescence $\Gamma$ increases (*e.g.*, *Triticum* — Peisker and Apel 1976, Feller and Erismann 1978; *Phaseolus* — Raggi 1978a, b, 1980; *Cyperus* and *Poa* — Imai and Murata 1979). No differences in $\Gamma$ could be detected between juvenile and adult ivy leaves (Bauer and Bauer 1980).

Due to the high temperature sensitivity of $\Gamma$, the differences between leaves of different ages are small at 20 °C and very substantial at 35 °C even if the ontogenetic course of $\Gamma$ is similar in shape at different leaf temperatures (*Phaseolus* — Tichá, unpublished; Fig. 7.10 b).

A rise in $\Gamma$ during leaf ontogeny was accelerated, *e.g.*, by the removal of flowers (*Capsicum* — Hall and Brady 1977) or infection of leaves (*Phaseolus* — Raggi 1978a, b, 1980).

---

→

**Fig. 7.10.** Carbon dioxide compensation concentration ($\Gamma$) is high in very young leaves and declines rapidly to a shorter or longer (according to life span of the leaf) stable level. In old leaves an increase in $\Gamma$ is found again (a — *Hevea brasiliensis* Muell. Arg., after Samsuddin and Impens 1979a; *Malus domestica* Borkh. — broken line — after Kennedy and Johnson 1981; first trifoliate leaflets of *Phaseolus vulgaris* L., after O'Toole et al. 1977; ninth leaf from the bottom of *Pisum sativum* L., after Bethlenfalvay and Phillips 1977). Leaf ontogeny is indicated in days or as leaf plastochron index. $\Gamma$ is extremely sensitive to temperature; even if the ontogenetic course of $\Gamma$ is similar at different leaf temperatures, the differences between young and old leaves are small at 20 °C, and substantial at 35 °C. The older the leaf, the more affected is $\Gamma$ by temperature (b — primary leaves of *Phaseolus vulgaris* L., Tichá, unpublished). $\Gamma$ in leaves of $C_3$ plants is higher, by about 10 times or more, than in $C_4$ plants; the values of $C_3$—$C_4$ intermediate plants are intermediate between these two limits. $\Gamma$ is thus often used as a quick indication of the $CO_2$ fixation type (see Section 7.2.6.2). Upper leaves usually have higher $\Gamma$ than leaves of middle insertion level in $C_3$ and $C_3$-$C_4$ intermediate plants; the tendency to an increase in $\Gamma$ in the lowest leaves can be seen in $C_4$ plants, too, as well as in $C_3$ and $C_3$-$C_4$ intermediate plants (c — *Moricandia arvensis* (L.) DC., plants at the end of the vegetative phase, from Apel et al.1978; *Pisum sativum* L., 40 d old plants, from Bethlenfalvay and Phillips 1977; *Populus euramericana* Hardtwalden, from Furukawa 1973a; *Zea mays* L., 60 d old plants, after Williams and Kennedy 1977, and 21 d old plants, after Crespo et al. 1979).

C$_4$ plants are characterized by a very low $CO_2$ compensation concentration; in comparison to C$_3$ plants, a similar but less pronounced trend of changes in $\Gamma$ during leaf ontogeny was found: a decrease of $\Gamma$ in young leaves followed by an increase in senescing leaves (*e.g.*, *Thyridolepis* and *Cenchrus* — Christie 1975;

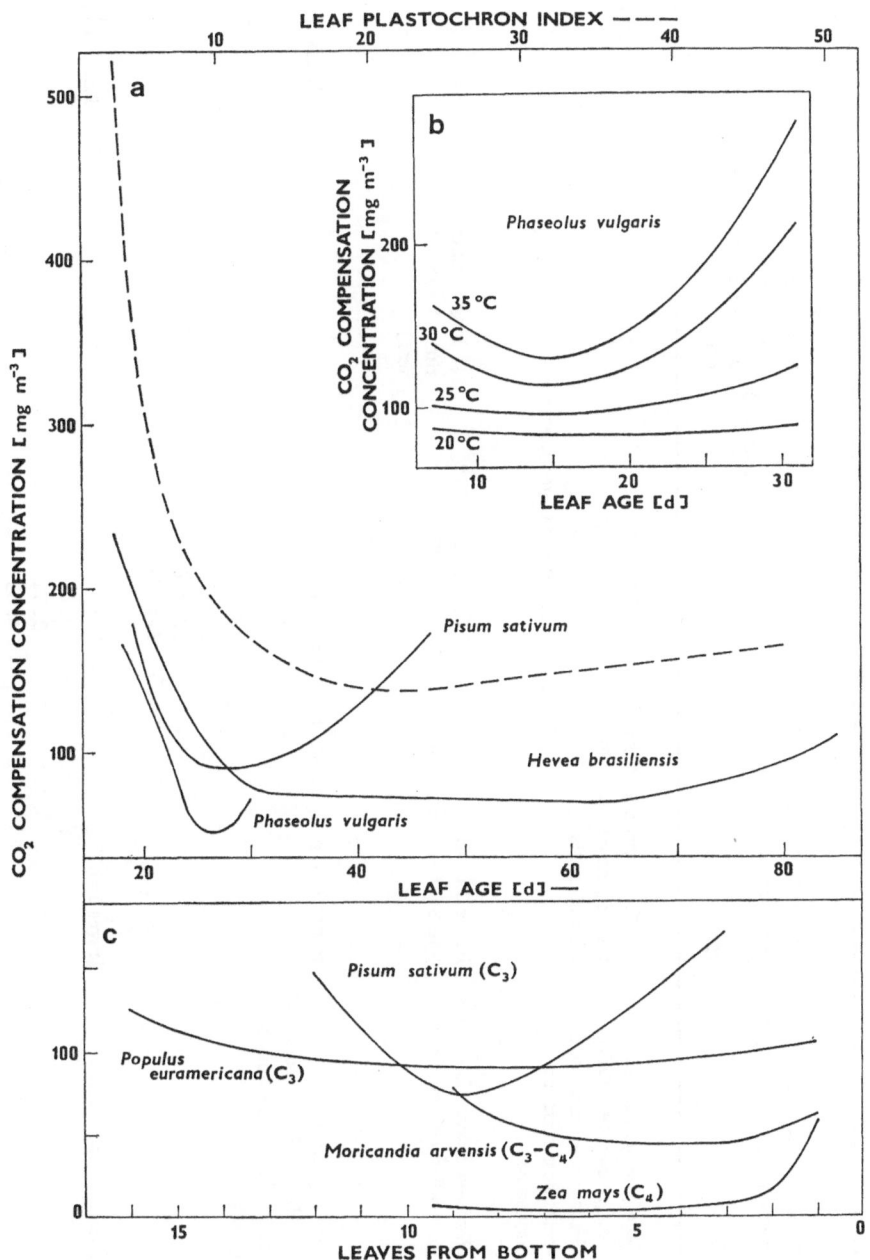

**Table 7.4** Changes in $CO_2$ compensation concentration ($\Gamma$) during leaf ontogeny and in leaves of different insertion levels. Explanations: $\downarrow$ decrease, $\uparrow$ increase.

| Plant species (and cultivation) (1) | Age of leaf (numbered from the oldest one) (2) | $\Gamma$ [mg m$^{-3}$] (data on line = ontogeny, data in column = insertion) (3) | Changes in $\Gamma$ induced by ageing or by insertion level (from upper to lower leaves) (4) | Author(s) and year of publication (5) |
|---|---|---|---|---|
| *Capsicum annuum* L. var. *grossum* cv. Market Giant (soil–coarse sand- -peat moss mixture, glasshouse) | leaf one internode above the first flower, 0 to 84 d after anthesis | 172...154...163 (fruiting plants) 172...217...199 (plants with removed flowers) | $\downarrow$, $\uparrow$ $\uparrow$, $\downarrow$ | Hall and Brady 1977 |
| *Hordeum vulgare* L. (controlled environment) | leaves 1 to 4, 6 to 36 d after planting | leaf 1: 123...101...118 2: $\pm$ 109 3: 136...101 4: 141...103 | $\downarrow$, long minimum, then $\uparrow$ (only 1st leaf) | Fair *et al.* 1972 |
| *Medicago sativa* L. cv. Hunter River (pots with sand, Hoagland solution, glasshouse) | leaves 2, 4, 6, 8, 10, 12, 14 | leaf 2: 72 4: 127 6: 127 8: 123 10: 127 12: 116 14: 138 | $\downarrow$ | Hodgkinson 1974 |
| *Nicotiana rustica* L. | leaves 1 to 7 | leaf 1: 83 2: 85 3: 71 4: 43 5: 34 6: 36 7: 34 | $\uparrow$ | Lurie *et al.* 1979 |

| Species (growth conditions) | Leaf/position | Values | Direction | Reference |
|---|---|---|---|---|
| *Phaseolus vulgaris* L. cv. Jantar and Harzgruß (pots with sand, nutrient solution, growth cabinet) | primary leaves, 7 to 20 d | 21% $O_2$: 306...95...270<br>2% $O_2$: 26...10...27 | ↓, ↑<br>↓, ↑ | Čatský *et al.* 1976, Čatský and Tichá 1979, Tichá *et al.* 1980, Peisker *et al.* 1981 |
| *Phaseolus vulgaris* L. cv. Harvester (soil-vermiculite, Hoagland solution, greenhouse) | 2nd trifoliate leaf, 1 to 7 weeks | 348...109...181 | ↓, ↑ | Smith *et al.* 1976 |
| *Pisum sativum* L. cv. Alaska (pots with vermiculite, growth chamber) | leaves 3, 6, 9, 12, 20 to 70 d from planting | leaf 3: 217...109...290<br>6: 136...105...226<br>9: 172...91...181<br>12: 167...109...136 | ↓, ↑<br>↓, ↑<br>↓, ↑<br>↓, ↑ | Bethlenfalvay and Phillips 1977 |
| *Populus deltoides* Marsh. (rooted cuttings, growth chamber, greenhouse, field) | leaf plastochron age from 5 to 35 (L.P.A. 29 to 35 = oldest leaves) | 471...109...580 | ↓, ↑ | Dickmann and Gjerstad 1973 |
| *Populus* × *euramericana* cv. Wisconsin-5 (pots, growth chambers) | leaf 8, LPI 4, 8, 19, 36 | 181...107...96...123 | ↓, ↑ | Dickmann and Gordon 1975 |
| *Quercus alba* L. (natural locality) | 21, 75, 90 and 100% of the final mature leaf size (24th April, 14th May, 26th May, 3rd June) on intact branches from upper fourth of the crown of 59 years old tree | 724...145...154...76 | ↓ | Dougherty *et al.* 1979 |
| *Vitis vinifera* L. cv. Sultana (rooted cuttings, field) | 6 positions on the shoot | 554<br>.<br>.<br>91 | ↓ | Kriedemann 1968 |

*Cyperus* and *Poa* — Imai and Murata 1979), or no changes in $\Gamma$ with ageing of leaves (*e.g. Atriplex* — Slatyer 1970; *Zea* — Morot-Gaudry *et al.* 1976) when short periods of leaf ontogeny were studied.

Estimations of $\Gamma$ in CAM plants are difficult because there is scarcely a time during the diurnal cycle when gas exchange approaches steady state. In addition to this, an endogenous circadian rhythm of $\Gamma$ was observed (*Bryophyllum* — Jones and Mansfield 1972; *Kalanchoë* — Allaway *et al.* 1974). During the trough of the daily oscillation (around midnight) $\Gamma$ is as low as 0 to 10 mg m$^{-3}$, during the maximum (around midday) it reaches 110 to 180 mg m$^{-3}$ (Kluge and Ting 1978) or 360 mg m$^{-3}$ (Black 1973). Hence, the variations in $\Gamma$ due to leaf ontogeny or leaf insertion in CAM plants are not yet known. Only Allaway *et al.* (1974) found in the fourth leaf from the apex in *Kalanchoë* a slow rise in $\Gamma$ from June to the end of July followed by a decrease towards November.

Submersed aquatic macrophytes have higher $\Gamma$ in winter (above 90 mg m$^{-3}$) than in summer (30 to 80 mg m$^{-3}$) (*Hydrilla* — Bowes *et al.* 1978, 1979, Holaday and Bowes 1980). In summer their $P_N$ increases by about 65% and respiration (photorespiration rate, but mainly "dark" respiration rate) decreases considerably. Furthermore, the RuBPC/PEPC ratio changes between winter and summer: low $\Gamma$ plants fix some $CO_2$ into $C_4$ acids, which can later be decarboxylated and re-fixed. This is advantageous in summer lake environments where the $CO_2$ levels are high at night but low during the day (Holaday and Bowes 1980).

Leaf insertion: The highest $\Gamma$ in $C_3$ plants is found in the upper leaves (*e.g. Vitis* — Kriedemann 1968; *Populus* — Furukawa 1973a; *Nicotiana* — Salin and Homann 1973; *Medicago* — Hodgkinson 1974; *Lycopersicon* — Khavari-Nejad 1980); intermediate or low values in the more or less numerous leaves of middle insertion levels, and the lowest or somewhat higher $\Gamma$ in the lower leaves (Table 7.4, Fig. 7.10 c).

Analysing insertion gradients of $\Gamma$ during the whole plant ontogeny, in young plants (25 d from sowing) a decrease in $\Gamma$ from the upper to lower leaves, in 40 d-old plants a decrease and increase, and in 50 or 60 d-old plants only an increase in $\Gamma$ from the upper towards lower leaves is found (*Pisum* — Bethlenfalvay and Phillips 1977). Thus, insertion gradients in $\Gamma$ change with plant age as the insertion gradients in $P_N$ do (*cf.* Section 7.2.2).

An increase in $\Gamma$ to values unusual for $C_4$ plants in lower maize leaves was reported (Williams and Kennedy 1977, Crespo *et al.* 1979). On the other hand, Imai and Murata (1979) found that senescent leaves of $C_4$ grasses retained their $C_4$ photosynthetic characteristics (low $\Gamma$) until near death.

In natural $C_3$–$C_4$ intermediate plants the differences in $\Gamma$ of leaves of different insertion levels resemble those of $C_3$ plants (*Moricandia* — Apel *et al.* 1978; Fig. 7.10 c). Also in these plants, possible changes in the type of carboxylation mechanism in leaves of different insertion levels were suggested (Raghavendra *et al.* 1978, Raghavendra 1980). According to other authors, there is no evidence that the carbon assimilation pathway in leaves of one plant may be essentially changed (Osmond *et al.* 1982).

Differences on the leaf blade: In the tip half of the upper tobacco leaves lower $\Gamma$ than in the base half was found (Homann 1975).

### 7.2.6.3 *Long Term $CO_2$ Effects ($CO_2$ Enrichment)*

$CO_2$ enrichment of air during plant growth substantially increases growth rate of plants and their $P_N$. Furthermore, morphological, histological, anatomical and cytological characteristics as well as biochemical and physiological processes are affected in plants grown at elevated $CO_2$ concentrations which may influence $P_N$ (for reviews see, *e.g.*, Madsen 1976, Lemon 1983).

The increase in $P_N$ due to the increased $CO_2$ concentration may be explained by the steeper $CO_2$ gradient between ambient atmosphere and carboxylation centres in the chloroplasts which results in a better $CO_2$ supply to the carboxylation centres. A long-term action of elevated $CO_2$ concentration does not affect the biosynthesis of RuBPCO in the leaves (*Brassica* — Andreeva *et al.* 1979). The leaves cannot therefore adapt to higher $CO_2$ concentrations at the level of RuBPCO. However, RuBPCO is present in leaves in a sufficient amount to process $CO_2$ during a short-term increase in $CO_2$ concentration resulting in a corresponding increase in $P_N$. The effect of $CO_2$ enrichment varies with leaf irradiance — the higher the irradiance the smaller the $CO_2$ effects. $CO_2$ enrichment is thus very important mainly under limiting irradiance (Andreeva *et al.* 1979). A further effect of $CO_2$ enrichment is the increased water-use-efficiency of $C_3$ plants and the suppressed photorespiration rate (Tinus 1974, Goudriaan and Ajtay 1979).

For primary leaves of *Phaseolus vulgaris* an attempt was made to compensate the lower $P_N$ in very young and senescent leaves by gradual $CO_2$ enrichment according to the ontogenetic course of $P_N$ (Tichá and Čatský 1982).

The effect of ambient $CO_2$ concentrations between 2170 and 4350 mg m$^{-3}$ on $P_N$ depends on the duration of $CO_2$ enrichment (*Cucumis* — Aoki and Yabuki 1977). $P_N$ of lower leaves decreased rapidly as the enrichment continued, although upper leaves retained $P_N$ higher than those of non-enriched plants (*Cucumis* — Aoki and Yabuki 1977). $CO_2$ enrichment from 580 to 2720 mg m$^{-3}$ stimulated mainly $P_N$ in the fifth uppermost leaf of cucumber; the insertion gradient with a maximum $P_N$ in the third or fourth leaf changed therefore under 2720 mg m$^{-3}$ $CO_2$ into a gradient with maximum $P_N$ in the uppermost fifth leaf (Frydrych 1976). On the contrary, soybean plants grown in air with 1810 mg m$^{-3}$ $CO_2$ had a lower $P_N$ than those grown in ambient air (545 mg m$^{-3}$ $CO_2$); only the uppermost leaf had a higher $P_N$ (Hofstra and Hesketh 1975). The insertion gradient in $P_N$ was not changed by $CO_2$ enrichment (from 545 to 1810 mg m$^{-3}$ $CO_2$) in tomato where $P_N$ of leaves 2 to 14 simultaneously increased about 1.5 times (Ito 1973).

## 7.2.7    Temperature

The temperature dependence of net photosynthetic rate in higher plants follows the well known optimum curve resulting from the differing temperature effects on the rates of carboxylation, photorespiration and partly light-inhibited "dark" respiration. There is an optimum temperature range, dependent upon species, ecotypes and environment within which $P_N$ is highest; as the temperature increases or decreases beyond this range, $P_N$ decreases until the limits are reached where $CO_2$ output equals intake. Beyond these limits, net $CO_2$ evolution occurs even in the light.

Most plants possess a relatively broad temperature optimum associated with the temperature conditions of the natural environment: for C$_4$ plants the range is 30 to 40 °C whereas for C$_3$ plants 10 to 25 °C (*cf.* Table 7.1; for review see Berry and Raison 1981).

The temperature of a leaf is determined by its energy budget:

$$R_{abs} = L_e + H + \lambda E + M$$

where $R_{abs}$ is the flux of absorbed radiation, $L_e$ is the flux of emitted radiation $H$ is the sensible heat loss, $E$ is transpiration rate, $\lambda$ is the latent heat of vapourization for water, and $M$ the energy release or storage by chemical reaction (photosynthesis and other metabolism) (see Campbell 1981 for details). The ontogenetic changes in the energy balance may thus contribute to the changes in $P_N$ under natural and controlled conditions.

In general, short-term or actual, and long-term temperature effects on $P_N$ can

$\rightarrow$

**Fig. 7.11.** Net photosynthetic rate ($P_N$) during leaf ontogeny is influced by many environmental factors. At different actual leaf temperatures, different absolute values of $P_N$ are obtained (according to the temperature optimum of photosynthesis), nevertheless the ontogenetic course in $P_N$ is maintained (a — primary leaves of *Phaseolus vulgaris* L., from Tichá *et al.* 1984). Similarly, when cultivating plants in different temperature regimes (long-term effects of temperature), the ontogenetic course of $P_N$ was also maintained but leaves from the high-temperature regime had a higher optimum temperature for $P_N$ at photosynthetic maturity, and a shorter life than leaves from the low-temperature regime (b — fourth leaves of *Festuca arundinacea* Schreb., from Woledge and Jewiss 1969). Water supply is another important factor affecting $P_N$; see the time trends of $P_N$ during ageing of control leaves and leaves during water stress and recovery (minimum leaf water potential A...−1.1 MPa, B... −2.3 MPa, C... −5.7 MPa) (c — eighth leaves of *Panicum maximum* Jacq., after Ludlow 1975). The slopes of relationship between $P_N$ and leaf water potential vary in leaves of different insertion level (d — *Helianthus annuus* L., after Rawson 1979).

be distinguished. Information on ontogenetic changes in temperature dependence of $P_N$ is, however, rather rare in the literature.

Short-term effects: Temperature optimum of $P_N$ varies with leaf age, insertion level, and with the date of leaf unfolding (*Cucumis* — Iwakiri and Inayama 1975; *Morus* — Murakami 1982). Young and mature leaves have usually a high temperature optimum of $P_N$, but the optimum decreases as the leaves age (*Rubus* — Marks and Taylor 1978). The ontogenetic trend of $P_N$ is maintained at different temperatures, the absolute values of $P_N$ are, of course, modified according to the

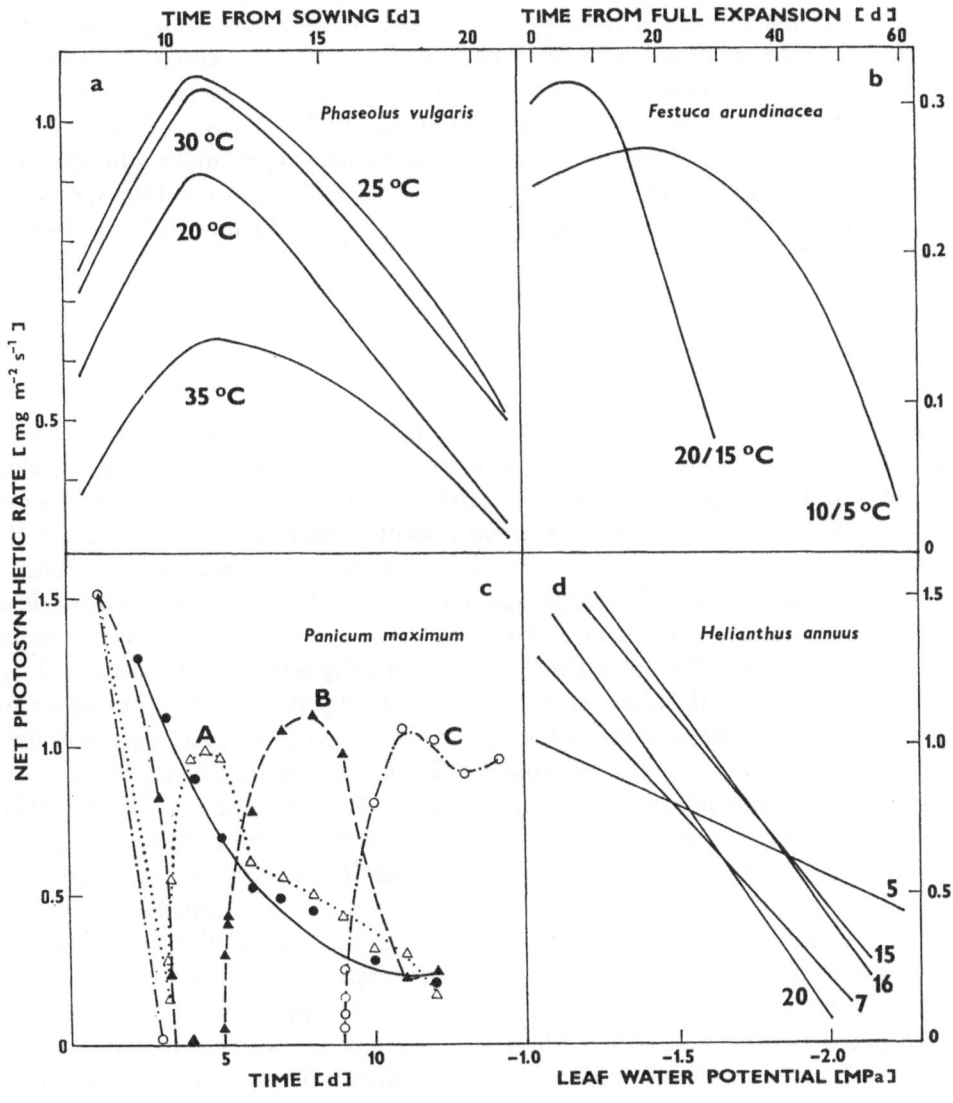

temperature optimum of photosynthesis; maximum $P_N$ characterizes the same stage of leaf ontogeny (*Phaseolus* — Tichá *et al*. 1984; Fig. 7.11 a).

Leaves of different insertion level were differently affected by temperature and vary in the ability to and duration of recovering after low or high temperature treatment (*Zea* — Thiagarajah *et al*. 1979).

Long-term effects: Plants and leaves grown in different temperature regimes vary in life span and have different photosynthetic characteristics, *e.g.* leaf area and morphology, contents of chlorophyll, components of electron transport chain and enzymes. Low temperatures limit the synthetic processes while high temperatures inactivate and denature the metabolically active compounds. This is reflected in the resulting $P_N$; nevertheless, in principle the common ontogenetic course of $P_N$ is maintained under most temperature regimes (*Dactylis* — Treharne and Cooper 1968; *Festuca* — Woledge and Jewiss 1969; *Oryza* — Uchida *et al*. 1980; Fig. 7.11 b). This holds even when the plants are grown under different root temperature (*Glycine* — Duke *et al*. 1979) or under a different combination of air temperature and relative air humidity (*Plectranthus* — Tichá 1970a, b, 1976b, 1984; Fig. 7.3).

## 7.2.8    Water Stress

Drought is probably the most inhibitory factor of the environment encountered by a plant during its life cycle and it is often responsible for reducing crop yield. Low leaf water potential influences the production of leaves through its effect on the leaf initiation in meristem and the subsequent enlargement, and hence, on the formation of the photosynthesizing area (*e.g. Nicotiana* — Clough and Milthorpe 1975). In addition, low leaf water potential can cause the loss of the existing leaf area if desiccation is prolonged and severe. Also the amount of radiant energy accepted by a wilting, flaccid leaf may be substantially less than that accepted by a horizontal turgid leaf. Water stress decreases $P_N$ by limiting $CO_2$ transport to the chloroplasts in the gaseous phase and in the liquid phase, and by restricting the biochemical activity in the chloroplasts. Thus, the influence of water stress on plant photosynthetic production is very complex (for a review see Slavík 1975).

The ontogenetic trend of leaf $P_N$, an increase to maximum followed by an obviously slower decrease, is usually accelerated due to water stress, especially the decrease in $P_N$ in senescent leaves. However, in the $C_4$ grass *Panicum maximum* the decline in $P_N$ with leaf ontogeny appeared to be suspended by water stress; after rewatering, leaves attained rates greater than those of unstressed leaves. Ageing recommenced after rewatering when photosynthetic activity started again (Ludlow 1975; Fig. 7.11c). The insertion gradients in $P_N$ in maize were not changed in their trends but only in absolute values when the plants were grown at 60 or 90% of soil moisture (Šesták and Václavík 1965). Similarly, the same onto-genetic trend of $P_N$ and only differences in absolute values were proposed by Sharpe and De Michele (1974) in their model for leaves affected by air humidity (90 and 60%). In wilting plants, however, $P_N$ of upper leaves (preferentially sup-plied with water) declined more slowly than in middle and lower leaves, the maximum $P_N$ being in water saturated plants in middle leaves, and in wilted plants in the uppermost leaf (*Brassica* — Čatský 1965; *Lolium* — Wardlaw 1969).

$P_N$ is usually very slightly affected by water stress until the "critical" value of the leaf water potential is reached, but then it decreases to zero in a narrow range of leaf water potentials. The changes in sensitivity of $P_N$ to water stress during leaf ontogeny have been studied very rarely; more often leaves of different insertion levels were compared. The range of leaf water potentials inducing the decrease in $P_N$ and the slope of dependence of $P_N$ on leaf water potential were similar for most leaf positions in kale (Čatský 1965) and sunflower plants, with the exception of the lowest leaves (Rawson 1979; Fig. 7.11d). The slope of the dependence of $P_N$ on leaf water potential was also similar for outer and inner leaves of both young and mature sugar-beet plants (Lawlor and Milford 1975) or during ageing of pine needles (El Aouni 1976). However, the decrease in $P_N$ was induced by the more negative values of the leaf water potential in inner rather than in outer sugar-beet leaves and by higher values of water saturation deficit for middle rather than for both young and old pine needles. The differences in sensitivity to water stress in leaves of different insertion levels are smaller for $P_N$ than for stomatal conductance (*cf.* Chapter 8), probably due to the above mentioned complexity of photosynthesis.

### 7.2.9  Oxygen Effect

An altered oxygen concentration produces a variety of physiological and bioche-mical responses in plants including photosynthesis and respiration (Forrester *et al.* 1966 a, b). The $O_2$ inhibition of $P_N$ in $C_3$ plants is connected with the dual function of RuBPCO: with respect to partial pressures of $CO_2$ or $O_2$, the carbo-xylation or oxygenation of RuBP is supported or inhibited. Under atmospheric $CO_2$ and $O_2$ concentrations both processes occur in parallel. The decrease in $O_2$ concentration down to 1 or less % means a preference for carboxylation, and hence an up to 50% increase in $P_N$. In contrast to this, increasing $O_2$ concentration above the atmospheric one inhibits carboxylation and stimulates oxygenation, *i.e.* also stimulates the glycollate pathway and photorespiration rate (*cf.* Chapter 9). With increasing $O_2$ concentration in $C_3$ plants the carboxylation efficiency (see Section 7.2.6.1) declines and $CO_2$ compensation concentration (see Section 7.2.6.2) in-creases. The $CO_2$ concentrating mechanism in $C_4$ plants enables to create at the site of operation of RuBPCO in the bundle sheath cells a relatively high internal $CO_2$ concentration favouring carboxylation. Carboxylation efficiency and $CO_2$

compensation concentrations in $C_4$ plants are not affected by changes in ambient $O_2$ concentration. Besides the $O_2$ effect on $P_N$ *via* the competitive inhibition between $CO_2$ and $O_2$ also a direct $O_2$ inhibition or effect on $P_N$ is supposed. Unfortunately, there are no data available on modifications of the ontogenetic course of $P_N$ by $O_2$ concentration or data on long-term effects of $O_2$ on the developmental pattern of leaf photosynthesis.

## 7.2.10  Mineral Nutrition

The effect of mineral nutrition on photosynthesis is very complex. Mineral nutrition affects the formation of the photosynthetic apparatus, determines the development of leaf area and structure and the chemical composition of leaves, especially that of the thylakoid membranes, it affects conductances for $CO_2$ transfer and membrane permeability. Mineral elements are not only necessary components for the synthesis of structural compounds of the cells, but they also directly participate in individual photosynthetic reactions (*e.g.* manganese in water splitting). Nevertheless, due to a rather slow distribution of absorbed mineral elements within the plant tissues and the possibility to act mostly only after a series of biosynthetic steps, usually long term effects of mineral nutrition on photosynthesis are observed in higher plants. Furthermore, the effect of a particular mineral element cannot be isolated from that of other elements (for a review see Nátr 1975).

The rate of $CO_2$ uptake is decreased by nitrogen, phosphorus, potassium or other mineral deficiency but the general trends in $P_N$ during leaf ontogeny and in insertion profiles seem to be unchanged as can be summarized from the rather sporadic literature (*Zea* — Meinl and Bellmann 1965; *Hordeum* — Nátr 1970, Dale 1972, Nátr *et al.* 1975, Ma and Hunt 1975, Passera 1978; *Triticum* — Cartwright *et al.* 1974, Migus and Hunt 1980; *Arachis* — Gallaher *et al.* 1976; *Trifolium* — Maag and Nösberger 1980). The effects of nitrogen on $P_N$ during leaf ontogeny were greater than those of phosphorus or potassium (*Triticum* — Osman *et al.* 1977). Trifoliate leaves of potassium-stressed soybean seedlings fixed 38% less $CO_2$ than did control leaves (Wells *et al.* 1979). Fully developed mature bean leaves were much less sensitive to potassium deficiency than were growing leaves (Ozbun *et al.* 1965).

## 7.2.11  True (Gross) Photosynthetic Rate and Leaf Ontogeny

The rate of $CO_2$ uptake by photosynthetic carboxylation at carboxylation sites in the chloroplast is understood to be the gross or true photosynthetic rate, $P_G$. $P_G$ can be estimated by summing $P_N$ and all respiration losses of the leaf (*cf.* Section 7.1). This calculation, however, cannot account for the $CO_2$ reassimilation, which certainly occurs within the leaf. Earlier, but also in several recent papers, $P_G$ was estimated as the sum of $P_N$ and $R_D$ neglecting the photorespiration rate and the inhibition of respiration rate in the light (*e.g.* *Triticum* — Sawada 1970; *Nicotiana* — Tanaka and Tatemichi 1971; cereals — Takeda and Udagawa 1976; *Colocasia* — Sato *et al.* 1978; *Amorphophallus* — Miura and Osada 1981). $P_G$ may be measured directly as the rate of $^{14}CO_2$ uptake in short-time exposures (*Triticum* — Thomas *et al.* 1978; *Helianthus* — Rawson and Constable 1980).

Similar ontogenetic changes in $P_G$ and $P_N$ are usually found: a rapid increase to a maximum is followed by a slower decline (*Phaseolus* — Čatský *et al.* 1976; *Triticum* — Thomas *et al.* 1978; *Helianthus* — Rawson and Constable 1980). An increase in $P_G$ with increasing leaf insertion to a maximum in middle leaves is followed by a decrease in the uppermost leaves (*Helianthus* — Rawson and Constable 1980). $P_G$ may be computed as a function of irradiance and leaf surface; the simulation for sunflower plants led to similar ontogenetic changes and insertion gradients in both $P_G$ and $P_N$ (Kobayashi 1975).

## 7.3    RESPIRATION RATE

Respiration ("dark", "mitochondrial") plays an important role in the carbon balance of the plant or leaf (*cf.* Section 10. 8). Dark respiration rate, *i.e.* the carbon dioxide efflux from the leaf in darkness, $R_D$, was one of the first parameters largely studied in leaf's gas exchange. Comprehensible but partial information on the ontogenetic changes in $R_D$ can therefore be derived from numerous ecological and ecophysiological studies, a part of which was summarized by Šesták and Čatský (1967c).

However, in the last decade or so respiration of a leaf has been analysed in more detail. It was stressed that $R_D$ of a leaf includes respiration rates in both chlorophyllous and achlorophyllous cells and tissues and is closely related to the photosynthetic rate and the amount of photosynthates formed in the previous light period. This effect of leaf preconditioning is reflected also in the rather long period, *e.g.* several hours, before a steady-state $R_D$ is attained. For example, $R_D$ immediately after light switching was reported up to twice that achieved later (*e.g.* *Cucumis* — Hopkinson 1964; *Nicotiana* — Wada *et al.* 1967, Rawson and Hackett 1974).

Moreover, respiration rates may be studied with respect to photosynthate utilization using the concept of growth and maintenance respiration (*cf.* McCree 1970, 1982a, b, Thornley 1971, 1977, Penning de Vries 1974, 1975b, de Wit *et al.* 1978, Lambers *et al.* 1983, *etc.*). Briefly, in growth respiration, $R_G$, photosynthates are used to provide energy required for the synthesis of structural and storage compounds. $R_G$ may be related directly to photosynthetic rate. In maintenance respiration, $R_M$, photosynthates are used in energy consuming processes that are necessary to maintain the plant structure. $R_M$ is proportional to leaf dry mass (*cf.* Fig. 7.12). The "structure maintenance" was further distinguished from "tool maintenance" which refers to energy use for maintenance of the tools of biosynthesis, such as RNA and protein turnover associated with biosynthetic enzymes (Penning de Vries *et al.* 1974, 1979).

There is also much uncertainty concerning the respiration rate in a photosynthesizing leaf (*cf.* Section 10.8). Photosynthetic, photorespiration and respiration rates cannot be determined simultaneously and independently in a range of $CO_2$ and $O_2$ concentrations (*cf.* Chapter 9). It is, therefore, uncertain whether and how much the respiration rate is reduced in the light or whether it proceeds at all in photosynthesizing cells.

All the above features of respiration and its measurement are reflected in respiration rates and can, therefore, explain many of the differences in the absolute values of $R_D$ and their changes during leaf ontogeny reported in the literature.


### 7.3.1 Dark Respiration Rate and Leaf Age

A very high $R_D$ in newly unfolded leaves decreasing rather rapidly to a more or less constant rate as the leaves expand and mature is the most common finding, both in $C_3$ and $C_4$ species (*e.g., Beta* — Nevins and Loomis 1970; *Dactylis* — Burris and Carson 1971; *Populus* — Dickmann 1971; *Glycine, Calopogonium, Pennisetum, Sorghum* — Ludlow and Wilson 1971; *Phaseolus* — Čatský *et al.* 1976; *Helianthus* — Horie 1977; *Larrea* — Syvertsen and Cunningham 1977; *Zea* — Williams and Kennedy 1977; *Populus* — Ceulemans and Impens 1979; *Larix, Pinus* — Gowin *et al.* 1980; *Oryza* — Sato and Kim 1980b; *Gossypium* — Constable and Rawson 1980). In senescent leaves, an increase in $R_D$ has usually been observed (*e.g., Perilla* — Hardwick *et al.* 1968; *Phaseolus* — Čatský *et al.* 1976; *Larix, Pinus* — Gowin *et al.* 1980; Figs. 7.12 a and 10.3).

---

→

**Fig. 7.12.** Total dark respiration rate ($R_D$) and its growth ($R_G$) and maintenance ($R_M$) components during leaf ontogeny and in leaves of different insertion level. $R_D$ decreases rapidly with leaf age and increases again with leaf senescence (*cf.* also Figs. 10.3 and 10.4). $R_G$ is proportional to the leaf growth rate, and $R_M$ to the leaf biomass. a: $R_D$ during leaf ontogeny in *Populus* cv. Unal 2 (*i.e. Populus deltoides* Marshal S.9-2 × *Populus nigra* L. cv. Ghoy). The dashed curve fitted by the authors is significant at P = 0.001. (Modified from Ceulemans and Impens 1979.) b: $R_D$ for each leaf of 75-d-old tobacco plants (*Nicotiana tabacum* L. cv. Mammoth) grown either with (o) or without (●) photoperiod extension (modified from Rawson and Hackett 1974). c: $R_D$, $R_G$ and $R_M$ during ontogeny of primary leaf of *Phaseolus vulgaris* L. cv. Harzgruß (Kaše and Čatský, unpublished). d: $R_D$, $R_G$, $R_M$ and plant growth rate during late ripening stages of rice plants (*Oryza sativa* L. cv. Yūkara) grown in nutrient solution (drawn from data of Yamaguchi 1978). e; f: $R_G$ and $R_M$ in the developing leaf of *Helianthus tuberosus* L. (modified from Kimura *et al.* 1978). $R_G$ and $R_M$ per leaf (e) in a large leaf (max. dry matter 812 mg) and the ratio $R_G/R_D$ (f) in large, middle-sized and small leaves (max. dry matter 812, 501 and 193 mg, respectively).

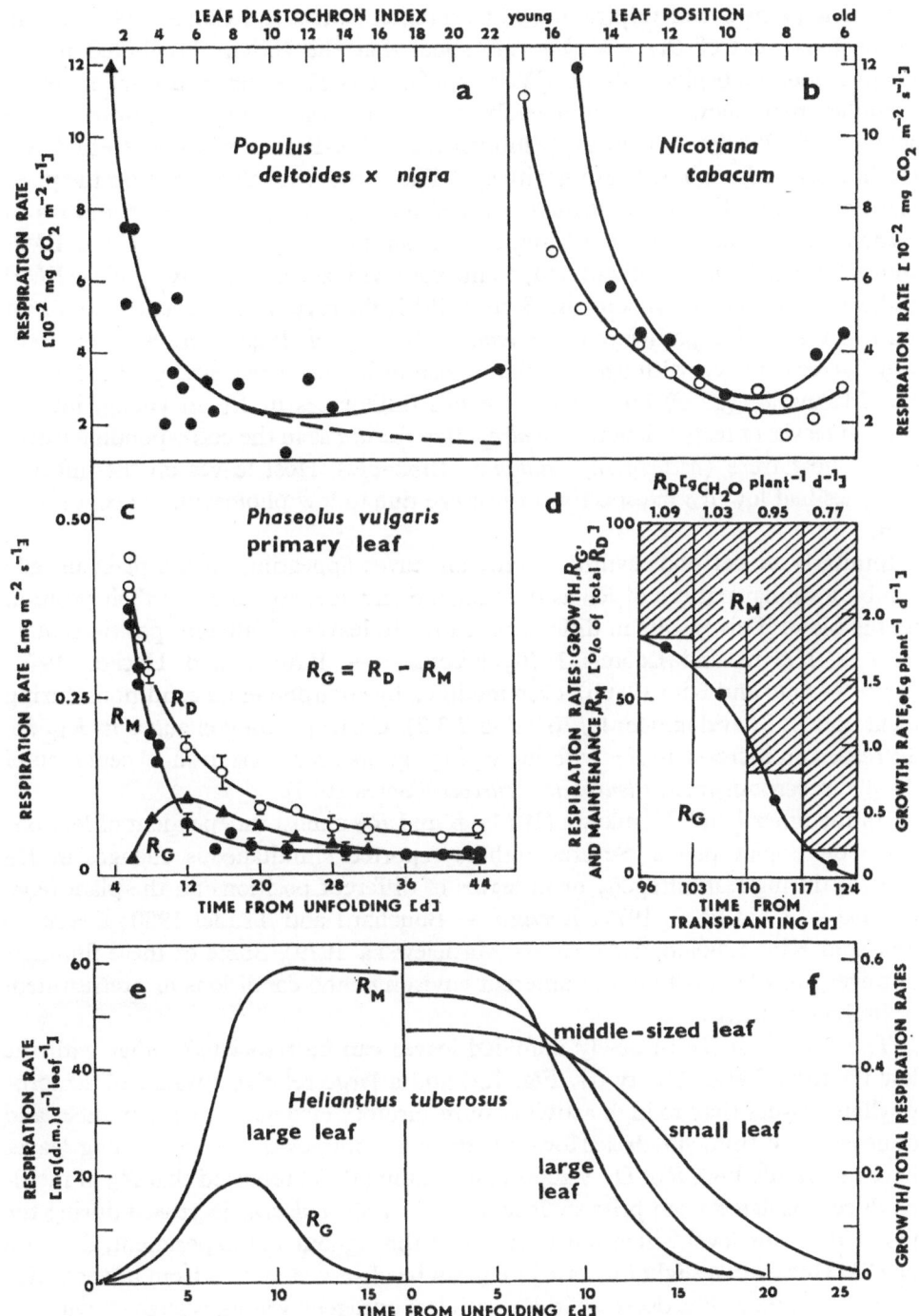

The decrease with age in total $R_D$ may be associated with similar changes in metabolic pathways of respiration. For example, Burris and Carson (1971) found in the basal part of *Dactylis glomerata* shoot that the activity attributed to the pentose monophosphate shunt (Warburg-Dickens-Horecker pathway) followed a similar trend showing a complete absence at 11 weeks of growth. However, the activity of glycolytic pathway (Embden-Meyerhof-Parnas scheme) shifted less predictably, although decreasing with age. The authors conclude that "as the plant matures, it would appear that there is a definite shift from the dual pathways of carbon metabolism (pentose monophosphate shunt and EMP) to the single EMP and/or storage". On the other hand, an increase with age in the activity of the EMP pathway (*Nicotiana* — Bloom and Stetten 1953; for review see Korkes 1956) or no shift in respiration pathways (*Nicotiana* — Baur *et al.* 1968) were also reported.

A decrease in $R_D$ with age was found also in leaves of parasitic plants. Kumar and Mukherjee (1969) found simultaneous differences in $R_D$ in young, mature and old leaves of leafy mistletoe *Dendrophthoe falcata* as in the corresponding leaves of the host trees (*Mangifera, Bauhinia, Mimusops*). Host leaves on the infected branches had low $R_D$ irrespective of leaf age due to low photosynthate concentration.

Similarly to net photosynthetic rate, all leaves appearing on the plant usually exhibit a similar pattern of $R_D$ as they expand and age, regardless of their position on the plant. This results in differences in $R_D$ in leaves of different positions (*e.g.*, *Beta* — Nevins and Loomis 1970; *Nicotiana* — Rawson and Hackett 1974; *Oryza* — Sato and Kim 1980b), even modified by environmental conditions during plant and leaf development (*cf.* Section 7.3.2). Corresponding changes in $R_D$, *i.e.* decrease and increase in $R_D$ with increasing age was reported in shoot segments of the pleurocarpous moss *Pleurozium schreberi* (Bates 1979).

On the other hand, Furukawa (1973a) found $R_D$ almost independent of leaf position on poplar plants. Several authors reported simultaneous changes in $R_D$ and $P_N$ during leaf ontogeny or in leaves of different positions on the plant (*e.g.*, *Glycine* — Chin et al. 1977; *Lactuca* — Bouchard and Trudel 1980; *Oryza* — Sato and Kim 1980a,b,c; *Malus* — Maciejewska 1979). Some of these findings, however, may be attributed to different environmental conditions of pretreatment of individual leaves.

The very high $R_D$ in newly unfolded leaves can be associated rather with the low photosynthetic activity (*cf.* Fig. 7.2) and a large relative amount of achlorophyllous tissues than to high activities of respiratory enzymes or large number and dimensions of mitochondria. However, there is some evidence that young leaves have inherently high $R_D$. De Villiers and Ashton (1977) reported that $R_D$ of mesophyll cells isolated from primary leaves of *Phaseolus vulgaris* decreased during the first 4 d to low level which was maintained throughout the experiment, *i.e.* until day 21. However, a slight increase in $R_D$ can be observed in cells from leaves which began to senesce. The decrease in $R_D$ with leaf ontogeny was associated — with the exception of the youngest leaf — with the decrease in protein, lipid and RNA syntheses in isolated cells. Zuo Bao-yu and Duan Xu-chuang (1978) found in younger flag leaf of wheat large and numerous mitochondria and $R_D$ twice as high in comparison with the 5[th] leaf which contained fewer and smaller mitochondria.

Only partial information is available on the maintenance ($R_M$) and growth components ($R_G$) of $R_D$: moreover, they are usually calculated for the whole leaf or even plant. During the first part of leaf ontogeny, much of respiration is associated with growth, but $R_M$ becomes equally important as leaf dry matter increases (*Hordeum* and *Zea* — Ryle *et al.* 1976; *Helianthus* — Horie 1977; *Oryza, Zea, Glycine* — Yamaguchi 1978). No reliable information on $R_M$ and $R_G$ is available for leaf senescence when massive mobilization and export of saccharides begins. It seems that $R_G$ decreases during this period (*cf.* Fig. 7.12 c to f). Using a dynamic model of sugar beet growth, Hunt and Loomis (1979) predicted $R_M$ and $R_G$ during crop development and confirmed the experimental findings.

### 7.3.2 Environmental Modifications of Ontogenetic Changes in Respiration

Respiration rate is influenced by environmental conditions, both actual or acting during the leaf or plant development.

Temperature effect on $R_D$ has been studied most frequently. The differences in $R_D$ with leaf age are almost unaffected by actual (measurement) temperature (*Gossypium* — Ludwig *et al.* 1965; *Lolium* — Woledge 1973; *Helianthus* — Horie 1977; *Zea* — Miedema and Sinnaeve 1980; *Oryza* — Sato and Kim 1980b). Only in young leaves, the optimum temperature for $R_D$ seems to be slightly shifted to a lower temperature (*e.g.* Ludwig *et al.* 1965). Also the growth temperature affects only the absolute values of $R_D$ but not the ontogenetic changes. For example, Gowin *et al.* (1980) did not find any special effect of growth temperature on $R_D$ of pine and larch seedlings. Similar rates were obtained in both species grown at 12, 17, 24 and 27 °C with the exception of pine seedlings grown at 12 °C where high $R_D$ often exceeded net photosynthetic rate.

Irradiance during leaf development influences $R_D$ as is generally known from studies on intact plants in a canopy: $R_D$ is usually higher in sun than in shade leaves irrespective of leaf age, and adaptations occur. Similar findings have been obtained if plants were grown under different radiation treatments (*e.g.*, *Malus* — Barden 1974; *Oryza* — Sato and Kim 1980c). In the population of rice plants in a paddy field, $R_D$ of leaves decreased proportionally to $P_N$ during leaf ontogeny and under increasing shading (Sato and Kim 1980d). The authors stressed that even the lowest, oldest leaf did not seem as "parasitic" as was often thought.

The extension of the photoperiod induced a shift of maximum $R_D$ along the leaf position to older leaves (*Nicotiana* — Rawson and Hackett 1974 — Fig. 7.12b).

Also the mineral supply or the content of mineral elements do not greatly affect the trend of the changes in $R_D$ with leaf ontogeny (*e.g. Beta* — Nevins and Loomis 1970; *Nicotiana* — Rawson and Hackett 1974).

It may be concluded from the above information and also from other, even partial data on the influence of further factors as $CO_2$, $O_2$, *etc.* on $R_D$ that the leaves of different ages have inherently different $R_D$; *i.e.* that ontogenetic changes in $R_D$ (similarly to $P_N$) are more plant-induced than environment-induced.


## 7.4    DRY MATTER ACCUMULATION

The assimilated carbon which is not lost by respiration processes nor by translocation represents an increase in leaf total dry matter. The accumulation of photosynthates is the basis of plant production important for the supply of food, raw materials and energy.

Measuring the rates of dry matter increment per unit time and leaf area is to a certain degree another possibility of estimating $P_N$ (the respective papers were treated in Sections 7.2.1 and 7.2.2). There are, however, more leaf characteristics connected with dry matter accumulation. Specific leaf mass (originally specific leaf weight, SLW), *i.e.* the ratio between leaf dry matter and leaf area [g m$^{-2}$], represents the distribution of leaf dry matter relative to leaf area. It shows the relative use of carbon for leaf thickening and leaf expansion. SLW is further used as a selection criterion for leaf photosynthesis, for SLW was found to be correlated with mean $P_N$ in genotypes, ecotypes or cultivars (*Glycine* — Dornhoff and Shibles 1970; *Avena* — Criswell and Shibles 1971).

Leaf density [g m$^{-3}$] represents the distribution of leaf dry matter relative to leaf volume.

In this section only dry matter characteristics of individual leaves during leaf ontogeny or with leaf position on the plant, and the carbon balance of a developing leaf area are treated. Growth analysis of whole plants is beyond the scope of this book.


### 7.4.1    Dry Matter Accumulation during Leaf Ontogeny

Total leaf dry matter: In general, dry matter of a leaf increases during ontogeny; after a small decline (*Sinapis* — Zerbe and Wild 1980a, b) or a short lag period in very young leaves a period of rapid dry matter increase starts followed by periods of slow increase and constant leaf dry matter (*Callistephus* — Cockshull

1966; *Lolium* — Silsbury 1970; *Plectranthus* — Tichá 1970a; *Hordeum* — Felippe and Dale 1973; *Lycopersicon* — Tanaka et al. 1974; *Vigna* — Schoch and Candelario 1974; *Phaseolus* — Naito *et al.* 1978; *Cucumis* — Horie *et al.* 1979; *Lycopersicon* — Ho and Shaw 1979; *Oryza* — Mae and Ohira 1981). In senescent leaves a slight decline in dry mass of the leaves may occur (*Nicotiana* — Kakie 1972; *Helianthus* — Judel *et al.* 1975; in *Lolium* a loss of 30% of their

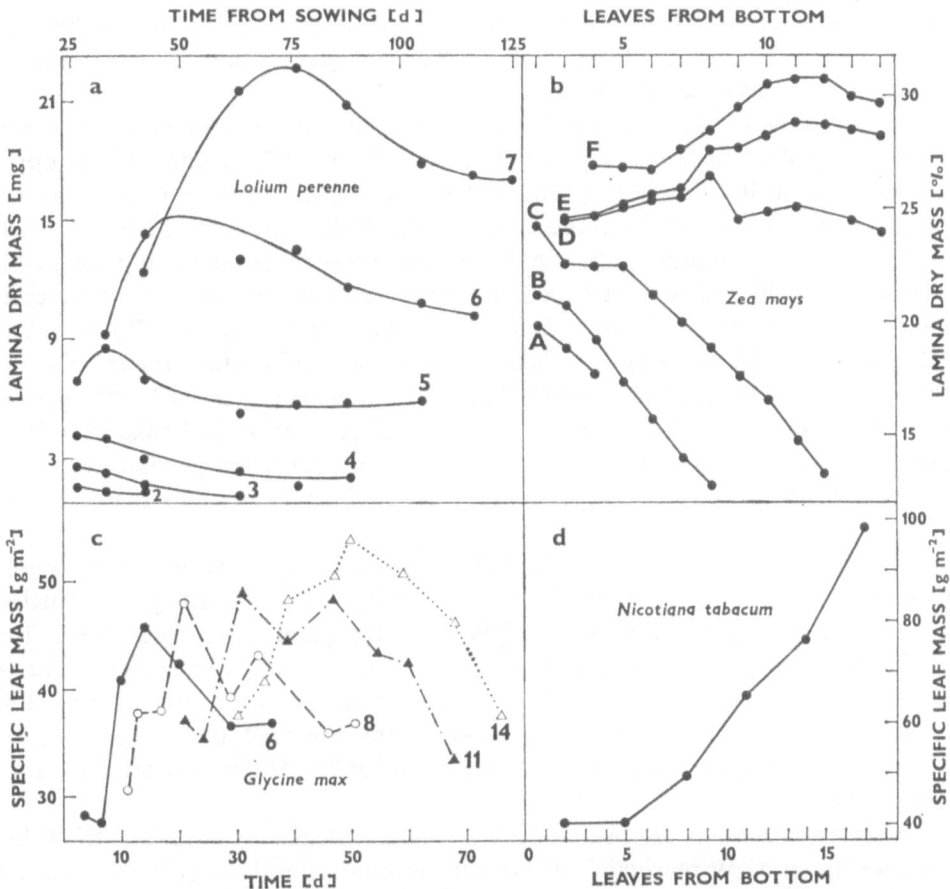

**Fig. 7.13.** Lamina dry matter of successive leaves increases rapidly during leaf ontogeny to a maximum followed by a slower decrease. The higher the sequential leaf number the higher the dry matter of the whole leaf (a — *Lolium perenne* L. from a closed community, after Robson and Deacon 1978). Insertion profiles of lamina dry matter per leaf vary with plant ontogeny (b — A to C: shooting with a difference of one week, D: tasseling, E: flowering, F: milky ripeness — *Zea mays* L., after Pethő 1967). Specific leaf mass of soybean leaves from nodi 6, 8, 11, and 14 (from bottom) changed in a similar way during leaf ontogeny; maximum specific leaf mass increased with increasing leaf position (c — *Glycine max* (L.) Merrill., after Lugg and Sinclair 1981). Specific leaf mass increased with increasing leaf insertion level in tobacco plants with 18 "harvestable" leaves grown in controlled environment (d — *Nicotiana tabacum* L., after Raper and Downs 1973).

mass — Robson and Deacon 1978; *Phaseolus* — Braber 1980; *Triticum* — Patterson *et al.* 1980; Fig. 7.13 a). In cotyledons, dry matter content is nearly stable (*Vigna* — Schoch and Candelario 1973) or decreases — after a short plateau — during almost the whole leaf life span (*Phaseolus* — Raafat *et al.* 1971; *Acacia* — Ashcroft and Murray 1979). Sheaths show ontogenetic changes in dry matter accumulation similar to leaf blades, but their maximum mass reaches only one third of that of the leaf lamina (*Lolium* — Robson and Deacon 1978). The ontogenetic courses of dry mass of successive leaves on a plant were very similar while their absolute dry masses differed substantially (*Lolium* — Robson and Deacon 1978; *Oryza* — Mae and Ohira 1981, see also Section 7.4.2).

The ontogenetic pattern of total leaf dry mass increment in individual leaves is modified by shading (*Hordeum* — Felippe and Dale 1973), namely of the upper leaves, even during the early primordial growth which depends entirely on the supply of photosynthates from the first leaf. The lack of immediate effect of shade on the second leaf indicates the use of stored reserves from the grain during early growth. The values of dry matter content during leaf ontogeny are not substantially modified by nitrogen deficiency (*Lolium* — Robson and Deacon 1978), however, they are affected by temperature (*Phaseolus* — Wilson and Ludlow 1968; *Lolium* — Silsbury 1970), sink (ear) removal (*Triticum* — Patterson and Brun 1980), application of kinetin, indole-3-acetic acid (*Sinapis* — Zerbe and Wild 1980a, b) or benzyladenine (*Phaseolus* — Naito *et al.* 1978); nevertheless, the general pattern of ontogenetic changes in dry matter content is maintained under all these treatments.

Specific leaf mass (SLW) initially declines because the leaf area rapidly increases (*Vigna* — Schoch and Candelario 1973; *Glycine* — Lugg and Sinclair 1981). The subsequent increase in SLW generally parallels the increase in leaf thickness, being followed by a decline in SLW in senescing leaves (*Nicotiana* — Hackett and Rawson 1974; *Glycine* — Lugg and Sinclair 1981; Fig. 7.13 c). The changes in $P_N$ are, however, not proportional to those of SLW during the ontogeny of individual leaves on a plant (*Beta* — Hodáňová 1975; *Malus* — Marini and Barden 1981).

Individual leaves on the plant follow a similar ontogenetic course of SLW but the maxima in SLW reached increase with leaf insertion level (*Glycine* — Lugg and Sinclair 1981; Fig. 7.13 c).

The cartography of the SLW distribution on leaf lamina showed the highest accumulation of dry matter relative to leaf area unit at the base and in the central part, and the lowest in the apex of a maize leaf (Baldy 1971), and the highest accumulation in the upper half, the lowest at the base of a tobacco leaf (Baldy and Le Buhan 1971).

## 7.4.2    Leaf Insertion and Dry Matter Accumulation

Total leaf dry matter: More dry matter is accumulated in leaves in the middle part of the plants than in upper leaves (*Lycopersicon* — Friis-Nielsen 1973) or lower leaves (*Callistephus* — Cockshull 1966; *Brassica* — Šesták and Čatský 1967a; *Zea* — Tanaka and Yamaguchi 1972; *Morus* — Satoh 1974; *Hoya* — Niemann et al. 1980). The maximum of leaf dry matter occurs at two thirds of plant height (*Zea* — Tanaka and Yamaguchi 1972). For $C_3$ grasses, only an increase in dry matter content towards the upper leaves was reported (*Lolium* — Robson and Deacon 1978).

Complicated insertion gradients in dry matter accumulation are found in conifers. In *Pinus*, e.g., current-year needles formed 47% of the total needle biomass and the contribution decreased with needle age (i.e. insertion). On the tree, the maximum biomass was in the middle whorls (4 to 7) (Ågren et al. 1980). During tree ontogeny, the total biomass of shoots increased rapidly, however, needles formed in the previous year only lost their dry matter (Chung and Barnes 1980a, b). The relative distribution of age classes of needles consistently shifts due to a combined result of changing tree size and needle longevity: in the smallest trees (7 years old), 93% of the needles are 1 year old whereas no 3 or 4-year-old needles are present; in the largest trees (39 years old) only 55% of the needles are 1-year old and a few 4-year-old needles are present (*Pinus* — Madgwick et al. 1977).

The insertion gradient in dry mass along a plant is rather varied, e.g., by age of the plant (*Zea* — Pethő 1967; *Brassica* — Šesták and Čatský 1967b; Fig. 7.13 b), by night-break (*Callistephus* — Cockshull 1966) or by water supply (*Lycopersicon* — Friis-Nielsen 1973), etc. Under water stress the differences between successive leaves enlarge (*Lycopersicon* — Friis-Nielsen 1973). In all examples, however, the general pattern of the insertion gradient is maintained.

Specific leaf mass: The common finding is that SLW increases with leaf insertion level (*Medicago* — Pearce et al. 1968, Ku and Hunt 1973; *Nicotiana* — Raper and Downs 1973; *Diplacus* — Gulmon and Chu 1981; *Pinus* — Coyne and Bingham 1982; Fig. 7.13 d), i.e. in upper leaves more photosynthates are stored relative to leaf area. However, in barley a decrease in SLW in the uppermost leaves was reported (Ma and Hunt 1975). The higher the insertion level of the leaf, the higher is the maximum SLW reached during leaf ontogeny (*Glycine* — Lugg and Sinclair 1981).

The insertion gradient in SLW is maintained also under the influence of different temperature regimes (*Medicago* — Ku and Hunt 1973), nitrogen supply (*Diplacus* — Gulmon and Chu 1981), or $CO_2$ enrichment (*Nicotiana* — Raper and Downs 1973). In light-unlimited desert plants, SLW changes very little with leaf insertion whereas in plants in a closed canopy SLW increases from lower to upper leaves (annuals – Mooney et al. 1981).

Leaf density: A similar trend to that of SLW with leaf insertion level is found (*Medicago* — Ku and Hunt 1973) with absolute values modified by the temperature regime.

### 7.4.3    Carbon Budget of the Developing Leaf

Translocation and partitioning of photosynthates are clearly of economic importance. During plant ontogeny, the young leaf is always a sink which imports intensively saccharides from seed or photosynthates from the other leaves. Maximum sink strength given as the product of sink activity and sink size was attained when 10 to 25% of leaf area reached full expansion (*Beta* — Fellows and Geiger 1974). Thereafter the sink strength declined rapidly and asymptotically to a near zero value at about 45% of the final area (*Beta* — Giaquinta 1978). During this period, however, the rapid decline in translocation into the leaf was offset by a rapid rise in $P_N$ of the sink leaf, maintaining a near constant relative rate of dry matter increase until the sink leaf had expanded to about 17% of its final area (*Phaseolus* — Swanson and Hoddinott 1978). However, it was not possible to control the rate of translocation to the sink leaf by varying its $P_N$ through short term exposures to the light or darkness (*Phaseolus* — Swanson and Hoddinot 1978). Generally, the leaf became self-sufficient at an early stage of development when the amount of imported carbon was only about one-sixth of that required for growth. The leaf became a net carbon exporter before half of its life span was reached when the amount of exported carbon was 27% of that fixed; this amount increased to 85% of exported carbon in senescing leaves (*Cucurbita* — Turgeon and Webb 1975; *Lycopersicon* — Ho and Shaw 1979; Fig. 7.14). Maximum carbon export occurred when a leaf had just attained maximum size. Translocation from a newly exporting leaf was primarily upward to developing leaves at the apex. As a leaf at any one position aged, the translocation pattern gradually shifted from upward to bi-directional and, finally, to a predominantly downward direction — into stem and roots. Maximum $P_N$ coincided with the downward shift of carbon export (*Populus* — Larson and Gordon 1969). Generally, translocation was about three quarters of $P_N$ (*Beta* — Terry and Mortimer 1972). In principal, photosynthates moved to all parts of the plant but preferentially to the sink closest to the photosynthesizing leaf (*Triticum* — Rawson and Hofstra 1969). The contributions of carbon from various sources changed throughout the period of expansion of a barley leaf (Anderson and Dale 1983).

Upper leaves which are characterized by higher $P_N$, utilize a larger portion of photosynthates in respiration and dry matter accumulation than middle and bottom leaves; therefore the export from upper leaves is smaller (*Trifolium* — Hoshino *et al.* 1971).

The greater photosynthetic capacity of apical than basal parts of leaves may account for the slightly greater basipetal than acropetal carbon translocation rate (*Zea* — Keith 1979).

The general pattern of translocation was not changed when the plants were fertilized and irrigated (*Pinus* — Ericsson 1979).

The changes in the biochemical composition of photosynthates during leaf development are discussed in Section 6.4.

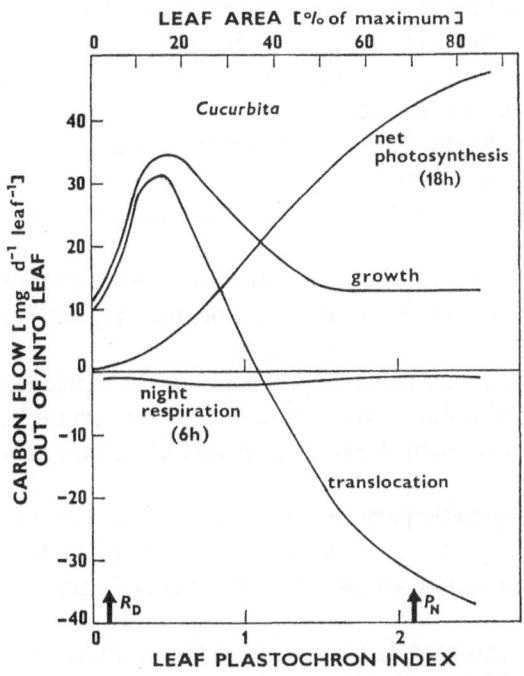

**Fig. 7.14.** Carbon balance of a leaf during leaf ontogeny. Values expressed as the mass of carbon exchanged over a 24-h period. Values for net photosynthesis and dark respiration were calculated for 18-h and 6-h periods, respectively. Arrows labelled with $R_D$ and $P_N$ indicate the maximum respiration and photosynthetic rates calculated on unit area basis, respectively (*Cucurbita pepo* L., after Turgeon and Webb 1975).

Several models of leaf growth which have appeared recently, are based on leaf carbon balance (*e.g.* Charles-Edwards 1979, Thornley *et al.* 1981, Laison and Thornley 1982); they confirm the experimental results.

## 7.5    CONCLUSIONS

An increase in net photosynthetic rate during leaf expansion, a maximum $P_N$ usually before reaching the maximum area of the leaf blade, and a decline in $P_N$ in mature and mainly in the senescing leaf is the usual ontogenetic course of $P_N$

measured as the rate of net $CO_2$ uptake or dry matter increase. A complex of structural, physiological, biochemical and environmental factors is involved in forming this common ontogenetic trend.

All leaves appearing on the plant usually exhibit a similar trend of $P_N$ as they expand and age, regardless of their position on the plant. The extent of $P_N$ values is, however, modified because subsequently formed leaves on a plant develop from different primordia at different phases of plant ontogeny and in various microenvironmental conditions, they are differently supplied with nutrients and water. This results in differences in $P_N$ in leaves of different insertion levels. These insertion gradients in $P_N$ change their character during plant development: in middle-aged plants a higher $P_N$ is mostly found in middle leaves than in the upper and lower ones, but in young and old plants the shape of the insertion gradient does not recall the shape of leaf ontogenetic changes (basipetal and acropetal gradients).

A very high $R_D$ in newly unfolded leaves decreasing rather rapidly to a more or less constant rate as the leaves expand and mature, and an increase in $R_D$ in senescent leaves is the most common finding.

Dry matter accumulation continues throughout the whole leaf life span: it starts with a short lag period followed by a rapid dry matter accumulation reaching a certain upper limit. In senescing leaves, the export of carbon may cause a decline in leaf dry matter.

The inherent ontogenetic patterns in $P_N$, $R_D$ and dry matter accumulation and the changes with leaf position on the plant are substantially modified by various internal and environmental factors, but the general trend described is mostly maintained.

The stage of leaf ontogeny is an important factor affecting $P_N$ and related characteristics and should be carefully considered during experimentation, in interpreting values from the literature, and drawing conclusions for photosynthetic production of leaves.

# 8 CONDUCTANCES FOR CARBON DIOXIDE TRANSFER IN THE LEAF

*J. Čatský, Jarmila Solárová, Jana Pospíšilová and Ingrid Tichá*

Carbon dioxide passes on the pathway from the atmosphere to the carboxylation sites in chloroplasts through a series of structures differing in physical, chemical and biological properties which more or less control its flow rate. Almost all the

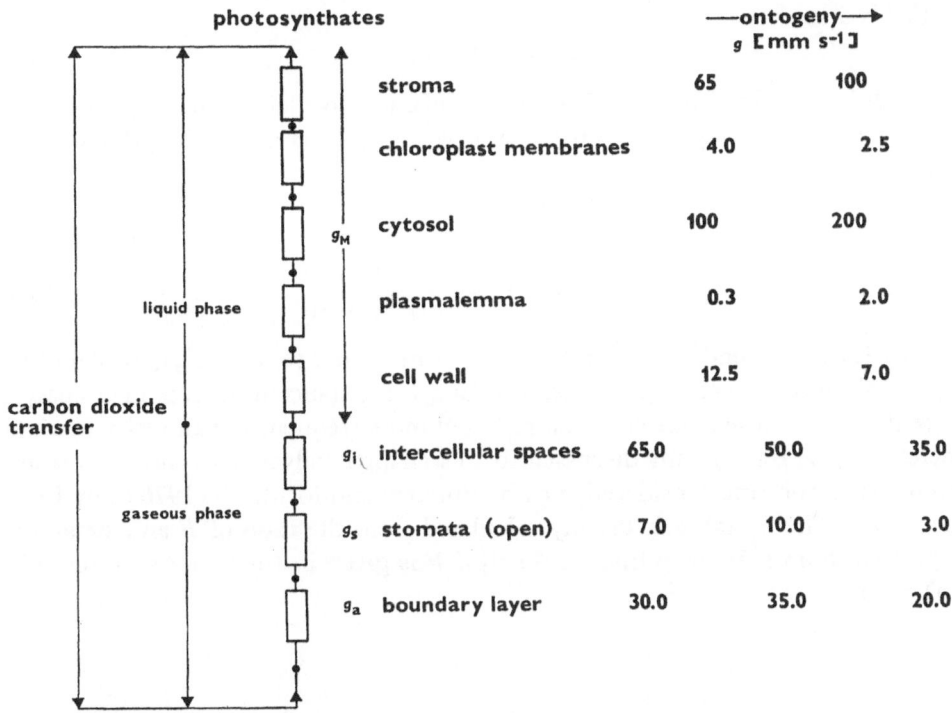

**Fig. 8.1.** Schematic representation of individual leaf structures and the respective conductances for CO₂ transfer on the pathway from the atmosphere to the reaction centres of chloroplasts (*cf.* also Nobel 1974, and Table 6.1 in Raven and Glidewell 1981). *Right*: typical values of conductances for CO₂ transfer in these structures and their ontogenetic changes (young and senescent, or young, mature and senescent leaves).

structures change considerably as the leaf ages (*cf.* Chapter 1) which is reflected in the changes of $CO_2$ flux through them. In principle, two main parts of the $CO_2$ transfer pathway may be distinguished (Fig. 8.1) — the transfer of molecular $CO_2$ in the gaseous phase and the transfer of bicarbonate and molecular $CO_2$ in the liquid phase. The conductances of the two parts have generally been studied in the last decade as elementary photosynthetic parameters (see Čatský and Tichá 1982, Tichá 1982, Solárová and Pospíšilová 1983 for reviews on ontogenetic changes). The analysis of $CO_2$ transfer inside the leaf is further complicated by the creation of $CO_2$ within the cell through respiration (Chapters 7 and 10) and in $C_3$ plants also through photorespiration (Chapter 9).

## 8.1    CARBON DIOXIDE TRANSFER IN THE GASEOUS PHASE

In the gaseous phase, $CO_2$ diffuses* along a concentration gradient, *i.e.* according to Fick's first law

$$F = - D \, dc/dl \tag{8.1}$$

where $F$ is $CO_2$ flux [mg m$^{-2}$ s$^{-1}$], $D$ is a diffusion coefficient [m$^2$ s$^{-1}$], and $c$ the $CO_2$ concentration [mg m$^{-3}$], $l$ pathway length [m]. Over a finite length $l$ we may write

$$F = g \Delta c \tag{8.2}$$

and

$$g = D/l = 1/r = \frac{1}{r_a + r_s + r_1} \tag{8.3}$$

where $g$ [m s$^{-1}$] is conductance and $r$ [s m$^{-1}$] is resistance to transfer, $r_a$ is boundary layer resistance, $r_s$ epidermal resistance, and $r_1$ resistance in intercellular spaces. Recently, the conductance has been reported more frequently than resistance and therefore, $g$ is preferentially discussed in this chapter (advantages and disadvantages of using $g$ or $r$ are considered, *e.g.*, by Burrows and Milthorpe 1976 or by Ludlow 1982). The overall $g$ is usually calculated from the ratio of $F$ and $\Delta c$ and is expressed in [m s$^{-1}$], or in [mol m$^{-2}$ s$^{-1}$] if $F$ is given in [mol m$^{-2}$ s$^{-1}$] and $c$ in [m$^3$ m$^{-3}$].

---

* The more realistic three-dimensional diffusion may be approximated by a one-dimensional model by determining the equivalent one-dimensional resistance components (Cooke and Rand 1980).

## 8.1.1　Conductance of Leaf Boundary Layer

Before entering the leaf, $CO_2$ diffuses through the boundary layer of air adhering to a leaf surface. The conductance of the boundary layer, $g_a$, is given by its thickness, $l_a$, i.e. mainly by the size, shape and surface quality of the leaf, and wind speed. Approximately, the mean $l_a$ is (Nobel 1974)

$$l_a = 4 \sqrt{L/u} \tag{8.4}$$

where $L$ is leaf length [m] and $u$ is wind speed [m s$^{-1}$]. Gates et al. (1968), from experiments with leaf models made of wet blotting papers showed boundary layer resistance, $r_a$,

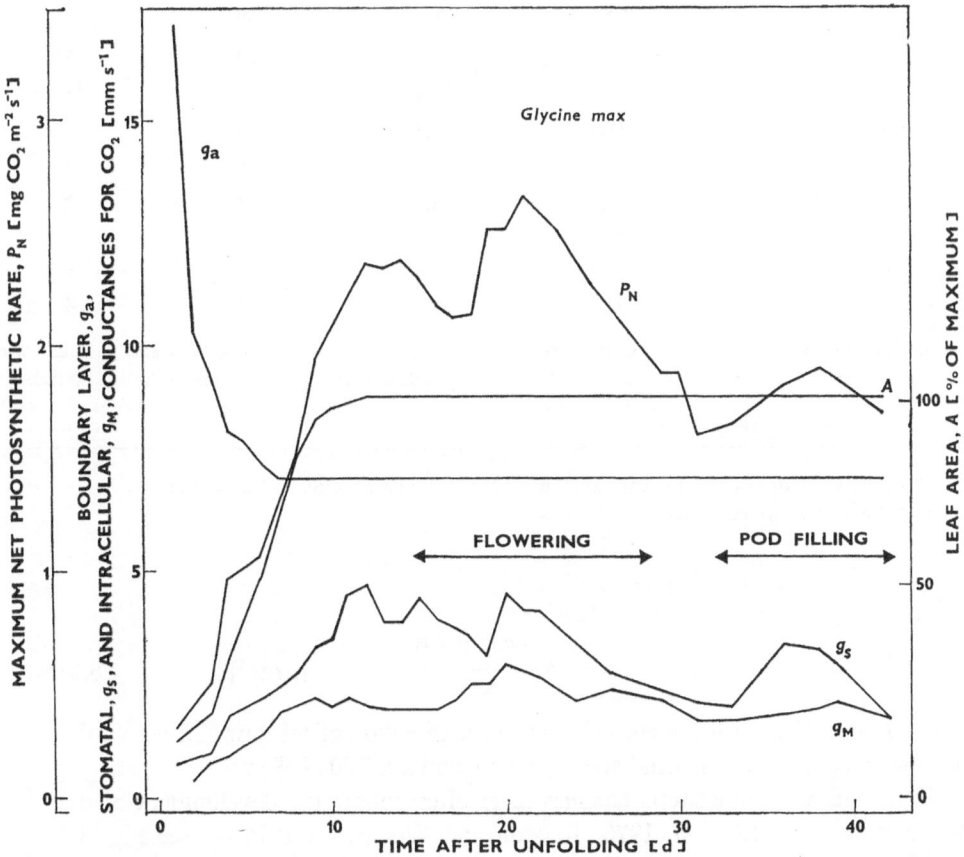

**Fig. 8.2.** Ontogenetic changes modified by developmental events (flowering, pod filling) in maximum net photosynthetic rate, $P_N$, and boundary-layer, stomatal and intracellular conductances for $CO_2$ transfer, $g_a$, $g_s$, $g_M$, respectively, of the fourth leaf on the main stem of *Glycine max* (L.) Merrill cv. Lee (air temperature 27.4 °C, vapour pressure deficit 1640 Pa, quantum irradiance 1800 μmol m$^{-2}$ s$^{-1}$, $CO_2$ concentration 325 cm$^3$ m$^{-3}$). (After Woodward and Rawson 1976.)

**Table 8.1** Timing of peaks of stomatal ($g_s$) and mesophyll ($g_M$) conductances, and net photosynthetic

| Gas exchange characteristics | Calopogonium mucunoides Desv.; | Glycine javanica L. cv. Cooper runner; | Glycine max (L.) Merrill cv. Lee; | Glycine wightii (R. Grah. ex Wight & Arn.) Verdcourt cv. Cooper; |
|---|---|---|---|---|
| | growth chamber | growth chamber | glasshouse | growth chamber |
| $g_s$ initial* [mm s$^{-1}$] | 7.6 | 4.2 | 1.6 | 4.2 |
| % $g_s$ max | 65 | 51 | 16 | 49 |
| $A$ initial* [cm$^2$] | 14.4 | 12.5 | 10 | 11.6 |
| % $A$ max | 43 | 13 | 8.5 | 14 |
| $g_s$ max [mm s$^{-1}$] | 11.7 | 8.3 | 10.3 | 8.6 |
| at % $A$ max | 100 | 74 | 100 | 84 |
| $g_M$ max [mm s$^{-1}$] | 3.6 | 2.5 | 4.6 | 2.6 |
| at % $A$ max | 100 | 74 | 100 | 84 |
| $P_N$ max [mg m$^{-2}$ s$^{-1}$] | 1.11 | 0.86 | 1.17 | 0.83 |
| at % $A$ max | 100 | 74 | 100 | 84 |
| Author(s) and year of publication | Ludlow and Wilson 1971 | Wilson and Ludlow 1970 | Woodward and Rawson 1976 | Ludlow and Wilson 1971 |

\* Initial value of $g_s$ or $A$, i.e. the value measured immediately after leaf unfolding.
\*\* 4 d after leaf appearance.

to be

$$r_a = k \frac{L^{0.20} \, W^{0.35}}{u^{0.55}} \qquad [\text{s m}^{-1}] \qquad (8.5)$$

where $L$ and $W$ are leaf dimensions [m] in the direction of air flow and across the air flow, respectively, $u$ is wind speed [m s$^{-1}$], and $k$ is 180 s$^{0.45}$ m$^{-1}$.

Therefore, the ontogenetic changes in $g_a$ often follow the development of the leaf area (e.g. Čatský et al. 1976, Rawson and Woodward 1976 — see Fig. 8.2 and Table 8.1). However, the changes in the roughness of the leaf surface (trichomes, folding, etc.) are similar or even more important than leaf dimensions. It is regrettable that the effect of ontogenetic changes in leaf roughness (cf. Chapter 1) on $r_a$ has not yet been determined.

rate ($P_N$) during expansion of leaves ($A$ — area of leaf blade) of different plant species.

| Gossypium hirsutum L. cv. Deltapine 16; | Helianthus annuus L. cv. Peredovik; | Helianthus annuus L. cv. Suncross 51; | Nicotiana tabacum L. cv. Mammoth**; | Phaseolus vulgaris L. cv. Jantar; | Sorghum almum Parodi cv. Crooble; |
|---|---|---|---|---|---|
| glasshouse | glasshouse | field | growth chamber | growth chamber | growth chamber |
| 0.5 | 3.8 (1.4/2.4) | | 2.8 | 5.6 | 10.00 |
| 4 | 25 (31/19) | | 23 | 63 | 100 |
| 0.1 | | | 3 or 6 | 3.3 | 52 |
| 5 | 7 | | 7 | 14.5 | 21 |
| 10.7 | 16.4 | 16.4 | 12.9 | 8.8 | 10.0 |
| 80–95 | 80 | after 100 | 78 | 46 | 22 |
| 4.4 | | 5.2 | 3.1 | 2.8 | 41.7 |
| 70 | | after 100 | 62 | 78 | 73 |
| 1.10 | 0.81 | 1.55 | 0.85 | 1.08 | 2.39 |
| 75–90 | 25 | 50–80 | 72 | 56 | 73 |
| Constable and Rawson 1980 | Rawson and Woodward 1976 | Rawson and Constable 1980 | Rawson and Woodward 1976 | Čatský et al. 1976 | Wilson and Ludlow 1970 |

## 8.1.2 Epidermal Conductance and its Components

The gas exchange between leaf tissues and ambient atmosphere depends on the epidermal conductance. Several attempts were reported to separate its two components, stomatal and cuticular, by transpiration curve analysis or artificial stomata closing or determination of diffusive conductance of the epidermis in the dark (Hygen 1951, 1953, Slavík 1958a, b, Ng 1978). The cuticular diffusive conductance reaches in herbaceous dicotyledons only 2—10%, in grass 6—10% and in broadleaf trees 2.5—3.5% of maximum epidermal diffusive conductance, i.e. when the stomata are fully open (for review see Larcher 1980). Ontogenetic changes in cuticular conductance induced by quantitative structural changes in the cuticle (cf. Section 1.3.2) and epidermis around the stomata are consistently lower than those resulting from ontogenetic changes in stomata dimensions and their reactivity.

This is reflected also in recent terminology (*e.g.* Jarvis and Morison 1981) using the term stomatal diffusive conductance with symbols $g_{ad}, g_{ab}, g_s$ for conductances of adaxial (upper), abaxial (lower) or both epidermes, respectively. This terminology is used also in this chapter.

### 8.1.3  Stomata and Stomatal Conductance

Stomatal diffusive conductance changes with stomata number, size, shape and aperture width varying in dependence on guard cells mobility, and internal and environmental factors. All epidermal and stomatal characteristics are not only species specific, but vary according to habitat and even among individuals according to leaf age and insertion level (Zalenskiï 1904, Miller 1931, Tichá 1982, *etc.*). In most plants the pore area (*i.e.* the sum of maximal area of stomatal pores) amounts to 0.5–1.2% of the leaf area, in plants of tropical forests up to 3%; succulents with low stomata density, evergreens and sclerophyllous plants are capable of opening stomata only slightly and the pore area thus reaches only 0.1–0.3%. Leaves of about 40% of plant species are amphistomatous, *i.e.* with stomata on both sides of the leaf, usually with more stomata on the lower side, but in cereals this relation is reversed. About 50% of all plant species, including broadleaf trees, have hypostomatous leaves with stomata only on the lower side. Leaves floating on water level are epistomatous with stomata only on the upper side, and some succulents have stomata on the surface of the whole aboveground part (Slavík 1974). Stomatal index is the ratio of the number of stomata per unit area to the sum of the number of stomata and epidermal cells per unit area (Salisbury 1927). All these characteristics are reflected in the diffusive conductances of leaf epidermes.

Brown and Escombe (1900) calculated the resistance to diffusion through stomata using approximations of stoma geometry. Since this time, several authors have used anatomical parameters of leaf stomata in estimating stomata resistance, $r_s$, or conductance, $g_s$, of different types of stomata (*e.g.* Penman and Schofield 1951, Lee and Gates 1964, Milthorpe and Penman 1967, Cowan and Milthorpe 1968, Parlange and Waggoner 1970) and applied this approach to different plant species and for different purposes (*e.g.* Brown and Rosenberg 1970, Wilson 1972, Lloyd and Woolhouse 1978, Yoshida 1978, Wild and Wolf 1980).

For more or less circular stomata, $g_s$ is calculated as (Penman and Schofield 1951)

$$g_s = \frac{nD}{\left( \dfrac{d}{\pi ab} + \dfrac{1}{2\sqrt{ab}} \right)} \quad [\text{mm s}^{-1}] \tag{8.6}$$

where $a$ and $b$ are the semi-length and semi-width of the stoma aperture, respectively, and $d$ is the depth of the stoma tube [mm], $n$ is the stomata density [mm$^{-2}$],

222

and $D$ the diffusion coefficient of $CO_2$ or water vapour in air, *i.e.* 1.4 or 2.4 mm$^2$ s$^{-1}$, respectively. For more elongated or slit-like stomata the formula (Parlange and Waggoner 1970)

$$g_s = \frac{nD}{\left(\dfrac{d}{\pi ab} + \dfrac{ln(4a/b)}{\pi a}\right)} \qquad [\text{mm s}^{-1}] \qquad (8.7)$$

is used.

Unfortunately, there is no information on the changes in the necessary stomata parameters during leaf ontogeny. However, the parameters were reported for leaves of different insertion levels: $g_s$ calculated for lower leaves of sugar beet was nearly constant, but much higher than in the uppermost immature leaves (Brown and Rosenberg 1970).

As stomata are the main means of gas exchange between inner leaf spaces and the ambient atmosphere, ontogenetic changes in their characteristics are very important parameters of leaf photosynthetic activity. In spite of the small stomata pores area they permit $CO_2$ uptake from the ambient atmosphere at a rate sufficient to support an adequate rate of photosynthesis for normal leaf and plant growth and development. On the other hand, their ability to close decreases the loss of water from leaves and plants in critical situations and thus prevent plant damage.

### 8.1.4 Ontogenetic Changes in Maximum Values of Stomatal Diffusive Conductance

The general ontogenetic trend of changes in maximum (for given experimental conditions) diffusive conductance of both leaf epidermes separately or together is similar to those in net photosynthetic rate, transpiration and mesophyll conductance, *i.e.* an increase to maximum followed by an obviously slower decrease (Figs. 8.2 and 8.3, Table 8.2). A similar trend is usually observed among leaves in the insertion gradient. The absolute values differ not only among plant species, but also among cultivars and in dependence on cultivation and measurement conditions (see Table 8.2).

Very soon after leaf unfolding stomatal conductance was measured only in leaves of a few plant species such as *Calopogonium muconoides*, *Glycine javanica*, *G. wightii*, *G. max*, *Gossypium hirsutum*, *Helianthus annuus*, *Nicotiana tabacum*, *Phaseolus vulgaris* and *Sorghum almum* (Wilson and Ludlow 1970, Ludlow and Wilson 1971, Fraser and Bidwell 1974, Čatský *et al.* 1976, Woodward and Rawson 1976, Constable and Rawson 1980). The same value of $g_s$ may be reached by leaves of different leaf area (3 and 6 cm$^2$) if differently irradiated during cultivation (4 d-old leaves of tobacco: 2.8 mm s$^{-1}$ — Rawson and Woodward 1976).

The stomata density increases only in the very early stages of leaf development. Then due to rapid expansion of leaf area the density of stomata decreases, though

**Table 8.2** Examples of changes in stomatal, $g_S$, and intracellular, $g_M$, conductances during leaf ontogeny (in lines) and in leaves of different insertion levels (in columns). Abbreviations and explanations: A — leaf area, IRGA — infra-red gas analyser, DP — diffusion porometer.

| Plant and cultivation | Leaf age [d] or insertion level (numbered from oldest leaf) | Method of measurement | Range of conductance [mm s⁻¹] | | Author(s) and year of publication |
|---|---|---|---|---|---|
| | | | stomatal | intracellular | |
| (1) | (2) | (3) | (4) | (5) | (6) |
| *Capsicum annuum* L. var. grossum cv. Market Giant (pots with mixture of soil, sand, peat and moss, glasshouse) | leaves inserted one node above the first flower, 0 to 84 d after anthesis | IRGA, open system | 1.7...3.3...2.0 | 2.0 oscillates 0.8 | Hall and Brady 1977 |
| *Glycine javanica* L. cv. Cooper runner (pots with alluvial fertilized clay-loam, growth chamber) | youngest leaf on 30 d old plant, 2 to 42 d; 1st to 18th leaves on 4–6 week old plants | IRGA, open system | 3.2...7.1...2.5 — 13.9 18th, 5.6 12th, 6.8 1st | 0.9...2.5...0.8 — 1.4 18th, 2.3 12th, 0.6 1st | Ludlow and Wilson 1971; Wilson and Ludlow 1970 |
| *Glycine max* (L.) Merrill cv. Lee (pots with mixture of sand and compost, glasshouse) | 4th leaf of main stem, 1 to 43 d | IRGA, open system | 1.6...9.8...3.1 | 0.3...4.0...0.5 | Woodward and Rawson 1976 |
| *Glycine wightii* (R. Grag. ex Wight & Arn.) Verdcourt cv. Cooper (pots with alluvial fertilized clay-loam, growth chamber) | youngest leaf on 30 d old plant, 2 to 42 d | IRGA, open system | 4.3...8.1...3.5 | 0.9...2.5...0.8 | Ludlow and Wilson 1971 |
| *Gossypium hirsutum* L. cv. Deltapine 16 (pots with river loam and peat moss, glasshouse) | leaves on 5th, 7th and 9th nodes | IRGA, open system | 0.6...10.7...1.1 | 0.5...4.0...1.0 | Constable and Rawson 1980 |

| Species (growing conditions) | Leaf description | Method | | | Reference |
|---|---|---|---|---|---|
| *Hevea brasiliensis* Muell. Arg. (pots with soil, glasshouse, then phytotron) | Leaf Blade Class Concept (9 classes according to angle between leaves and stem) | *Siemens* chamber, IRGA, open system | 8.3...28.0...24.5 | 10.0...1.7...2.1 | Samsuddin and Impens 1979b |
| *Lolium perenne* L. (pots with perlite, nutrient solution, growth chamber) | 5th leaf after $A_{max}$ | IRGA, open system | 2.9...0.2 | 2.6...0.1 | Woledge 1972, 1977 |
| | 5th to 8th leaves from the base | | 2.6 8th<br>2.8 7th<br>4.2 6th<br>4.8 5th | 1.3 8th<br>1.7 7th<br>1.4 6th<br>1.4 5th | |
| *Medicago sativa* L. cv. Hunter River (pots with sand plus nutrient solution, inoculation with *Rhizobium*, growth chamber) | 2nd leaf, 10 to 80 d of regrowth; 2nd to 14th leaves after 49 d of regrowth | IRGA, open system | 30.7... 8.9 | 3.7... 0.7 | Hodgkinson 1974 |
| | | | 13.3 14th<br>33.3 12th<br>16.0 4th<br>25.0 2nd | 3.8 14th<br>3.1 12th<br>1.8 4th<br>0.9 2nd | |
| *Morus alba* L., plants from rooted cuttings (pots with mixture of sand, loam and peat, glasshouse) | the uppermost (12th—14th) leaves, 0 to 30 d; | DP, IRGA, open system | 1.3...0.7...1.3 | 1.4...1.5...1.1 | Kriedemann *et al.* 1976 |
| | leaves 7th to 23th; 0 to 40 d after shoot pruning | | 8.2 23th<br>15.6 15th<br>8.8 7th | 0.4 23th<br>1.3 17th<br>0.7 7th | Satoh *et al.* 1977 |
| *Nicotiana rustica* L. (pots) | 1st to 7th leaves on 4 month plants | IRGA, open system | 2.7 7th<br>8.8 5th<br>1.7 1st | 0.5 7th<br>0.7 5th<br>0.1 1st | Lurie *et al.* 1979 |
| *Nicotiana tabacum* L. cv. Mammoth 17 L (pots with mixture of sand, peat and loam, plus nutrient solution, growth chamber) | 12th leaf from 10 to 100% of maximal leaf area | IRGA, open system, differential psychrometer | 2.8...16.3...2.3 | 0.6...3.0...0.6 | Rawson and Woodward 1976 |

225

**Table 8.2** (continued)

| Plant and cultivation | Leaf age [d] or insertion level (numbered from oldest leaf) | Method of measurement | Range of conductance [mm s⁻¹] stomatal | intracellular | Author(s) and year of publication |
|---|---|---|---|---|---|
| (1) | (2) | (3) | (4) | (5) | (6) |
| *Phaseolus vulgaris* L. cv. Jantar (pots, sand and nutrient solution, growth chamber, spring and autumn experiments) | primary leaves, 30 (spring) or 13 (autumn) d after unfolding | DP, IRGA, open system | spring: 5.0...5.9...0.3 autumn: 5.5...8.8...2.8 | 0.8...3.3...0.7 | Solárová 1973, 1980, Čatský et al. 1976, Čatský and Tichá 1980 |
| *Phaseolus vulgaris* L. cv. Pencil Pod Black Wax (pots with mixture of sandy loam, peat moss and turface, growth chamber) | primary, 1st and 2nd trifoliate leaves, 28 d from primary leaf unfolding | IRGA, closed system, wet-bulb psychrometer | 2.8...6.0...2.5 0.8...4.3...1.6 2.7...4.1...1.3 | 0.6...1.1...0.5 0.5...1.4...0.5 0.6...1.0...0.8 | Fraser and Bidwell 1974 |
| *Sorghum almum* Parodi cv. Crooble (pots with alluvial fertilized clay-loam, growth chamber) | youngest leaf on 30 d old plant, 2 to 30 d; 1st to 8th leaves on 4—6 weeks plants | IRGA, open system | 9.5...1.6  3.2 8th 7.1 6th 2.5 1st | 13.5...41.8...4.0  1.5 8th 33.3 5th 1.6 1st | Ludlow and Wilson 1971 |
| *Triticum aestivum* L. (pots with John Innes Compost No. 1, growth chamber) | 4th leaf, 4 to 28 d from leaf emergence | IRGA, open system | 3.7...12.2 | 0.6...0.7...0.3 | Osman and Milthorpe 1971 |
| *Triticum aestivum* L. cv. Kalyansona (pots with | flag leaf from anthesis | IRGA, open | 4.5...6.4...3.3 | 3.1...1.3 | Rawson et al. 1976 |

<table>
<tr><td>Vigna luteola (Jacq.) Benth. cv. Dalrymple (pots with alluvial fertilized clay-loam, growth chamber)</td><td>1st to 14th leaves on 4–6 week plants</td><td>IRGA, open system</td><td>4.2 14th<br>12.5 11th<br>2.6 1st</td><td>0.9 14th<br>4.5 7th<br>0.6 1st</td><td></td></tr>
<tr><td>mixture of perlite and vermiculite, nutrient solution, glasshouse, later growth chamber</td><td>to harvest (40 d)</td><td>system</td><td></td><td></td><td>Ludlow and Wilson 1971</td></tr>
</table>

new stomata develop from stomata mother cells over a rather long period of time (*cf.* Chapter 1). This "dilution" of stomata density is at the beginning of leaf expansion largely compensated by the increasing size of individual stomata, later (mainly at the end of leaf enlargement) stomatal area per unit leaf area decreases (Solárová 1973, 1980, Rawson and Craven 1975). These changes in the quantitative anatomical characteristics of leaf epidermes are accompanied apparently by an antagonistic sharp increase in stomatal conductance of both leaf sides to a maximum, reached obviously earlier by $g_{ad}$. This increase in $g_s$ may be explained only to some extent by an increase in stomatal pore size and distance between individual stomata, and mainly, by the gradual development of the capability of stomata to open, this being connected with an increase in elasticity of guard cell walls. Some stomata begin to open only after reaching their maximal size (Solárová 1973, 1980, Rawson and Craven 1975).

The development of $g_s$ mentioned above, its maximal value, and time when it is reached depend on the plant species and conditions of plant and leaf development. Any factors influencing the leaf development and its final area, such as irradiance, water availability, *etc.*, affect also the development of stomata (Tichá 1982, see Chapter 1) and in consequence also $g_s$ (Rawson and Craven 1975). The assumption that the changes in $g_s$ are firmly associated with changes in net photosynthetic and transpiration rates and intracellular conductance, and that peaks of these parameters are attained at maximal leaf area, was confirmed for *Festuca arundinacea* (Jewiss and Woledge 1967), *Calopogonium muconoides* (Ludlow and Wilson 1971), and *Glycine max* (Rawson and Woodward 1976). A contradictory assumption that maximum $g_s$ is not necessarily coupled with maximum net photosynthetic rate was confirmed for *Gossypium hirsutum* (Constable and Rawson 1980), *Helianthus annuus* (Rawson and Constable 1980), *Phaseolus vulgaris* (Čatský *et al.* 1976), and *Sorghum almum* (Wilson and Ludlow 1970). In other plant species the maximum values could not be determined or were not evident as the intervals between individual measurements were rather long. Differences between two experiments with *Helianthus annuus* confirm the effect of cultivar, environmental conditions and/or manner of cultivation on $g_s$ development (Rawson and Woodward 1976, Rawson and Constable 1980).

Ontogenetic changes in $g_s$ depend also on the life span of the leaf (Fig. 8.4). In leaves with a short life span, especially when the leaf ontogeny proceeds only during the vegetative phase of plant development, for instance in primary leaves of *Phaseolus vulgaris* (Solárová 1973, 1980, Čatský *et al.* 1976), *Helianthus annuus*, *Nicotiana tabacum* (Rawson and Woodward 1976) and *Gossypium hirsutum* (Constable and Rawson 1980), a rapid decrease in $g_s$ is obvious. On the contrary, $g_s$ of

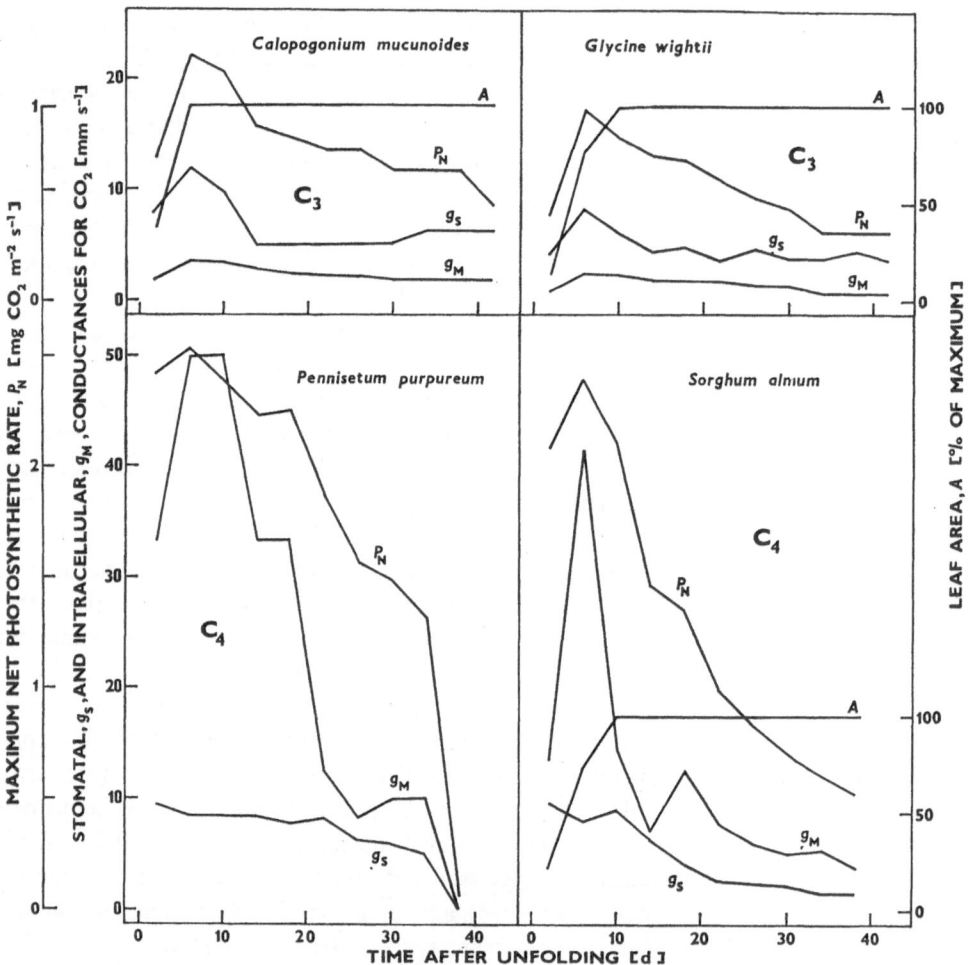

**Fig. 8.3.** Changes in leaf area, *A*, net photosynthetic rate, $P_N$, stomatal, $g_s$, and intracellular, $g_M$, conductances for $CO_2$ transfer during leaf ontogeny (leaf temperature 30 °C, vapour pressure deficit 17 mm, illuminance 100 klx, $CO_2$ concentration 300 $cm^3$ $m^{-3}$). Comparison of plants with different carbon fixation pathways: $C_3$ — *Calopogonium mucunoides* Desv. and *Glycine wightii* (R. Grah. *ex* Wight & Arn.) Verdcourt cv. Cooper, and $C_4$ — *Pennisetum purpureum* Schum. cv.Q5088 and *Sorghum almum* Parodi cv. Crooble. (After Wilson and Ludlow 1970, Ludlow and Wilson 1971.)

leaves with a long life span (leaves of trees and shrubs, and needles) reaches maximum value during leaf expansion or contemporarily with maximum leaf area, and then its value is steady or very slowly decreases in dependence on changes in environmental factors, *e.g.*, water availability, plant ontogeny stage, or pathological agents. The rapid decrease of $g_S$ in hypostomatous leaves of *Betula alleghaniensis* and *Fagus grandifolia* starts with senescence associated with a decrease

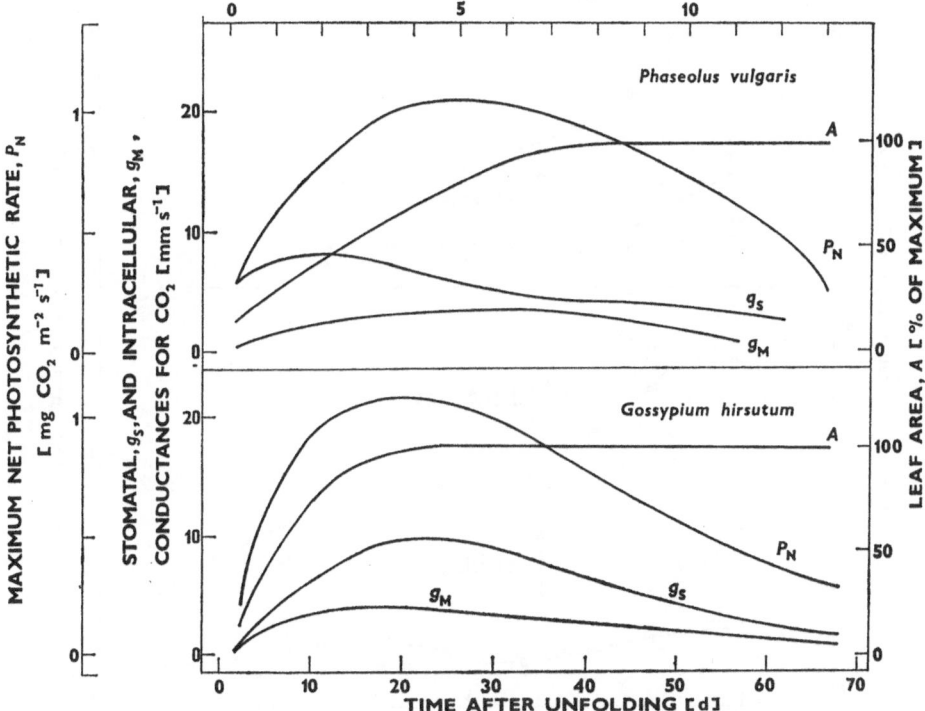

**Fig. 8.4.** Changes in leaf area, $A$, net photosynthetic rate, $P_N$, and stomatal, $g_S$, and intracellular, $g_M$, conductances for $CO_2$ transfer during leaf ontogeny of *Phaseolus vulgaris* L. cv. Jantar and *Gossypium hirsutum* L. cv. Deltapine 16. Changes were similar in spite of different leaf life span: primary leaves of *Phaseolus vulgaris*, eighth leaf on the main stem of *Gossypium hirsutum*. (After Čatský *et al.* 1976, Constable and Rawson 1980).

in chlorophyll content and conspicuous leaf colour change (Gee and Federer 1972). It is regrettable that most authors reporting the changes in $g_S$ of deciduous woody plants considered mainly the position of the leaf in the crown, *i.e.* the height from the ground (Gee and Federer 1972, Cline and Campbell 1976, Federer 1976, 1980, Federer and Gee 1976, Turner and Heichel 1977, Roberts *et al.* 1981) but did not follow the ontogenetic changes in $g_S$ of the same leaf. This is due mostly to the relatively high portion of leaves which are attacked by insects or damaged in other ways and unsuitable for remeasurement (Federer 1976).

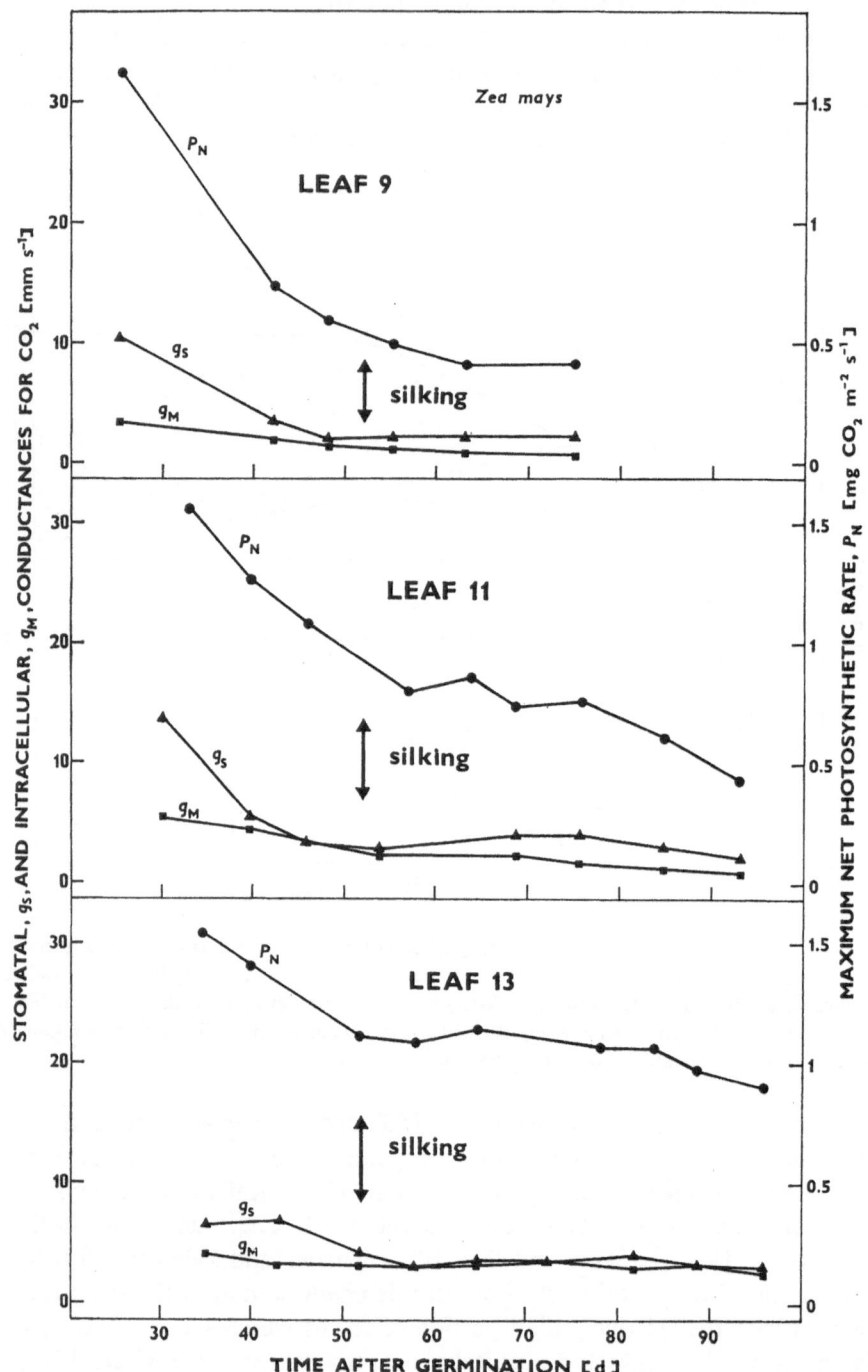

In the course of plant ontogeny, changes in $g_s$ depend on both leaf and plant age and stage of development, as was shown for soybean by Woodward and Rawson (1976): $g_s$ of the fourth secondary leaf increased to a maximum reached at the same time as leaf area reached maximum and the following decrease was interrupted by two peaks: the first of them was lower and coincided with plant flowering, the second, higher, with pod filling. At the same time maxima of net photosynthetic and transpiration rates and intracellular conductance were observed. These additional peaks are probably induced by growth regulators produced in roots or reproductive tissues rather than by an effect of sink size. Similar modifications of the intrinsic ontogenetic pattern of $g_s$ of all developed French bean leaves were observed in the time of subsequent leaf emergence or flowering (Fraser and Bidwell 1974). Also the silking of maize plants affected the decrease in $g_s$, $g_M$ and $P_N$ induced by leaf ageing (Thiagarajah et al. 1981 — Fig. 8.5).

Artificial prolongation of leaf life span or leaf physiological activity by plant decapitation or by keeping the same leaf irradiance by moving the secondary leaves aside resulted in a retardation of $g_s$ decrease (*Phaseolus vulgaris* — Solárová 1973, 1980).


### 8.1.5    Leaf Insertion Level
and Stomatal Diffusive Conductance

In many plant species primary leaves differ from the subsequent ones. Even if they are not different morphologically, they are smaller and thinner than the middle leaves; the upper, youngest leaves are again smaller and thinner than the leaves in the middle insertion level. Also the anatomical structure of leaves of different insertion levels changes (Chapter 1) — with higher insertion level, the size of both epidermal and guard cells decreases, and the stomata density increases in addition to other quantitative anatomical characteristics (Zalenskiï 1904).

Changes in physiological parameters along the insertion gradient (Table 8.2) are similar to those during ontogeny of one leaf, *i.e.* when $g_s$ of all leaves on one plant is determined at the same time, a sharp increase to maximum and a slower decrease along the stem from the top (youngest) leaves to the bottom (old) ones is observed. Thus the highest values of $g_s$ are reached mostly in leaves of the upper part of the middle portion of the leaf set. An exceptional distribution of $g_{ab}$ was

←

**Fig. 8.5.**   Changes in net photosynthetic rate, $P_N$, and stomatal, $g_s$, and intracellular, $g_M$, conductances for $CO_2$ transfer during ontogeny of *Zea mays* L. cv. Harrow 691 leaves. Comparison of mature leaves of different insertion levels (leaves 9, 11 and 13 from the base) and the effect of development of whole plant (arrows indicate date of flowering). As measurements began approximately when each leaf matured, increase of every parameter at the beginning of leaf development is not mentioned. (After Thiagarajah et al. 1981.)

found for the evergreen shrub *Lyonia lucida* in which the third leaf from the bottom of the shoot had the maximum value of $g_{ab}$, and the youngest three leaves the minimum one. This may be caused by differences in leaf colour (the youngest leaves are red), anatomical structure, water relations or sensitivity to water stress (Schlesinger and Chabot 1977). Also on twigs of *Carpinus betulus* the maximum $g_{ab}$ was found in the leaves near the base (Eliáš 1979).

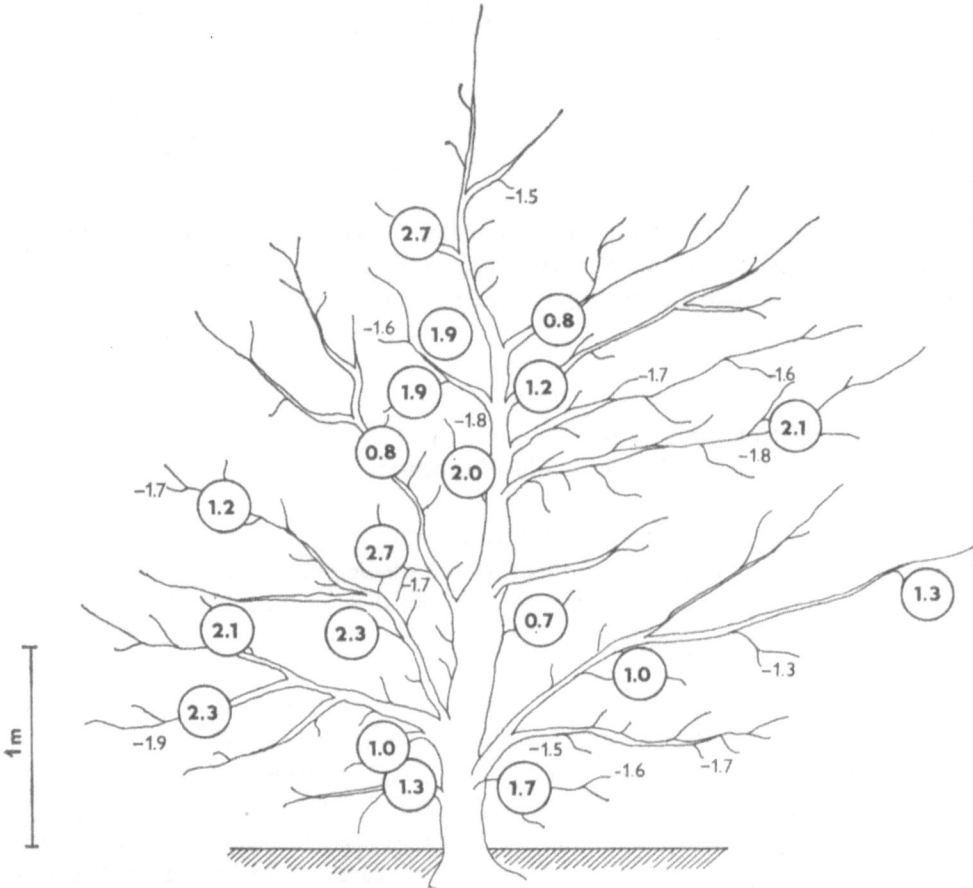

**Fig. 8.6.** The spatial variation of stomatal conductance, $g_s$ ([mm s$^{-1}$], encircled numbers) and leaf water potential $\Psi_w$ [MPa] in seven-years old trees of *Prunus armeniaca* L. cv. Ungarische Beste grown on SW-exposed walls near Innsbruck (640 m above sea level). All values were obtained between 14.30 and 15.30 local time. (After Körner 1981).

Only a few researchers pursued all individual leaves of a plant at various stages of plant development (Ishihara *et al.* 1971a, b, 1978c, Ishihara 1979). More papers report $g_s$ of many or all leaves at different insertion levels at only one phase of plant development (Solárová 1965, Frommhold 1972, Graham and Ulrich 1972,

Čatský *et al.* 1973, Kanemasu *et al.* 1973/1974, Hodgkinson 1974, Jordan *et al.* 1975, Denmead and Millar 1976, Millar and Denmead 1976, Jones 1977, Schlesinger and Chabot 1977, Woledge 1977, Düring 1978, Lurie *et al.* 1979), other papers report $g_s$ in different canopy layers (Impens *et al.* 1967, Brun *et al.* 1973, Stigter 1974, Lof 1976, Hatfield and Carlson 1978) or in different parts of the tree crown (Fig. 8.6; Garland and Branson 1977, Beadle *et al.* 1978, Eliáš 1979, Körner 1981), but most papers compare only $g_s$ of upper, middle and bottom leaves, or only two of these leaf classes.

A smooth trend of changes is observed usually only in unstressed plants grown in homogeneous optimal environmental conditions. In canopies, tree crowns, *etc.*, gradients of internal factors complicate or change the general gradient or course. Thus irradiance decreases with the depth of the canopy and accelerates the decrease in $g_s$ through the insertion profile. Contrary to this, a higher $g_s$ in the lower canopy layer may be conditioned by a relatively strong irradiance in some places of the canopy or parts of the leaf blade (Woledge 1972, Kanemasu *et al.* 1973/1974, Nagarajah 1975a, b, Hatfield and Carlson 1978).

### 8.1.6 Stomatal Conductance and Leaf Blade Heterogeneity

Little attention has been paid to the heterogeneity of the leaf blade. Kanemasu *et al.* (1973/1974) measured $g_s$ in the tip, centre, and base of leaves of *Sorghum bicolor* at different insertion levels in the stand. At the leaf base the decrease in $g_s$ with the insertion level begins from the youngest leaf, in the centre from the second leaf, and at the tip even from the third leaf from the apex. At all insertion levels $g_{ad}$ and $g_{ab}$ and contemporarily differences between $g_{ad}$ and $g_{ab}$ were lowest at the leaf base and highest at the tip. Also Liu and Zobel (1980) found in the palm *Elaeis guineensis* the highest $g_s$ near the tip of the leaflet and of the leaf, in the upper layers of leaflets on a single leaf, and on young (but not youngest) leaves.

### 8.1.7 Ontogenetic Changes in Responses of Stomatal Diffusive Conductance to Environmental and Internal Factors

During leaf ontogeny not only anatomical characteristics of the stomatal apparatus (see Chapter 1) and maximum values of stomatal conductance may vary, but also the reactivity of stomata to environmental and internal factors. Unfortunately only a few papers deal with this problem and they are not easily comparable as each of them concerns the response of $g_s$ (eventually $r_s$ or stomatal aperture) to another factor, and leaf ontogeny is characterized either by leaf age, leaf insertion, level in the canopy, plant growth stages, or only by the seasonal pattern.

Maximum values of $g_s$ reached during the day usually increase as the leaves expand, reach a broad plateau in nearly matured or mature leaves, and decrease during leaf senescence. Similarly, the diurnal maximum of $g_s$ is higher in upper than lower leaves. Maximum daytime $g_s$ in almost fully expanded leaves is accompanied by the minimum night-time $g_s$, *i.e.*, at this phase of leaf development the stomata open more widely and close more tightly than at other stages (Turner 1974a, Burrows and Milthorpe 1976). During the leaf senescence the stomata neither open as widely nor close as tightly, *e.g.*, $g_s$ of old needles of larch remains high at night and this lack of stomatal control promotes the autumn defoliation (Benecke *et al.* 1981). During the leaf expansion period the stomata close as tightly as in mature leaves (Burrows and Milthorpe 1976). Similarly, stomata on the upper leaves of *Xanthium strumarium* close more tightly in the dark than stomata on the lower leaves (Krizek and Milthorpe 1973). On the other hand, Jodo (1973) found a pre-dawn opening of stomata on the upper and the middle leaves of tobacco.

Also the mean daytime stomatal opening and the period for which the stomata are open decrease from the upper to the lower leaves of tobacco, sorghum and maize (Jodo 1973, Turner and Begg 1973) or from the upper to lower whorls on one year old shoots of pine (Troeng and Linder 1982b). The time peak in the stomatal opening appears earlier in the lower than in the upper leaves of tobacco (Jodo 1973) or sorghum (Kanemasu *et al.* 1973/1974) but later in the lower than in the upper leaves of cotton (Ackerson 1981).

After reaching the diurnal maximum opening the stomata on the lower leaves of rice, sorghum and maize close sooner than stomata on the upper leaves (Ishihara *et al.* 1971a, b, 1978a, b, Kanemasu *et al.* 1973/1974, Turner 1975, Ishihara 1979). The differences in the diurnal course of $g_s$ in relation to leaf insertion in rice were more expressive on sunny than cloudy days (Ishihara *et al.* 1971a, b), in plants with a partially excised root system (Ishihara *et al.* 1978b), or in plants transported from the interior to the border of a paddy field (Ishihara *et al.* 1978c). In addition, the lower leaves of tobacco showed a midday closure of stomata even under mild soil moisture stress (Jodo 1973).

The highest $g_s$ on the top of the crown of pine occurred only in the morning but due to a high evaporative demand in the middle of the day $g_s$ was lower at the top than at lower levels (Hellkvist *et al.* 1980).

In rice maximum $g_s$ during the day in the levels of corresponding insertion was practically the same from the tillering stage to the heading stage, but after the heading stage it became lower and also the stomata closed sooner in the afternoon (Ishihara *et al.* 1971a, b).

When environmental conditions are favourable, the diurnal course of $g_s$ is related to changes in irradiance, and therefore the observed differences in diurnal courses of $g_s$ during leaf ontogeny may be, at least partially, considered as its different response to irradiance. However, Kanemasu and Tanner (1969) observed the "diurnal course" of $g_s$ in both the upper and the lower leaves in plants exposed to constant irradiance.

### 8.1.7.2 *Response of Stomatal Conductance to Irradiance*

The basic character of dependence of $g_s$ on irradiance was similar in leaves of different ages in alfalfa (Hodgkinson 1974) and French bean (Čatský and Tichá 1980, Solárová 1980), in leaves of different insertion in maize (Čatský *et al.* 1973) and wheat (Denmead and Millar 1976) or in the upper leaves of cotton during the different growth stages (Ackerson *et al.* 1977), even if the absolute values of $g_s$ were different. Values of $g_s$ under a certain irradiance increase after unfolding, reach maximum in nearly matured leaves, and decrease during senescence, or decrease from the upper to the lower leaves (for review see Turner 1974b). The hyperbola expressing the relation of $g_s$ to irradiance had more marked quasilinear parts in young and mature leaves than in senescing bean leaves (Čatský and Tichá 1980, Solárová 1980 — see Fig. 8.7). Only in nearly emerged and senescent leaves of maple and oak (Turner and Heichel 1977) or French bean (Solárová 1980) $g_s$ did not respond to changes in irradiance.

Consequently, different levels of irradiance probably took part only in the ontogenetic changes in maximum $g_s$ per day but other changes in the diurnal course of $g_s$ (*e.g.*, decrease in the afternoon in lower leaves) were induced rather by the ontogenetic changes in the response of $g_s$ to the transient leaf water deficit than to irradiance.

Ontogenetic changes in $g_s$ were similar in tobacco plants grown under high or low irradiance, only the maximum $g_s$ per day was reached a little sooner in plants grown under high irradiance (Rawson and Woodward 1976). Decrease in $g_s$ with the depth of canopy was observed in both sunlit and shaded parts of the maize leaves, but it was more marked in the shaded parts (Stigter and Lammers 1974, Stigter *et al.* 1977). Ageing of leaves also delayed the beginning of stomata opening and decreased the opening rate after switching on the light (Solárová and Pospíšilová 1984).

### 8.1.7.3 *Response of Stomatal Conductance to Water Stress*

With the decreasing leaf water potential the stomatal closure proceeded from the lowest to the uppermost leaves in sugar beet (Lawlor and Milford 1975) or in cotton (Jordan *et al.* 1975, Ackerson 1980, Radin 1981 — Fig. 8.8). As the values of leaf water potential in individual leaves on a plant usually did not differ significantly, critical water potential for stomatal closure was higher in the lower leaves

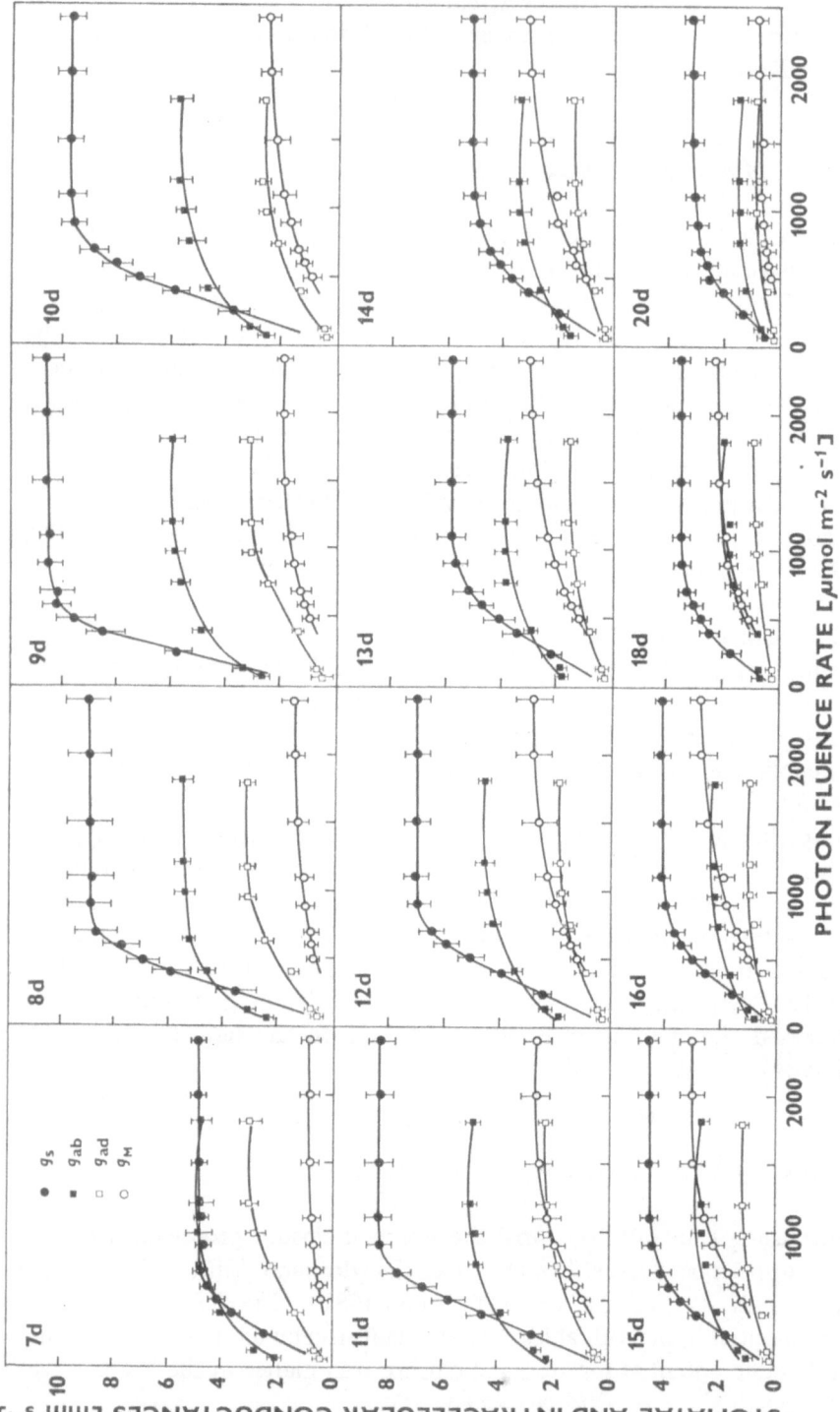

**Fig. 8.7.** Effect of quantum irradiance on stomatal, $g_S$, and intracellular, $g_M$, conductances for $CO_2$ transfer. Comparison during ontogeny (7 to 20 d from sowing) of primary leaves of *Phaseolus vulgaris* L. cv. Jantar. Stomatal conductance of abaxial, $g_{ab}$ (■), and adaxial, $g_{ad}$ (□), epidermes was measured with diffusion porometer *Lambda LI-60*, stomatal conductance of both epidermes together, $g_S$ (●), and intracellular, $g_M$ (○), conductances were calculated from the measurement of transpiration and photosynthetic rates in an open gas exchange system with IRGA. (Modified from Čatský and Tichá 1980, Solárová 1980.)

than in the upper ones. Similarly, the stomata closed at higher values of water potential in the lower level of the canopy than in the higher one in tobacco, sorghum, maize (Turner 1974a) and Sitka spruce (Beadle et al. 1978). On the other hand, in *Citrus sinensis* $g_s$ of young leaves began to decrease at higher leaf water potential than $g_s$ of old (over-wintered) leaves (Syvertsen et al. 1981) and in water stressed tea plants the upper leaves exhibited lower $g_s$ than the lower leaves (Sandanam et al. 1981). Reduced water potential caused an increase in the abscisic acid level in all leaves, and in cotton, rice, millet and wheat the upper leaves exhibited a higher level of abscisic acid than the lower leaves (Jordan et al. 1975, Ackerson 1980, Quarrie and Henson 1981). Similarly, the amount of abscisic acid accumulated in response to water stress decreased as the leaves of rice, millet, French bean and wheat aged (Eze et al. 1981, Quarrie and Henson 1981). Thus the ontogenetic pattern of abscisic acid accumulation in response to decreased water potential did not correspond to the pattern of stomatal closure induced by decreased water potential.

Certain differences in reponse of $g_s$ to leaf water potential between the upper and the lower leaves of cotton were observed also in experiments where irradiance incident on both types of leaves was equalized (Jordan et al. 1975). Thus the different response of stomata to leaf water potential on the upper and lower leaves was associated with physiological ageing and only partially with irradiance during leaf development.

A higher critical leaf water potential for stomatal closure was observed in the older sorghum plants (Jones and Rawson 1979), and a higher decrease in $g_s$ corresponding to the same decrease in water potential was found in the older oat plants (Sandhu and Horton 1977). On the other hand, the stomata of wheat were found less responsive to decreased water potential as the plants aged (Frank et al. 1973) or as the flag leaf aged (Jones 1977). In Sitka spruce the critical water potential varied with the season, being higher in July than in October and November (Beadle et al. 1978).

Although the critical values of leaf water potential changed with leaf age, insertion or plant age, the basic character of the response of $g_s$ to water potential was usually the same; only Lawlor and Milford (1975) determined in sugar beet grown in moist air a linear decrease in $g_s$ with decreasing water potential in mature plants and a curvilinear one in young plants; in plants grown in dry air a linear decrease in $g_s$ in both mature and young plants was observed.

With decreasing soil moisture, $g_s$ decreased similarly in both young and old leaves of *Coffea arabica* (Nunes 1976). As a result of water stress, $g_s$ in the middle leaves of soybean decreased earlier during the day and to a greater extent than in the upper leaves (Stevenson and Shaw 1971), and $g_s$ in leaves of younger soybean and sunflower plants decreased less than in leaves of older plants (Sionit and Kramer 1976, Silvius et al. 1977). Due to a decrease in soil moisture, $g_s$ of grape-vine leaves decreased more in June than in July and August (Hofäcker 1976).

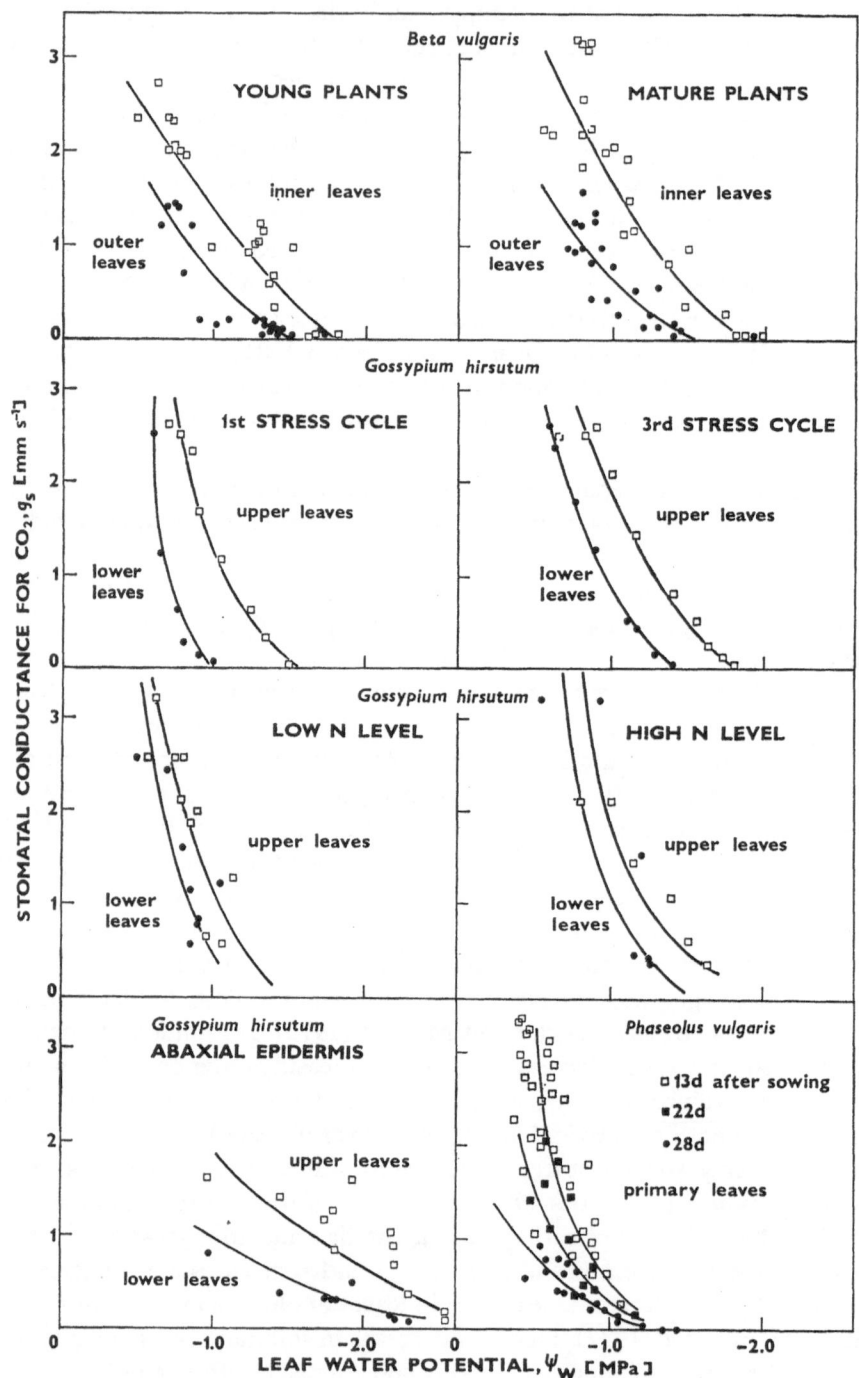

After rewatering, $g_s$ recovered more quickly in younger than in the older wheat plants (Frank *et al.* 1973). As the number of stress cycles increased, the stomata of cotton became less responsive to leaf water potential, especially the stomata of the upper leaves (Ackerson 1980, Fig. 8.8). On the contrary, in oat the earlier stress contributed to the adverse effect of the later stress (Sandhu and Horton 1977).

Decrease in $g_s$ due to salinity treatment was more marked in the mature than in the young leaves of tobacco (West and Black 1978).

A lateral and longitudinal gradient in stomatal sensitivity to air humidity or osmotic potential of the root medium was found in the leaf blade of *Commelina communis* (Maier-Maercker 1981). The response significantly increased from the centre towards the margin of the leaf.

### 8.1.7.4 *Response of Stomatal Conductance to Temperature*

Only a seasonal pattern of response of $g_s$ to temperature was described (Giurgevich and Dunn 1978, 1979) and it is not easy to distinguish whether the observed changes (*e.g.* higher $g_s$ values at specific temperatures in the tall form of the grass *Spartina alternifolia* in September and lower $g_s$ in December and March, or in the short form higher $g_s$ in May and lower in March) were caused by ontogeny of plants or by changes in environmental conditions as the dependence of $g_s$ on temperature was very sensitive to conditions during the days preceding the measurement.

### 8.1.7.5 *Effects of $CO_2$, $SO_2$, and Nitrogen Supply*

Growing of the soybean plants in $CO_2$-enriched air lead to a decrease in their $g_s$, especially in the upper leaves (Hofstra and Hesketh 1975). Fumigation with $SO_2$ had the greatest effect on $g_s$ of middle leaves of *Populus tremuloides* and the decrease in $g_s$ of upper and particularly lower leaves was much less pronounced (Kimmerer and Kozlowski 1981). Also the effect of high doses of nitrogen (the

---

←

Fig. 8.8. Influence of leaf water potential, $\Psi_w$, on stomatal conductance, $g_s$. Comparison of inner (younger) and outer (older) leaves in young and mature plants of *Beta vulgaris* L. cv. Sharpe's Klein (after Lawlor and Milford 1975); upper (younger) and lower (older) leaves of *Gossypium hirsutum* L. cv. Stoneville 213 plants during first and third stress cycle (after Ackerson 1980); upper and lower leaves of *Gossypium hirsutum* L. cv. Deltapine 61 plants affected by different nitrogen nutrition (after Radin 1981); conductance of abaxial epidermis of upper and lower leaves of *Gossypium hirsutum* L. cv. Stoneville 213 plants grown under equal irradiance during stress cycle (after Jordan *et al.* 1975); primary leaves of *Phaseolus vulgaris* L. cv. Harzgruss of different age (after Pospíšilová and Solárová 1984).

wider stomatal aperture from the morning until reaching the maximum aperture) was more pronounced in the upper than in the lower leaves of rice at all growth stages (Ishihara *et al.* 1978a).

### 8.1.8 Intercellular Conductance

Conductance for $CO_2$ in the intercellular air spaces, $g_i$, is relatively large (*cf.* Chapter 1). In addition to the diffusion coefficient it depends on the pathway length. The pathway length for $CO_2$ is usually taken as about equal to the pathway length for $H_2O$, however, the recent experimental evidence suggests that the pathway for $CO_2$ is longer than that for $H_2O$ (for review see Cooke and Rand 1980). Intercellular conductance can, only with difficulty, be measured and therefore its changes during leaf ontogeny are not known. Only the ontogenetic changes in leaf structure (*cf.* Chapter 1) suggest that the $CO_2$ pathway becomes longer during leaf ontogeny and therefore a decrases in $g_i$ may be supposed.

## 8.2 CARBON DIOXIDE TRANSFER IN THE LIQUID PHASE

After reaching a cell on the interface between intercellular spaces and the mesophyll, $CO_2$ is dissolved in water in pores of the cell wall. In the liquid phase different processes participate in the transport of $CO_2$ in both molecular and bicarbonate forms. Also this transport, however, is usually regarded — for practical reasons — as a diffusion and the conductance in the whole liquid phase is expressed in units of diffusive conductance. Diffussion coefficient in the liquid phase is about $10^4$ times lower than in the gaseous phase. Nevertheless, the liquid phase conductance, *i.e.* intracellular conductance (formerly called mesophyll conductance) is of the same order as the conductance in the gaseous phase because of a fairly short transfer pathway and a large internal surface, *i.e.* the interface between the two phases.

### 8.2.1 The Overall Intracellular Conductance

The intracellular conductance, $g_M$, is usually studied as an overall residual conductance (or resistance) which includes both $CO_2$ transfer and carboxylation (see Jarvis 1971 and Raven and Glidewell 1981 for details):

$$r_M = \frac{\Delta c}{F} - r_s - r_a \qquad [\text{s m}^{-1}] \qquad (8.8)$$

and

$$g_M = 1/r_M \; [\text{m s}^{-1}]$$

where $r_M$, $r_s$, $r_a$ are intracellular, stomatal and boundary layer resistances, respectively, $\Delta c$ is the difference in $CO_2$ concentration [mg m$^{-3}$] in the ambient air and after carboxylation, and $F$ is the net $CO_2$ influx [mg m$^{-2}$ s$^{-1}$] (or net photosynthetic rate, $P_N$). As $CO_2$ concentration after carboxylation is often regarded as equal zero, $\Delta c$ is then substituted by the ambient $CO_2$ concentration. The value of residual conductance and the accuracy of its determination is limited by determining a series of $CO_2$ and water vapour parameters, and also by calculation procedures used (cf. also Meidner 1975, Cooke and Rand 1980, Čatský and Tichá 1982). This may contribute to some extent to the variation in $g_M$ values reported in the literature.

Besides the assessment as residual conductance, $g_M$ as a whole or its components may be studied by several indirect approaches (cf. Section 8.2.3.2). For studies with an ontogenetic aspect the approaches deriving $g_M$ from the biochemical properties of the carboxylation enzymes and the geometry of cells and leaves seem to be promising (e.g. Sinclair et al. 1977, Raven and Glidewell 1981).

## 8.2.2 Ontogenetic Changes in the Overall Intracellular Conductance

Under non-limiting environmental conditions, the shape of the curve relating $g_M$ to both leaf age or leaf insertion level follows the shape of a similar curve for net photosynthetic rate modified by the ontogenetic changes in $g_s$ and $g_a$ (Fig. 8.9). The most common finding is that $g_M$ and $g_s$ change essentially in parallel as leaves age (Wilson and Ludlow 1970, Ludlow and Wilson 1971, Woodward and Rawson 1976, O'Toole et al. 1977, Table 8.1) with the result that the $CO_2$ concentration in the intercellular spaces remains constant (Davis and McCree 1978, Goudriaan and van Laar 1978, Constable and Rawson 1980). However, maximum $g_M$ before (e.g. Constable and Rawson 1980) or after maximum $g_s$ (e.g. Čatský et al. 1976, Čatský and Tichá 1980) were also reported. The decrease in net photosynthetic rate with leaf age is, therefore, accounted for by a decrease in both $g_s$ and $g_M$. However, the correlation between $g_M$ and net photosynthetic rate is usually higher than that between $g_s$ and net photosynthetic rate (e.g. Aslam et al. 1977). The higher correlation between net photosynthetic rate and $g_M$ and lower values of $g_M$ suggest that changes in this conductance are more important than changes in $g_s$ in determining the changes in net photosynthetic rate with age. Aslam et al. (1977) support this conclusion by the plot of $r_s$ against $r_M$ which shows that a doubling of $r_M$ is not matched by an equal doubling of $r_s$. Also Woledge (1972) reported in ryegrass that both $g_M$ and $g_s$ decreased as the leaf aged, but the former decreased faster. Samsuddin and Impens (1979b) found similar ontogenetic changes in *Hevea* using a newly developed Leaf Blade Class Concept describing the physiological age of leaves.

The ontogenetic or insertion changes in $g_M$ are most often only slightly modified by various environmental and internal factors during growth or during measurement (*e.g.* water potential — Adams *et al.* 1977, vernalization — Woledge 1979,

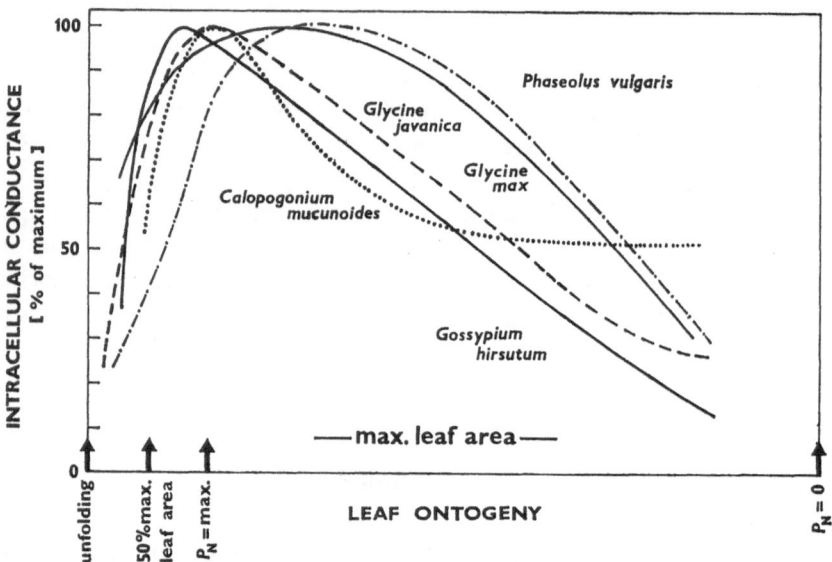

**Fig. 8.9.** Generalized course of intracellular conductance, $g_M$, during leaf ontogeny (expressed in % of maximum) of *Calopogonium mucunoides* Desv. (...), *Glycine javanica* L. (---), *Glycine max* (L.) Merrill (—), *Gossypium hirsutum* L. (——), and *Phaseolus vulgaris* L. (—·—) plants. (Calculated from Wilson and Ludlow 1970, Ludlow and Wilson 1971, Čatský *et al.* 1976, Constable and Rawson 1980.)

gibberellic acid — Little and Loach 1975, irradiance — Woledge 1972, Rawson and Woodward 1976, Čatský and Tichá 1980, high level of nutrition — Rawson and Woodward 1976, drought — Moorby *et al.* 1975, pollution – Srivastava *et al.* 1975, *etc.*). No considerable differences in the ontogenetic changes in $g_M$ were found in 15 cultivars of cassava (Aslam *et al.* 1977). Generally, stress factors such as drought usually accelerate leaf senescence and thus the decrease in $g_M$ (Hanscom and Ting 1977). A similar response may be induced by fruit development (Woodward and Rawson 1976). Fraser and Bidwell (1974) postulated on the basis of extensive experimental work with French bean that there is an intrinsic pattern of gas exchange for each leaf, but this is modified by events that take place during development in other parts of the plant, such as subsequent leaf emergence or flowering.

Similar findings were reported for evergreen plants with leaf life span lasting several years. Slow decrease in $g_M$ with ageing was found in current and 1-year old needles of balsam fir (*e.g.* Little and Loach 1975), but $g_M$ of creosotebush leaves was not affected by leaf age once the leaf had fully expanded (Syvertsen and Cunningham 1977). A lower $g_M$ was found in juvenile ivy leaves than in adult ones (Bauer and Bauer 1980) and in rather new than old leaves of evergreen sclerophylls (Dunn 1975).

Although differences in $g_M$ between successive leaves on the same plant do not necessarily reflect quantitatively the difference due to age (Saeki 1959, *etc.*), changes in $g_M$ (and other photosynthetic parameters) during leaf ontogeny and with leaf number are similar (*e.g.* Ludlow and Wilson 1971) and reflecting the changes in net photosynthetic rate often reported in the literature (see Šesták and Čatský 1967c, and Chapter 7). A similar trend may be found also in changes of $g_M$ in the course of the vegetation period (*e.g.* Giurgevich and Dunn 1978, 1979, Linder and Troeng 1980), but they are modified by the changing canopy environment (*e.g.* shading — Woledge 1977, 1978, temperature — Moore *et al.* 1973, radiation — Taylor and Pearcy 1976, drought — Shmat'ko *et al.* 1979) or internal factors (*e.g.* flowering and fruiting — Frank and Barker 1976, Woodward and Rawson 1976, Hall and Brady 1977, Thiagarajah *et al.* 1981, *etc.*).

### 8.2.2.1 Comparison of the Ontogenetic Changes in Intracellular Conductance of $C_3$ and $C_4$ Plants

Leaves of $C_4$ plants have a higher net photosynthetic rate than leaves of $C_3$ plants under a wide range of environmental conditions and at various ages and positions on the plant. This is associated with, and may be caused by, a higher $g_M$, which appears to result from a lack of photorespiration and one or more of the following: possession of the $C_4$ pathway, carbonic anhydrase activity, and higher transport component of intracellular conductance, resulting from leaf, cellular and subcellular structure (Slatyer 1970, Wilson and Ludlow 1970). The intracellular conductance in $C_4$ plants is of a different kind than in $C_3$ plants. As $CO_2$ transported from the atmosphere is carboxylated with the help of phosphoenolpyruvate carboxylase (PEPC), $CO_2$ concentration at the carboxylation site may be considerably lower than in $C_3$ plants. Thus the very low overall $r_M$ (high $g_M$) found in $C_4$ plants may include only a very small chemical, or carboxylation resistance (see Section 8.2.4.2) representing only the carboxylation catalysed by PEPC.

The general trend of ontogenetic changes in $g_M$ reported for $C_4$ plants is similar to that described for $C_3$ plants. Also in leaves of $C_4$ plants of different insertion levels a similar trend of $g_M$ was found as in leaves of $C_3$ plants (*e.g.* Slatyer 1970).

## 8.2.3 Components of Intracellular Conductance

Even if attempts to study individual components of $g_M$ may be misleading (*cf.* Tenhunen *et al.* 1980), the following sections summarize some useful findings resulting from the ontogenetic aspects of these studies together with the necessary basic information on the individual concepts.

### 8.2.3.1 *Transport and Carboxylation Components of Intracellular Conductance*

The sites where $CO_2$, evolved in photorespiration and respiration of achlorophyllous cells, enters the main $CO_2$ pathway several components of $g_M$ are distinguished. Although these components of $g_M$ cannot be reliably localized on certain cell structures, several attempts to distinguish the conductance for $CO_2$ transfer, $g_M$, from the chemical, "carboxylation" conductance, $g_x$, were reported. Most frequently the curve relating net photosynthetic rate to irradiance or to the ambient $CO_2$ concentration was analysed by means of a model expressing both the $CO_2$ transfer and biochemical processes (*e.g.* Chartier 1970, Chartier *et al.* 1970, Jones and Slatyer 1972).

Although this approach may be criticized for several reasons (*e.g.* Jarvis 1971, Laïsk 1977) it may provide some information on the trend of ontogenetic changes in two main components of $g_M$. Using Chartier's model, Čatský and Tichá (1980) reported that the transfer component was the main portion (*ca.* 60–75%) of the overall $r_M$ during the whole ontogeny of the primary French bean leaves. In spite of the large variation, the increase in $g_x$ after the end of leaf expansion was associated with an increase in the rates of Hill reaction and non-cyclic photophosphorylation (*Phaseolus vulgaris* — Zima *et al.* 1981).

Also $g_x$ and $g_m$, similarly to $g_M$, may be modified by environmental factors. For example, Prioul (1971) reported that $g_x$ and $g_m$ decreased irrespective of leaf position with decreasing irradiance during the growing of ryegrass plants.

### 8.2.3.2 *Transport Component Calculated from Leaf Anatomy*

Another approach to assess the changes in the relative significance of the two main components of $g_M$ has been based on the analysis of anatomical and cytological parameters of the $CO_2$ pathway inside the leaf. Laïsk *et al.* (1970) and Rakhi (1971) estimated the transport component of $g_M$ as

$$g_{m \, (anat)} = DS\beta kL^{-1} \quad [\text{mm s}^{-1}] \qquad (8.9)$$

where $D$ is the diffusion coefficient for $CO_2$ in water [mm$^2$ s$^{-1}$], $S$ is the ratio of internal to external leaf surface area [$mm^2$ $mm^{-2}$], $\beta$ is the solubility coefficient of $CO_2$ in water [g $g^{-1}$], $k$ is the part of the internal leaf surface covered with chloroplasts, and $L$ is the mean distance [mm] of chloroplasts from the mesophyll cell surface (= mean length of the $CO_2$ pathway in the liquid phase from the interface to carboxylation sites). Using this formula, Tichá and Čatský (1977) found in French bean a brief decrease in $g_{m\ (anat)}$ after leaf unfolding followed by a slow increase. Comparing $g_{m\ (anat)}$ with $g_M$ determined by infra-red gas analysis the authors conclude that the leaf structure may limit $CO_2$ transfer in mature leaves with the near-maximum net photosynthetic rates.

### 8.2.4 Transport and Carboxylation Conductances and Enzyme Activities

The ontogenetic changes in net photosynthetic rate, mainly the decrease in net photosynthetic rate as the leaf ages, are usually associated with a reduction in the activity of enzymes involved in photosynthesis, e.g. in ribulose-1,5-bisphosphate carboxylase, RuBPC (see Chapter 6), and carbonic anhydrase, CA. For $C_3$ plants, it may be suggested that the activity of CA is associated with the transport component, $g_m$, and the activity of RuBPC with the carboxylation component, $g_x$ (cf. Section 7.2.6). Such relationships were reported for ryegrass by Reyss and Prioul (1975) and for aspen by Tsel'niker et al. (1981). Nevertheless, comparing net photosynthetic rate and other parameters of $CO_2$ exchange found in vivo with enzyme activities measured in vitro may lead to unrealistic conclusions. On the other hand, many of recent models of $CO_2$ exchange in leaves explain in detail the dependence of $CO_2$ exchange rate on enzyme activity and amount under different conditions (see Tenhunen et al. 1980, Raven and Glidewell 1981).

#### 8.2.4.1 Carbonic Anhydrase

After dissolving in water in pores in the cell wall, $CO_2$ is transported mainly as $HCO_3^-$. However, the conductance for the transport of ions across a lipid membrane is very low, e.g. $2 \times 10^{-4}$ mm $s^{-1}$ for $HCO_3^-$ transfer (Nobel 1974). It was shown in model experiments that $CO_2$ is transported across plasmalemma and chloroplast membranes presumably in molecular form, the conductance for which may reach ca. 2 to 3 mm $s^{-1}$. The diffusion may be facilitated by carbonic anhydrase (carbonate dehydrase, carbonate hydro-lyase, E.C. 4.2.1.1) catalysing the slow reaction ($t_{1/2} = 14$ s) $CO_2 + H_2O \rightleftharpoons H_2CO_3$ which was found in $C_3$ plants, first in cytosol and later also in chloroplasts (cf. review of Poincelot 1979).

Information on ontogenetic changes in CA activity is very rare: in young cotton leaves it was similar at any $CO_2$ level, while in adult and senescing leaves, the CA activity depended on $CO_2$ concentration (Chang 1975). Under atmospheric $CO_2$ concentration, an increase in CA activity to a maximum (appearing probably later than the maximum of net photosynthetic rate) and a decrease was found during leaf ontogeny, and a two-peak curve along the insertion gradient of leaves (Fig. 8.10). Under $CO_2$ concentration of 1000 cm³ m⁻³ the ontogenetic changes were less pronounced, and a one-peak curve was found in leaves of different insertion levels (Chang 1975). Moorby *et al.* (1975) found a decrease in the activity of CA with the age of potato leaves from 5 to 21 d .

### 8.2.4.2 *Carboxylation Conductance and Ribulose-1,5-bisphosphate Carboxylase*

It was proposed by Monteith (1963) that the "chemical", carboxylation component of $g_M$ expressed as carboxylation conductance, $g_x$, is proportional, under saturating irradiance, to the carboxylation activity. Then, if carboxylation is simply expressed as $CO_2 + A \xrightarrow{k} ACO_2$, $g_x$ is given (Acock *et al.* 1971) by

$$g_x = hkA \qquad [\text{mm s}^{-1}] \qquad (8.10)$$

where $h$ is mesophyll thickness [mm], A is acceptor concentration [mol m⁻³], and k is the rate constant [m³ mol⁻¹ s⁻¹].

Recently, it has been shown that the concept of $g_x$ can be confusing (Tenhunen *et al.* 1980), even if some authors use it. There is, of course, no direct evidence of ontogenetic changes in $g_x$ but the ontogenetic changes in the overall enzymatic capacity for photosynthesis and mainly in RuBPC (see Chapter 6) may indicate similar changes in $g_x$ (*cf.* also Ludlow and Wilson 1971).

### 8.2.5 Excitation Conductance and Photosynthetic Efficiency

Monteith (1963) proposed that under low irradiance the carboxylation rate is not limited by the rate of $CO_2$ transfer to carboxylation sites in chloroplasts but by the rate of energy transfer, *i.e.* by the rate of photochemical reactions. This limitation to net photosynthetic rate may be expressed by excitation resistance, $r_e$, or conductance, $g_e$ (*cf.* also Woolhouse 1967/1968, Ludlow 1970, Jarvis 1971, Ludlow and Wilson 1971)

$$g_e = \alpha I / c_c \, [\text{m s}^{-1}] \qquad (8.11)$$

where $\alpha$ is the photosynthetic efficiency [mg mol⁻¹], $I$ is irradiance [mol m⁻² s⁻¹] and $c_c$ is the $CO_2$ concentration [mg m⁻³] at the chloroplast surface. With some limitation, data in the literature on ontogenetic changes in $\alpha$ may therefore be interpreted in terms of excitation conductance (see Section 7.2.5).

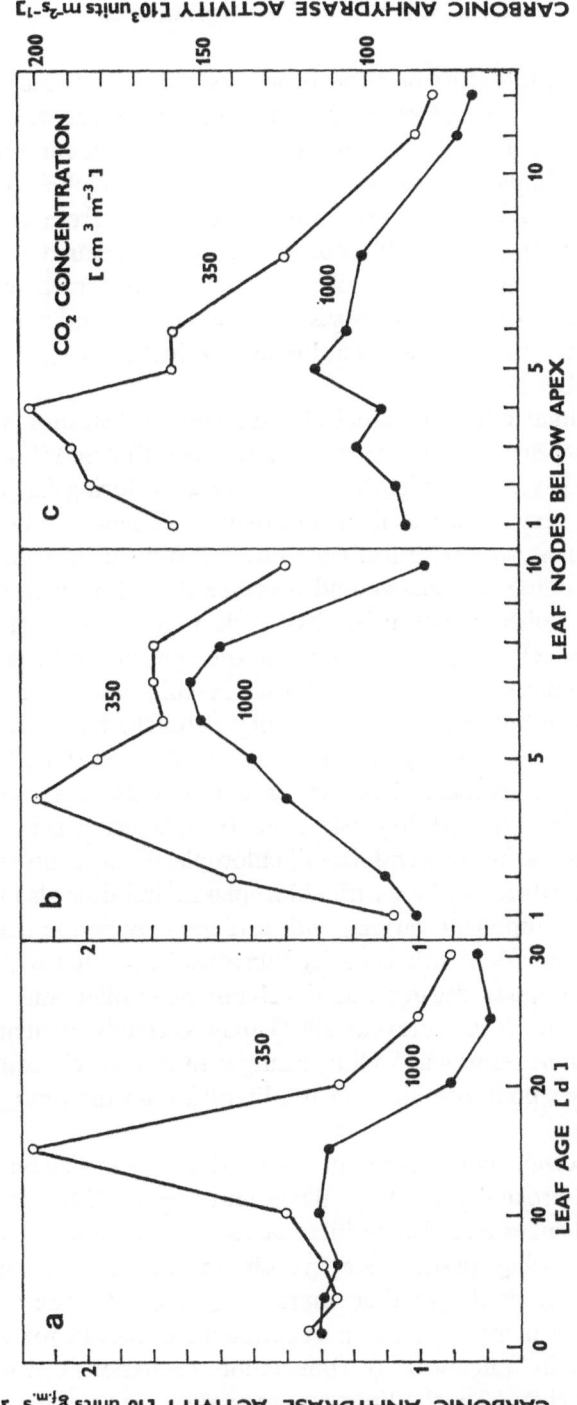

**Fig. 8.10.** Changes in carbonic anhydrase activity in *Gossypium hirsutum* L. leaves during ontogeny (a) and in leaves of different insertion levels (b, c) under two $CO_2$ concentrations (b, c — calculated on mass and area basis, respectively). (Modified and recalculated from Chang 1975.)

## 8.2.6   CO₂ Transport on Individual Cell Structures

The lack of respective literature on these problems calls for an analysis of the physico-chemical properties and cytological parameters of cellular structures. It may provide information yielding rather minimum than realistic conductances which may be several times higher in leaves *in situ* (*cf.* Chapter 1; Nobel 1974, Raven and Glidewell 1981). The main structures limiting $CO_2$ flux from the intercellular spaces to carboxylation sites in chloroplasts are plasmalemma and chloroplast membranes (envelope and thylakoid) the conductance of which may reach 3 to 4 mm s⁻¹ or less. The other structures such as the cell wall and mainly cytosol and stroma of chloroplasts may have a much higher conductance, *e.g.* between 10 and 100 mm s⁻¹ (*cf.* Nobel 1974).

There is some evidence from cytological and anatomical studies (see Chapter 1) that the internal leaf surface may vary with leaf age (Turrell 1965). The ratio between internal and external leaf surfaces ($S$) increases during leaf ontogeny, just after leaf unfolding, very rapidly at first, and then more slowly. The increase in $S$ yields an increase in $g_{m (anat)}$ with leaf age. The cell walls of mesophyll cells may thicken during leaf ontogeny; this should decrease the cell wall conductance. On the other hand, vacuolization of mesophyll cells may decrease the distance of chloroplasts from the cell wall and thus increase the cytosol conductance. A similar response may be induced by starch formation in chloroplasts. The number of chloroplasts per cell in both palisade and spongy parenchama tissue increases 2 to 10 times during leaf ontogeny (*e.g.* Possingham and Saurer 1969, Whatley 1980, Wild and Wolf 1980). Average diameter of chloroplasts increases during leaf development 3 to 4 times (*e.g.* Holowinsky *et al.* 1965, Więckowski 1967a, Whatley 1980). The increase in the number and sizes of chloroplasts may contribute to a better coverage of the internal leaf surface with chloroplasts, and thus also to an increase in $g_{m (anat)}$. Another component varying with leaf age may be the flattening of the chloroplasts upon irradiation: this capacity increases in parallel with the number of lamellae per chloroplast during leaf development (Miller and Nobel 1972). Chloroplast movements (*e.g.* Zurzycki 1961) may certainly contribute to $CO_2$ supply of carboxylation centres as well as changes in cytosol viscosity, streaming, *etc.* There is, however, no information in the literature on the ontogenetic trends of these components.

During leaf senescence, chloroplasts markedly decrease in size and the internal thylakoid system is broken down (*e.g.* Ikeda and Ueda 1964, Barton 1966; *cf.* Chapter 2). These changes and also an increase in apparent free space (*e.g.* Jacoby *et al.* 1973) in senescing tissues, may result in an increase in the overall transport conductance, similar to that found in $g_{m (anat)}$ for the whole palisade cells. Nevertheless, the increase in transport conductance may be masked by a more pronounced decrease in "chemical" (carboxylation, excitation) conductance.

There is enough evidence in the literature to suggest that the intracellular conductance and processes contributing to it are associated significantly with the ontogenetic changes in $CO_2$ exchange of the leaf. The purpose of recent studies has been to find ways how to control the individual processes and properties involved in the $CO_2$ pathway in the liquid phase in order to control photosynthesis and photosynthetic production.


## 8.3   CONCLUSIONS

The ontogenetic changes in carbon dioxide transfer in the leaf are mostly accounted for by ontogenetic changes in stomatal and intracellular conductances. Under non-limiting environmental conditions both conductances usually increase as the leaves expand, reach a broad maximum in nearly mature or mature leaves, and decrease during leaf senescence. The common finding is that stomatal and intracellular conductances change essentially in parallel as the leaves age, however maximum intracellular conductance before or after maximum stomatal conductance was also reported. Ontogenetic changes in both conductances depend also on the life span of the leaf, plant age and stage of development and on other factors affecting leaf growth. Although changes in physiological parameters between successive leaves on the same plant do not necessarily reflect quantitatively the differences due to age, they are qualitatively similar and, therefore, a sharp increase in value of stomatal and intracellular conductance to maximum (reached mostly in the upper part of the middle portion of the leaf set) followed by a slower decrease is found along the insertion gradient. Also under stress conditions the ontogenetic pattern of stomatal and intracellular conductances remains the same, or even more expressive than under non-limiting conditions as the nearly-matured or matured leaves and upper leaves on the stem are usually less sensitive to the negative influence of environmental factors than the other leaves.

# 9 PHOTORESPIRATION DURING LEAF ONTOGENY

*J. Čatský and Ingrid Tichá*

## 9.1 PHOTORESPIRATION AND PHOTOSYNTHESIS

Three principal oxygen uptake processes in leaves are presently recognized. These are the oxygenase reaction of ribulose-1,5-bisphosphate carboxylase/oxygenase (RuBPCO) and the associated metabolism of phosphoglycollate, the Mehler reaction, and the oxygen uptake associated with "dark" respiration which may continue in the light (*cf.* Sections 7.3 and 10.8; Canvin *et al.* 1980). The oxidation of carbon compounds to $CO_2$ in the light via glycollate metabolism, induced by the activity of the photosynthetic system, is known as photorespiration. It could include the direct oxidation of organic acids by the two photosystems, the production of higher concentrations of oxidizable substrates, the increase in the level of substrates available for mitochondrial oxidation, and the oxygen dependent oxidation of ribulose-1,5-bisphosphate (RuBP) by RuBPCO (Schaedle 1975). In brief (Heber and Krause 1980), glycollate is formed in the chloroplast at the expense of sugar-phosphate intermediates of the Calvin cycle: oxygen competes with $CO_2$ at the active sites of RuBPCO. Instead of being carboxylated, RuBP is oxygenated to form phosphoglycerate and phosphoglycollate. The latter is hydrolysed to glycollate. After export from the chloroplast, the glycollate undergoes a series of reactions taking place in the peroxisome, mitochondrion, and, finally, again in the chloroplast. There, phosphoglycerate is produced, linking glycollate metabolism to the Calvin cycle. A peculiar feature of photorespiration is that, in contrast to mitochondrial, dark respiration, it does not evolve but consumes energy.

Whereas the mechanism of photorespiration has been the subject of numerous studies, rather few data are available on the associated gas fluxes in tissues, individual organs or whole plants, although such gas exchange studies initiated the discovery of the main expressions of photorespiration, *i.e.* the loss of $CO_2$ and the uptake of oxygen in light (Gerbaud and André 1980). The most conspicuous gas exchange expressions of photorespiration in $C_3$ species are the oxygen dependence of net photosynthetic rate (and also dry matter production) and the $CO_2$ compensation concentration (*cf.* Section 7.2.6.2).

The exact role of photorespiration is still not known, although several functions have been proposed (*cf.* Chollet and Ogren 1975, Heber and Krause 1980). Most frequently the protection of the photosynthetic apparatus in the case of various stresses has been discussed, *e.g.* the protection of chloroplasts against photooxidative destruction under high irradiance and low $CO_2$ concentration (*e.g.* Osmond and Björkman 1972). Photorespiration could be viewed as a scavenging process, acting to convert undesirable by-products of photosynthesis. According to some authors the photorespiratory conversion of phosphoglycollate back to phosphoglycerate serves only to salvage at least 75% of the carbon lost from the Calvin cycle because of the oxygenation of RuBP is an unavoidable side reaction to carboxylation (*cf.* Andrews and Lorimer 1978). Some authors suggest that photorespiration is involved in nitrogen metabolism because the energy liberated during photorespiration may be used to drive the reduction of nitrate to ammonia in the peroxisome (*cf.* Bassham *et al.* 1981, Canvin 1981). On the other hand, photorespiration is frequently regarded as a wasteful process (*e.g.* Zelitch 1971).

Photorespiration has a considerable bearing on the photosynthetic production at least in C3 species (*cf.* Zelitch 1971, Kumar and Singh 1979, Keys and Whittingham 1981, Moore 1981, Keys *et al.* 1982). In several, mainly arid environments, conventional photosynthesis of C3 plants has drawbacks which some tropical C4 plants overcome in different ways — by an additional biochemical pathway, a distinctive morphology and differing biochemical capabilities between their two main classes of photosynthetic cells. The virtual absence of photorespiration in C4 species has stimulated the efforts of plant scientists to endow C3 plants with C4 characteristics. There is, however, not much optimism that this can be done in the near future. Also the probability of finding plants, *e.g.* mutants, with high net photosynthetic rate and low photorespiration rate under natural conditions appears to be very low because the two processes seem integrally and tightly coupled (*cf.* McCashin and Canvin 1979).

## 9.2    MEASURES OF PHOTORESPIRATION RATE

There has been an explosion of interest in measuring photorespiration rate, $R_L$, over the past decade. Although the mechanism of photorespiration has been fairly well recognized (*cf.* Tolbert 1979), the studies of its rate still yield variable data (*cf.* Ludlow and Jarvis 1971, Gibbs and Latzko 1979). In light, at least three simultaneous processes occur in plant tissue — photosynthesis, respiration and photorespiration. Photorespiration rate must, therefore, be measured indirectly, and as a consequence, its rate is always somewhat in doubt (Schaedle 1975).

All methods commonly used in the studies of $R_L$ are unsatisfactory from one viewpoint or another (see Ludlow and Jarvis 1971 for discussion). The method using the $CO_2$ burst following irradiation (post-illumination burst — PIB) seems to be the least suitable while the most suitable methods would be those which measure the $CO_2$ efflux into $CO_2$-free air and the $CO_2$ compensation concentration $\Gamma$, and to which a model can be applied to calculate intercellular re-assimilation. Non-isotopic methods, however, do not allow a measurement of $R_L$ during photosynthesis, moreover, they do not yield estimates relevant to natural $CO_2$ concentration because $R_L$ varies with ambient $CO_2$ concentration being closely linked with photosynthesis (Ludlow and Jarvis 1971, Gerbaud and André 1980). On the other hand, the isotopic methods using carbon isotopes give the $CO_2$ efflux from the plant, but are inevitably biased by the recycling of $CO_2$. Consequently, the method used for the determination of $R_L$ should be taken into account when summarizing and interpreting information from the literature, particularly on the ontogenetic changes in $R_L$.

## 9.3 ONTOGENETIC CHANGES
### IN PHOTORESPIRATION RATE IN C₃ PLANTS

Irrespective of the method used, $R_L$ of most higher $C_3$ plants ranges from 0.1 to 0.5 mg m$^{-2}$ s$^{-1}$ (Table 9.1). In most cases, the rates reported in the literature exceed dark respiration rates by factors of 1.2 to 4, and are 10 to 50% of net photosynthetic rates (*cf.* Canvin 1979, Zelitch 1979, Canvin *et al.* 1980). Re-assimilation inside the cell and tissues may at a guess be 30 to 40% of $R_L$ measured (Canvin *et al.* 1980).

As photosynthesis and photorespiration are tightly coupled, similar ontogenetic changes may be expected in photorespiration as is generally known in photosynthesis (Chapter 7). The $CO_2$ evolution in light, however, originates from photorespiration and any mitochondrial respiration that occurs in the light. The source of $CO_2$ can hardly be distinguished. As the dark respiration rate is relatively small in adult leaves of $C_3$ species most of $CO_2$ evolved in the light represents photorespiration, with the limitation that any $CO_2$ released and immediately refixed within the tissue is not measured (*cf.* Section 9.2).

In spite of numerous reports on $R_L$ as affected by various internal and environmental factors, complete data on ontogenetic changes in $R_L$ are rather rare in the literature. The most complete information may be derived from reports of $CO_2$ compensation concentration which has been summarized in Chapter 7 (Fig. 7.10). On the other hand, only few papers report $R_L$ measured by one or more methods throughout the whole leaf life (Cornic *et al.* 1970, Furukawa 1973b, Kisaki *et al.* 1973, Hodgkinson 1974, Čatský *et al.* 1976, Thomas *et al.* 1978, Samsuddin and Impens 1979a, Tichá *et al.* 1980, Kennedy and Johnson 1981; Table 9.1, Figs. 9.3 and 10.3). Irrespective of the method used, the most common finding is an increase in $R_L$ to a maximum near to (usually after) the maximum of net photosynthetic rate followed by a usually slower decrease (Figs. 9.1, 9.2, 9.4). The position of maximum may be affected to some extent by mitochondrial respiration which may influence $R_L$ — besides inducing an error in determination — by changing the carbon dioxide concentrations and fluxes inside the leaf. Cornic *et al.* (1970) explained the decrease in net photosynthetic rate in ageing mustard leaves by the influence of photorespiration acting mainly by increasing the internal re-assimilation.

The ontogenetic changes in $R_L$ expressed in carbon dioxide fluxes have been significantly correlated to changes in some parameters of photorespiratory mecha-

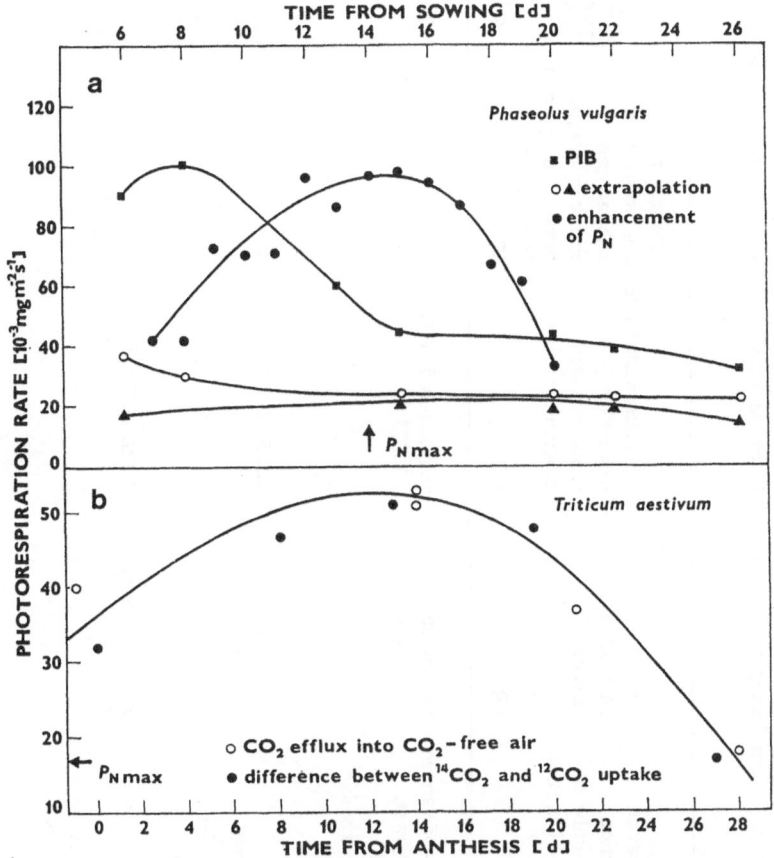

**Fig. 9.1.** Ontogenetic changes in photorespiration rate measured by different methods. a: Enhancement of net photosynthetic rate in 2% oxygen, post-illumination burst of $CO_2$ in the dark (PIB), and $CO_2$ efflux into $CO_2$-free air estimated by extrapolation of the $CO_2$ dependence of net photosynthetic rate to zero $CO_2$ concentration ($\bigcirc$ — $R_{L \text{ extr}}$, or $\blacktriangle$ — $R_{L \text{ extr}}$ minus 0.25 $R_D$) in primary leaves of *Phaseolus vulgaris* L. (Recalculated from Čatský *et al.* 1976 and from Kaše and Čatský, unpublished.) b: $CO_2$ efflux into $CO_2$-free air and $CO_2$ evolution in the light estimated from the difference between $^{14}CO_2$ and $^{12}CO_2$ uptake by the same leaf of *Triticum aestivum* L. (Recalculated from Thomas *et al.* 1978.)

nism. Studying the differences in photorespiratory and photosynthetic characteristics in tobacco leaves of different positions on a stalk, Kisaki *et al.* (1973) found maximum $R_L$ and maximum activities of glycollate oxidase from peroxisomes and

**Table 9.1** Photorespiration rate, $R_L$, during leaf ontogeny and in leaves of different insertion levels.

Abbreviations: $c_a$ — ambient $CO_2$ concentration; $c_i$ — internal $CO_2$ concentration; IRGA — infra-red gas analyser; LPI — leaf plastochron index; max — maximum; min — minimum; $P_G$ — gross photosynthetic rate; $P_N$ — net photosynthetic rate; PIB — post illumination burst; $r_M$ — intracellular resistance to $CO_2$ transfer; $\Gamma$ — $CO_2$ compensation concentration; $\downarrow$ — decrease; $\uparrow$ — increase. (2) The leaves are numbered from the base of the plant. Where the leaves are numbered from the top of the plant and the overall number of leaves on the plant is not indicated, the symbols leaf $n$, leaf $(n+1)$, etc. are used.

| Plant species and cultivation | Age of leaf or leaf insertion (leaves numbered from the oldest one) | Method of measurement | Photorespiration rate [mg m⁻² s⁻¹], if not indicated otherwise | Changes in photorespiration rate induced by ageing or insertion level (from upper to lower leaves) | Author(s) and year of publication |
|---|---|---|---|---|---|
| (1) | (2) | (3) | (4) | (5) | (6) |
| *Capsicum annuum* var. *grossum* cv. Market Giant (soil-coarse sand-peat moss mixture 1:1:1, glasshouse) | leaf one node above the first flower, 0 to 84 d after its anthesis | extrapolation of $P_N$ to $c_i = 0$ | 0.190...0.085...0.120 ...0.075 | $\downarrow, \uparrow, \downarrow$ | Hall and Brady 1977 |
| *Glycine max* (L.) Merrill cv. Harosoy (pots with sand, growth chamber) | leaf 7, 12–16 d and 31–34 d after flowering | extrapolation of $P_N$ to $c_i = 0$ | 0.255...0.147 | $\downarrow$ | Sambo *et al.* 1977 |
| *Hevea brasiliensis* Muell. Arg. (controlled environment) | attached leaves, 17 to 84 d (Leaf Blade Class 5 to 9 + 50 d) | extrapolation of $P_N$ to $c_a = 0$ | 0.002...0.050...0.039 | $\uparrow, \downarrow$ (max d 44) | Samsuddin and Impens 1979a |
| *Malus domestica* Borkh., one-year old seedlings (pots, glasshouse) | intact leaves, LPI 7.9 to 14.9 | | 0.505...0.122...0.131 | $\downarrow, \uparrow$ | Kennedy and Johnson 1981 |

| Species (conditions) | Leaves / age | Method | Values | Trend | Reference |
|---|---|---|---|---|---|
| *Medicago sativa* L. cv. Hunter River (glasshouse) | leaves 2 to 14 | open system, IRGA, $CO_2$ evolution into $CO_2$-free air (1), or $R_L = \dfrac{2\Gamma}{r_M}$ (2) | (1): 0.135...0.060 (2): 0.350...0.080 | ↓ ↓ | Hodgkinson 1974 |
| *Phaseolus vulgaris* L. cv. Pencil Pod Black Wax (growth chamber) | primary leaf and leaves 1 to 3, 0 to 26 d | radiometry ($^{12}CO_2$, $^{14}CO_2$), $R_L = P_G - P_N$, IRGA | leaf 3: 0.106...0.111...0.106 leaf 2: 0.106...0.050...0.072 leaf 1: 0.111...0.038...0.181 ...0.038 primary leaf: 0.050...0.100...0.033... ...0.067 | ↑, ↓ (max d 22) ↓, ↑ (min d 19) ↓, ↑, → (min d 17, max d 21) ↑, ↓, ↑ (max d 13, min d 20) | Fraser and Bidwell 1974 |
| *Phaseolus vulgaris* L. cv. Jantar and Harzgruß (growth chamber) | primary leaf, 7 to 20 d from sowing | closed system, IRGA, extrapolation $P_N/c_a$ to $c_a = 0$ and from the enhancement of $P_N$ in 2% $O_2$ as compared with 21% $O_2$ | 0.205...0.500...0.195 0.280...0.180...0.190 | ↑, ↓ (max d 15) ↓, ↑ | Čatský et al. 1976, Čatský and Tichá 1979, 1980, Tichá et al. 1980 |
| *Populus euramericana* (glasshouse) | leaves $n$ to ($n + 13$) | open system, IRGA, $CO_2$ evolution in $CO_2$-free air and PIB (cut leaves) | $CO_2$ output in $CO_2$-free air: 0.039...0.096...0.028 PIB: 0.111...0.139...0.049 | ↑, ↓ max leaf ($n + 9$) ↑, ↓, max leaf ($n + 8$) | Furukawa 1973b |
| *Populus* × *euramericana* cv. Wisconsin-5 (growth chamber) | leaf 8 at LPI 4, 8, 19, 36 | closed system, IRGA, $CO_2$ evolution into $CO_2$-free air | 0.158...0.058...0.081 | ↓, ↑ (min at LPI 19) | Dickmann and Gordon 1975 |

**Table 9.1** (continued)

| Plant species and cultivation | Age of leaf or leaf insertion (leaves numbered from the oldest one) | Method of measurement | Photorespiration rate [mg m$^{-2}$ s$^{-1}$], if not indicated otherwise | Changes in photorespiration rate induced by ageing or insertion level (from upper to lower leaves) | Author(s) and year of publication |
|---|---|---|---|---|---|
| (1) | (2) | (3) | (4) | (5) | (6) |
| *Sinapis alba* L. (phytotron) | leaves 7 and 13 (4 to resp. 46 d) | IRGA, PIB and $CO_2$ evolution in $CO_2$-free air | leaf 13: 0.111...0.178...0.052 leaf 7: 0.089...0.145...0.096 | ↑, ↓ (max d 7) ↑, ↓ (max d 10) | Cornic *et al.* 1970 |
| *Triticum aestivum* L. cv. Carola (controlled environment) | intact flag leaf, 0, 16 and 27 d after anthesis | open system, IRGA; extrapolation of $P_N$ to $c_i = 0$ at 21% $O_2$ | 0.170...0.097 | ↓ | Peisker and Apel 1976 |
| *Triticum aestivum* L. cv. Maris Huntsman (field) | intact flag leaf, 0 to 30 d from anthesis | radiometry: difference between $^{14}CO_2$ and $^{12}CO_2$ uptake, $CO_2$ evolution into $CO_2$-free air | 0.154...0.229...0.088 | ↑, ↓ (max d 14 from anthesis) | Thomas *et al.* 1978 |

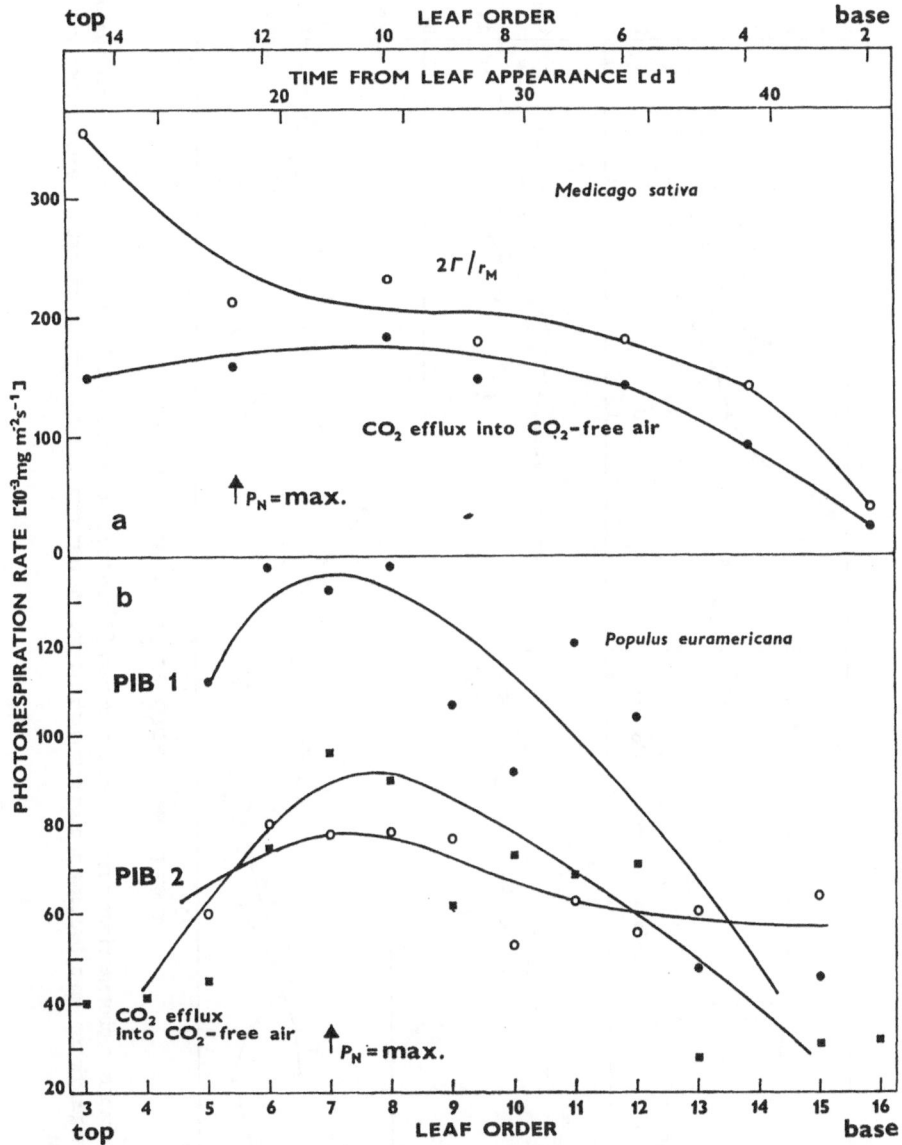

**Fig. 9.2.** Photorespiration rate measured or calculated by different methods in leaves of different insertion level. a: Photorespiration rate in *Medicago sativa* L. measured as $CO_2$ efflux into $CO_2$-free air or calculated as $CO_2$ flux at the $CO_2$ compensation concentration — $2\Gamma/r_M$ according to Ludlow and Jarvis (1971). (Recalculated from Hodgkinson 1974.)

b: Photorespiration rate in *Populus euramericana* measured as $CO_2$ efflux into $CO_2$-free air or as post-illumination bursts of $CO_2$ in the dark. However, two bursts in very young and very old leaves could not be clearly distinguished. (Recalculated from Furukawa 1973b.)

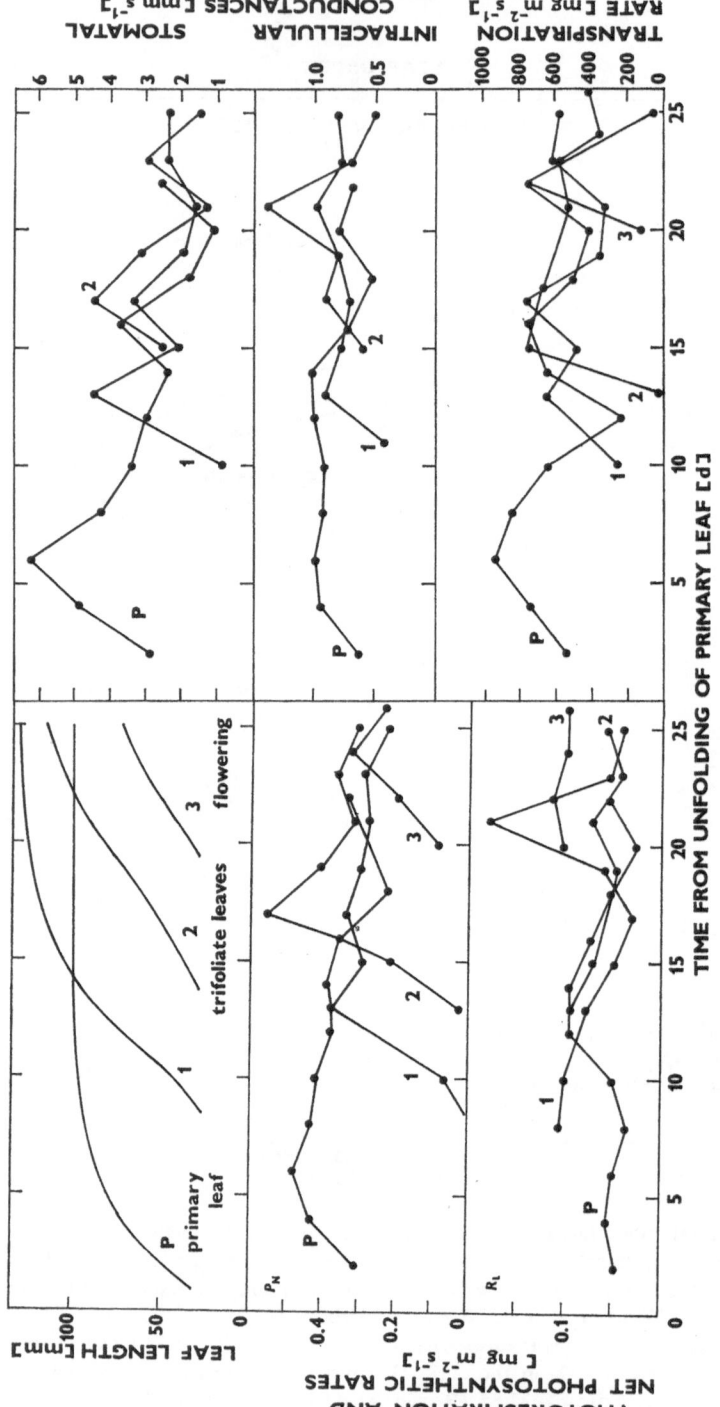

**Fig. 9.3.** Ontogenetic changes in photorespiration rate, $R_L$, of the primary (P) and the first, second, and third secondary leaves of *Phaseolus vulgaris* L. as compared with ontogenetic courses of leaf length, net photosynthetic rate, $P_N$, transpiration rate, stomatal, and intracellular conductances (from Fraser and Bidwell 1974).

phosphoglycollate phosphatase from the chloroplast in the third youngest leaf (Fig. 9.4). Similar, but two-peak changes in glycollate oxidase and phosphoglycollate phosphatase were reported during the ontogeny of the primary leaf of barley (Passera and Albuzio 1977). On the other hand, Salin and Homann (1971, 1973) found that young citrus and tobacco leaves photorespired less, had lower activities

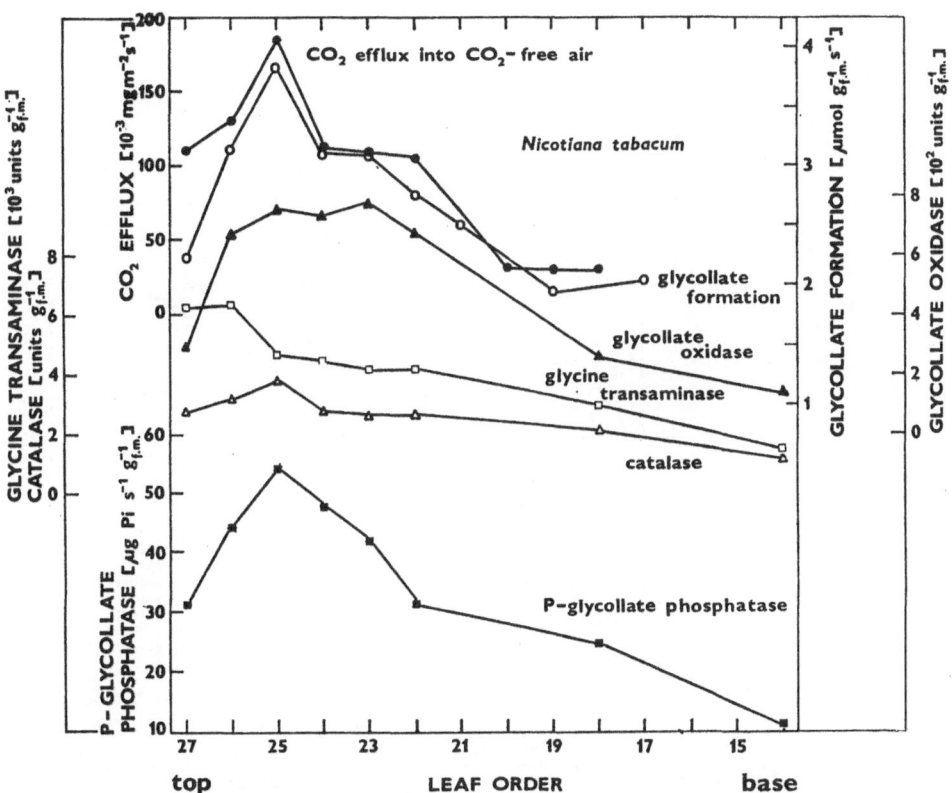

**Fig. 9.4.** Photorespiration rate measured as $CO_2$ efflux into $CO_2$-free air, glycollate formation and the activity of some peroxisomal (glycollate oxidase, E.C.1.1.3.1, catalase, E.C.1.11.1.6, glycine transaminase or glutamate-glyoxylate aminotransferase, E.C.2.6.1.4) and chloroplast (P-glycollate phosphatase, E.C.3.1.3.18) enzymes connected with glycollate metabolism, in leaves of *Nicotiana tabacum* L. of different insertion level. (Recalculated from Kisaki *et al.* 1973.)

of photorespiratory enzymes and a slower synthesis of glycollate than older leaves. The "young" leaves used by Salin and Homann were not fully unfolded and had fewer photosynthetic pigments than older leaves; this explains the apparent disagreement with the findings of Kisaki *et al.* (1973).

Good correlation has usually been found between the ontogenetic changes in $R_L$ and the ratio between the oxygenase and carboxylase activities of RuBPCO (*cf.*

Chapter 6). Passera (1975) reported that $R_L$ of barley leaves increased with plant age as a consequence of a higher oxygenation activity of RuBPCO. However, according to Thomas *et al.* (1978) $R_L$ declined much less than the net photosynthetic rate as wheat leaves aged. On the other hand, the oxygenase activity of RuBPCO declined at the same rate as the carboxylase activity, so that the carboxylase to

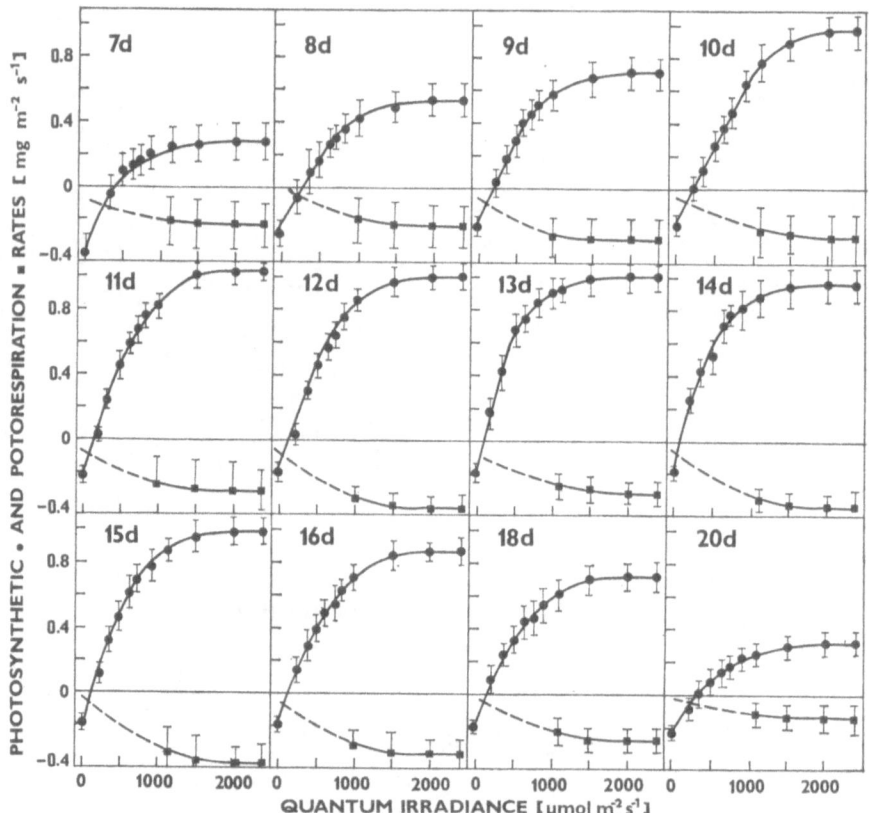

**Fig. 9.5.** Modification of the ontogenetic changes [d from unfolding] in photorespiration ($R_L$) and net photosynthetic ($P_N$) rates by irradiance in primary leaves of *Phaseolus vulgaris* L. $R_L$ was determined as enhancement of $P_N$ in 2% oxygen. Means ± standard mean errors. (Modified from Čatský and Tichá 1980.)

oxygenase ratio remained constant. The authors concluded that the change in the relative rates of photosynthesis and photorespiration *in vivo* did not reflect a change in carboxylase to oxygenase activities of RuBPCO *in vitro*.

The ontogenetic course of $R_L$ is modified — similarly to net photosynthetic rate — by individual developmental events, *e.g.* unfolding of new leaves, flowering, *etc.* (Fraser and Bidwell 1974). Hodgkinson (1974) reported that both net photosynthetic and photorespiration rates of lucerne leaves declined as leaves aged and increased after partial defoliation.

The shape of the ontogenetic curve of $R_L$ is modified also by different environmental factors, *e.g.* irradiance (Carpenter and Hanover 1974, Fig. 9.5), mineral elements (Ozbun *et al.* 1965, Kabaki *et al.* 1979), growth regulators (Poskuta *et al.* 1975), *etc.* Ozbun *et al.* (1965) reported that potassium deficiency accelerated $R_L$ indicated by $CO_2$ release ($^{18}O_2$ and $^{13}CO_2$ method) at both low and high irradiances, regardless of the age of leaves of *Phaseolus vulgaris*. However, rather a decrease in photorespiratory activity under phosphorus deficiency was reported (Kabaki *et al.* 1979) in young and old rice leaves when oxygenase and carboxylase activities of RuBPCO were measured.

Similarly to photosynthesis, the ontogenetic decrease in $R_L$ may be stimulated by pathogenic organisms. It seems, however, that mild infection with some pathogens does not change significantly $R_L$ in ageing leaves before the appearance of visible damage of leaves, at least in 21% $O_2$ (*e.g. Uromyces* on *Phaseolus* — Raggi 1978b).

In leaves of evergreen plants, similar ontogenetic changes may be expected in $R_L$ as those reported for net photosynthetic rate (Chapter 7). Nilsen and Mortensen (1978) demonstrated that $R_L$ (measured as $P_N$ enhancement in 1% oxygen) increased with leaf age in spruce plants 3 to 6 months old.

## 9.4   ONTOGENETIC CHANGES IN LIGHT RESPIRATION IN $C_4$ SPECIES

Zero or very low photorespiration rates have frequently been reported in $C_4$ plants. However, any $CO_2$ production in light in these species may be masked by an intensive recycling of $CO_2$ in $C_4$ photosynthesis. Moreover, the $CO_2$ concentration inside the bundle-sheath cells is too high (due to the "$CO_2$ pump" in mesophyll cells) to enable to distinguish photorespiratory $CO_2$ evolution. Thus, higher $R_L$ reported in senescent leaves may be connected with the operation of both $C_3$ and $C_4$ photosynthesis in the respective leaves. According to Imai and Murata (1979), $R_L$ in senescent leaves of $C_4$ grasses (*Sorghum, Zea, Panicum, Digitaria, Pennisetum, etc.*) was as low as in mature leaves (*i.e. ca.* 0.010 mg m$^{-2}$ s$^{-1}$) while in $C_4$ Cyperaceae (*e.g. Cyperus microria*) it greatly increases (from *ca.* 0.025 to 0.060 mg m$^{-2}$ s$^{-1}$) with senescence. Thus, the senescent leaves of grasses retained their $C_4$ characteristics until near death, although their $CO_2$-fixing ability became very low. On the other hand, the photorespiratory activity in $C_4$ plants may be guessed from the activity of the enzymes involved in the glycollate metabolism. In maize, glycollate oxidase activity calculated on a protein basis decreased with leaf position: the lowest, 1st leaf showed an activity 60% higher than the upper 11th leaf (Soldatini 1979).

In the $C_4$ species *Portulaca oleracea* (Kennedy 1976), young and mature leaves were unaffected by changes in irradiance and oxygen concentration, and exhibited typical $C_4$ plant light/dark $^{14}CO_2$ evolution ratios. Senescent leaves, on the other hand, had photorespiration rates similar to those of $C_3$ plants: they were affected by the absence of $CO_2$ and the presence of 100% oxygen and nitrogen as in $C_3$ plants.

## 9.5    CONCLUSIONS

As photorespiration is tightly coupled with photosynthesis, ontogenetic changes in the photorespiration rate in leaves of $C_3$ species are similar to ontogenetic changes in the net photosynthetic rate. Irrespective of the method used for the assessment of the photorespiration rate, it increases during leaf ontogeny to a maximum, usually after maximum net photosynthetic rate, and then slowly decreases. The changes in photorespiration rate in leaves of different ages and positions are similar, taking into account the features discussed in Chapter 7.

The ontogenetic changes in photorespiration rate expressed in carbon dioxide fluxes are significantly correlated with changes in some parameters of photorespiratory mechanism, *e.g.* the activities of photorespiratory enzymes, the ratio between carboxylase and oxygenase activities of RuBPCO, the concentration of glycollate, *etc.* The ontogenetic course of photorespiration rate is modified by individual developmental events, *e.g.* unfolding of new leaves, flowering, *etc.*, by artificial defoliation, or by environmental factors.

In $C_4$ species, measurable photorespiration rates may be found in very young or senescent leaves. This feature may be connected with the operation of both $C_3$ and $C_4$ photosynthesis in the respective leaves.

# 10 INTEGRATION OF PHOTOSYNTHETIC CHARACTERISTICS DURING LEAF DEVELOPMENT

*Z. Šesták, Ingrid Tichá, J. Čatský, Jarmila Solárová, Jana Pospíšilová and Danuše Hodáňová*

## 10.1 GENERAL CONSIDERATIONS

The formation of leaf structure, the development of the chloroplast as a basic unit of photosynthesis, the synthesis of pigment complexes and the components of the electron transport chain in the thylakoid as the place of the photochemical reactions of photosynthesis have been touched on or thoroughly analysed by many authors. Unfortunately, the major part of the existing literature contains the determination of usually only one or two characteristics which belong to the complex mosaic called "the photosynthetic apparatus". These findings have already been described in the previous Chapters. The papers with a more synthetic aspect are as rare as those that thoroughly analyse the whole leaf life span or the complete leaf insertion gradient.

Due to the complexity of the problem of obtaining plants with leaves of optimum photosynthetic production, a steady search for important characteristics of the photosynthetic apparatus which may easily be determined, is necessary. Such characteristics are certainly significant for the current work in plant selection. Of course, proposals for new agricultural plants as perspective prototypes for the future means of selection require modelling of physiological processes in plants, the formation of plant ideotypes. These ideotypes must include the optimum development of a plant, and mainly optimum ontogeny of shape and functions of a leaf as the principal organ of photosynthesis.

This is why the role of basic photosynthetic characteristics of various levels is often questioned, *e.g.*, whether it is better for a plant to have more chloroplasts with less chlorophyll per chloroplast or *vice versa*, whether many small or few large stomata are optimal, whether to have more leaves with a shorter life span or *vice versa*, whether larger or smaller leaves are better, whether the photosynthetic characteristics of $C_4$ plants are really advantageous, *etc.* To answer them is not always easy, even if we take into account concrete climatic conditions.

To say that for optimum photosynthetic activity of a leaf all its photosynthetic characteristics must be optimal throughout leaf ontogeny is of as little value as to say that the existing photosynthetic mechanisms are ineffective and should be completely changed. Of course, present day knowledge does not permit a definite analysis of ontogeny of the photosynthetic apparatus and its functions nor does it find the exact timing of bottle-necks in the leaf life span. Thus the goal of this Chapter is to show some examples of the comparison of the time courses of two or more photosynthetic characteristics, and to point out the role of their interaction in the final photosynthetic balance. Finally, an attempt will be made to synthesize present day knowledge on the ontogeny of the photosynthetic apparatus. This task will be very difficult, as even the data, though reliable, were obtained in different plant species and cultivars, grown under various environmental conditions, the determinations were obtained under different external conditions by methods of differing accuracy and standard error, and the results were calculated on different bases; only rarely the primary aspect of the experiments was a purely ontogenetic one and thus very different plants were taken for the "control" ones.

## 10.2    LEAF STRUCTURE AND PHOTOSYNTHESIS

Leaf structure represents a skeleton in which physiological processes take place including photosynthesis with its photophysical, photochemical and biochemical reactions. The formation of this structural basis for the functioning of photosynthesis and its ontogenetic changes are thus the first assumption of the performance of photosynthesis during leaf development. But photosynthesis is a very complex process and structural changes are only one of numerous events which affect the photosynthetic performance of the leaf during its development.

Leaf area enlargement serves as a common indication of leaf growth and development (*cf.* Section 1.2). Leaf lamina shows a 15 to 30-fold increase in area following leaf unfolding. Growth in length and area of the leaf is often sigmoid when plotted against time. Laminar extension is due to cell division and cell enlargement, and cell extension may continue, *e.g.*, in the palisade cells even after expansion in area has ceased.

Comparing the sigmoid leaf area enlargement and the course of net photosynthetic rate during leaf development, several patterns are found (Fig. 10.1). The most general trend of the ontogenetic changes in net photosynthetic rate is a rapid increase to a maximum followed by a slower decline. Maximum leaf area is preceded by the maximum net photosynthetic rate which is attained at about 35 to 90% of the maximum leaf area (*e.g.*, 35 to 55% in *Solanum* — Borzenkova and Nefedova 1981, 37% in *Nicotiana* — Rawson and Hackett 1974, 50 to 80% in *Helianthus* — Rawson and Constable 1980, 70% in *Cucurbita* — Turgeon and Webb 1975, 70% in *Phaseolus* — Čatský et al. 1976, 75 to 90% *Gossypium* — Constable and Rawson 1980; Fig. 10.1). The peak for net photosynthetic rate is reached at 35 to 40% of final laminar length (*Beta* — Fellows and Geiger 1974). This type of interrelationship between net photosynthetic rate and leaf area during leaf ontogeny is common in mesophytic plants. In some plant species, the decline in net photosynthetic rate after the maximum was reached, is interrupted by a second or more maxima in net photosynthetic rate during the leaf life span (Fig. 10.1). These secondary maxima are ascribed to various events in the plant development such as unfolding of further leaves, flowering, *etc.* (*e.g.*, *Phaseolus* — Fraser and Bidwell 1974, Šesták *et al.* 1975; *Glycine* — Woodward and Rawson 1976; *Capsicum* — Hall and Brady 1977). Leaf area expansion in these species is sigmoid and not

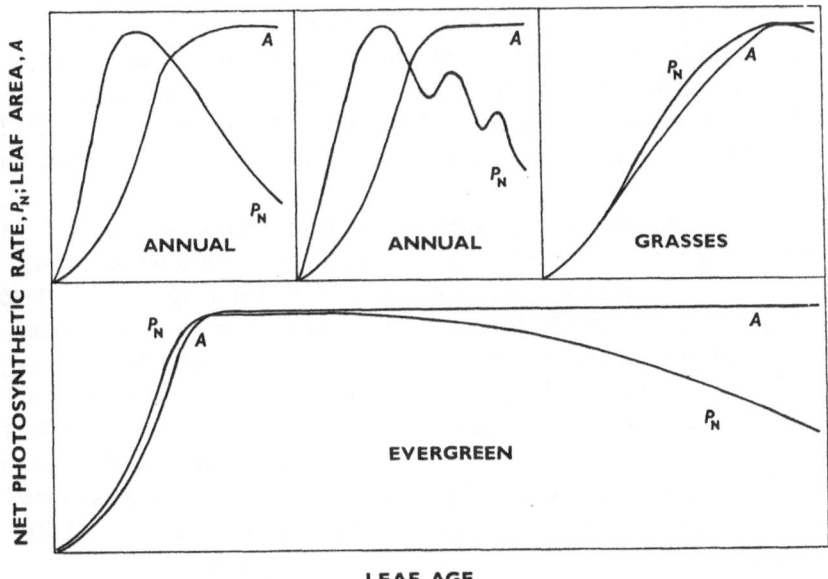

**Fig. 10.1.** Comparison of ontogenetic patterns in net photosynthetic rate ($P_N$) and leaf area ($A$) in different plant species. In annuals maximum $P_N$ at 35 to 90% of maximum $A$ is attained. Two or more maxima in $P_N$ may occur due to various events in the plant development. In other species (*e.g.* some grasses), $P_N$ and $A$ continue to rise during the whole leaf life span, and maximum $P_N$ is reached at maximum $A$. In evergreen plants there is a rapid, nearly parallel increase in $P_N$ and $A$; during the very long leaf duration minimum changes in $A$ and a very slow decline in $P_N$ are observed. (Redrawn after Mokronosov 1981.)

affected by these events. In leaves of some grasses net photosynthetic rate continuously increases throughout the leaf life span in parallel with the leaf area growth. Maximum net photosynthetic rate is here attained at maximum leaf area (*e.g.*, *Festuca* — Jewiss and Woledge 1967; Fig. 10.1). Leaves of deciduous and mainly evergreen plants are characterized by long leaf duration; leaf area expansion and development of net photosynthetic rate activity are distributed on a longer time scale. There is a rapid and nearly parallel increase in leaf area and net photosynthetic rate in a young leaf; the achieved maximum values of both parameters are maintained for a long period, and in senescing leaves net photosynthetic rate decreases very slowly (Fig. 10.1).

A rather irregular relation has been found between photosynthesis and leaf thickness in several grasses and herbs (El-Sharkawy and Hesketh 1965) although Wilson and Cooper (1967, 1969) report a strong negative relationship between the mean cross-sectional area per mesophyll cell and net photosynthetic rate in *Lolium*. The increase in leaf thickness during leaf ontogeny may result in a higher concentration of compartments of the photosynthetic apparatus (more chloroplasts) and in lengthening the effective optical path of radiation inside the leaf (*cf*. Chapter 4) but also in longer diffusion pathways for $CO_2$ from the ambient air into carboxylation centres. The enlargement of the internal leaf surface during leaf ontogeny (from 10—12% to 30—50% — *e.g*. Tichá and Čatský 1977, Mokronosov 1981) may contribute to a better $CO_2$ transfer as more parallel $CO_2$ fluxes are possible. More chloroplasts in the mesophyll cells can occupy a larger internal leaf surface.

Stomata are the component of the leaf epidermis which provides the leaf with the opportunity to change both the partial pressure of $CO_2$ at the sites of carboxylation and the rate of transpiration ( *cf*. Sections 10.7 and 10.10). The number of stomata per unit leaf area in a young leaf increases by stomata mother cell division. After that a permanent decline in stomata density occurs accompanied by an enlargement of stomata sizes (*cf*. Chapter 1). The increase in stomata size may compensate, to some extent, the decline in stomata density so that the relative index of the actual area of stomata pores may be similar during leaf ontogeny and throughout the leaf blade (*e.g*. Slavík 1963). The ontogenetic course of stomata density correlates with the changes in stomatal conductance (*cf*. Chapter 8 and Section 10.7) during leaf development.

Photosynthetic activity of a leaf is correlated also with further anatomical parameters of the leaf mesophyll as, *e.g*., the number of chloroplasts per leaf or per leaf area unit, the total chloroplast surface, and the size, surface area and volume of mesophyll cells (El-Sharkawy and Hesketh 1965, Mokronosov 1978, 1981, Nobel and Longstreth 1981, Tsel'niker 1978). Note that the term leaf mesophyll was originally used for all leaf structures between both epidermes of a dorsiventral leaf, *e.g*. including palisade parenchyma, spongy parenchyma and conducting and supporting tissue. Recently, as mesophyll cells the outer concentric layer of cells around the bundle sheaths cells in the leaf structure of $C_4$ plants is described, and from this some confusion may arise.

In the course of leaf ontogeny the number of chloroplasts per cell increases (*cf*. Chapter 1 and Figs. 1.8 a and 10.2), but per unit leaf area it declines 1.5 to 3 times (Fig. 10.2). The reason for this is that the elongation of palisade cells proceeds more rapidly than chloroplast division; the relative volume of chloroplasts per cell declines from 15 to 35% in young leaves to 3 to 6% in mature leaves (Mokronosov 1981). Thus, when the photosynthetic activity per chloroplast does not increase, this anatomical factor may contribute to the decline in net photosynthetic rate after reaching the maximum net photosynthetic rate (Fig. 10.2). Chloroplast growth in size during leaf ontogeny represents a longer pathway for $CO_2$ to the carboxylation sites. On the other hand, due to the proceeding vacuolization of mesophyll cells chloroplasts are located in the narrow layer of cytosol nearest the

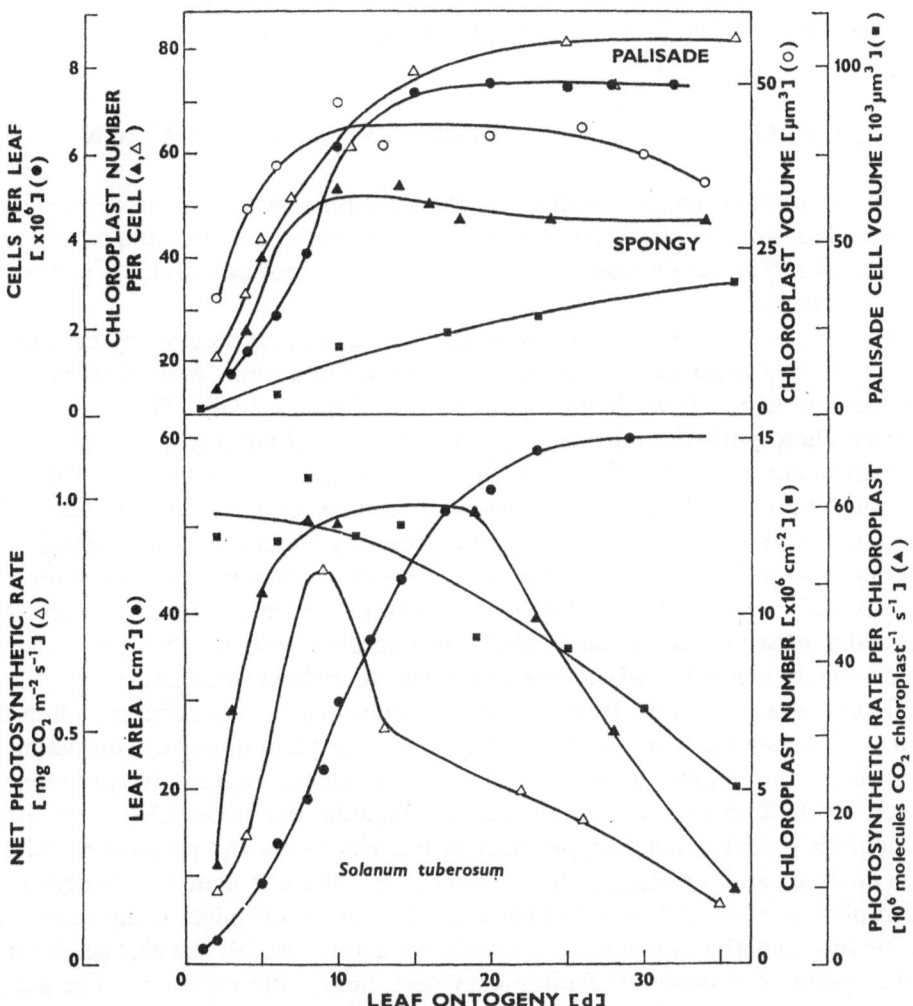

**Fig. 10.2.** Some anatomical and morphological parameters as number of cells per leaf (●), chloroplast numbers per palisade (△) and spongy parenchyma (▲) cells, chloroplast (○) and palisade cell (■) volumes (*top*), leaf area (●), and number of chloroplasts per leaf area unit (■) compared with net photosynthetic rate per leaf area unit (△) and per chloroplast (▲) (*bottom*) during potato leaf ontogeny. The diminished net photosynthetic rate after the maximum net photosynthetic rate was reached, can partly be explained by the rapid decline of chloroplast number per leaf area unit (1.5 to 3 times per cm²) which is not compensated by an increase in the photosynthetic activity of the chloroplast itself (*Solanum tuberosum* L., after Mokronosov 1981).

cell sufrace. This may partially compensate for the longer $CO_2$ pathway (Tichá and Čatský 1977).

Bearing in mind the importance of leaf anatomical compartmentation for the functioning of the $C_4$ type of $CO_2$ fixation, the $C_4$ syndrome, it is surprising that only little information on ontogenetic changes in leaf anatomy in $C_4$ plants is available till now. This gap in our knowledge must be closed in the future.

## 10.3 CHLOROPLAST FORMATION AND DEGENERATION

As a chloroplast is the basic unit of higher plant photosynthesis, its development and mainly formation of substructure and composition of the thylakoid membranes (see Chapter 2) is closely associated with the development of individual photosynthetic reactions.

The onset of activities of Photosystems 1 and 2 is made possible by the formation in the thylakoid membrane of the photosynthetic unit, even if its primary composition is far from being optimum (see Section 10.4). The rather large primary photosynthetic unit has many molecules of chlorophyll, mainly present in forms *in vivo* serving in the light-harvesting complexes, but the biosynthesis of reaction centre complexes, electron transporters and enzymes still continues. The formation of grana substructure is then a necessary condition for the functioning of Photosystem 2. The phase of reaching optimum structure and composition of the thylakoid apparatus is reflected in maximal photochemical activities (the expression of which in the final balance of photosynthesis is often prevented by increased resistances to carbon dioxide transfer to carboxylation sites).

Chloroplasts are also the first organelles to show evidence of degeneration in the course of senescence which is induced by the whole plant ontogeny (formation of new leaves, flowers, *etc.*) as well as microclimatic factors (water stress, shading of leaves, *etc.*). The senescence phenomena include the accumulation of large osmiophilic globules, the disappearance of free ribosomes and polysomes, chloroplast swelling, and, finally, the dissolution of envelope and complete distortion of chloroplast structure. The accompanying changes in chemical composition are closely related with those of chloroplast ultrastructure, namely the changes in types and amounts of proteins (including enzymes), lipids, and pigments. The steady balance of synthetic and degradative processes shifts during leaf development in favour of the latter. Rapid reaction of chloroplasts to various interfering effects and abrupt climatic factors, such as plant decapitation, temperature shocks, changes in leaf irradiance, *etc.*, which often results in the reversion of the degradation processes, shows the important role of growth regulators in the developmental processes.

Examples for the above-mentioned dependences of structure and function may be found in the literature, even if their number is not high enough to allow a final decision. Thus in potato the highest organization of chloroplast substructure (densely packed lamellae, large grana) and the highest concentrations of ATP and ATPase were found in leaves of the middle tiers during bud formation; in these leaves the highest net photosynthetic rates and the highest assimilation numbers were found (Kislyakova *et al.* 1967). The leaves of *Populus deltoides* reached maximum area 10 to 12 d after leaf unfolding: at this time the chloroplasts were larger, their thylakoid system was well organized, the number of membranes per granum was increased, the net photosynthetic rate and cyclic photophosphorylation activities were maximum. Chlorophyll amount for the same leaf area unit peaked later (30 to 40 d). 60 d leaves contained almost maximum amounts of chlorophyll, their chloroplasts were densely packed with thylakoids and contained many plastoglobules, but their photosynthetic and photophosphorylation rates were already half or less of the peak value. The decline in all characteristics continued to the end of leaf life at *ca.* 130 d (Hernández-Gil and Schaedle 1973). In tobacco, the photosynthetic rates and chlorophyll content per leaf area unit and the ratio of monogalactosyl diacylglyceroles to digalactosyl diacylglyceroles declined from young through mature to old yellowing leaves. Already the young leaves contained chloroplasts with a well developed thylakoid system; at maturity the size of plastoglobules and starch grains increased, while during yellowing the swelling and destruction of stroma lamellae was observed. The activities of Photosystems 1 and 2 per chlorophyll unit and their ratio declined from young to mature leaves, but in old yellowing leaves they abruptly increased, especially with regard to Photosystem 2; the cyclic and non-cyclic photophosphorylation rates reached maximum in mature leaves (Wolińska 1976).

In C4 plants different relationships may be observed in chloroplasts of mesophyll and bundle sheath cells. In maize very young bundle sheath chloroplasts have well-developed grana, but with the increase in chlorophyll amount they lose their grana and concurrently also their Hill reaction activity (Downton and Pyliotis 1971). Also their Photosystem 1 activity, much lower than that of the mesophyll chloroplasts, declines with the development of chloroplast ultrastructure. In mesophyll chloroplasts the peak Photosystem 1 and NADP reduction activities are in chloroplasts with the best developed ultrastructure (Andersen *et al.* 1972).

## 10.4   CHLOROPHYLL AND PHOTOSYNTHETIC ACTIVITIES

The general rule of photosynthesis is "Without chlorophyll there is no photosynthetic activity". And thus from the time of Willstätter and Stoll (1918), who proposed the idea of "assimilation number" as the photosynthetic activity of a unit chlorophyll amount, the quantitative relationship of chlorophyll amount and photosynthetic rate has often been studied.

In the early phases of leaf development or during greening of an etiolated leaf, chlorophyll formation starts earlier than the formation of photosystems. Photosystem 1 may be formed earlier than Photosystem 2, the formation of which requires appressed thylakoids. Nevertheless, the formation of appressed thylakoids and grana proceeds rather early after leaf vernation or irradiation. (These changes are reflected also in the absorption of radiant energy — see Kirk and Goodchild 1972.) $CO_2$ fixation starts later and develops more slowly than $O_2$ evolution (barley, maize — Lüttge et al. 1974) probably due to the light-dependent synthesis of enzymes of carbon cycles (namely RuBPC). The formation of chlorophyll is connected with changes in the amounts of chlorophylls a and b and proteins and in the ratio of in vivo forms of chlorophyll a; this is reflected in the shape of absorption curves in vivo (cf. Więckowski 1967a, 1969, Šesták 1972) as well as in the contents of individual proteins in thylakoids. (The total amount of proteins changes usually similarly to the content of chlorophyll in a leaf — cf. Woolhouse 1967, Gamaleï and Kulikov 1977, while the relative contents of individual proteins are very different in, e.g., young inner and old outer leaves of Lactuca sativa and are in relation to the contents of the light-harvesting chlorophyll-protein complex as well as to the substructure of the thylakoid membrane — Henriques and Park 1976.)

During leaf expansion, the formation of chloroplast ultrastructure, chlorophyll accumulation and synthesis of other components of the photosynthetic apparatus proceed almost in parallel, which is reflected in a proportional increase of the net photosynthetic rate. The phase of photosynthetic maturity is then followed by senescence, during which the rate of photosynthesis declines more rapidly than chlorophyll degradation proceeds.

Based on this general character of ontogenetic changes, Więckowski (1969) proposed a scheme with six phases of changes of assimilation numbers: the first two phases are characterized by zero values (embryonic leaf primordia with plastid initials; chlorophyll formation preceding the existence of an active photosynthetic apparatus). Phase 3 of rapid increase in assimilation numbers is followed by their slower decline during early leaf development (phase 4; e.g. Phaseolus — Więckowski 1967a). Phase 5 (according to Więckowski steady or slightly declining assimilation numbers, in our experience often increasing ones) characterizes photosynthetically mature leaves, while the steady decline (phase 6) is typical for leaf senescence and is often found in the leaf insertion gradient (e.g. Nicotiana — Šesták 1963a).

The validity of this scheme is confirmed by comparing the majority of results of those authors, who measured the photosynthetic rate under optimum conditions, especially under saturating irradiance. In the majority of cases only leaves in phases 5 and 6 were used for experiments and thus either an increase followed by a decrease or a steady decline were observed (e.g. Nicotiana — Wada et al. 1967, Fleck-Gerndt 1971a; Dactylis — Treharne et al. 1968; Spirodela — Gaponenko and Stazhetskiï 1969; Zea — Šesták and Bartoš 1963, Šesták and Václavík 1965, Kupka and Truong Quang Tan 1975; Phaseolus — Bolhár-Nordenkampf 1975;

*Vitis* — Kriedemann 1968; *Solanum* — Frier 1977; *Brassica* — Šesták 1966a, b; *Picea* — Khodasevich *et al.* 1979). The character of changes of assimilation numbers was modified by factors which affected leaf growth, *e. g.* temperature (Treharne *et al.* 1968), photoperiod (Frier 1977), plant decapitation (Iordanov and Popov 1967).

The rather large variation in assimilation numbers among plant species and genotypes, in plants grown under various environmental conditions, and during leaf and plant ontogeny lead to various explanations. The surplus of chlorophyll in the photosynthetic apparatus has been postulated together with its limiting function only under feeble irradiance (chlorophyll as "Schwachlichtfaktor" — Gabrielsen 1948). Nevertheless, Gabrielsen's results reached by illuminances up to 11 klx were not confirmed at higher irradiances (Šesták 1963b). The review of literature results shows always a better agreement between photosynthetic rates and chlorophyll content during leaf ontogeny in measurements under optimum conditions (*cf.* Table 22.32 in Šesták and Čatský 1967c; see also Baker and Hardwick 1975a, 1976).

The discrepancy is often caused by reaching maximum photosynthetic rate earlier in leaf ontogeny than reaching maximal chlorophyll amount (*e.g. Zea* — Šesták and Bartoš 1963, Šesták and Václavík 1965; *Populus* — Hernández-Gil and Schaedle 1973, Kuno 1980; *Perilla* — Hardwick *et al.* 1968); in some plants (especially those having thick leaves) this relation for the same leaf area unit may be reversed (*e.g. Brassica* — Šesták 1966b; *Theobroma* — Baker and Hardwick 1973; *Medicago* — Okubo *et al.* 1975b). For plants of a given population and age a linear relation with net photosynthetic rate may be found for a rather wide range of chlorophyll (*a* + *b*) concentrations when leaves in the insertion gradient are tested (*e.g. Nicotiana* — Šesták and Čatský 1962, Šesták 1963a, b; *Brassica* — Šesták 1963a, b, 1966a, b; *Zea* — Šesták 1963b, Šesták and Bartoš 1963; *Beta* — Šesták 1966a, b; *Medicago, Trifolium* — Okubo *et al.* 1975a). The dependence is always better when chlorophyll *a* rather than chlorophyll (*a* + *b*) is used for calculation (Šesták 1966a, Okubo *et al.* 1975a). The photosynthetic rate per chlorophyll amount is in direct relation to the increment of chlorophyll fluorescence induced by the infiltration of leaf tissue with the inhibitor of Photosystem 2, dichlorodiphenyl-1,5-dimethyl urea (leaves of five different ages of *Ipomoea* — Kulandaivelu and Daniell 1980). A more detailed analysis has shown (*Zea* — Šesták and Václavík 1965; *Beta* — Šesták 1966a; *Populus* — Kuno 1980) that the photosynthetic efficiency of a unit chlorophyll amount is higher in young than older leaves. This difference was explained by the changes in relative contents of chlorophylls *a* and *b* and various *in vivo* forms of chlorophyll *a*, the ineffectiveness of some of which causes the increase in the chlorophyll compensation point (Šesták 1966a) with leaf senescence.

Recent explanations of the relation of chlorophyll and photosynthesis use the concept of the photosynthetic unit (chlorophyll unit) as a basic unit of the photosynthetic apparatus. Of course, the present understanding of the unit is more a statistical aid than an idea of a functional complex discretely localized in the thylakoid membrane. The size of the unit for the whole electron transport chain is given by the amount of chlorophyll molecules per one molecule of the reaction centre of Photosystem 1, *i.e.* P700. Its size increases and then declines during the ontogeny of the primary leaf of *Phaseolus* (Šesták and Demeter 1976) or only increases with age (*Zea* — Keresztes and Faludi-Dániel 1973; *Mertensia* — Harvey 1980). Even on the area of one leaf blade the size of the photosynthetic unit increases from the youngest base to the old tissues of the leaf tip (7 d plants of *Zea* — Baker and Leech 1977). The capacity of the photosynthetic apparatus and its changes including the ontogenetic ones shall just be related to the number of photosynthetic units: *e.g.* this number increases with age of the cotton leaf (of the 7th–8th node) to 10 d and then declines (Khodzhaev *et al.* 1978). Another concept to the same effect distinguishes the chlorophyll complexes collecting, transporting and concentrating radiant energy, and photosynthetic processing complexes (electron carriers and enzymes) — then the size of the complex being minimum determines the capacity of the photosynthetic apparatus.

## 10.5 ACTIVITIES OF PHOTOCHEMICAL REACTIONS AND THE PHOTOSYNTHETIC RATE

The literature showing the changes in photochemical activities during leaf ontogeny (see Chapter 5) does not give any uniform picture. The onset of photochemical activities of photosynthesis is certainly bound with the formation of minimum primary structure and composition of the thylakoid membrane. The reaching of maximum activities of individual photoreactions is probably connected with the reaching of optimum thylakoid and grana substructure (*cf.* Section 10. 3). The development of photoreaction activities may generally parallel the development of the photosynthetic unit (see Section 10.4). In some cases the ontogenetic courses of some of these activities agree with that of the photosynthetic rate (*e.g.* cyclic photophosphorylation — Hernández-Gil and Schaedle 1973, Hill reaction — Volodarskiĭ *et al.* 1978).

Recently the indirect estimations of photochemical activities by means of fluorescence or luminescence characteristics have often been used for these comparisons: thus Bystrykh and Matorin (1975) found an agreement in the developmental changes of the net photosynthetic rate and electron transport activity determined as delayed luminescence in the vegetative phases of sunflower ontogeny, but later flowering interfered with this interrelationship. In cocoa leaves, the increase in leaf area, chlorophyll content and $O_2$ evolution rate to maximum was accompanied by the increase in the ratio of variable to maximum fluorescence ($F_v/F_m$) at 695 nm (estimating the probability of photon utilization for primary photochemistry of

Photosystem 2), while the ratio $F_v/F_m$ at 735 and 695 nm (estimating the fraction of energy within Photosystem 1 from Photosystem 2) and the ratio of variable to minimum fluorescence ($F_v/F_o$) at 735 nm (which monitors the energy transfer from Photosystem 2 to Photosystem 1) reached a sharp maximum prior to the onset of the linear phase of leaf expansion; the ratio of fluorescence emission at 685 and 735 nm continuously declined during leaf development (Baker and Miranda 1981).

In contrast to some authors who found maxima of individual photochemical activities in leaves of different ages (*e.g.* Wolińska 1976), ontogenetic courses with peaks of all photochemical activities in the same phases of leaf development were also found: During ontogeny of primary leaves of *Phaseolus vulgaris* the activities of Photosystems 1 and 2 and non-cyclic photophosphorylation per unit chlorophyll amount peaked simultaneously two to three times (Šesták *et al.* 1975, 1977, 1978b). The first peak preceding the phase of maximum net $CO_2$ input and even more the reaching of the maximum chlorophyll amount may be connected with the optimum small size of the photosynthetic unit in young leaves (Šesták and Demeter 1976). The second peak coincides with the optimum ultrastructure of chloroplasts. The third peak appearing in some cases prior to the end of the leaf life span is connected with senescence processes (antenna chlorophyll degradation preceding the degradation of the electron transport chain). Similar time courses with more peaks during the leaf life may be found in the literature (*Phaseolus* — Heyes and Dale 1971; *Cucurbita* — Harnischfeger 1974; *Triticum* — Volodarskiï *et al.* 1978; *Sinapis* — Wild *et al.* 1981b). Nevertheless, due to the combination of resistances to $CO_2$ transfer these rather sharp peaks in photochemical activities are usually reflected in ontogenetic courses of net $CO_2$ influx as small secondary peaks only (Šesták *et al.* 1975).

## 10.6  ACTIVITIES OF PHOTOSYNTHETIC ENZYMES AND PHOTOSYNTHETIC RATE

The relationship of activity of carboxylation reactions or carboxylation enzymes with the photosynthetic rate is often discussed with different conclusions (see Chapter 6); the data are certainly influenced by the methods and conditions of determination, modes of expression, *etc.* Even if there were no theoretical reason for the discrepancy between these two processes, the resistances limiting the carbon dioxide transfer to carboxylation sites (see Chapter 8), respiratory (Chapter 7) and photorespiratory processes (Chapter 9) might affect their relationship.

The leaf ontogenetic course of activity of the enzyme ribulose-1,5-bisphosphate carboxylase (RuBPC) may more or less agree with the changes in net photosynthetic rate (per chloroplast in the leaves of the 7th tier of potato — Mokronosov and Nekrasova 1977; per leaf in pea leaves — Gordon *et al.* 1978) or precede it (per area of leaves of the 7th tier of potato — Borzenkova and Nefedova 1981). Peak RuBPC activity may appear in the leaf with maximum chlorophyll amount (tobacco — Kawashima and Mitake 1969) or subsequently (*Pisum* — Gordon *et al.* 1978; *Theobroma* — Baker and Hardwick 1973). The maximum ratio of RuBPC activity to net photosynthetic rate in *Theobroma* is reached five days after reaching the maximum leaf area (Baker and Hardwick 1973). The ontogenetic peak in RuBPC activity is preceded by peaks of activities of ribulose-5-phosphate isomerase and NADH-glyoxylate reductase (*Pisum* — Gordon *et al.* 1978, *Perilla* — Batt and Woolhouse 1975), coincides with that of phosphoribulokinase, and is followed by those of glyceraldehyde-3-phosphate dehydrogenase and phosphoglycerate kinase (third leaf pair in *Perilla* — Batt and Woolhouse 1975). The activity of phosphoenolpyruvate carboxylase peaks much earlier than the amount of chlorophyll (*Zea* — Möller *et al.* 1977).

## 10.7 PHOTOSYNTHESIS AND CONDUCTANCES FOR CARBON DIOXIDE TRANSFER

The rate of photosynthesis saturated with radiant energy has often been expressed as a function of conductances in the carbon dioxide transfer pathway (Gaastra 1959, Jarvis and Morison 1981). The main limitations to diffusive conductance in this pathway lie within the stomatal pore and in the liquid phase (see Chapter 8). In comparison with these limiting conductances, conductance of leaf air boundary layer is rather high and only slight changes take place with leaf expansion and senescence. The stomata (*cf.* Section 10.2) are situated in the leaf surface in a position where they most effectively control the efflux of water vapour and influx of carbon dioxide. Their main role is, probably, to minimize plant water loss, especially in cases of reduced water uptake and high evaporative demand. Contemporaneously they also limit carbon gain. Changes in stomatal conductance ($g_s$) as well as intracellular conductance ($g_M$) are thus associated with changes in net photosynthetic rate ($P_N$) and transpiration ($E$); they all reflect the pattern of leaf expansion (Fig. 10.3 — *cf.* Wilson and Ludlow 1970, Ludlow and Wilson 1971, Fraser and Bidwell 1974, Čatský *et al.* 1976, Rawson and Woodward 1976, Woodward and Rawson 1976, Constable and Rawson 1980). In dicotyledonous plants considerable leaf expansion occurs after unfolding; simultaneously the length of stomata guard cells increases, number of stomata per leaf area unit decreases and stomata capability to fully open develops. In consequence of these changes stomatal conductance increases, reaches its maximum value mostly before the leaves are fully expanded, and thereafter declines. More complicated changes in net photosynthetic rate ($P_N$), transpiration rate ($E$), $g_s$ and $g_M$ of grass leaves are conditioned by their growth pattern as they have an intercalar meristem and, hence, the apical parts of leaf laminae are fully expanded soon after the leaf appears while the extension of basal parts carries on. Changes in $g_M$ are conditioned by quantitative anatomical changes, mesophyll cell walls and plasmalemma aging, and by changes in photochemical and biochemical activities (*cf.* Sections 10.5 and 10.6). During leaf ageing, stomatal as well as mesophyll conductances decrease; the decrease of the latter is more pronounced and, thus, it most effectively limits the $CO_2$ transfer. The decrease in stomatal conductance may be conditioned by increasing the substomatal concentration of $CO_2$ (Meidner and Mansfield 1965) and/or by the loss of stomata ability to open fully (Solárová 1973) because of the loss of stomata cell wall elasticity.

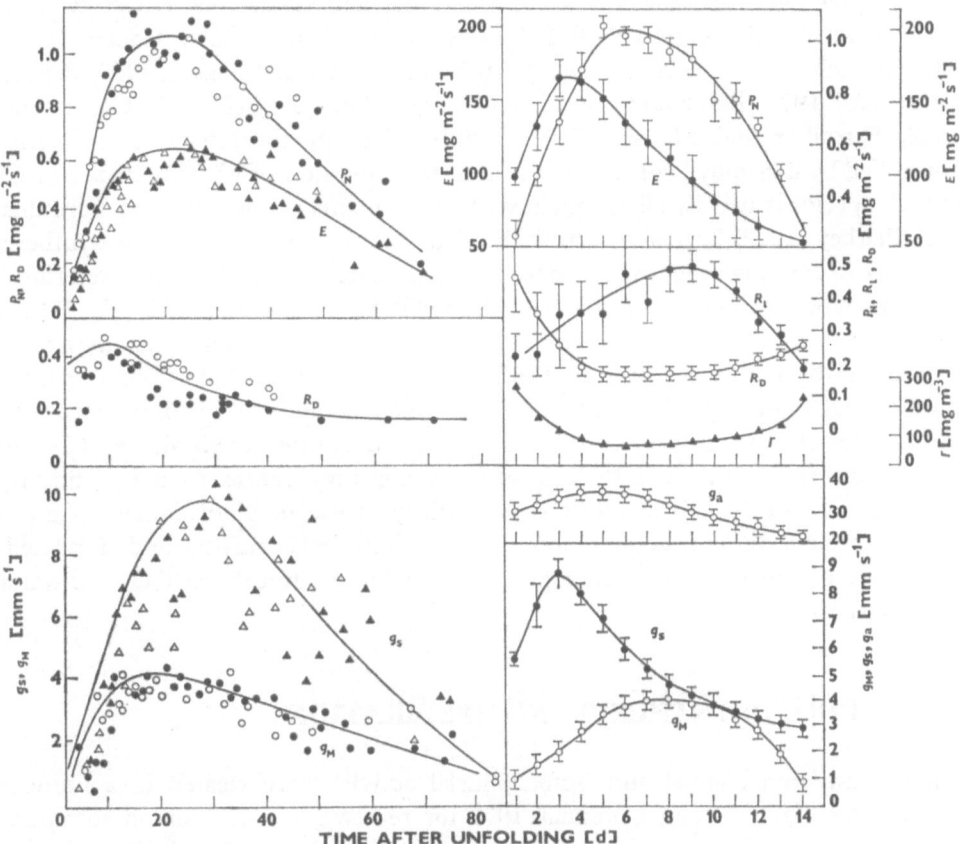

**Fig. 10.3.** Changes in the parameters of carbon dioxide and water vapour exchange in leaves of different ages. Rates of net photosynthesis ($P_N$), respiration in darkness ($R_D$), photorespiration ($R_L$), transpiration ($E$), $CO_2$ compensation concentration ($\Gamma$), and boundary layer ($g_a$), stomatal ($g_s$) and intracellular ($g_M$) conductances are compared (*left*) for leaves of *Gossypium hirsutum* L. in the winter and spring/summer experiments (full and open points, respectively; means of two to eight leaves pooled for all leaf positions; recalculated from Constable and Rawson 1980), and (*right*) for the ontogeny of the primary leaf of *Phaseolus vulgaris* L. (means ± standard mean errors; redrawn from Čatský *et al.* 1976).

Interrelationship of $g_s$ and $g_M$ values depends predominantly on the type of carbon fixation; changes in this relation during leaf ontogeny are rather small. Generally, the C4 species have in most cases $g_M/g_s$ ratio $> 1$ and the C3 species $< 1$. During leaf ontogeny the $g_M/g_s$ ratio changes in a loop form being the highest at the beginning and end of the leaf life span (*cf.* Fig. 10.3). Contrary to the conclusions of some experiments using both C3 and C4 plants (van den Driessche *et al.* 1971, McPherson and Slatyer 1973, Pasternak and Wilson 1973, Homann 1975) according to which $g_s$ is the prime controlling factor in photosynthesis, more experimental data (Gaastra 1959, Bierhuizen and Slatyer 1964, Beardsell *et al.* 1973a, b, Fraser and Bidwell 1974, Srivastava *et al.* 1975, Vignes and Calmés 1975, Čatský *et al.* 1976, Rawson and Woodward 1976, Woodward and Rawson 1976, Constable and Rawson 1980) and also recent leaf photosynthesis models (Gaastra 1959, Peisker 1976, Laïsk 1977, Berry and Farquhar 1978, Hall 1979, Farquhar and Caemmerer 1982, Farquhar and Sharkey 1982) confirm that photosynthesis is primarily controlled by mesophyll conductance. Stomatal limitation of photosynthesis in C3 and C4 species is usually slight, mainly when the plants are not stressed. Farquhar and Sharkey (1982) have shown that the slight stomatal control of photosynthesis does not in any way reduce the importance of stomata, for only by being well tuned to the metabolism and environment of the plant can they achieve this role. Hence the authors conclude that the current challenge in guard cell physiology is to determine the nature of this fine tuning. Optimal stomatal behaviour may save only a small amount of water (expressed per unit carbon fixed) in comparison with the amount of water transpired through stomata uniformly open all day (Cowan and Farquhar 1977). Nevertheless, such saving may represent a significantly increased rate of growth or reduced probability of leaf or plant damage (Cowan 1982). To solve these problems, more studies, both experimental and simulated, are needed on the ontogenetic changes in stomatal and intracellular $CO_2$ and water vapour transfer.

## 10.8    PHOTOSYNTHESIS AND RESPIRATION

Chloroplast, peroxisomal and mitochondrial activities are related in a number of ways (*cf.* Graham and Chapman 1979 for review). The role of photorespiration in the overall photosynthetic activity and carbon balance of the leaf was discussed in Chapter 9. However, also the interactions between photosynthesis and respiration may contribute to the ontogenetic variation in the carbon balance of the leaf. The interaction between the two simultaneously operating processes has only rarely been studied because of methodological difficulties. Nevertheless, the meagre information obtained suggests, but mostly does not prove, some conclusions.

It seems that respiration rate is more or less depressed in the light (*cf.* also

Section 7.3). The effect of light on respiration is mediated through the photosynthetic or photorespiratory mechanism because no significant inhibition was found when photosynthesis was absent (*e.g.* inhibited by DCMU, in etiolated leaves, *etc.*). However, the most effective in respiration depression is the long-wave (above 680 nm) component of radiation which is less effective in photosynthetic $CO_2$ uptake. Photophosphorylation, enhanced by long-wave radiation, may continuously deprive respiration of ADP, thereby suppressing oxidative phosphorylation (*cf.* Mangat *et al.* 1974, Richter 1978).

There is some biochemical evidence, that the degree of inhibition of respiration by light depends on leaf age. In rice, Nishida (1962) found that the tricarboxylic acid cycle was initially inhibited in the light, but subsequently it could operate in younger leaves at a rate comparable with that in the dark, while in mature leaves it was still inhibited. This difference may be related to higher rates of synthetic processes in young leaves.

Similar age-dependent light inhibition of respiration followed from the analysis of a simple model of leaf $CO_2$ exchange: Peisker *et al.* (1981) reported an increase (from *ca.* 40 to *ca.* 75%) in light inhibition of respiration rate during expansion of primary leaves of *Phaseolus vulgaris*, followed by a decrease (to *ca.* 20%) during leaf maturity and senescence.

Some information is available on the indirect effect of photosynthesis on the components of respiration rate ($R_D$), *i.e.* maintenance and growth respiration (*cf.* Section 7.3). The growth respiration ($R_G$) seems mainly responsible for the ontogenetic changes — with the exception of leaf senescence — in the rate of respiration in the dark measured as $CO_2$ efflux from the leaf and often reported in the literature (Figs. 7.12 and 10.3). $R_G$ in the unfolding leaf (*e.g. Helianthus* — Kimura *et al.* 1978) is driven by reserve substances supplied from the seed or photosynthates from the older leaves with higher, near-maximum net photosynthetic rates. During unfolding, the major transport saccharide — saccharose — is already detectable in the blade and its content increases steadily during leaf expansion (*Glycine* — Silvius *et al.* 1978). However, the increase in saccharose content, accompanied by corresponding increases in saccharose phosphate synthetase (E.C. 2.4.1.14) and translocation rate, is usually associated with the decrease in $R_D$. The other major transport saccharides, stachyose and raffinose, appear later than saccharose, *i.e.* when about one-fifth of the final blade size is reached and intralaminar phloem transport from the tip to the base has begun (*e.g. Cucurbita* — Turgeon and Webb 1975).

In the period of leaf growth and during the photosynthetic maturity of the leaf, $R_G$ is still high, and later gradually decreases with the decrease in the rate of leaf growth (Fig. 7.12). In this period, $R_D$ and the synthetic activities of the leaf decrease whereas the content of storage starch does not change. This suggests that the starch accumulation is not entirely controlled by the energy demands of the leaf (*Glycine* — Silvius *et al.* 1978).

In many species, the maximum net photosynthetic rate coincides with the peak in leaf extension rate (*cf.* Rawson and Hackett 1974). Hence, $R_D$ seems to be associated with the supply of photosynthates (Fig. 10.4) but a significant correlation

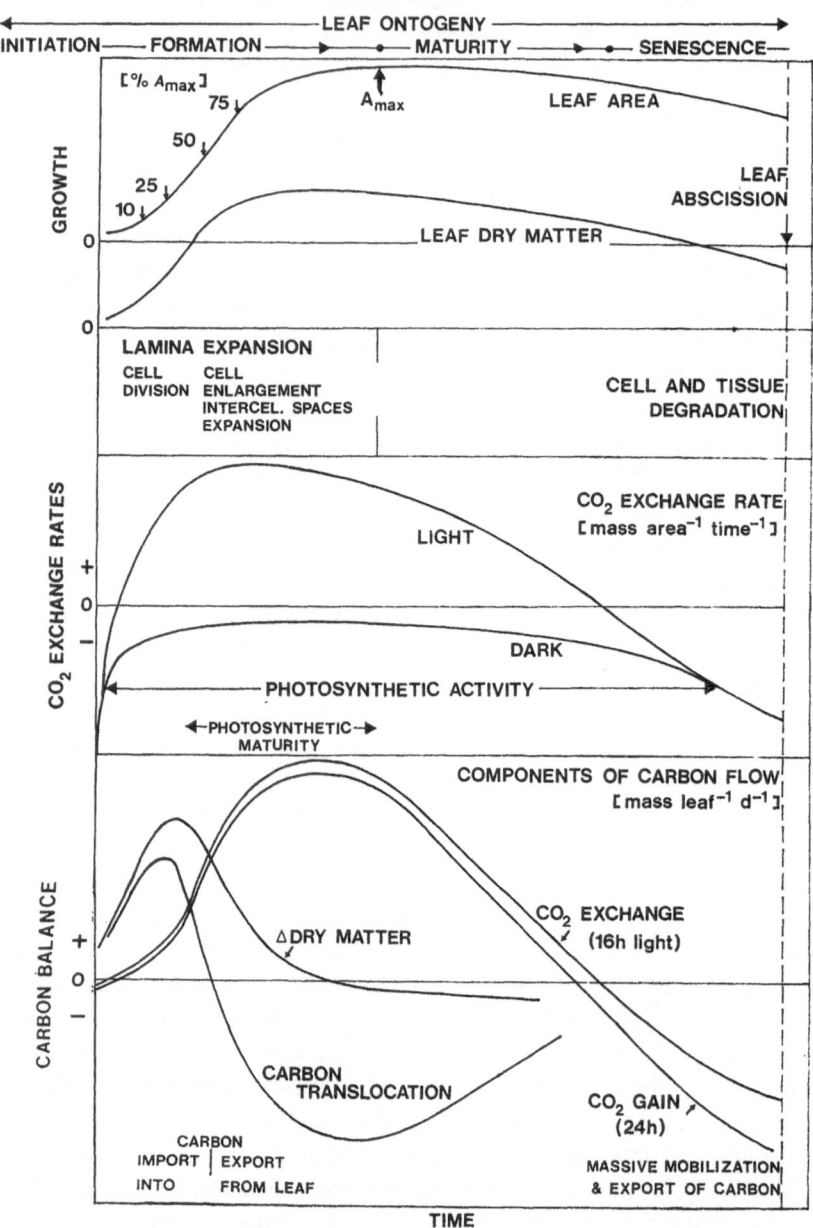

**Fig. 10.4.** The idealized ontogenetic pattern of carbon balance and related parameters in a leaf typical for the majority of C₃ and C₄ species. (From experimental data of Dickmann 1971, Turgeon and Webb 1975, and Čatský *et al.* 1976.)

between $R_D$ and any of the transport or storage saccharides can hardly be found.

In the second half of the leaf life span, $R_D$ approaches the rate of maintenance respiration, excepting the period of senescence. In that time, the leaf serves as a source of carbon and mineral elements for newly developing leaves and flowers or for storage. The rapid carbon translocation is consistent with an increase in $R_D$ which can be directly attributed to a functional change in leaf saccharide economy (Lloyd 1980).

The ontogenetic changes in the components of the leaf's carbon balance under different environmental conditions may be better understood by means of mathematical models (e.g. Horie 1977, Hunt and Loomis 1979). For example, Hunt and Loomis (1979) used the relation between photosynthesis and respiration described by Thornley (1970) as

$$R_D = (1 - Y_G)P_G + M_R Y_G W$$

where $R_D$ = dark respiration rate [kg d$^{-1}$], $Y_G$ = true growth yield [kg(dry m. synthesized) kg$^{-1}$(dry m. substrate)], $P_G$ = gross photosynthetic rate [kg d$^{-1}$], $M_R$ = maintenance respiratory coefficient [kg kg$^{-1}$ d$^{-1}$] and $W$ = dry matter [kg], all the variables being expressed for the whole plant, leaf, or another organ.

In 1977, Thornley proposed a more complex dynamic model (cf. also McCree 1982a) which was, however, not yet used for assessing the ontogenetic changes in $R_G$ and $R_M$.

Penning de Vries (1972, 1975a, b) developed a quantitative biochemical basis for determining $Y_G$. From $Y_G$, the respiratory costs of synthesis are derived as respiration coefficient for growth $G_R$ [kg kg$^{-1}$]:

$$G_R = (1 - Y_G)/Y_G .$$

For example, the following values were reported for leaves of Gossypium, Hordeum, Lolium and Triticum, respectively (cf. Horie 1977, review in Hunt and Loomis 1979):

$Y_G = 0.65;$     $0.7;$     $0.8;$     $0.75;$     decrease during leaf ontogeny
$G_R = 0.54;$     $0.43;$     $0.25;$     $0.33;$     decrease during leaf ontogeny
$M_R = 0.0264;$     $0.03;$     $0.014;$     $0.02;$     increase during leaf ontogeny

The relationship between photosynthesis and respiration rates is striking mainly in adult leaves. The ratio between the total leaf $R_D$ and the total leaf $P_N$ is fairly stable over a long period (e.g. Nicotiana — Tatemishi 1968, Rawson and Hackett 1974; Helianthus — Horie and Udagawa 1971). This finding may prove $R_D$ as a useful selection criterion and assist the attempt to increase crop productivity by genetic manipulation of dark respiration and to increase the efficiency of growth and/or maintenance processes (Wilson 1975).

## 10.9   PHOTOSYNTHESIS AND DRY MATTER ACCUMULATION. THE CARBON BALANCE OF THE LEAF

Total dry matter accumulation during leaf development is one aspect of leaf growth analogous to the increase in leaf area; it represents the assimilated carbon which is not lost by respiration processes nor translocated out of the leaf. The accumulation of dry matter is the basis of leaf and also plant production; all changes in photosynthetic characteristics with leaf age are finally reflected in the dry matter production of the leaf.

All accumulated dry matter of the leaf originates from photosynthesis, but, however, no simple relationship between net photosynthetic rate expressed as $CO_2$ uptake, and total dry matter accumulation of a leaf during ontogeny exists. This is not surprising when bearing in mind that $CO_2$ uptake is on the input, and dry matter accumulation on the output of the very complex system of photosynthesis where numerous processes are involved. The main inputs of the system are carbon dioxide, water and radiation. The rate of $CO_2$ uptake by a leaf is partly governed by the conductances for $CO_2$ transfer in the gaseous (mainly stomata) and liquid phases (intracellular conductances) (*cf.* Chapter 8), and also partly by the amount and level of activity of the carboxylating enzymes which accept the $CO_2$ molecules into the photosynthetic reduction cycle (*cf.* Chapter 6). As the result of photosynthetic reactions in the chloroplast, photosynthates are produced which may either be translocated into other leaves or plant parts, lost by respiration processes necessary for growth and maintenance of the leaf (*cf.* Section 10.8), or they are accumulated as leaf dry matter or biomass.

While the net photosynthetic rate during leaf ontogeny increases rapidly to a maximum followed by a slower decline, the general pattern of total leaf dry matter increment of the developing leaf is S-shaped with a substantial middle phase (*e.g. Lolium* — Silsbury 1970; *Cucurbita* — Turgeon and Webb 1975). Even after full leaf area expansion, leaves may continue to add dry matter to their structure. In the period between full expansion and death, total dry matter may fall by about 30 % (*Lolium* — Robson and Deacon 1978) because the leaf is a source of photosynthates which are translocated to other leaves or plant parts. The described pattern of total leaf dry matter accumulation is valid for all the successive leaves, the total dry matter of which generally increases with leaf insertion level (*Lolium* — Robson and Deacon 1978).

In the young leaf the losses by respiration (*cf.* Section 10.8) are enormous for the costs for construction (growth) are large. Net photosynthetic rate is still low, the leaf area small, and the photosynthetic systems are under development. Thus, the young leaf is a sink for photosynthates translocated from other leaves or seeds, and its dry matter accumulation is small. Before maximum leaf area is reached and growth of the leaf terminated, net photosynthetic rate and photosynthate production in the leaf rapidly increase and the leaf becomes a source of photosynthates which may be used in dry matter accumulation or translocated to other leaves. The fully expanded leaf has no further demand on photosynthates for growth, but a small residual supply is required to support cell wall and enzyme turnovers.

Dry matter accumulation is thus a substantial part of the carbon balance of the leaf. The original idea that the ratio of carbon to total dry matter is constant throughout the ontogeny of the leaf, at about 42 to 45% (*Cucurbita* — Turgeon and Webb 1975) has recently been replaced by the finding that % carbon content changes during the ontogeny of an organ. The decline of carbon to dry matter ratio during leaf development (*e.g. Lycopersicon* — Ho and Shaw 1977) may be due to both increased mineral accumulation and decreased carbon accumulation. The dry matter determinations are therefore not accurately proportional to carbon content, but do give a reasonable estimate of the investment of the plant into new material.

## 10.10   THE RATIO BETWEEN PHOTOSYNTHETIC AND TRANSPIRATION RATES

In many environments and situations water stress restricts photosynthetic rate and thus also crop production. In these cases the ratio of carbon dioxide fixed in photosynthesis to water vapour lost by transpiration ($P_N/E$ — kg $CO_2$ . $kg^{-1}$ $H_2O$ or mol $CO_2$ $mol^{-1}$ $H_2O$) or the ratio of dry matter produced (or crop yield) to amount of water used in transpiration (WUE[*] — water use efficiency — kg dry matter $kg^{-1}$ $H_2O$) becomes an important determinant of the total production of leaf, plant and crop. Other ratios used previously, *i.e.* transpiration coefficient or transpiration ratio, are only the reciprocal values of $P_N/E$ or WUE. The water use efficiency at any instant depends both on the environment and on a range of plant characters such as leaf morphology, *etc.* Some sections of the water vapour transfer from the transpiring surface (mesophyll cell walls) inside the leaf into the external atmosphere during transpiration are identical, but opposite in direction, with the carbon dioxide transfer pathway from the ambient air to the carboxylation centres in the chloroplasts during photosynthesis. Transfer of both carbon dioxide and water vapour in the common part of the pathway occurs by diffusion, the rates of which depend congruently on stomatal conductance and differently on different diffusion constant values, on $CO_2$ and water vapour concentration gradients (Jarvis 1971, Raven and Glidewell 1981 — *cf.* Section 8.2.1), ratio of photon fluence rate and temperature, and capacity of carboxylation centres.

Changes in stomatal aperture need not necessarily affect transpiration and photosynthetic rates to the same extent because photosynthesis is influenced, in addition to the conductances associated with diffusion of $CO_2$ through the stomata and air in intercellular spaces, by conductances to $CO_2$ molecules in the liquid phase and

---

[*] In some papers, the symbol WUE is used for the ratio of $P_N/E$.

by the carboxylation process itself. Consequently, linearity of the relation between transpiration and photosynthesis is quite fortuitous (Cowan and Troughton 1971). Due to differences in the ratio of mesophyll and stomatal conductances (*cf.* Section 10.7) $P_N/E$ and thus WUE in $C_4$ plants is considerably higher than in $C_3$ plants.

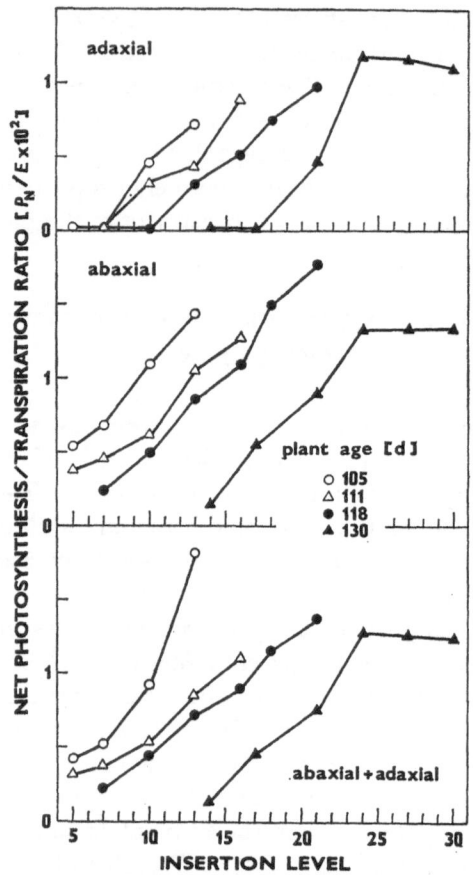

**Fig. 10.5.** Net photosynthesis/transpiration ratio ($P_N/E$) of the abaxial, adaxial, and abaxial + adaxial surfaces, respectively, in tobacco leaves of different insertion levels during plant ontogeny. (From Václavík 1974.)

Maximum efficiency of water use during the day is reached when stomatal conductance is synchronized with the $CO_2$ requirement so that the gain ratio $[(\partial P_N/\partial g_s)/(\partial E/\partial g_s)]$ remains constant. These partial differentials describe the sensitivity of photosynthetic rate ($P_N$) and transpiration rate ($E$) to changes in stomatal conductance ($g_s$) (Cowan and Farquhar 1977, Farquhar *et al.* 1978, Hall and Schulze 1980). During leaf ontogeny or in leaves of different insertion both the maximum stomatal conductance (*e.g.* Slatyer and Bierhuizen 1964, Holmgren *et al.* 1965,

Wilson and Ludlow 1970, Ludlow and Wilson 1971, Čatský *et al.* 1976, Rawson and Woodward 1976, Woodward and Rawson 1976, Constable and Rawson 1980) and the mesophyll conductance (*e.g.* Wilson and Ludlow 1970, Jarvis 1971, Ludlow and Wilson 1971, Čatský *et al.* 1976, Rawson and Woodward 1976, Woodward and Rawson 1976, Constable and Rawson 1980) can change simultaneously or differently (for details see Chapter 8). As the transpiration rate depends mainly on changes in stomatal conductance, the changes of transpiration during ontogeny of the leaf are relatively smaller than those of photosynthesis, which are influenced by changes in both stomatal and mesophyll conductances: this is reflected also in the $P_N/E$ ratio (*e.g.* Václavík 1973, 1974, 1975, 1980, Rawson and Woodward 1976, Woodward and Rawson 1976, André *et al.* 1978, Constable and Rawson 1980; *cf.* Fig. 10.5). Thus in pigeon pea, peanut, sunflower and cotton leaves WUE improved as the leaves expanded, and then it declined significantly as they aged (Rawson and Constable 1980). Unfortunately, many environmental or internal factors which reduce transpiration (*e.g.* by decreasing epidermal conductance, raising leaf reflectivity) also tend to reduce photosynthesis (*cf.* Solárová *et al.* 1980) irrespective of plant and leaf age. There are, however, some factors which generally increase net photosynthetic rate without any effect on transpiration rate (*e.g.* high intracellular conductance, low respiration rate), and others (*e.g.* low cuticular conductance, high air humidity) which minimize transpiration rate with little effect on photosynthesis (Jones 1976) and which thus positively influence $P_N/E$. These factors further complicate the ontogenetic pattern of $P_N/E$ or WUE, especially under natural conditions.

## 10.11  CONCLUSIONS

Not only leaf size and morphology, but also metabolic activities change in the course of leaf ontogeny. These changes have often been neglected by biochemists and in plant physiological and agricultural studies, although there are over 3000 papers which describe differences in some photosynthetic characteristic with respect to leaf age. Analysis of this vast literature (the selection of which is presented here) and the results of the authors' own experiments enabled to sum up the courses of individual photosynthetic characteristics during leaf ontogeny.

Leaf development starts with the formation of all components of the leaf blade ensuring not only its mechanical properties necessary for keeping its position in the canopy, resisting the adverse factors of environment, *etc.*, but also by producing structures which enable the functioning of all biochemical processes supplying biomass necessary to maintain the continuity of the plant. The expansion of the leaf blade is accompanied by increment in amount, size and substructure of cells, chloroplasts, mitochondria, peroxisomes, *etc.*, by the formation of leaf tissues and internal structure (*cf.* Fig. 10.2). Compartmentation of a chloroplast is characterized by the formation of its ultrastructure, the development of the thylakoid system being accompanied by the synthesis of photosynthetic pigments, electron transport chain components and carbon cycle enzymes. The increasing capacity of chloroplasts to perform primary photochemical functions (*i.e.* radiant energy harvesting and its utilization to produce $NADPH_2$ and ATP in the electron transport chain) and the biochemical reactions using $NADPH_2$ and ATP to $CO_2$ fixation and reduction to primary photosynthates, starts soon to be limited by barriers in the path of $CO_2$ as substrate for carboxylation from air to the carboxylation centres.

Optimum balance of the photophysical, biochemical and transport processes is usually reached prior to attaining the maximum leaf blade size, in a phase called photosynthetic maturity (Šesták and Čatský 1962). The formative processes continue at a slower rate to the reaching of maximum leaf size, and then, often after a plateau (especially extended in evergreen plants — *cf.* Fig. 10.1), the degradative processes increasingly more precede the synthetic ones. This phase of senescence continues to the end of leaf life, ending in cell and tissue degradation.

The concurrent respiratory processes supply with released energy and substances the growth and maintenance of the leaf. In the initial phases of development, the leaf is supplemented by saccharides transported from seed or older leaves, and therefore the "dark" respiration must be very active. Beginning with the photosynthetic maturity, the leaf starts to export photosynthates; this reduces its own dry matter accumulation. The photorespiratory processes are connected with the functioning of the enzyme ribulose-1,5-bisphosphate carboxylase/oxygenase and therefore their ontogenetic course is near to that of the net photosynthetic rate (*cf.* Fig. 10.3).

Among the resistances to $CO_2$ transfer to the carboxylation centres in the chloroplast, the intracellular (mesophyll) resistance seems to be the most important. Its limiting function erases the effects of transitory peaks in activities of photosystems, photophosphorylating mechanisms or carboxylation enzymes that are connected with changes in the substructure and composition of thylakoid membranes. These peaks may appear only as shoulders on the generally smooth curve of net $CO_2$ uptake (*cf.* Fig. 10.1).

The ontogenetic patterns of photosynthetic characteristics are generally repeated in allleaves successively formed on a plant. Nevertheless, the final blade size and the peak photosynthetic activities reached increase with higher leaf insertion to *ca.* one to two thirds of plant age (in herbs) and in leaves formed later they decline again (*cf.* Figs. 7.5 and 10.6). Thus the curves representing photosynthetic characteristics in a leaf insertion gradient change their character during plant ontogeny. Only in middle-aged plants, leaf insertion gradients recall to some extent the changes during ontogeny of an individual leaf. Therefore, studying insertion gradients which are, of course, measured more easily, is not a good substitution for studying leaf ontogeny. Unfortunately, the studies on leaf insertion gradients are often marked by the term "leaf age studies".

The above ontogenetic changes depend also on the structural and biochemical differences connected with the pathway of photosynthetic carbon metabolism. The $C_3$, $C_4$ and CAM plants differ often in their response to environmental factors, and, hence, the quantitative expression of their ontogenetic courses of photosynthetic characteristics may be differently affected by irradiance, $CO_2$ supply, temperature, water deficit, enzymatic inhibitors, *etc.* These responses may be different with

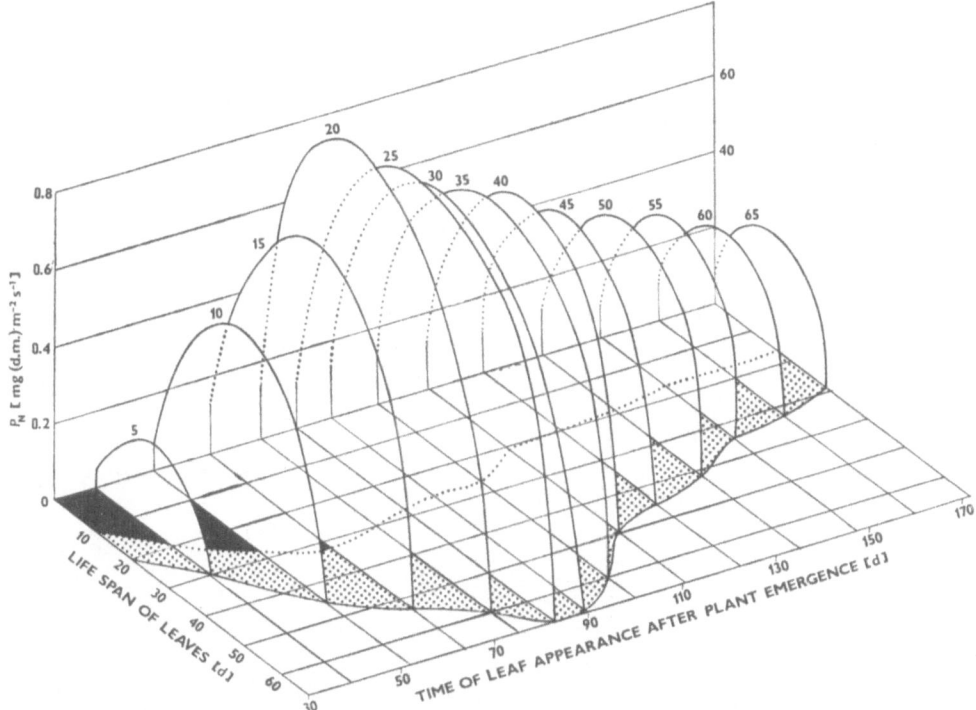

**Fig. 10.6.** Net photosynthetic rate measured as dry matter increment ($P_N$) of the successive sugar beet leaves (the order of leaf sequence is indicated by the figures at individual curves) as related to their life span, leaf area expansion (black area) and duration of the period after the leaf attained the maximum area (dotted area). (From Hodáňová 1981.)

respect to leaf age; they are especially expressed in the insertion gradients in leaf canopies, with irradiance as the main modifying factor. Under "optimum" environmental conditions maximum values of photosynthetic characteristics occur usually in nearly mature leaves or in the middle to upper leaves on the stalk. These leaves are usually less sensitive to negative influences of external factors and thus also under "stress" conditions the ontogenetic patterns remain the same or even more expressive than under "optimum" conditions.

In the course of leaf ontogeny all photosynthetic characteristics undergo continuous changes which are finally reflected in the dry matter production of the plant. This is why leaf ontogeny is not only interesting from the point of view of plant physiology and biochemistry, but it should be taken into account also in designing the ideotypes for plant selection and in proposing models of highly productive agricultural plant communities and forests.

# 11 REFERENCES

The form of references, abbreviations of journal titles, and transliterations of Russian, Ukrainian, Belorussian and Serbian characters are made according to Šesták, Z., Čatský, J. (ed.): Photosynthesis Bibliography. Vol. 1 to 11. Dr. W. Junk, B. V., Publishers, The Hague 1973—1983.

Abdullaev, Kh. A., Usmanov, P. D., Tageeva, S.V.: Sistema fotosinteticheskikh membran i evolyutsiya khloroplastov. [System of photosynthetic membranes and evolution of chloroplasts.] — Zh. obshch. Biol. 40: 43–59, 1979.

Abrahamsen, M., Mayer, A. M.: Photosynthetic and dark fixation of $^{14}CO_2$ in detached soybean cotyledons. — Physiol. Plant. 20: 1–5, 1967.

Abrams, G. J. von, Pratt, H. K.: Interaction of naphthaleneacetic acid and kinetin in the senescence of detached leaves. — Plant Physiol. 41: 1525–1530, 1966.

Ackerson, R. C.: Stomatal response of cotton to water stress and abscisic acid as affected by water stress history. — Plant Physiol. 65: 455–459, 1980.

Ackerson, R. C.: Osmoregulation in cotton in response to water stress. II. Leaf carbohydrate status in relation to osmotic adjustment. — Plant Physiol. 67: 489–493, 1981.

Ackerson, R. C., Krieg, D. R., Haring, C. L., Chang, N.: Effects of plant water status on stomatal activity, photosynthesis, and nitrate reductase activity of field grown cotton. — Crop Sci. 17: 81–84, 1977.

Acock, B., Thornley, J. H. M., Warren Wilson, J.: Photosynthesis and energy conversion. — In: Wareing, P. F., Cooper, J. P. (ed.): Potential Crop Production. A Case Study. Pp. 43–75. Heinemann Educational Books, London 1971.

Adams, J. A., Johnson, H. B., Bingham, F. T., Yermanos, D. M.: Gaseous exchange of *Simmondsia chinensis* (jojoba) measured with a double isotope porometer and related to water stress, salt stress, and nitrogen deficiency. — Crop Sci. 17: 11–15, 1977.

Adamson, H., Hiller, R. G.: Chlorophyll synthesis in the dark in angiosperms. — In: Akoyunoglou, G. (ed.): Photosynthesis. Vol. V. Pp. 213–221. Balaban Int. Sci. Serv., Philadelphia 1981.

Adedipe, N. O., Hunt, L. A., Fletcher, R. A.: Effects of benzyladenine on photosynthesis, growth and senescence of the bean plant. — Physiol. Plant. 25: 151–153, 1971.

Aérov, I. L.: Nekotorye osobennosti fotosinteticheskogo apparata v svyazi s rostom i razvitiem soi na raznykh fotoperiodakh. [Some characteristics of the photosynthetic apparatus in relation to growth and development of soybean in different photoperiods.] — In: Puti Povysheniya Intensivnosti i Produktivnosti Fotosinteza. Pp. 144–151. Naukova Dumka, Kiev 1966.

Aérov, I. L., Likholat, D. A.: Izmeneniya pigmentnoĭ sistemy u raznykh po vozrastu i raspolozheniyu list'ev yabloni v techenie vegetatsii. [Changes in pigment system in apple leaves of different age and insertion level during vegetation.] — In: Puti Povysheniya Intensivnosti i Produktivnosti Fotosinteza. Vol. 2. Pp. 176–184. Naukova Dumka, Kiev 1967.

Aerov, I. L., Manuïl's'kiĭ, V. D.: Vyvchennya stanu khlorofilu v lystkakh za dopomogoyu spektrofotometriĭ. [Spectrophotometric studies of chlorophyll state in leaves.] — Dopovidi Akad. Nauk URSR 29 B: 740–743, 1967.

Ågren, G. I., Axelsson, B., Flower-Ellis, J. G. K., Linder, S., Persson, H., Staaf, H., Troeng, E.:

Annual carbon budget for a young Scots pine. — Ecol. Bull. (Stockholm) 32 [Persson, T. (ed.): Structure and Function of Northern Coniferous Forests — An Ecosystem Study]: 307–313, 1980.

Akoyunoglou, G.: The effect of age on the phytochrome-mediated chlorophyll formation in dark--grown bean leaves. — Physiol. Plant. 23: 29–37, 1970.

Akoyunoglou, G. (ed.): Photosynthesis. Vol. 5. Chloroplast Development. — Balaban Int. Sci. Serv., Philadelphia 1981.

Akoyunoglou, G., Argyroudi-Akoyunoglou, J. H.: Effect of intermittent and continuous light on the chlorophyll formation in etiolated plants at various ages. — Physiol. Plant. 22: 288–295, 1969.

Akoyunoglou, G., Argyroudi-Akoyunoglou, J. H.: $CO_2$-assimilation by etiolated bean leaves exposed to intermittent light. — In: Forti, G., Avron, M., Melandri, A. (ed.): Photosynthesis, Two Centuries after Its Discovery by Joseph Priestley. Vol. 3. Pp. 2427–2436. Dr. W. Junk N. V. Publ., The Hague 1972.

Akoyunoglou, G., Argyroudi-Akoyunoglou, J. H. (ed.): Chloroplast Development. — Elsevier/ North-Holland Biomed. Press, Amsterdam — New York — Oxford 1978.

Akoyunoglou, G., Michalopoulos, G.: The relation between the phytylation and the 682→672 nm shift in vivo of chlorophyll a. — Physiol. Plant. 25: 324–329, 1971.

Akoyunoglou, G., Siegelman, H. W.: Protochlorophyllide resynthesis in dark-grown bean leaves. — Plant Physiol. 43: 66–68, 1968.

Akulovich, N. K., Godnev, T. N., Orlovskaya, K. I.: Osobennosti spektral'nykh izmeneniĭ protokhlorofill-(id)-golokhroma ètiolirovannykh list'ev v protsesse ego formirovaniya. [Features of spectral transformation of protochlorophyll(ide) holochrome of etiolated leaves during its formation.] — Dokl. Akad. Nauk SSSR 191: 1406–1409, 1970.

Akulovich, N. K., Orlovskaya, K. I., Parshikova, T. A.: Vzaimosvyaz' sostoyaniya i funktsii form protokhlorofillovogo pigmenta ètiolirovannykh rasteniĭ. [Interrelation of shape and function of forms of the protochlorophyll pigment in etiolated plants.] — In: Formirovanie Pigmentnogo Apparata Fotosinteza. Pp. 3–29. Nauka i Tekhnika, Minsk 1973.

Akulovich, N. K., Orlovskaya, K. I., Parshikova, T. A.: Kharakteristika protsessov nakopleniya protokhlorofilla i obrazovaniya ego spektral'nykh form v razvivayushchikhsya ètiolirovannykh rasteniyakh. [Characteristic of processes of protochlorophyllide accumulation and formation of its spectral forms in developing etiolated plants.] — In: Shlyk, A. A. (ed.): Khlorofill. Pp. 168–179. Nauka i Tekhnika, Minsk 1974.

Akulovich, N. K., Raskin, V. I.: Formirovanie protokhlorofill-golokhroma v ètiolirovannykh list'yakh i ego fotoprevrashchenie v khlorofill-golokhrom. [Formation of protochlorophyll-holochrome in etiolated leaves and its phototransformation in chlorophyll-holochrome.] — In: Problemy Biosinteza Khlorofillov. Pp. 5–52. Nauka i Tekhnika, Minsk 1971.

Al-Abbas, A. H., Barr, R., Hall, J. D., Crane, F. L., Baumgardner, M. F.: Spectra of normal and nutrient-deficient maize leaves. — Agron. J. 66: 16–20, 1974.

Alexander, A. G., Kumar, A.: Relationships of chlorophyll and enzyme gradients to sucrose content of sugarcane leaves. — J. Agr. Univ. Puerto Rico 58: 406–417, 1974.

Aliev, D. A., Azizov, I. V.: Fotokhimicheskaya aktivnost' izolirovannykh khloroplastov iz raznykh assimilyatsionnykh organov ozimoĭ pshenitsy. [Photochemical activity of isolated chloroplasts from different assimilatory organs of winter wheat.] — Izv. Akad. Nauk azerb. SSR, Ser. biol. Nauk 1975 (5): 33–37, 1975.

Aliev, D. A., Safarov, S. A.: Raspredelenie khlorofilla a i b v list'yakh razlichnykh sortov pshenitsy po vertikali posevov. [Distribution of chlorophyll a and b in the leaves of different wheat cultivars along crop height.] — Izv. Akad. Nauk azerb. SSR, Ser. biol. Nauk 1978 (5): 3–7, 1978.

Allaway, W. G., Austin, B., Slatyer, R. O.: Carbon dioxide and water vapour exchange parameters of photosynthesis in a Crassulacean plant, Kalanchoë daigremontiana. — Aust. J. Plant Physiol. 1: 397–405, 1974.

Allen, L. H., Jr., Gausman, H. W., Allen, W. A.: Solar ultraviolet (UV) radiation in terrestrial plant communities. — J. environm. Qual. 4: 285–294, 1975.

Allen, W. A., Gausman, H. W., Richardson, A. J.: Mean effective optical constants of cotton leaves. – J. opt. Soc. Amer. 60: 542–547, 1970.

Allen, W. A., Gausman, H. W., Richardson, A. J., Cardenas, R.: Water and air changes in grapefruit, corn, and cotton leaves with maturation. — Agron. J. 63: 392–394, 1971.

Allen, W. A., Gausman, H. W., Richardson, A. J., Thomas, J. R.: Interaction of isotropic light with a compact plant leaf. — J. opt. Soc. Amer. 59: 1376–1379, 1969.

Allen, W. A., Gausman, H. W., Richardson, A. J., Wiegand, C. L.: Mean effective optical constants of thirteen kinds of plant leaves. — Appl. Opt. 9: 2573–2577, 1970.

Allen, W. A., Richardson, A. J.: Interaction of light with a plant canopy. — J. opt. Soc. Amer. 58: 1023–1028, 1968.

Alleweldt, G., Eibach, R., Rühl, E.: Untersuchungen zum Gaswechsel der Rebe. I. Einfluss von Temperatur, Blattalter und Tageszeit auf Nettophotosynthese und Transpiration. — Vitis 21: 93–100, 1982.

Allison, J. C. S., Weinmann, H.: Effect of absence of developing grain on carbohydrate content and senescence of maize leaves. — Plant Physiol. 46: 435–436, 1970.

Ames, I. H., Pivorun, J. P.: A cytochemical investigation of a chloroplast inclusion. — Amer. J. Bot. 61: 794–797, 1974.

Amiri, Z.: Contribution à l'étude des pigments chlorophylliens de la feuille de Tabac. — Ann. SEITA-DEE Sect. 2 1971 (8): 73–121, 1971.

Ampofo, S. T., Moore, K. G., Lovell P. H.: The role of the cotyledons in four *Acer* species and in *Fagus sylvatica* during early seedling development. — New Phytol. 76: 31–39, 1976.

Andersen, K. S., Bain, J. M., Bishop, D. G., Smillie, R. M.: Photosystem II activity in agranal bundle sheath chloroplasts from *Zea mays*. — Plant Physiol. 49: 461–466, 1972.

Anderson, L. S., Dale, J. D.: The sources of carbon for developing leaves of barley. — J. exp. Bot. 34: 405—414, 1983.

André, M., Daguenet, A., Massimino, J., Massimino, D., Richaud, C.: Le laboratoire $C_{23}A$. Un outil au service de la physiologie de la plante entière II.— Possibilités de la mini-informatique et premiers résultats. — Ann. agron. 30: 153–166, 1979.

André, M., Massimino, D., Daguenet, A.: Daily patterns under the life cycle of a maize crop. I. Photosynthesis, transpiration, respiration. — Physiol. Plant. 43: 397–403, 1978.

Andreenko, S. S., Titova, Z. V.: Kolichestvennye izmeneniya khlorofilla v list'yakh prorostkov kukuruzy pri raznoĭ temperature v zone korneĭ. [Quantitative changes in chlorophyll content in leaves of maize seedlings at different temperature in the root zone.] — Dokl. Akad. Nauk SSSR 116: 157–160, 1957.

Andreeva, T. F., Avdeeva, T. A.: Belok "fraktsii 1" i fotosinteticheskaya aktivnost' list'ev. ["Fraction 1" protein and photosynthetic activity of leaves.] — Fiziol. Rast. 17: 225–233, 1970.

Andreeva, T. F., Avdeeva, T. A.: Adaptatsiya fotosinteza $C_3$- i $C_4$-rasteniĭ k usloviyam vneshneĭ sredy. [Adaptation of photosynthesis of $C_3$- and $C_4$-plants to conditions of environment.] — Fiziol. Biokhim. kul't. Rast. 8: 236–241, 1976.

Andreeva, T. F., Strogonova, L. E., Stepanenko, S. Yu., Maevskaya, S. N., Protasova, N. N., Murashov, I. N.: Zavisimost' aktivnosti fotosinteticheskogo apparata i rostovykh protsessov ot intensivnosti sveta i kontsentratsii $CO_2$ pri dlitel'nom vozdeĭstvii ètikh faktorov. [Dependence of the activity of the photosynthetic apparatus and growth processes on light intensity and $CO_2$ level in long-term action.] — Fiziol. Rast. 26: 1156–1162, 1979.

Andrews, A. K., Svec, L. V.: Photosynthetic activity of soybean pods at different growth stages compared to leaves. — Can. J. Plant Sci. 55: 501–505, 1975.

Andrews, T. J., Lorimer, G. H.: Photorespiration — still unavoidable? — FEBS Lett. 90: 1–9, 1978.

Aoki, M., Yabuki, K.: Studies on the carbon dioxide enrichment for plant growth. VII. Changes in dry matter production and photosynthetic rate of cucumber during carbon dioxide enrichment. — Agr. Meteorol. 18: 475–485, 1977.

Aoki, M., Yabuki, K., Totsuka, T.: Remote sensing of the physiological functions of plants by infrared color aerial photography (I): Relations between leaf reflectivity ratio, bi-band ratio and photosynthetic function of leaves in several woody plants. – Res. Rep. nat. Inst. environm. Studies 11 (Studies on the Effects of Air Pollutants on Plants and Mechanisms of Phytotoxicity): 225–237, 1980.

Apel, P.: Leitbündeldichte und Stomatafrequenz von Gramineen-Arten mit $C_3$-beziehungsweise $C_4$-pathway der Photosynthese. – Kulturpflanze 27: 91–95, 1979.

Apel, P., Lehmann, C. O.: Variabilität und Sortenspezifität der Photosyntheserate bei Sommergerste. — Photosynthetica 3: 255–262, 1969.

Apel, P., Peisker, M.: Pflanzenarten mit intermediärer Merkmalsausprägung in bezug auf den $C_3$- und $C_4$-pathway der Photosynthese. — Kulturpflanze 26: 49–66, 1979.

Apel, P., Tichá, I., Peisker, M.: $CO_2$-Kompensationspunkt von *Moricandia arvensis* (L.) DC. bei Blättern unterschiedlicher Insertionshöhe und bei verschiedenen $O_2$-Konzentrationen. — Biochem. Physiol. Pflanzen 172: 547–552, 1978.

Appiano, A., D'Agostino, G., Pennazio, S.: Development of dimorphic chloroplasts in a $C_4$ dicotyledon, *Gomphrena globosa* L., in relation to plastochron age. — J. submicr. Cytol. 11: 479–488, 1979.

Argyroudi-Akoyunoglou, J. H., Akoyunoglou, G.: Photoinduced changes in the chlorophyll *a* to chlorophyll *b* ratio in young bean plants. — Plant Physiol. 46: 247–249, 1970.

Armond, P. A., Staehelin, L. A., Arntzen, C. J.: Spatial relationship of photosystem I, photosystem II, and the light-harvesting complex in chloroplast membranes. — J. Cell Biol. 73: 400–418, 1977.

Arnon, D. I., Tsujimoto, H. Y., Tang, G.M.-S.: The oxygenic and anoxygenic photosystems of plant photosynthesis: An updated concept of light-induced electron and proton transport and photophosphorylation. — In: Akoyunoglou, G. (ed.): Photosynthesis. Vol. II. Pp. 7—18. Balaban Int. Sci. Serv., Philadelphia 1981.

Arntzen, C. J.: Dynamic structural features of chloroplast lamellae. — In: Sanadi, D. R., Vernon, L. P. (ed.): Current Topics in Bioenergetics. Vol. VIII. Pp. 111–160. Academic Press, New York – San Francisco – London 1978.

Arntzen, C. J., Dilley, R. A., Neumann, J.: Localization of photophosphorylation and proton transport activities in various regions of the chloroplast lamellae. — Biochim. biophys. Acta 245: 409–424, 1971.

Artamonov, V. I., Kuramagomedov, M. K.: Vliyanie gibberellina na razrushenie khlorofilla v intaktnykh rasteniyakh i vysechkakh iz list'ev fasoli. [Effect of gibberellin on the breakdown of chlorophyll in intact bean plants and leaf discs.] — Nauch. Dokl. vyssh. Shkoly, biol. Nauki 16 (6): 79–84, 1973.

Asada, K., Takahashi, M.–A., Tanaka, K., Nakano, Y.: Formation of active oxygen and its fate in chloroplasts. – In: Hayaishi, O., Asada, K. (ed.): Biochemical and Medical Aspects of Active Oxygen. Pp. 45–63. Japan Scientific Society Press, Tokyo 1977.

Ashby, E.: Studies in the morphogenesis of leaves. II. The area, cell size and cell number of leaves of *Ipomoea* in relation to their position on the shoot. — New Phytol. 47: 177–195, 1948.

Ashby, E., Wangermann, E.: Studies in the morphogenesis of leaves. IV. Further observations on area, cell size and cell number of leaves of *Ipomoea* in relation to their position on the shoot. — New Phytol. 49: 23–35, 1950.

Ashcroft, W. J., Murray, D. R.: The dual functions of the cotyledons of *Acacia iteaphylla* F. Muell. (*Mimosoideae*). — Aust. J. Bot. 27: 343–352, 1979.

Aslam, M., Lowe, S. B., Hunt, L. A.: Effect of leaf age on photosynthesis and transpiration of cassava (*Manihot esculenta*). — Can. J. Bot. 55: 2288–2295, 1977.

Aubuchon, R. R., Thompson, D. R., Hinckley, T. M.: Environmental influences on photosynthesis within the crown of a white oak. — Oecologia 35: 295–306, 1978.

Auclair, D., Gaudillère, J.-P.: Propriétés photosynthétiques de feuilles d'*Abies alba* (Mill.) et de *Picea abies* (L.). — Compt. rend. Acad. Sci. Paris, Sér. C 280: 905–908, 1975.

Auld, B. A., Dennett, M. D., Elston, J.: The effect of temperature changes on the expansion of individual leaves of *Vicia faba* L. — Ann. Bot. 42: 877–888, 1978.

Aussenac, G., Ducrey, M.: Etude bioclimatique d'une futaie feuillue (*Fagus silvatica* L. et *Quercus sessiliflora* Salisb.) de l'Est de la France. I. Analyse des profils microclimàtiques et des caractéristiques anatomiques et morphologiques de l'appareil foliaire. — Ann. Sci. forest. 34: 265–284, 1977.

Austin, R. B., Morgan, C. L., Ford, M. A., Bhagwat, S. G.: Flag leaf photosynthesis of *Triticum aestivum* and related diploid and tetraploid species. — Ann. Bot. 49: 177–189, 1982.

Avdeeva, T. A., Andreeva, T. F.: Nitrogen nutrition and activities of $CO_2$-fixing enzymes and glyceraldehyde phosphate dehydrogenase in broad bean and maize. — Photosynthetica 7: 140–145, 1973.

Avery, G. S., Jr.: Structure and development of the tobacco leaf. — Amer. J. Bot. 20: 565–592, 1933.

Axelsson, L.: The photostability of different chlorophyll forms in dark grown leaves of wheat III. Dependence on age of the plants. — Physiol. Plant. 41: 217–222, 1977.

Axelsson, L., Klockare, B., Sundqvist, C.: Oak seedlings grown in different light qualities. II. Photostability of early forms of chlorophyll(ide). — Physiol. Plant. 51: 314–320, 1981.

Azcón-Bieto, J., Farquhar, G. D., Caballero, A.: Effects of temperature, oxygen concentration, leaf age and seasonal variations on the $CO_2$ compensation point of *Lolium perenne* L. Comparison with a mathematical model including non-photorespiratory $CO_2$ production in the light. — Planta 152: 497–504, 1981.

Baker, N. R., Butler, W. L.: Development of the primary photochemical apparatus of photosynthesis during greening of etiolated bean leaves. — Plant Physiol. 58: 526–529, 1976.

Baker, N. R., Hardwick, K.: Biochemical and physiological aspects of leaf development in cocoa (*Theobroma cacao*). I. Development of chlorophyll and photosynthetic activity. — New Phytol. 72: 1315–1324, 1973.

Baker, N. R., Hardwick, K.: A model for the development of photosynthetic units in *Cocoa* leaves — In: Avron, M. (ed.): Proceedings of the Third International Congress on Photosynthesis. Vol. III. Pp. 1897–1906. Elsevier, Amsterdam–Oxford–New York 1975a.

Baker, N. R., Hardwick, K.: Biochemical and physiological aspects of leaf development in cocoa (*Theobroma cacao*). III. Changes in soluble sugar content and sucrose synthesizing capacity. — New Phytol. 75: 519–524, 1975b.

Baker, N. R., Hardwick, K.: Development of the photosynthetic apparatus in cocoa leaves. — Photosynthetica 10: 361–366, 1976.

Baker, N. R., Hardwick, K., Jones, P.: Biochemical and physiological aspects of leaf development in cocoa (*Theobroma cacao*) II. Development of chloroplast ultrastructure and carotenoids. — New Phytol. 76: 513–518, 1975.

Baker, N. R., Leech, R. M.: Development of photosystem I and photosystem II activities in leaves of light-grown maize (*Zea mays*). — Plant Physiol. 60: 640–644, 1977.

Baker, N. R., Miranda, V.: Development of primary photosynthetic processes in leaves grown under a diurnal light regime. — In: Akoyunoglou, G. (ed.): Photosynthesis. Vol. V. Pp. 367–376. Balaban Int. Sci. Serv., Philadelphia 1981.

Baldy, C. M.: Analyse de la photosynthèse du maïs dans les conditions naturelles par une méthode gravimétrique. — Oecol. Plant. 6: 101–113, 1971.

Baldy, C. M., Le Buhan, J.-P.: Répartition de la photosynthèse nette dans les feuilles de Tabac. — Photosynthetica 5: 421–423, 1971.

Balegh, S. E., Biddulph, O.: The photosynthetic action spectrum of the bean plant. — Plant Physiol. 46: 1–5, 1970.

Baranov, A. A., Dorokhov, B. L., Shiryaeva, G. A., Saakov, V. S.: Vliyanie neblagopriyatnykh temperaturnykh uslovii̇ na tonkuyu strukturu spektra pigment-lipoproteidnogo kompleksa list'ev. [Effect of adverse temperature conditions on the ultrastructure of spectra of pigment-lipoproteid complex of leaves.] — Izv. Akad. Nauk mold. SSR, Ser. khim. biol. Nauk 1974 (5): 29–36, 1974.

Bardat, F., Bourneuf, P., d'Harlingue, A., Bahl, J., Monéger, R.: Analyse des quinones chloroplastiques. Application à quelques espéces appartenant aux genres *Triticum* et *Aegilops*. – Physiol. vég. 19: 197–210, 1981.

Barden, J. A.: Net photosynthesis, dark respiration, specific leaf weight, and growth of young apple trees as influenced by light regime. — J. amer. Soc. hort. Sci. 99: 547–551, 1974.

Barr, R., Arntzen, C. J.: The occurrence of $\delta$-tocopherylquinone in higher plants and its relation to senescence. — Plant Physiol. 44: 591–598, 1969.

Barr, R., Crane, F. L.: Comparative studies on plastoquinones. V. Changes in lipophilic chloroplast quinones during development. — Plant Physiol. 45: 53–55, 1970.

Barrera, R., Wernsman, E. A.: Trichome type, density, and distribution on the leaves of certain tobacco varieties and hybrids. — Tobacco Sci. 10: 157–161, 1966.

Bartels, F.: Strukturelle Veränderungen der Chloroplasten-Thylakoide in Palisadenzellen von *Peperomia metallica* im Licht-Dunkelwechsel. — Protoplasma 72: 27–41, 1971.

Barton, R.: Fine structure of mesophyll cells in senescing leaves of *Phaseolus*. — Planta 71: 314 to 325, 1966.

Barton, R.: The production and behaviour of phytoferritin particles during senescence of *Phaseolus* leaves. — Planta 94: 73–77, 1970.

Basiouny, F. M., Van, T. K., Biggs, R. H.: Some morphological and biochemical characteristics of $C_3$ and $C_4$ plants irradiated with UV-B. — Physiol. Plant. 42: 29–32, 1978.

Bassham, J. A.: Increasing crop production through more controlled photosynthesis. – Science 197: 630–638, 1977.

Bassham, J. A., Larsen, P. O., Lawyer, A. L., Cornwell, K. L.: Relationships between nitrogen metabolism and photosynthesis. — In: Bewley, J. D. (ed.): Nitrogen and Carbon Metabolism. Pp. 135–163. Martinus Nijhoff/Dr. W. Junk Publ., The Hague–Boston–London 1981.

Bassi, R., Passera, C.: Effect of growth conditions on carboxylating enzymes of *Zea mays* plants. — Photosynthesis Res. 3: 53–58, 1982.

Bassman, J. H., Dickmann, D. I.: Effects of defoliation in the developing leaf zone on young *Populus* × *euramericana* plants. I. Photosynthetic physiology, growth, and dry weight partitioning. — Forest Sci. 28: 599–612, 1982.

Baszyński, T.: Distribution of plastid quinones and pigments along the leaf in different stages of plastid development of *Zea mays* L.seedlings. — Ann. Univ. Mariae Curie-Skłodowska, Sect. C 26: 187–198, 1971.

Bates, J. W.: The relationship between physiological vitality and age in shoot segments of *Pleurozium schreberi* (BRID.) MITT. — J. Bryol. 10: 339–351, 1979.

Batt, T., Woolhouse, H. W.: Changing activities during senescence and sites of synthesis of photosynthetic enzymes in leaves of the labiate, *Perilla frutescens* (L.) Britt. — J. exp. Bot. 26: 569–579, 1975.

Bauer, H., Bauer, U.: Photosynthesis in leaves of the juvenile and adult phase of ivy (*Hedera helix*). — Physiol. Plant. 49: 366–372, 1980.

Baur, J. R., Halliwell, R. S., Langston, R.: Pathways of carbohydrate catabolism in senescent and non-senescent tobacco leaves. — Physiol. Plant. 21: 45–51, 1968.

Bazhanova, N. V., Gasparyan, O. B., Oganesyan, D. A.: Nekotorye dannye, kharakteriziruyushchie pigmentnyi̇ apparat prorostkov kukuruzy, vyrashchennoi̇ v kamere iskusstvennogo klimata dlya zelenoi̇ podkormki zhivotnykh. [Characteristics of pigment apparatus of maize seedlings grown in an air-conditioned cabinet for use in green feeding of animals.] — Soobshch. Inst. agrokhim. Problem i Gidroponiki Akad. Nauk arm. SSR 7 (Issledovaniya v Oblasti Gidroponiki):69–78, 1967.

Beadle, C. L., Turner, N. C., Jarvis, P. G.: Critical water potential for stomatal closure in Sitka spruce. — Physiol. Plant. 43: 160—165, 1978.

Beardsell, M. F., Mitchell, K. J., Thomas, R. G.: Effects of water stress under contrasting environmental conditions on transpiration and photosynthesis in soybean. — J. exp. Bot. 24: 579–586, 1973a.

Beardsell, M. F., Mitchell, K. J., Thomas, R. G.: Transpiration and photosynthesis in soybean. Effects of temperature and vapour pressure deficit. — J. exp. Bot. 24: 587–595, 1973b.

Bedenko, V. P.: Fotosintez i Produktivnost' Pshenitsy na Yugo-Vostoke Kazakhstana. [Photosynthesis and Productivity of Wheat in the South-East of Kazakhstan.] — Nauka Kaz. SSR, Alma-Ata 1980.

Bendall, D. S.: Development of photosynthetic electron transport in greening barley. — Biochem. Soc. Trans. 5: 84–88, 1977.

Benecke, U., Schulze, E.-D., Matyssek, R., Havranek, W. M.: Environmental control of $CO_2$ – assimilation and leaf conductance in Larix decidua Mill. I. A comparison of contrasting natural environments. — Oecologia 50: 54–61, 1981.

Bergmann, H., Lerch, G., Müntz, K.: Über die physiologische Wirkung von Sonne und Schatten auf Coffea arabica L. "caturra" unter Freilandbedingungen in Kuba. II. Stoffwechsel von Jungpflanzen. — Z. Pflanzenphysiol. 63: 444–460, 1970.

Berner, E., Jr.: Studies in the nitrogen metabolism of barley leaves. Part I. The assimilation of nitrate. Part II. The effect of nitrate and ammonium on respiration and photosynthesis. – Physiol. Plant. (Suppl. VI): 1–56, 1971.

Berry, J., Farquhar, G.: The $CO_2$ concentrating function of $C_4$ photosynthesis. A biochemical model. — In: Hall, D. O., Coombs, J., Goodwin, T. W. (ed.): Proceedings of the Fourth International Congress on Photosynthesis. Pp. 119–131. Biochem. Soc., London 1978.

Berry, J. A., Raison, J. K.: Response of macrophytes to temperature. — In: Lange, O. L., Nobel, P. S., Osmond, C. B., Ziegler, H. (ed.): Physiological Plant Ecology I. Responses to the Physical Environment. Pp. 277–338. Springer-Verlag, Berlin–Heidelberg–New York 1981.

Bertsch, A., Domes, W.: $CO_2$-Gaswechsel amphistomatischer Blätter 1. Der Einfluss unterschiedlicher Stomaverteilung der beiden Blattepidermen auf den $CO_2$-Transport. — Planta 85: 183–193, 1969.

Bethlenfalvay, G. J., Phillips, D. A.: Ontogenetic interactions between photosynthesis and symbiotic nitrogen fixation in legumes. — Plant Physiol. 60: 419–421, 1977.

Bezuglov, V. K., Chernysheva, L. M.: Vozrastnaya dinamika sostoyaniya vody v list' yakh i fotokhimicheskaya aktivnost' khloroplastov pri razlichnom vodosnabzhenii rastenii gorokha. [Age dynamics of water state in leaves and photochemical activity of chloroplasts at different water supply of pea plants.] — In: Sbornik Aspirantskikh Rabot. Estestvennye Nauki. Biologiya. Pp. 47–55. Izdat. kazan. Univ., Kazan' 1974.

Bierhuizen, J. F., Slatyer, R. O.: Photosynthesis of cotton leaves under a range of environmental conditions in relation to internal and external diffusive resistances. — Aust. J. biol. Sci. 17: 348–359, 1964.

Biscoe, P. V., Incoll, L. D., Littleton, E. J., Ollerenshaw, J. H.: Barley and its environment VII. Relationships between irradiance, leaf photosynthetic rate and stomatal conductance. — J. appl. Ecol. 14: 293–302, 1977.

Bishop, D. G., Andersen, K. S., Smillie, R. M.: Lamellar structure and composition in relation to photochemical activity. — In: Hatch, M. D., Osmond, C. B., Slatyer, R. O. (ed.): Photosynthesis and Photorespiration. Pp. 372–381. Willey-Interscience, New York–London–Sydney–Toronto 1971.

Biswal, U. C., Singhal, G. S., Mohanty, P.: Dark stress induced senescence of barley leaves: Changes in chlorophyll a fluorescence of isolated chloroplasts. — Indian J. exp. Biol. 17: 262 to 264, 1979.

Björkman, O.: Responses to different quantum flux densities. — In: Lange, O .L., Nobel, P. S.,

Osmond, C. B., Ziegler, H. (ed.): Physiological Plant Ecology I. Responses to the Physical Environment. Pp. 55–107. Springer Verlag, Berlin –Heidelberg–New York 1981.

Black, C. C., Jr.: Photosynthetic carbon fixation in relation to net $CO_2$ uptake. – Annu. Rev. Plant Physiol. 24: 253–286, 1973.

Blenkinsop, P. G., Dale, J. E.: The effects of shade treatment and light intensity on ribulose-1,5-diphosphate carboxylase activity and Fraction I protein level in the first leaf of barley. — J. exp. Bot. 25: 899–912, 1974.

Bloom, B., Stetten, De W., Jr.: Pathways of glucose catabolism. — J. amer. chem. Soc. 75: 5446, 1953.

Boardman, N. K.: Chloroplasts — structure and photosynthesis. — Bot. Monogr. (Oxford) 14: 85–104, 442–483, 1977a.

Boardman, N. K.: Development of chloroplast structure and function. — In: Trebst, A., Avron, M. (ed.): Photosynthesis I. Pp. 583–600. Springer-Verlag, Berlin–Heidelberg–New York 1977b.

Boardman, N. K.: Comparative photosynthesis of sun and shade plants. — Annu. Rev. Plant Physiol. 29: 355–377, 1977c.

Boardman, N. K., Anderson, J. M., Hiller, R. G., Kahn, A., Roughan, P. G., Treffry, T. E., Thorne, S. W.: Biosynthesis of the photosynthetic apparatus during chloroplast development in higher plants. — In: Forti, G., Avron, M., Melandri, A. (ed.): Photosynthesis, Two Centuries after Its Discovery by Joseph Priestley. Vol. 3. Pp. 2265–2287. Dr. W. Junk N. V. Publ., The Hague 1972.

Boardman, N. K., Anderson, J. M., Kahn, A., Thorne, S. W., Treffry, T. E.: Formation of photosynthetic membranes during chloroplast development. — In: Boardman, N. K., Linnane, A. W., Smillie, R. M. (ed.): Autonomy and Biogenesis of Mitochondria and Chloroplasts. Pp. 70–84. North-Holland Publ. Comp., Amsterdam–London 1971.

Boardman, N. K., Björkman, O., Anderson, J. M., Goodchild, D. J., Thorne, S. W.: Photosynthetic adaptation of higher plants to light intensity: Relationship between chloroplast structure, composition of the photosystems and photosynthetic rates. — In: Avron, M. (ed.): Proceedings of the Third International Congress on Photosynthesis. Vol. III. Pp. 1809–1827. Elsevier, Amsterdam — Oxford — New York 1975.

Bocharov, E. A., Dzhanumov, D. A., Klimov, S. V.: Sostav fosfolipidov i galaktolipidov v khloroplastakh kontrastnykh po morozoustoĭchivosti sortov ozimykh kul'tur. [Composition of phospholipids and galactolipids in chloroplasts of winter crop cultivars of contrasting hardiness.] — Nauch. Tr. nauch.-issled. Inst. sel'. Khoz. tsentr. Raĭonov nechernozem. Zony 41 (Sozdanie Sortov Zernovykh Kul'tur Intensivnogo Tipa): 114–121, 1977.

Boffey, S. A., Ellis, J. R., Selldén, G., Leech, R. M.: Chloroplast division and DNA synthesis in light-grown wheat leaves. — Plant Physiol. 64: 502–505, 1979.

Bogacheva, I. I., Golubkova, B. M., Kislyakova, T. E.: Struktura i funktsii khloroplastov v ontogeneze kartofelya. [Structure and function of chloroplasts during potato ontogenesis.] — In: Ontogenez Vysshikh Rastenii. Pp. 94–100. Izd. Akad. Nauk arm. SSR, Erevan 1970.

Bolhár-Nordenkampf, H. R.: Die Veränderungen des Chlorophyllgehaltes in ontogenetisch verschiedenen Blättern von Phaseolus vulgaris var. nanus L. nach Behandlung mit Atrazin. — Biochem. Physiol. Pflanzen 167: 41–64, 1975.

Bolle-Jones, E. W.: Variations of chlorophyll and soluble sugar in oil palm leaves in relation to position, time of day and yield. — Oléagineux 23: 505–511, 1968.

Bonhomme, R., Varlet Grancher, C., Chartier, M., Artis, P.: Utilisation de l'énergie solaire par une culture de Vigna sinensis IV. Influence de l'âge et des eclairements passés sur le potentiel photosynthétique des feuilles cotylédonaires. — Ann. agron. 28: 159–169, 1977.

Borzenkova, R. A., Bortnikova, I. F.: Svetozavisimost' deĭstviya kinetina v protsesse khloroplastogeneza. [Light dependence of kinetin action during chloroplast ontogeny.] — Fiziol. Rast. 25: 254–261, 1978.

Borzenkova, R. A., Mokronosov, A. T.: Rol' fitogormonov v biogeneze khloroplastov. [Role of phytohormones in chloroplast biogenesis.] – Fiziol. Rast. 23: 490–496, 1976.

Borzenkova, R. A., Nefedova, O. A.: Formirovanie fotosinteticheskogo apparata i soderzhanie endogennykh tsitokininov v ontogeneze lista kartofelya. [Formation of photosynthetic apparatus and content of endogenous cytokinins during potato leaf ontogeny.] — Fiziol. Rast. 28: 825–833, 1981.

Boschetti, A.: Biogenese der Chloroplasten und Mitochondrien. VIII. — Gustav Fischer Verlag, Stuttgart–New York 1978.

Bottrill, D. E., Possingham, J. V.: The effect of mineral deficiency and leaf age on the nitrogen and chlorophyll content of spinach chloroplasts. – Biochim. biophys. Acta 189: 80–84, 1969.

Bouchard, M., Trudel, M. J.: Évolution et incidence des échanges de CO₂ pour évaluer l'accumulation de poids sec chez la laitue pommée durant la stade plantule. — Can. J. Plant Sci. 60: 577–583, 1980.

Bourdu, R.: Discussion sur les caractéristiques structurales et infra-structurales des feuilles en fonction de l'appartenance aux trois types métaboliques. — Physiol. vég. 14: 551–561, 1976.

Bourdu, R., Delay, C., Darmanaden, J., Halpern, S.: Ontogénie et développement des structures foliaires et chloroplastiques de Lactuca sativa. — Physiol. vég. 13: 265–289, 1975.

Bourdu, R., Krivitzky, M.: Sur le rapport chlorophylle a/chlorophylle b des chloroplastes dimorphiques de Zea mays L. – Compt. rend. Acad. Sci. Paris, Sér. D 275: 1039–1042, 1972.

Bowes, G., Holaday, A. S., Haller, W. T.: Seasonal variation in the biomass, tuber density, and photosynthetic metabolism of Hydrilla in three Florida lakes. — J. aquat. Plant Manage. 17: 61–65, 1979.

Bowes, G., Holaday, A. S., Van, T. K., Haller, W. T.: Photosynthetic and photorespiratory carbon metabolism in aquatic plants. — In: Hall, D. O., Coombs, J., Goodwin, T. W. (ed.): Proceedings of the Fourth International Congress on Photosynthesis. Pp. 289–298. Biochem. Soc., London 1978.

Boyer, J. S.: Differing sensitivity of photosynthesis to low leaf water potentials in corn and soybean. — Plant Physiol. 46: 236–239, 1970.

Bozarth, C. S., Kennedy, R. A., Schekel, K. A.: The effects of leaf age on photosynthesis in rose. — J. amer. Soc. hort. Sci. 107: 707—712, 1982.

Braber, J. M.: Catalase and peroxidase in primary bean leaves during development and senescence. — Z. Pflanzenphysiol. 97: 135–144, 1980.

Bradbeer, J. W.: Plastid development in primary leaves of Phaseolus vulgaris. — J. exp. Bot. 22: 382–390, 1971.

Bradbeer, J. W.: Chloroplasts — structure and development. — In: Smith, H. (ed.): The Molecular Biology of Plant Cells. Bot. Monogr. Vol. 14. Pp. 64–84. Blackwell sci. Publ., Oxford 1977.

Bradbeer, J. W.: Development of photosynthetic function during chloroplast biogenesis. —In: Hatch, M. D., Boardman, N. K. (ed.): The Biochemistry of Plants. Vol. 8. Pp. 423–472. Academic Press, New York–London–Toronto–Sydney–San Francisco 1981.

Bradbeer, J. W., Ireland, H. M. M., Smith, J. W., Rest, J., Edge, H. J. W.: Plastid development in primary leaves of Phaseolus vulgaris. VII. Development during growth in continuous darkness. — New Phytol. 73: 263–270, 1974.

Brady, C. J., Patterson, B. D., Heng Fong Tung, Smillie, R. M.: Protein and RNA synthesis during ageing of chloroplasts in wheat leaves. — In: Boardman, N. K., Linnane, A. W., Smillie, R. M. (ed.): Autonomy and Biogenesis of Mitochondria and Chloroplasts. Pp. 453–456. North-Holland Publ. Comp., Amsterdam–London 1971.

Brady, C. J., Scott, N. S.: The persistance of plastid polyribosomes and Fraction 1 protein synthesis in ageing wheat leaves. — In: Bogorad, L., Weil, J. H. (ed.): Acides Nucléiques et Synthèse des Protéines chez les Végétaux. Coll. Int. C.N.R.S. No. 261. Pp. 387–393. Édit. CNRS, Paris 1977a.

Brady, C. J., Scott, N. S.: Chloroplast polyribosomes and synthesis of fraction 1 protein in the developing wheat leaf. — Aust. J. Plant Physiol. 4: 327–335, 1977b.

Brandt, A. B., Tageeva, S. V.: Opticheskie Parametry Rastitel'nykh Organizmov. [Optical Parameters of Plant Organisms.] — Nauka, Moskva 1967.

Brangeon, J.: Compared ontogeny of the two types of chloroplasts of *Zea mays*. — J. Microscop. 16: 233–242, 1973.

Brangeon, J.: Effect of irradiance on granal configurations of *Zea mays* bundle sheath chloroplasts. — Photosynthetica 7: 365–372, 1973.

Brangeon, J., Mustárdy, L. A.: The ontogenetic assembly of intra-chloroplastic lamellae viewed in 3-dimension. — Biol. cell. 36: 71–80, 1979.

Branton, D.: Fracture faces of frozen membranes. — Proc. nat. Acad. Sci. USA 55: 1048–1056, 1966.

Branton, D., Bullivant, S., Gilula, N. B., Karnovsky, M. J., Moor, H., Mühlenthaler, K., Northcote, D. H., Packer, L., Satir, B., Satir, P., Speth, V., Staehelin, L. A., Steere, R. L,. Weinstein, R. S.: Freeze-etching nomenclature. — Science 190: 54–56, 1975.

Briant, R. E.: An analysis of the effects of gibberellic acid on tomato leaf growth. — J. exp. Bot. 25: 764–771, 1974.

Bricker, T. M., Newman, D. W.: Changes in the chlorophyll-proteins and electron transport activities of soybean (*Glycine max* L., cv. Wayne) cotyledon chloroplasts during senescence. — Photosynthetica 16: 239–244, 1982.

Britz, S. J.: Chloroplast and nuclear migration. — In: Haupt, W., Feinleib, M. E. (ed.): Physiology of Movements. Pp. 170–205. Springer-Verlag, Berlin–Heidelberg–New York 1979.

Brouwer, R.: The influence of the suction tension of the nutrient solutions on growth, transpiration and diffusion pressure deficit of bean leaves (*Phaseolus vulgaris*). — Acta bot. neerl. 12: 248–261, 1963.

Brown, H. T., Escombe, F.: Static diffusion of gases and liquids in relation to the assimilation of carbon and translocation in plants. — Phil. Trans. roy. Soc. London, Ser. B 193: 223–291, 1900.

Brown, K. W., Rosenberg, N. J.: Influence of leaf age, illumination, and upper and lower surface differences on stomatal resistance of sugar beet (*Beta vulgaris*) leaves. — Agron. J. 62: 20–24, 1970.

Brun, L. J., Kanemasu, E. T., Powers, W. L.: Estimating transpiration resistance. — Agron. J. 65: 326–328, 1973.

Bryant, F. D., Latimer, P., Seiber, B. A.: Changes in total light scattering and absorption caused by changes in particle conformation — A test of theory. — Arch. Biochem. Biophys. 135: 109–121, 1969.

Buchanan, B. B.: Role of light in the regulation of chloroplast enzymes. — Annu. Rev. Plant Physiol. 31: 341–374, 1980.

Buchanan-Bollig, I. C., Kluge, M., Lüttge, U.: PEP-carboxylase activities and the regulation of CAM: Effects of extraction procedures and leaf age. — Z. Pflanzenphysiol. 97: 457—470, 1980.

Bucke, C., Hallaway, M.: The distribution of plastoquinone *C* and the seasonal variation in its level in young leaves of *Vicia faba* L. — In: Goodwin, T. W. (ed.): Biochemistry of Chloroplasts. Vol. I. Pp. 153–157. Academic Press, London–New York 1966.

Bulley, N. R., Nelson, C. D., Tregunna, E. B.: Photosynthesis: Action spectra for leaves in normal and low oxygen. — Plant Physiol. 44: 678–684, 1969.

Bunce, J. A., Patterson, D. T., Peet, M. M., Alberte, R. S.: Light acclimation during and after leaf expansion in soybean. — Plant Physiol. 60: 255–258, 1977.

Bünning, E., Sagromsky, H.: Die Bildung des Spaltöffnungsmusters in der Blattepidermis (mit Anmerkungen über weitere Musterbildungen). — Z. Naturforsch. 3 B: 203–216, 1948.

Burris, J. S., Carson, E. W.: Carbohydrate metabolism in orchardgrass (*Dactylis glomerata* L.) as affected by age. — Crop Sci. 11: 171–174, 1971.

Burrows, F. J., Milthorpe, F. L.: Stomatal conductance in the control of gas exchange. — In: Kozlowski, T. T. (ed.): Water Deficits and Plant Growth. Vol. IV. Pp. 103–152. Academic Press, New York–San Francisco–London 1976.

Buschmann, C., Lichtenthaler, H. K.: Hill-activity and P700 concentration of chloroplasts isolated from radish seedlings treated with δ–indoleacetic acid, kinetin or gibberellic acid. — Z. Naturforsch. 32C: 798–802, 1977.

Buschmann, C., Prehn, H.: *In vivo* studies of radiative and non-radiative de-excitation processes of pigments in *Raphanus* seedlings by photoacoustic spectroscopy. — Photobiochem. Photobiophys. 2: 209–215, 1981.

Butler, R. D.: The effect of light intensity on stem and leaf growth in broad bean seedlings. — J. exp. Bot. 14: 142–152, 1963.

Butler, R. D.: The fine structure of senescing cotyledons of cucumber. — J. exp. Bot. 18: 535–543, 1967.

Butler, W. L.: Absorption spectroscopy *in vivo*: theory and application. — Annu. Rev. Plant Physiol. 15: 451–470, 1964.

Butler, W. L.: Development of photosynthetic systems 1 and 2 in a greening leaf. — Biochim. biophys. Acta 102: 1–8, 1965.

Butterfass, T.: Patterns of Chloroplast Reproduction. (Cell Biology Monographs, Vol. 6.) — Springer-Verlag, Wien–New York 1979.

Bystrykh, E. E., Matorin, D. N.: Funktsional'naya aktivnost' fotosinteticheskogo apparata v ontogeneze rastenii podsolnechnika. [Functional activity of the photosynthetic apparatus during ontogenesis of sunflower plants.] — Sel'skokhoz. Biol. 10: 230–236, 1975.


Caemmerer, S. von, Farquhar, G. D.: Some relationships between the biochemistry of photosynthesis and the gas exchange of leaves. — Planta 153: 376–387, 1981.

Caldwell, M. M.: Plant response to solar ultraviolet radiation. — In: Lange, O. L., Nobel, P. S., Osmond, C. B., Ziegler, H. (ed.): Physiological Plant Ecology I. Responses to the Physical Environment. Pp. 169–197. Springer-Verlag, Berlin–Heidelberg–New York 1981.

Callow, J. A.: Ribosomal RNA, Fraction I protein synthesis, and ribulose diphosphate carboxylase activity in developing and senescing leaves of cucumber. — New Phytol. 73: 13–20, 1974.

Cameron, R. J.: Light intensity and the growth of *Eucalyptus* seedlings. I. Ontogenetic variation in *E. fastigata*. — Aust. J. Bot. 18: 29–43, 1970a.

Cameron, R. J.: Light intensity and the growth of *Eucalyptus* seedlings II. The effect of cuticular waxes on light absorption in leaves of *Eucalyptus* species. — Aust. J. Bot. 18: 275–284, 1970b.

Campbell, G. S.: Fundamentals of radiation and temperature relations. — In: Lange, O. L., Nobel, P. S., Osmond, C. B., Ziegler, H. (ed.): Physiological Plant Ecology I. Responses to the Physical Environment. Pp. 11–40. Springer-Verlag, Berlin–Heidelberg–New York 1981.

Campbell, R.: Electron microscopy of the development of needles of *Pinus nigra* var. *maritima*. — Ann. Bot. 36: 711–720, 1972.

Campbell, W. H., Black, C. C.: The relationship of $CO_2$ assimilation pathways and photorespiration to the physiological quantum requirement of green plant photosynthesis. — BioSystems 10: 253–264, 1978.

Canvin, D. T.: Photorespiration: Comparison between $C_3$ and $C_4$ plants. — In: Gibbs, M., Latzko, E. (ed.): Photosynthesis II. Pp. 368–396. Springer-Verlag, Berlin–Heidelberg–New York 1979.

Canvin, D. T.: Photorespiration and nitrogen metabolism. — In: Bewley, J. D. (ed.): Nitrogen and Carbon Metabolism. Pp. 178–194. Martinus Nijhoff/Dr. W. Junk Publ., The Hague–Boston–London 1981.

Canvin, D. T., Berry, J. A., Badger, M. R., Fock, H., Osmond, C. B.: Oxygen exchange in leaves in the light. — Plant Physiol. 66: 302–307, 1980.

Carlson, R. E., Yarger, D. N., Shaw, R. H.: Factors affecting the spectral properties of leaves with special emphasis on leaf water status. — Agron. J. 68: 486–489, 1971.

Carpenter, S. B., Hanover, J. W.: Comparative growth and photosynthesis of black walnut and honey locust seedlings. — Forest Sci. 20: 317–324, 1974.

Cartwright, B., Kriedemann, P. E., Bowen, G. D., Torokfalvy, E., Cunningham, R. B.: Effects of phosphate nutrition on photosynthesis in wheat leaves. — In: Bieleski, R. L., Ferguson, A. R.,

Cresswell, M. M. (ed.): Mechanisms of Regulation of Plant Growth. Bull. 12. Pp. 145–150. Roy. Soc. New Zealand, Wellington 1974.

Casadoro, G., Rascio, N.: Relatinship between plastids with a membrane-bound body and cell differentiation gradients in belladonna. — Caryologia 30: 189–198, 1977.

Casadoro, G., Rascio, N.: Thylakoid membranes in sunflower and in other plants. — J. Ultrastruct. Res. 65: 30–35, 1978a.

Casadoro, G., Rascio, N.: Chloroplast ontogenesis in *Lippia citriodora* L. — In: Proceedings of the 9th International Congress on Electron Microscopy. Vol. II. Pp. 416–417. Toronto 1978b.

Casadoro, G., Rascio, N.: Plastid ultrastructural features in the various tissues of sunflower leaves. — Cytobios 24: 157–166, 1979a.

Casadoro, G., Rascio, N.: Patterns of thylakoid system formation. — J. Ultrastruct. Res. 69: 307–315, 1979b.

Casadoro, G., Rascio, N., Paganelli Cappelletti, E. M.: Membrane-bound plastidial inclusions in belladonna (*Atropa belladonna* L. L.). – Biol. cell. 29: 61—66, 1977.

Čatský, J.: Water saturation deficit and photosynthetic rate as related to leaf age in the wilting plant. — In: Slavík, B. (ed.): Water Stress in Plants. Pp. 203–209. Publ. House Czechosl. Acad. Sci., Prague 1965. Dr. W. Junk, The Hague 1965.

Čatský, J., Chartier, P., Djavanchir, A.: Assimilation nette, utilisation de l'eau et microclimat d'un champ de maïs. IV. — Évolution diurne de la résistance stomatique et du déficit de saturation des feuilles; conséquences sur la fixation du $CO_2$. — Ann. agron. 24: 287–305, 1973.

Čatský, J., Nováková, J., Šesták, Z.: Daily carbon dioxide balance and its changes with age in a fodder cabbage plant grown in controlled conditions. — Photosynthetica 1 : 215–218, 1967.

Čatský, J., Tichá, I.: $CO_2$ compensation concentration in bean leaves: Effect of photon flux density and leaf age. — Biol. Plant. 21: 361–364, 1979.

Čatský, J., Tichá, I.: Ontogenetic changes in the internal limitations to bean-leaf photosynthesis. 5. Photosynthetic and photorespiration rates and conductances for $CO_2$ transfer as affected by irradiance. — Photosynthetica 14: 392–400, 1980.

Čatský, J., Tichá, I.: Photosynthetic characteristics during ontogenesis of leaves. 6. Intracellular conductance and its components. — Photosynthetica 16: 253–284, 1982.

Čatský, J., Tichá, I., Solárová, J.: Ontogenetic changes in the internal limitations to bean-leaf photosynthesis. 1. Carbon dioxide exchange and conductances for carbon dioxide transfer. — Photosynthetica 10: 394–402, 1976.

Ceulemans, R., Impens, I.: Study of $CO_2$ exchange processes, resistances to carbon dioxide and chlorophyll content during leaf ontogenesis in poplar. — Biol. Plant. 21: 302–306, 1979.

Ceulemans, R., Impens, I., Moermans, R.: The response of $CO_2$ exchange rate to photosynthetic photon flux density for several *Populus* clones under laboratory conditions. — Photosynthesis Res. 1 : 137–142, 1980.

Chabot, J. F., Chabot, B. F.: Developmental and seasonal patterns of mesophyll ultrastructure in *Abies balsamea*. — Can. J. Bot. 53: 295–304, 1975.

Chaïka, M. T., Savchenko, G. E.: Metabolizm pigmentov v protsesse razvitiya zelenogo lista. [Pigment metabolism in the course of development of a green leaf.] — In: Formirovanie Pigmentnogo Apparata Fotosinteza. Pp. 105–129. Nauka i Tekhnika, Minsk 1973.

Chaïka, M. T., Savchenko, G. E.: Fotoregulyatsiya biosinteza khlorofilla v protsesse razvitiya khloroplasta. [Photoregulation of chlorophyll biosynthesis and chloroplast development.] — In: Fotoregulyatsiya Metabolizma i Morfogeneza Rastenii. Pp. 120–134, 251. Nauka, Moskva 1975.

Chaïka, M. T., Savchenko, G. E.: Biosintez Khlorofilla v Protsesse Razvitiya Plastid. [Chlorophyll Biosynthesis in the Course of Plastid Development.] — Nauka i Tekhnika, Minsk 1981.

Chain, R. K., Arnon, D. I.: Quantum efficiency of photosynthetic energy conversion. — Proc. nat. Acad. Sci. USA 74: 3377–3381, 1977.

Chang, C. W.: Carbon dioxide and senescence in cotton plants. — Plant Physiol. 55: 515–519, 1975.

Charles-Edwards, D. A.: A model of leaf growth. — Ann. Bot. 44: 523–535, 1979.

Chartier, P.: A model of $CO_2$ assimilation in the leaf. — In: Prediction and Measurement of Photosynthetic Productivity. Pp. 307–315. Pudoc, Wageningen 1970.

Chartier, P., Chartier, M., Čatský, J.: Resistances for carbon dioxide diffusion and for carboxylation as factors in bean leaf photosynthesis. — Photosynthetica 4: 48–57, 1970.

Chartier, P., Morrot-Gaudry, J. F., Bethenod, O., Thomas, D. A.: The net assimilation of $C_3$ and $C_4$ plants as influenced by light and carbon dioxide, and an analysis of the role of the gene *opaque* 2 in young maize. — In: Landsberg, J. J., Cutting, C. V. (ed.): Environmental Effects on Crop Physiology. Pp. 125–136. Acad. Press, London–New York –San Francisco 1977.

Chin, H. F., Neales, T. F., Wilson, J. H.: The effects of cotyledon excision on growth and leaf senescence in soya-bean plants. — Ann. Bot. 41: 771–777, 1977.

Ching, T. M., Hedtke, S., Garay, A. E.: Energy state of wheat leaves in ammonium nitrate-treated plants. — Life Sci. 16: 603–610, 1975.

Chollet, R., Ogren, W. L.: Regulation of photorespiration in $C_3$ and $C_4$ species. – Bot. Rev. 41: 137–179, 1975.

Chonan, N.: [Studies on the photosynthetic tissues in the leaves of cereal crops. I. The mesophyll structure of wheat leaves at different levels of the shoot.] — Tohoku J. agr. Res. 16: 1—12, 1965.

Chonan, N.: [Studies on the photosynthetic tissues in the leaves of cereal crops. III. The mesophyll structure of rice leaves inserted at different levels of the shoot.] — Proc. Crop Sci. Soc. Jap. 36: 291–296, 1967.

Christie, E. K.: Physiological responses of semiarid grasses. 4. Photosynthetic rates of *Thyridolepis mitchelliana* and *Cenchrus ciliaris* leaves. — Aust. J. agr. Res. 26: 459–466, 1975.

Chub, A. I.: Izmenenie sostava assimilyatov u soi pod vliyaniem molibdena. [Changes in photosynthates composition in soybean induced by molybdenum.] — Fiziol. Biokhim. kul't. Rast. 7: 607–610, 1975.

Chuchalin, A. I., Eroshin, N. S., Tikhomirov, A. A., Anistratova, N. A., Shur, L. A., Shilenko, M. P.: Soderzhanie pigmentov i opticheskie svoĭstva list'ev pshenitsy v usloviyakh intensivnoĭ svetokul'tury rasteniĭ. [Pigment concentration and optical properties of leaves of wheat grown in intense light.] — Izv. sibir. Otd. Akad. Nauk SSSR, Ser. biol. Nauk 1977 (2): 38–43, 1977.

Chugunova, N . G., Chermnykh, L. N., Kosobryukhov, A. A., Karpilova, I. F., Chermnykh, R. M.: Vzaimosvyaz' rostovykh protsessov i fotosinteza v ontogeneze lista ogurtsov pri deĭstvii ponizhennoĭ nochnoĭ temperatury. [Interelationship of growth processes and photosynthesis during cucumber leaf ontogeny under low night temperature.] — Fiziol. Rast. 27: 1101–1109, 1980.

Chung, H. H., Barnes, R. L.: Photosynthate allocation in *Pinus taeda*. II. Seasonal aspects of photosynthate allocation to different biochemical fractions in shoots. — Can. J. Forest Res. 10: 338–347, 1980a.

Chung, H. H., Barnes, R. L.: Photosynthate allocation in *Pinus taeda*. III. Photosynthate economy: its production, consumption and balance in shoots during the growing season. — Can. J. Forest Res. 10: 348–356, 1980b.

Clark, J. B., Lister, G. R.: Photosynthetic action spectra of trees II. The relationship of cuticle structure to the visible and ultraviolet spectral properties of needles from four coniferous species. — Plant Physiol. 55: 407–413, 1975.

Clayton, R. K.: Photosynthesis. Physical Mechanisms and Chemical Patterns. — Cambridge Univ. Press, Cambridge–London–New York–New Rochelle–Melbourne–Sydney 1980.

Cline, R. G., Campbell, G. S.: Seasonal and diurnal water relations of selected forest species. — Ecology 57: 367–373, 1976.

Clough, B. F., Milthorpe, F. L.: Effects of water deficit on leaf development in tobacco. — Aust. J. Plant Physiol. 2: 291–300, 1975.

Cockshull, K. E.: Effects of night-break treatment on leaf area and leaf dry weight in *Callistephus chinensis*. — Ann. Bot. 30: 791–806, 1966.

Constable, G. A., Rawson, H. M.: Effect of leaf position, expansion and age on photosynthesis, transpiration and water use efficiency of cotton. — Aust. J. Plant Physiol. 7: 89–100, 1980.

Cooke, J. R., Rand, R. H.: Diffusion resistance models. — In: Hesketh, J. D., Jones, J. W. (ed.): Predicting Photosynthesis for Ecosystem Models. Vol. I. Pp. 93–121. CRC Press, Boca Raton 1980.

Cornic, G., Mousseau, M., Monteny, B.: Importance de la photorespiration dans le bilan photosynthétique au cours de la croissance foliaire. — Oecol. Plant. 5: 355–363, 1970.

Cowan, I. R.: Regulation of water use in relation to carbon gain in higher plants. — In: Lange, O. L., Nobel, P. S., Osmond, C. B., Ziegler, H. (ed.): Physiological Plant Ecology II. Water Relations and Carbon Assimilation. Pp. 589–613. Springer-Verlag, Berlin–Heidelberg–New York 1982.

Cowan, I. R., Farquhar, G. D.: Stomatal function in relation to leaf metabolism and environment. — In: Jennings, D. H. (ed.): Integration of Activity in the Higher Plant. Pp. 471–505. Cambridge Univ. Press, Cambridge–London–New York–Melbourne 1977.

Cowan, I. R., Milthorpe, F. L.: Plant factors influencing the water status of plant tissues. — In: Kozlowski, T. T. (ed.): Water Deficits and Plant Growth. Vol. I. Pp. 137–193. Academic Press, New York–London 1968.

Cowan, I. R., Troughton, J. H.: The relative role of stomata in transpiration and assimilation. — Planta 97: 325–336, 1971.

Cowart, F. F.: Apple leaf structure as related to position of the leaf upon shoot and to type of growth. — Proc. Soc. hort. Sci. 32: 145–148, 1936.

Coyne, P. I., Bingham, G. E.: Variation in photosynthesis and stomatal conductance in an ozone-stressed ponderosa pine stand: light response. — Forest Sci. 28: 257–273, 1982.

Crespo, H. M., Frean, M., Cresswell, C. F., Tew, J.: The occurrence of both $C_3$ and $C_4$ photosynthetic characteristics in a single *Zea mays* plant. — Planta 147: 257–263, 1979.

Criswell, J. G., Shibles, R. M.: Physiological basis for genotypic variation in net photosynthesis of oat leaves. — Crop Sci. 11: 550–553, 1971.

Crookston, R. K.: The structure and function of $C_4$ vascular tissue — some unanswered questions. — Ber. deut. bot. Ges. 93: 71–78, 1980.

Cunninghame, M. E., Bowes, B. G., Hillman, J. R.: An ultrastructural study of foliar senescence in *Taxus baccata* L. — Ann. Bot. 43: 527–528, 1979.

Cutler, J. M., Rains, D. W., Loomis, R. S.: The importance of cell size in the water relations of plants. — Physiol. Plant. 40: 255–260, 1977.

D'Agostino, G., Pennazio, S.: An ultrastructural study of the senescence induced in *Gomphrena globosa* leaves by mineral deficiency. — J. submicroscop. Cytol. 13: 373–384, 1981.

Dale, J. E.: Leaf growth in *Phaseolus vulgaris* 2. Temperature effects and the light factor. — Ann. Bot. 29: 293–308, 1965.

Dale, J. E.: Growth and photosynthesis in the first leaf of barley. The effect of time of application of nitrogen. — Ann. Bot. 36: 967–979, 1972.

Dale, J. E.: The Growth of Leaves. — Edward Arnold Ltd., London 1982.

Dale, J. E., Heyes, J. K.: A virescent mutant of *Phaseolus vulgaris*; growth, pigment and plastid characters. — New Phytol. 69: 733–742, 1970.

Daley, L. S., Dailey, F., Criddle, R. S.: Light activation of ribulose bisphosphate carboxylase. Purification and properties of the enzyme in tobacco. — Plant Physiol. 62: 718–722, 1978.

Dalgarn, D., Miller, P., Bricker, T., Speer, N., Jaworski, J. G., Newman, D. W.: Galactosyl transferase activity of chloroplast envelopes from senescent soybean cotyledons. — Plant Sci. Lett. 14: 1–6, 1979.

Damsz, B.: Biometric analysis of the thylakoid system development during chloroplast ontogenesis in leaves of two orchids, *Coelogyne cristata* and *Cymbidium insigne*. — Biochem. Physiol. Pflanzen 174: 802–810, 1979.

D'Aoust, A. L., Canvin, D. T.: Caractéristiques du $^{14}CO_2$ dégagé à la lumière et à l'obscurité par des feuilles de haricot, de radis, de tabac et de tournesol pendant et après une photosynthèse en présence de $^{14}CO_2$. — Physiol. vég. 12: 545–560, 1974.

Dassiou, C., Akoyunoglou, G.: Effect of light on the RuDP-carboxylase activity in etiolated plants of different ages. — In: Metzner, H. (ed.): Progress in Photosynthesis Research. Vol. III. Pp. 1631–1635. Tübingen 1969.

Davies, B. H.: Carotenoids. — In: Goodwin, T. W. (ed.): Chemistry and Biochemistry of Plant Pigments. 2$^{nd}$ Ed. Vol. 2. Pp. 38–165. Academic Press, London–New York–San Francisco 1976.

Davis, S. D., McCree, K. J.: Photosynthetic rate and diffusion conductance as a function of age in leaves of bean plants. — Crop Sci. 18: 280–282, 1978.

Deering, R. A.: Ultraviolet radiation and nucleic acid. — Sci. Amer. 207: 135–144, 1962.

De Greef, J. A., Caubergs, R.: Studies on greening of etiolated seedlings I. Elimination of the lag phase of chlorophyll biosynthesis by a pre-illumination of the embryonic axis in intact plants. — Physiol. Plant. 26: 157–165, 1972.

Dei, M.: Inter-organ control of greening in etiolated cucumber cotyledons. — Physiol. Plant. 43: 94–98, 1978.

De Jong, D. W.: The influence of growth conditions and leaf maturity in relation to the chlorogenic acid stimulation of glycollate oxidase from tobacco leaves. — Can. J. Bot. 52: 209–215, 1974.

De Jong, D. W., Woodlief, W. G.: Chloroplast properties as causative factors in the growth performance of pale-yellow tobacco hybrid. — Tobacco Sci. 18: 105–107, 1974. Tobacco 176 (18): 29–31, 1974.

De Jong, D. W., Woodlief, W. G.: High-speed, low pressure liquid chromatography of chloroplast pigments from tobacco mutants. — J. agr. Food Chem. 26: 1281–1288, 1978.

De Jong, D. W., Woodlief, W. G.: Some factors influencing tobacco leaf senescence. — Beitr. Tabakforsch. 10: 48–56, 1979.

Denmead, O. T., Millar, B. D.: Field studies of the conductance of wheat leaves and transpiration. — Agron. J. 68: 307–311, 1976.

Denne, M. P.: Leaf development in Trifolium repens. — Bot. Gaz. 127: 202–210, 1966.

Dennett, M. D., Auld, B. A.: The effects of position and temperature on the expansion of leaves of Vicia faba. L. — Ann. Bot. 46: 511–517, 1980.

Dennett, M. D., Auld, B. A., Elston, J.: A description of leaf growth in Vicia faba L. — Ann. Bot. 42: 223–232, 1978.

Dennett, M. D., Elston, J., Milford, J. R.: The effect of temperature on the growth of individual leaves of Vicia faba L. in the field. — Ann. Bot. 43: 197–208, 1979.

Dennis, D. T., Stubbs, M., Coultate, T. P.: The inhibition of Brussels sprout leaf senescence by kinins. — Can. J. Bot. 45: 1019–1024, 1967.

Dennison, D. S.: Phototropism. — In: Haupt, W., Feinleib, M. E. (ed.): Physiology of Movements. Pp. 506–566. Springer-Verlag, Berlin–Heidelberg–New York 1979.

Detchon, P., Possingham, J. V.: Ribosomal-RNA distribution during leaf development in spinach. — Phytochemistry 11: 943–947, 1972.

de Villiers, O. T., Ashton, F. M.: Metabolic activity of isolated leaf cells of Phaseolus vulgaris in relation to leaf development. — Plant Physiol. 59: 1072—1075, 1977.

De Wit, C. T. et al.: Simulation of Assimilation, Respiration and Transpiration of Crops. — PUDOC, Wageningen 1978.

Dezhi, L.: Aktivnost' oksidazy glikolevoï kisloty i soderzhanie khlorofilla v list'yakh kukuruzy. [Activity of glycollate oxidase and chlorophyll content in maize leaves.] — Acta bot. Acad. Sci. hung. 20: 221–226, 1974.

Dickmann, D. I.: Photosynthesis and respiration by developing leaves of cottonwood (Populus deltoides Bartr.). — Bot. Gaz. 132: 253–259, 1971.

Dickmann, D. I., Gjerstad, D. H.,: Application to woody plants of a rapid method for determining leaf $CO_2$ compensation concentration. — Can. J. Forest Res. 3: 237–242, 1973.

Dickmann, D. I., Gordon, J. C.: Incorporation of $^{14}C$-photosynthate into protein during leaf development in young *Populus* plants. — Plant Physiol. 56: 23–27, 1975.

Dickson, R. E., Larson, P. R.: $^{14}C$ fixation, metabolic labeling patterns, and translocation profiles during leaf development in *Populus deltoides*. — Planta 152: 461–470, 1981.

Diepenbrock, W., Geisler, G.: Untersuchungen zur Bedeutung der Fruchtwand der Rapsschote als Organ der Assimilatbildung und als Stickstoffreservoir für die Samen. — Z. Acker-Pflanzenbau 146: 54–67, 1978.

Döbel, P.: Untersuchung der Wirkung von Streptomycin-, Chloramphenicol- und 2-Thiouracil-Behandlung auf die Plastidenentwicklung von *Lycopersicum esculentum* Miller. — Biol. Zentralbl. 82: 275–295, 1963.

Dockerty, A., Lord, J. M., Merrett, M. J.: Development of ribulose-1,5-diphosphate carboxylase in castor bean cotyledons. — Plant Physiol. 59: 1125–1127, 1977.

Dodge, J. D.: Changes in chloroplast fine structure during the autumnal senescence of *Betula* leaves. — Ann. Bot. 34: 817–824, 1970.

Doley, D., Yates, D. J.: Gas exchange of Mitchell grass [*Astrebla lappacea* (Lindl.) Domin] in relation to irradiance, carbon dioxide supply, leaf temperature and temperature history. — Aust. J. Plant Physiol. 3: 471–487, 1976.

Domes, W.: Unterschiedliche $CO_2$-Abhängigkeit des Gasaustausches beider Blattseiten von *Zea mays*. — Planta 98: 186–189, 1971.

Domes, W., Bertsch, A.: $CO_2$-Gaswechsel amphistomatischer Blätter 2. Ein Vergleich von diffusivem $CO_2$-Austausch der beiden Blattepidermen von *Zea mays* mit dem im Porometer gemessenen viscosen Volumfluss. — Planta 86: 84–91, 1969.

Dontschev, T., Lossner, G.: Der Einfluss von Chlorcholinchlorid (CCC) auf das Wachstum und den Blattfarbstoffgehalt von *Phaseolus vulgaris* L. und *Sinapis alba* L. — Arch. Phytopathol. Pflanzenschutz 12: 49–56, 1976.

Dornhoff, G. M., Shibles, R. M.: Varietal differences in net photosynthesis of soybean leaves. — Crop Sci. 10: 42–45, 1970.

Dornhoff, G. M., Shibles, R.: Soybean leaf net $CO_2$-exchange as influenced by experimental environment and leaf age. — Iowa State J. Res. 48:311–317, 1974.

Dornhoff, G. M., Shibles, R.: Leaf morphology and anatomy in relation to $CO_2$-exchange rate of soybean leaves. — Crop Sci. 16: 377–381, 1976.

Dougherty, P. M., Teskey, R.O., Phelps, J. E., Hinckley, T. M.: Net photosynthesis and early growth trends of a dominant white oak (*Quercus alba* L.). — Plant Physiol. 64: 930–935, 1979.

Downton, J., Slatyer, R. O.: Variation in levels of some leaf enzymes. — Planta 96: 1–12, 1971.

Downton, W. J. S., Pyliotis, N. A.: Loss of photosystem II during ontogeny of sorghum bundle sheath chloroplasts. — Can. J. Bot. 49: 179–180, 1971.

Downton, W. J. S., Törökfalvy, E.: Photosynthesis in developing asparagus plants. — Aust. J. Plant Physiol. 2: 367–375, 1975.

Draper, S. R.: Lipid changes in senescing cucumber cotyledons. — Phytochemistry 8: 1641–1647, 1969.

Drury, S., Park, R. B.: The effect of leaf senescence on quantum efficiencies of photosynthetic light reactions. — Plant Physiol. 43 (Suppl.): S-29, 1968.

Duke, S. H., Schrader, L. E., Henson, C. A., Servaites, J. C., Vogelzang, R. D., Pendleton, J. W.: Low root temperature effects on soybean nitrogen metabolism and photosynthesis. — Plant Physiol. 63: 956–962, 1979.

Dunaeva, S. E.: Ul'trastruktura khloroplastov pshenitsy v svyazi s vozrastom lista. [Ultrastructure of wheat chloroplasts in connection with leaf growth.] — Tsitologiya 21: 5–11, 1979.

Dunn, E. L.: Environmental stresses and inherent limitations affecting $CO_2$ exchange in evergreen sclerophylls in mediterranean climates. — In: Gates, D. M., Schmerl, R. B. (ed.): Perspectives of Biophysical Ecology. Pp. 159–181. Springer-Verlag, Berlin–Heidelberg–New York 1975.

Dunstone, R. L., Gifford, R. M., Evans, L. T.: Photosynthetic characteristics of modern and primitive wheat species in relation to ontogeny and adaptation to light. — Aust. J. biol. Sci. 26: 295–307, 1973.

Düring, H.: Untersuchungen zur Umweltabhängigkeit der stomatären Transpiration bei Reben. II. Ringelungs- und Temperatureffekte. — Vitis 17: 1–9, 1978.

Düring, H.: Stomatafrequenz bei Blättern von Vitis-Arten und -Sorten. — Vitis 19: 91–98, 1980.

Dutton, J. E., Rogers, L. J., Haslett, B. G., Takruri, I. A. H., Gleaves, J. T., Boulter, D.: Comparative studies on the properties of two ferredoxins from Pisum sativum L. — J. exp. Bot. 31: 379–391, 1980.

Dutzmann, S., Forche, S., Döll, G., Kranz, J.: Licht- und rasterelektronenmikroskopischer Vergleich von Blattoberflächen verschiedener Sommergerstensorten. — Z. Pflanzenkrankh. Pflanzenschutz 88: 518–524, 1981.

Duysen, M. E., Freeman, T. P., Zabrocki, R. D.: Light and the correlation of chloroplast development and coupling of photophosphorylation to electron transport. — Plant Physiol. 65: 880–883, 1980.


Ehara, Y., Misawa, T.: Occurrence of abnormal chloroplasts in tobacco leaves infected systemically with the ordinary strain of cucumber mosaic virus. — Phytopathol. Z. 84: 233–252, 1975.

Ehleringer, J.: Leaf absorptances of Mohave and Sonoran Desert plants. — Oecologia 49: 366–370, 1981.

Ehleringer, J., Björkman, O.: Pubescence and leaf spectral characteristics in a desert shrub, Encelia farinosa. — Oecologia 36: 151–162, 1978.

Ehleringer, J. R., Mooney, H. A.: Leaf hairs: effects on physiological activity and adaptive value to a desert shrub. — Oecologia 37: 183–200, 1978.

Eilam, Y., Butler, R. D., Simon, E. W.: Ribosomes and polysomes in cucumber leaves during growth and senescence. — Plant Physiol. 47: 317–323, 1971.

El Aouni, M. H.: Action du déficit hydrique interne sur les mouvements stomatiques, la transpiration et la photosynthèse nette d'aiguilles excisées de Pin noir d'Autriche (Pinus nigra Arn.). Évolution avec l'âge foliaire. — Photosynthetica 10: 403–410, 1976.

El Aouni, M. H., Mousseau, M.: Relation d'échange de CO₂ chez les aiguilles du Pin noir d'Autriche (Pinus nigra Arn.) avec l'âge, la teneur en chlorophylle et réassimilation. — Photosynthetica 8: 78–86, 1974.

Eliáš, P.: Stomatal activity within the crowns of tall deciduous trees under forest conditions. — Biol. Plant. 21: 266–274, 1979.

Eliáš, P., Huzulák, J.: Hustota prieduchov v korune javora pol'ného (Acer campestre L.). [The density of stomata in the crown of Acer campestre L.] — Acta Musei silesiae, Ser. dendrol. 24: 129–135, 1975.

Eliáš, P., Kozinka, V.: Stomata in the leaves of Asperula odorata L. and Pulmonaria officinalis L. subsp. maculosa (Hayne) Gams. — Biológia (Bratislava) 31: 33–40, 1976.

Eller, B. M.: Die optische Eigenschaften der Blätter von Rhododendron ferrugineum L. und Alnus viridis (Chaix) DC. — Ber. schweiz. bot. Ges. 85: 25–30, 1975.

Eller, B. M.: Die strahlungsökologische Bedeutung von Epidermisauflagen. — Flora 168: 146–192, 1979.

Eller, B. M., Glättli, R., Flach, B.: Optische Eigenschaften und Pigmente von Sonnen- und Schattenblättern der Rotbuche (Fagus silvatica L.) und der Blutbuche (Fagus silvatica cv. Atropunicea). — Flora 171: 170–185, 1981.

Eller, B. M., Willi, P.: Die Bedeutung der Wachsausblühungen auf Blättern von Kalanchoë pumila L. Baker für die Absorption der Globalstrahlung. — Flora 166: 461–474, 1977.

Ellis, M. A., Ferree, D. C., Spring, D. E.: Photosynthesis, transpiration, and carbohydrate content of apple leaves infected by Podosphaera leucotricha. — Phytopathology 71: 392–395, 1981.

Elmore, C. D., Hesketh, J. D., Muramoto, H.: A survey of rates of leaf growth, leaf ageing and leaf photosynthetic rates among and within species. — J. Arizona Acad. Sci. 4: 215 – 219, 1967.

El-Sharkawy, M., Hesketh, J.: Photosynthesis among species in relation to characteristics of leaf anatomy and $CO_2$ diffusion resistances. — Crop Sci. 5: 517–521, 1965.

Erickson, R. O.: Relative elemental rates and anisotropy of growth in area: a computer programme. — J. exp. Bot. 17: 390–403, 1966.

Erickson, R. O., Michelini, F. J.: The plastochron index. — Amer. J. Bot. 44: 297–305, 1957.

Ericsson, A.: Effects of fertilization and irrigation on the seasonal changes of carbohydrate reserves in different age-classes of needle on 20-year-old Scotch pine trees (Pinus silvestris). — Physiol. Plant. 45: 270–280, 1979.

Esau, K.: Anatomy of Seed Plants. 2nd ed. — John Wiley and Sons, New York – Santa Barbara – London – Sydney – Toronto 1977.

Evans, L. S., Ting, I. P.: Ozone sensitivity of leaves: Relationship to leaf water content, gas transfer resistance, and anatomical characteristics. — Amer. J. Bot. 61: 592–597, 1974.

Evans, L. T., Rawson, H. M.: Photosynthesis and respiration by the flag leaf and components of the ear during grain development in wheat. — Aust. J. biol. Sci. 23: 245–254, 1970.

Even-Chen, Z., Sachs, R. M.: Photosynthesis as a function of short day induction and gibberellic acid treatment in Bougainvillea "San Diego Red". — Plant Physiol. 65: 65–68, 1980.

Eze, J. M. O., Dumbroff, E. B., Thompson, J. E.: Effects of moisture stress and senescence on the synthesis of abscisic acid in the primary leaves of bean. — Physiol. Plant. 51: 418–422, 1981.

Fair, P., Tew, J., Cresswell, C.: The effect of age and leaf position on carbon dioxide compensation point ($\Gamma$), and potential photosynthetic capacity, photorespiration and nitrate assimilation in Hordeum vulgare L. — J. South Afr. Bot. 28: 81–95, 1972.

Fair, P., Tew, J., Cresswell, C. F.: Enzyme activities associated with carbon dioxide exchange in illuminated leaves of Hordeum vulgare L. I. Effects of light period, leaf age, and position, on carbon dioxide compensation point ($\Gamma$). — Ann. Bot. 37: 831–844, 1973.

Fair, P., Tew, J., Cresswell, C. F.: Enzyme activities associated with carbon dioxide exchange in illuminated leaves of Hordeum vulgare L. IV. The effect of light intensity on the carbon dioxide compensation point. — Ann. Bot. 38: 45–52, 1974.

Farkas, G. L., Rajháthy, T.: Untersuchungen über die xeromorphischen Gradienten einiger Kulturpflanzen. — Planta 45: 535–548, 1955.

Farquhar, G. D., Caemmerer, S. von, Berry, J. A.: A biochemical model of photosynthetic $CO_2$ assimilation in leaves of $C_3$ species. — Planta 149: 78–90, 1980.

Farquhar, G. D., Caemmerer, S. von: Modelling of photosynthetic response to environmental conditions. — In: Lange, O. L., Nobel, P. S., Osmond, C. B., Ziegler, H. (ed.): Physiological Plant Ecology II. Water Relations and Carbon Assimilation. Pp. 549–587. Springer-Verlag, Berlin – Heidelberg – New York 1982.

Farquhar, G. D., Dubbe, D. R., Raschke, K.: Gain of the feedback loop involving carbon dioxide and stomata. Theory and measurement. — Plant Physiol. 62: 406–412, 1978.

Farquhar, G. D., Sharkey, T. D.: Stomatal conductance and photosynthesis. — Annu. Rev. Plant Physiol. 33: 317–345, 1982.

Federer, C . A.: Differing diffusive resistance and leaf development may cause differing transpiration among hardwoods in spring. — Forest Sci. 22: 359–364, 1976.

Federer, C. A.: Paper birch and white oak saplings differ in responses to drought. — Forest Sci. 26: 313–324, 1980.

Federer, C. A., Gee, G. W.: Diffusion resistance and xylem potential in stressed and unstressed northern hardwood trees. — Ecology 57: 975–984, 1976.

Fedtke, C.: Effects of the herbicide methabenzthiazuron on the physiology of wheat plants. — Pestic. Sci. 4: 653–664, 1973.

Feierabend, J.: Der Einfluss von Cytokininen auf die Bildung von Photosyntheseenzymen in Roggenkeimlingen. — Planta 84: 11–29, 1969.

Felippe, G. M., Dale, J. E.: The uptake of $^{14}CO_2$ by developing first leaves of barley and partition of the labelled assimilates. — Ann. Bot. 36: 411–418, 1972.

Felippe, G. M., Dale, J. E.: Effects of shading the first leaf of barley plants on growth and carbon nutrition of the stem apex. — Ann. Bot. 36: 45–56, 1973.

Feller, U., Erismann, K. H.: Veränderungen des Gaswechsels und der Aktivitäten proteolytischer Enzyme während der Seneszenz von Weizenblättern (Triticum aestivum L.). — Z. Pflanzenphysiol. 90: 235–244, 1978.

Fellows, R. J., Geiger, D. R.: Structural and physiological changes in sugar beet leaves during sink to source conversion. — Plant Physiol. 54: 877–885, 1974.

Fenchuk, T. D.: Intensivnost' reaktsii Khilla v khloroplastakh gorokha v zavisimosti ot vozrasta list'ev i usloviĭ reaktsionnoĭ sredy. [Hill reaction rate in pea chloroplasts in relation to leaf ageand conditions of the reaction medium.] — Nauch. Dokl. vyssh. Shkoly, biol. Nauki 15 (3): 76–80,. 1972

Fenchuk, T. D.: Deĭstvie khloristogo ammoniya na élektrontransportnuyu aktivnost' izolirovannykh khloroplastov gorokha v zavisimosti ot vozrasta list'ev i intensivnosti osveshcheniya reaktsionnoĭ smesi. [Effect of ammonium chloride on the electron transport activity of isolated pea chloroplasts in dependence on leaf age and irradiation of the reaction mixture.] — In: Botanika (Issledovaniya). Vol. 22. Pp. 212–218. Nauka i Tekhnika, Minsk 1980.

Ferguson, C. H. R., Simon, E. W.: Membrane lipids in senescing green tissues. — J. exp. Bot. 24: 307–316, 1973.

Field, C., Mooney, H. A.: Leaf age and seasonal effects on light, water and nitrogen use efficiency in a California shrub. — Oecologia 56: 348—355, 1983.

Fleck-Gerndt, G.: Untersuchungen über die Photosynthese alternder Blätter. I. Beziehungen zwischen Chlorophyllgehalt, Stickstoffhaushalt und Assimilationsintensität. — Biol. Zentralbl. 90: 479–506, 1971a.

Fleck-Gerndt, G.: Untersuchungen über die Photosynthese alternder Blätter. II. Der Einbau von Radiokohlenstoff in die Assimilate. — Biol. Zentralbl. 90: 723–743, 1971b.

Flinn, A. M.: Regulation of leaflet photosynthesis by developing fruit in the pea. — Physiol. Plant. 31: 275–278, 1974.

Floyd, B. W., Noble, R. D.: Intraseasonal variation in chlorophyll content and chloroplast ultrastructure of selected plant species in a deciduous forest. — Can. J. Bot. 58: 1504–1519, 1980.

Ford, M., Black, M., Chapman, J. M.: Inter-organ synergism and the control of chlorophyll accumulation in sunflower (Helianthus annuus) cotyledons. — J. exp. Bot. 28: 926–934, 1977.

Forrester, M. L., Krotkov, G., Nelson, C. D.: Effect of oxygen on photosynthesis, photorespiration and respiration in detached leaves. I. Soybean. — Plant Physiol. 41: 422–427, 1966a.

Forrester, M. L., Krotkov, G., Nelson, C. D.: Effect of oxygen on photosynthesis, photorespiration and respiration in detached leaves. II. Corn and other monocotyledons. — Plant Physiol. 41: 428–431, 1966b.

Forti, G.: Electron transport and the structure of the thylakoid system. — In: Akoyunoglou, G. (ed.): Photosynthesis. Vol. II. Pp. 3–6. Balaban Int. Sci. Services, Philadelphia 1981.

Frank, A. B.: Effect of leaf age and position on photosynthesis and stomatal conductance of forage grasses. — Agron. J. 73: 70–74, 1981.

Frank, A. B., Barker, R. E.: Rates of photosynthesis and transpiration and diffusive resistance of six grasses grown under controlled conditions. — Agron. J. 68: 487–490, 1976.

Frank, A. B., Power, J. F., Willis, W. O.: Effect of temperature and plant water stress on photosynthesis, diffusion resistance, and leaf water potential in spring wheat. — Agron. J. 65: 777–780, 1973.

Fraser, D. E., Bidwell, R. G. S.: Photosynthesis and photorespiration during the ontogeny of the bean plant. — Can. J. Bot. 52: 2561–2570, 1974.

Freeman, B. A., Platt-Aloia, K., Mudd, J. B., Thomson, W. W.: Ultrastructural and lipid changes associated with the aging of *Citrus* leaves. — Protoplasma 94: 221–233, 1978.

French, C. S.: Various forms of chlorophyll *a* in plants. — In: The Photochemical Apparatus — Its Structure and Function. Brookhaven Symposia in Biology No. 11. Pp. 65—73. Upton 1958.

French, C. S.: Changes with age in the absorption spectrum of chlorophyll *a* in a diatom. — Arch. Mikrobiol. 59: 93–103, 1967.

Frey-Wyssling, A., Ruch, F., Berger, X.: Monotrope Plastiden-Metamorphose. — Protoplasma 45: 97–114, 1955.

Frier, V.: The relationship between photosynthesis and tuber growth in *Solanum tuberosum* L. — J. exp. Bot. 28: 999–1007, 1977.

Friis-Nielsen, B.: Growth, water and nutrient status of plants in relation to patterns of variations in concentrations of dry matter and nutrient elements in base-to-top leaves I. Distribution of contents and concentrations of dry matter in tomato plants under different growth conditions. — Plant Soil 39: 661–673, 1973.

Frimmel, G.: Die Variation der Grösse und der Dichte von Spaltöffnungen bei fünfzehn Sommerweizensorten. — Angew. Bot. 51: 333–342, 1977.

Frommhold, I.: Ontogenetische und funktionelle Entwicklung der Stomata von Hafer (*Avena sativa* L.). – Biochem. Physiol. Pflanzen 162: 410–416, 1971.

Frommhold, I.: Tagesverlauf der Stomataaperturen unterschiedlich alter Pflanzen bzw. Blätter von Hafer (*Avena sativa* L.). — Biochem. Physiol. Pflanzen 163: 216–224, 1972.

Frydrych, J.: Photosynthetic characteristics of cucumber seedlings grown under two levels of carbon dioxide. — Photosynthetica 10: 335–338, 1976.

Fuchs, M., Schulze, E.-D., Fuchs, M. I.: Spacial distribution of photosynthetic capacity and performance in a mountain spruce forest of Northern Germany. II. Climatic control of carbon dioxide uptake. — Oecologia 29: 329–340, 1977.

Fukshansky, L.: On the theory of light absorption in non-homogeneous objects. The sieve effect in one-component suspensions. — J. math. Biol. 6: 177–196, 1978.

Fukshansky, L.: Optical properties of plants. — In: Smith, H. (ed.): Plants and the Daylight Spectrum. Pp. 21–40. Academic Press, London–New York–San Francisco 1981.

Furukawa, A.: Carbon dioxide compensation points in poplar plant. — J. jap. Forest. Soc. 55(3): 95–99, 1973a.

Furukawa, A.: Photosynthesis and respiration in poplar plant in relation to leaf development. — J. jap. Forest. Soc. 55 (4): 119–123, 1973b.

Gaastra, P.: Photosynthesis of crop plants as influenced by light, carbon dioxide, temperature, and stomatal diffusion resistance. — Meded. Landbouwhogeschool (Wageningen) 59 (13): 1–68, 1959.

Gabrielsen, E. K.: Effects of different chlorophyll concentrations on photosynthesis in foliage leaves. — Physiol. Plant. 1: 5–37, 1948.

Gallaher, R. N., Brown. R. H., Ashley, D. A., Jones, J. B., Jr.: Photosynthesis of, and $^{14}CO_2$-photosynthate translocation from calcium-deficient leaves of crops. — Crop Sci. 16: 116–119, 1976.

Gamaleĭ, Yu. V.: Prodolzhitel'nost' zhizni khloroplastov v kletkakh mezofilla listopadnykh i vechnozelenykh rasteniĭ. [The length of life of chloroplasts in mesophyll of deciduous and evergreen plants.] — Tsitologiya 17: 1243–1248, 1975.

Gamaleĭ, Yu. V., Kulikov, G. V.: Vozrastnye izmeneniya kletok mezofilla listopadnykh i vechnozelenykh rasteniĭ. [Ontogenetic changes in mesophyll cells in deciduous and evergreen plants.] — Tsitologiya 19: 15–20, 1977.

Gamaleĭ, Yu. V., Kulikov, G. V.: Razvitie Khlorenkhimy Lista. [Development of Leaf Chlorenchyma.] — Nauka, Leningrad 1978.

Gaponenka, V. I., Baleva, E. F., Shaŭchuk, S. M.: Abnaŭlenne khlarafilu i asimilyatsyĭnyya liki listsyaŭ yachmenyu roznaga ŭzrostu. [Chlorophyll regeneration and assimilation numbers of barley leaves of various age.] — Vestsi Akad. Navuk belarus. SSR, Ser. biyal. Navuk 1977 (4): 47–52, 140, 1977.

Gaponenka, V. I., Baleva, E. F., Shaŭchuk, S. M.: Abnaŭlenne khlarafilu i inténsiŭnasts' fotasintezu listsyaŭ kukuruzy roznaga ŭzrostŭ. [Chlorophyll recovery and photosynthetic rate of maize leaves of different age.] — Vestsi Akad. Navuk belarus. SSR, Ser. biyal. Navuk 1978 (1): 21–25, 138, 1978.

Gaponenka, V. I., Shaŭchuk, S. M., Baleva, E. F.: Dasledavanne suvyazi pamizh abnaŭlennem khlarafilu i aktyŭnastsyu fotasintétychnaga aparatu ŭ antageneze listsyaŭ kukuruzy. [Relation between chlorophyll regeneration and activity of the photosynthetic apparatus during maize leaves ontogeny.] — Vestsi Akad. Navuk belarus. SSR, Ser. biyal. Navuk 1980 (1): 67–72, 1980.

Gaponenko, V. I.: Vliyanie Vneshnikh Faktorov na Metabolizm Khlorofilla. [Effect of External Factors on Chlorophyll Metabolism.] — Nauka i Tekhnika, Minsk 1976.

Gaponenko, V. I., Baleva, E. F., Shevchuk, S. N.: Obnovlenie khlorofilla — kharakternaya cherta fotosinteziruyushchego apparata. [Chlorophyll regeneration — a characteristic of photosynthesizing apparatus.] — In: Shlyk, A. A. (ed.): Khlorofill. Pp. 298–310, 411–412. Nauka i Tekhnika, Minsk 1974.

Gaponenko, V. I., Baleva, E. F., Shevchuk, S. N.: Korrelyatsiya mezhdu obnovleniem khlorofilla i assimilyatsionnymi chislami raznykh po vozrastu zon list'ev kukuruzy. [Correlation between chlorophyll turnover and assimilation numbers of different age zones of maize leaves.] — Dokl. Akad. Nauk belorus. SSR 21: 749–752, 1977.

Gaponenko, V. I., Nikolaeva, G. N., Stanishevskaya, E. M., Lositskaya, T. V., Shevchuk, S. N.: O labilizatsii khlorofilla v zelenom rastenii. [Labilization of chlorophyll in a green plant.] — In: Metabolizm i Stroenie Fotosintet--icheskogo Apparata. Pp. 134–143. Nauka i Tekhnika, Minsk 1970.

Gaponenko, V. I., Stazhetskiĭ, V.: Izmenenie intensivnosti fotosinteza i soderzhaniya khlorofilla u ryaski v svyazi s vozrastom i usloviyami osveschcheniya. [Change in photosynthetic rate and chlorophyll content in duckweed as related to age and illumination conditions.] — Fiziol. Rast. 16: 993–1001, 1969.

Garland, J. A., Branson, J. R.: The deposition of sulphur dioxide to pine forest assessed by a radioactive tracer method. — Tellus 29: 445–454, 1977.

Gates, D. M.: Biophysical Ecology. — Springer-Verlag, New York–Heidelberg–Berlin 1980.

Gates, D. M., Alderfer, R., Taylor, S. E.: Leaf temperature of desert plants. — Science 159: 994–995, 1968.

Gausman, H. W.: Photomicrographic record of light reflected at 850 nanometers by cellular constituents of Zebrina leaf epidermis. — Agron. J. 65: 504–505, 1973a.

Gausman, H. W.: Reflectance, transmittance, and absorptance of light by subcellular particles of spinach (Spinacia oleracea L.) leaves. — Agron. J. 65: 551–553, 1973b.

Gausman, H. W.: Reflectance of leaf components. — Remote Sens. Environ. 6: 1–9, 1977.

Gausman, H. W., Allen, W. A., Cardenas, R.: Reflectance of cotton leaves and their structure. — Remote Sens. Environ. 1: 19–22, 1969a.

Gausman, H. W., Allen, W. A., Cardenas, R., Richardson, A. J.: Relation of light reflectance to histological and physical evaluations of cotton leaf maturity. — Appl. Optics 9: 545–552, 1970.

Gausman, H. W., Allen, W. A., Cardenas, R., Richardson, A. J.: Effects of leaf nodal position on absorption and scattering coefficients and infinite reflectance of cotton leaves, Gossypium hirsutum L. — Agron. J. 63: 87–91, 1971a.

Gausman, H. W., Allen, W. A., Cardenas, R., Richardson, A. J.: Reflectance discrimination of cotton and corn at four growth stages. — Agron. J. 65: 194–198, 1973.

Gausman, H. W., Allen, W. A., Escobar, D. E., Rodriguez, R. R., Cardenas, R.: Age effects of cotton leaves on light reflectance, transmittance and absorptance and on water content and thickness. — Agron. J. 63: 465–469, 1971b.

Gausman, H. W., Allen, W. A., Myers, V. I., Cardenas, R.: Reflectance and internal structure of cotton leaves, *Gossypium hirsutum* L. — Agron. J. 61: 374–376, 1969b.

Gausman, H. W., Cardenas, R.: Effect of soil salinity on external morphology of cotton leaves. — Agron. J. 60: 566–567, 1968.

Gausman, H. W., Menges, R. M., Richardson, A. J., Walter, H., Rodriguez, R. R., Tamez, S.: Optical parameters of leaves of seven weed species. — Weed Sci. 29: 24–26, 1981.

Gausman, H. W., Rodriguez, R. R., Escobar, D.E.: Ultraviolet radiation reflectance, transmittance and absorptance by plant leaf epidermises. — Agron. J. 67: 719–724, 1975.

Gautam, O. P.: Chlorophyll development in wheat. I. Chlorophyll development during the ontogeny of the wheat plant and its relation to plant vigour. — Indian J. Plant Physiol. 6: 44–59, 1963.

Gay, A. P., Hurd, G. R.: The influence of light on stomatal density in the tomato. — New Phytol. 75: 37–46, 1975.

Gee, G. W., Federer, C. A.: Stomatal resistance during senescence of hardwood leaves. — Water Resour. Res. 8: 1456–1460, 1972.

Gej, B.: Zmiany w zawartości chlorofilu *a* i *b* w liściach różnego wieku niektórych roślin dwuliściennych. [Changes in chlorophyll *a* and *b* content in leaves of different ages in some dicotyledonous plants.] — Acta Soc. Bot. Pol. 35: 209–224, 1966.

Gej, B.: Changes in $^{14}CO_2$ fixation rate in different-aged leaves after decapitation of certain annual plants. — Bull. Acad. pol. Sci., Sér. Sci. biol., Cl. II, 18: 585–589, 1970.

Gej, B.: Changes in $^{14}CO_2$ absorption rates by the successive leaves in buckwheat and white mustard plants of various age. — Acta Soc. Bot. Pol. 40: 599–614, 1971.

Gej, B.: Charakterystyka fizjologiczna poszczególnych liści sałaty traktowanej symazyną. [Physiological characteristics of individual leaves of lettuce treated with simazine.] — Roczn. Nauk roln. A 101 (4): 7–19, 1976.

Genchev, S.: Prouchvane v"rkhu plastidnite pigmenti pri luka (*Allium cepa* L.). III. V"rkhu vr"zkata na khlorofila s katalazata, peroksidazata i askorbinovata kiselina v listata na luka. [Plastid pigments in onion (*Allium cepa* L.) III. Relationship of chlorophyll content and catalase, peroxidase and ascorbic acid in onion leaves.] — Fiziol. Rast. (Sofiya) 1: 25–31, 1970.

Genkel', P. A., Morozova, R. S.: Élektronno-mikroskopicheskoe issledovanie khloroplastov *Bellis perennis* v svyazi s perekhodom v sostoyanie zimnego pokoya. [Electron-microscópic study of *Bellis perennis* chloroplasts in connection with transition to winter period.] — Fiziol. Rast. 4: 509–513, 1957.

Genkel', P. A., Morozova, R. S.: Élektronno-mikroskopischeskoe issledovanie khloroplastov *Bellis perennis* v vesenniï period. [Electron-microscopic study of *Bellis perennis* chloroplasts during the autumn period.] — Fiziol. Rast. 6: 575–578, 1959.

Gerbaud, A., André, M.: Effect of $CO_2$, $O_2$, and light on photosynthesis and photorespiration in wheat. — Plant Physiol. 66: 1032–1036, 1980.

Giaquinta, R.: Source and sink leaf metabolism in relation to phloem translocation. Carbon partitioning and enzymology. — Plant Physiol. 61: 380–385, 1978.

Gibbs, M., Latzko, E.: Introduction. — In: Gibbs, M., Latzko, E. (ed.): Photosynthesis II. Pp. 1–5. Springer-Verlag, Berlin–Heidelberg–New York 1979.

Gibbs, S. P.: The chloroplasts of some algal groups may have evolved from endosymbiotic eukaryotic algae. — Ann. New York Acad. Sci. 361: 193–208, 1981.

Gindel, I.: Stomata constellation in the leaves of cotton, maize and wheat plants as a function of soil moisture and environment. — Physiol. Plant. 22: 1143–1151, 1969.

Girnth, C., Bergfeld, R., Kasemir, H.: Phytochrome-mediated control of grana and stroma thylakoid formation in plastids of mustard cotyledons. — Planta 141: 181–198, 1978.

Giurgevich, J. R., Dunn, E. L.: Seasonal patterns of $CO_2$ and water vapor exchange of *Juncus roemerianus* Scheele in a Georgia salt marsh. — Amer. J. Bot. 65: 502–510, 1978.

Giurgevich, J. R., Dunn, E. L.: Seasonal patterns of $CO_2$ and water vapor exchange of the tall and short height forms of *Spartina alterniflora* Loisel in a Georgia salt marsh. — Oecologia 43: 139–156, 1979.

Givnish, T. J.: Ecological aspects of plant morphology: leaf form in relation to environment. — In: Sattler, R. (ed.): Theoretical Plant Morphology. Pp. 83–142. Leiden Univ. Press, Leiden 1978.

Glater, R. B., Solberg, R. A., Scott, F. M.: A developmental study of the leaves of *Nicotiana glutinosa* as related to their smog-sensitivity. — Amer. J. Bot. 49: 954—970, 1962.

Gloser, J.: Characteristics of $CO_2$ exchange in *Phragmites communis* Trin. derived from measurements *in situ*. — Photosynthetica 11: 139–147, 1977.

Godnev, T. N., Khodasevich, E. V., Arnautova, A. I.: O kharaktere sezonnykh izmeneniĭ v soderzhanii i sootnoshenii pigmentov u khvoĭnykh v estestvennykh usloviyakh v svyazi s temperaturoĭ vozdukha. [Character of seasonal changes in pigment content and ratio in coniferous plants in natural conditions in relation to air temperature.] — Fizol. Rast. 16: 102–105, 1969.

Godnev, T. N., Shabel'skaya, Ė. F.: K voprosu o formirovanii plastidnogo apparata v ontogeneze lista sakharnoĭ svekly v estestvennykh usloviyakh. [Formation of the plastid apparatus during ontogenesis of sugar beet leaf in natural conditions.] — Dokl. Akad. Nauk belorus. SSR 10: 987–990, 1966.

Gœdheer, J. C.: Les changements du spectre d'absorption et de fluorescence au cours du verdissement et du vieillissement des plastes. – In: Sironval, C. (ed.): Le Chloroplaste, Croissance et Vieillissement. Pp. 77–85. Mason, Paris 1967.

Gœdheer, J. C., van der Cammen, J.C.J.M.: Protochlorophyll(ide) and chlorophyll(ide) fluorescence lifetime and other properties in etiolated and greening leaves. — In: Akoyunoglou, G. (ed.): Photosynthesis. Vol. V. Pp. 39–44. Balaban Int. Sci. Serv., Philadelphia 1981.

Golinka, P. I.: Vliyanie obrezki vinogradnykh kustov na razvitie fotosinteticheskogo apparata list'ev. [Effect of cutting of vine shrubs on the development of photosynthetic apparatus of leaves.] —Fiziol. Rast. 13: 607—613, 1966.

Golod, M. G.: Fotokhimichna aktyvnist' khlorofilu. [Photochemical activity of chlorophyll.] — In: Pytannya Ėksperymental'noĭ Botaniki. Pp. 35–40. Naukova Dumka, Kyïv 1964.

Goodwin, T. W.: Studies in carotenogenesis. 24. The changes in carotenoid and chlorophyll pigments in the leaves of deciduous trees during autumn necrosis. — Biochem. J. 68: 503–511, 1958.

Goodwin, T. W.: The Biochemistry of the Carotenoids. Vol. I. — Champan and Hall, London–New York 1980.

Gordon, A. J., Hesketh, J. D., Peters, D. B.: Soybean leaf photosynthesis in relation to maturity classification and stage of growth. — Photosynthesis Res. 3: 81–93, 1982.

Gordon, K. H. J., Peoples, M. B., Murray, D. R.: Ageing-linked changes in photosynthetic capacity and in fraction I protein content of the first leaf of pea *Pisum sativum* L. — New Phytol. 81: 35–42, 1978.

Goryshina, T. K.: Structural and functional features of the leaf assimilatory apparatus in plants of a forest-steppe oakwood. I. Leaf plastid apparatus in plants of various forest strata. — Acta oecol., Oecol. Plant. 1: 47–54, 1980a.

Goryshina, T. K.: Structural and functional features of the leaf assimilatory apparatus in plants of a forest-steppe oakwood. II. Seasonal dynamics of the leaf plastid apparatus in the herbaceous understory. — Acta oecol., Oecol. Plant. 1: 201–208, 1980b.

Goryshina, T. K., Laverycheva, I. G.: Ob ul'trastrukture khloroplastov v vesennikh i letnikh list'yakh dvukh travyanistykh vidov lesostepnoĭ dubravy v svyazi s sezonnoĭ dinamikoĭ svetovogo rezhima. [Chloroplast ultrastructure in spring and summer leaves of two grass species of forest-steppe oak forest in relation to seasonal dynamics of radiation regime.] – Bot. Zh. 65: 1150–1156, 1980.

Goryshina, T. K., Zabotina, L. N., Pruzhina, E. G.: Osobennosti assimilyatsionnykh tkaneĭ i plastidnogo apparata lista v raznykh chastyakh krony u nekotorykh drevesnykh porod v lesostepnoĭ dubrave. [Characteristics of assimilatory tissues and plastid apparatus in leaves of various parts of crown in some tree species of the oakwood.] — Vest. leningr. Univ. 1979 (3): 67–76, 125, 1979.

Goryshina, T. K., Zabotina, L. N., Pruzhina, E. G.: O mezostrukture fotosinteticheskogo apparata vetrenitsy dubravnoï (*Anemone nemorosa* L.) v raznykh mestoobitaniyakh. [Mesostructure of the photosynthetic apparatus of *Anemone nemorosa* L. in different localities.] — Ekologiya 1981 (1): 19–26, 1981.

Goudriaan, J.: A family of saturation type curves, especially in relation to photosynthesis. — Ann. Bot. 43: 783–785, 1979.

Goudriaan, J., Ajtay, G. L.: The possible effect of increased $CO_2$ on photosynthesis. — In: Bolin, B., Degens, E. T., Kempe, S., Ketner, P. (ed.): The Global Carbon Cycle. Pp. 237–249. John Wiley & Sons, Chichester 1979.

Goudriaan, J., van Laar, H. H.: Relations between leaf resistance, $CO_2$-concentration and $CO_2$-assimilation in maize, beans, lalang grass and sunflower. — Photosynthetica 12: 241–249, 1978.

Gowin, T., Lourtioux, A., Mousseau, M.: Influence of constant growth temperature upon the productivity and gas exchange of seedlings of Scots pine and European larch. — Forest Sci. 26: 301–309, 1980.

Graham, D., Chapman, E. A.: Interactions between photosynthesis and respiration in higher plants. —In: Gibbs, M., Latzko, E. (ed.): Photosynthesis II. Pp. 150–162. Springer-Verlag, Berlin–Heidelberg–New York 1979.

Graham, R. D., Ulrich, A.: Potassium deficiency-induced changes in stomatal behavior, leaf water potentials, and root system permeability in *Beta vulgaris* L. — Plant Physiol. 49: 105–109, 1972.

Granick, S.: Plastid structure, development and inheritance. — In: Ruhland, W. (ed.): Handbuch der Pflanzenphysiologie. Vol. I. Pp. 507–564. Springer-Verlag, Berlin–Göttingen–Heidelberg 1955.

Greenway, H., Winter, K., Lüttge, U.: Phosphoenolpyruvate carboxylase during development of crassulacean acid metabolism and during a diurnal cycle in *Mesembryanthemum crystallinum*. — J. exp. Bot. 29: 547–559, 1978.

Griffiths, W. T., Morgan, N. L., Mapleston, R. E.: Chlorophyll synthesis and the development of photosynthetic activity. — In: Bücher, T., Neupert, W., Sebald, W., Werner, W. (ed.): Genetics and Biogenesis of Chloroplasts and Mitochondria. Pp. 111–118. North-Holland Publ. Co., Amsterdam–New York–Oxford 1976.

Groen, J.: Photosynthesis of *Calendula officinalis* L. and *Impatiens parviflora* DC., as influenced by light intensity during growth and age of leaves and plants. — Meded. Landbouwhogeschool 73(8): 1–128, 1973.

Gruber, P. J., Becker, W. M., Newcomb, E. H.: The development of microbodies and peroxisomal enzymes in greening bean leaves. — J. Cell Biol. 56: 500–518, 1973.

Gulmon, S. L., Chu, C. C.: The effects of light and nitrogen on photosynthesis, leaf characteristics, and dry matter allocation in the chaparral shrub, *Diplacus aurantiacus*. — Oecologia 49: 207–212, 1981.

Gupta, B.: Correlation of tissues in leaves 2. Absolute stomatal numbers. — Ann. Bot. 25: 61–77, 1961.

Gupta, R. K., Woolley, J. T.: Spectral properties of soybean leaves. — Agron. J. 63: 123–126, 1971.

Guretskaya, F. S.: O razlichiyakh v stroenii list' ev raznykh yarusov. [Differences in structure of leaves of different order.] — Bot. Zh. 11: 19–27, 1954.

Gyldenholm, A. O.: Macromolecular physiology of plastids. V. On the nucleic acid metabolism during chloroplast development. — Hereditas 59: 142–168, 1968.

Haberlandt, G.: Physiologische Pflanzenanatomie. 6. Aufl. — W. Engelmann, Leipzig 1924.

Hackett, C.: An exploration of the carbon economy of the tobacco plant. I. Inferences from a simulation. — Aust. J. biol. Sci. 26: 1057–1071, 1973.

Hackett, C., Rawson, H. M.: An exploration of the carbon economy of the tobacco plant. II.

Patterns of leaf growth and dry matter partitioning. — Aust. J. Plant Physiol. 1: 271–281, 1974.

Hall, A. E.: A model of leaf photosynthesis and respiration for predicting carbon dioxide assimilation in different environments. — Oecologia 143: 299–316, 1979.

Hall, A. E., Schulze, E.-D.: Stomatal response to environment and possible interrelation between stomatal effects on transpiration and $CO_2$ assimilation. — Plant Cell Environ. 3: 467–474, 1980.

Hall, A. J., Brady, C. J.: Assimilate source-sink relationships in *Capsicum annuum* L. II. Effects of fruiting and defloration on the photosynthetic capacity and senescence of the leaves. — Aust. J. Plant Physiol. 4: 771–783, 1977.

Hall, J. L. (ed.): Electron Microscopy and Cytochemistry of Plant Cells. — Elsevier/North Holland Biomedical Press, Amsterdam–Oxford–New York 1978.

Hall, N. P., Keys, A. J., Merrett, M. J.: Ribulose-1,5-diphosphate carboxylase protein during flag leaf senescence. — J. exp. Bot. 29: 31–37, 1978.

Hannan, R. V.: Leaf growth and development in the young tobacco plant. — Aust. J. biol. Sci. 21: 855–870, 1968.

Hanscom III, Z., Ting, I. P.: Physiological responses to irrigation in *Opuntia basilaris* Engelm. & Bigel. — Bot. Gaz. 138: 159–167, 1977.

Haraguchi, N., Shimizu, S.: Photosynthetic activities in tobacco plants. II. Relationship between photosynthetic activity and chlorophyll content. — Bot. Mag. (Tokyo) 83: 411–418, 1970.

Hardwick, K., Wood, M., Woolhouse, H. W.: Photosynthesis and respiration in relation to leaf age in *Perilla frutescens* (L.) Britt. — New Phytol. 67: 79–86, 1968.

Harnischfeger, G.: Chloroplast degradation in ageing cotyledons of pumpkin. — J. exp. Bot. 24: 1236–1246, 1973.

Harnischfeger, G.: Studies on chloroplast degradation *in vivo* II. Effects of aging on Hill activity of plastids from *Cucurbita* cotyledons. — Z. Pflanzenphysiol. 71: 301–312, 1974.

Harris, J. B.: Development of a tubular apparatus in chloroplasts of ageing *Cyphomandra* leaves. — Cytobios 21: 151–164, 1978.

Harris, J. B., Naylor, A. W.: Changes in the etiolated tobacco leaf during greening. II. Composition of subcellular fractions. – Tobacco Sci. 12: 25–33, 1968a.

Harris, J. B., Naylor, A. W.: Changes in the etiolated tobacco leaf during greening. III. The effect of organic metabolites and light intensity on chlorophyll and carotenoid production. — Tobacco Sci. 12: 170–176, 1968b.

Harris, J. B., Naylor, A. W.: Changes in the etiolated tobacco leaf during greening. IV. Carbon dioxide fixation. — Tobacco Sci. 14: 69–72, 1970.

Harris, W. M.: Ultrastructural observations on the mesophyll cells of pine leaves. — Can. J. Bot. 49: 1107–1109, 1971.

Harte, C., Hansen, H.: Das Spaltöffnungsmuster in der Blattepidermis einer Standardsippe und einiger Mutanten von *Antirrhinum majus* L. — Biol. Zentralbl. 90: 1–26, 1971.

Harvey, G. W.: Seasonal alteration of photosynthetic unit sizes in three herb layer components of a deciduous forest community. — Amer. J. Bot. 67: 293—299, 1980.

Hatch, M. D.: $C_4$-pathway photosynthesis in *Portulaca oleracea* and the significance of alanine labelling. — Planta 125: 273—279, 1975.

Hatch, M. D., Kagawa, T., Craig, S.: Subdivision of $C_4$-pathway species based on differing $C_4$ acid decarboxylating systems and ultrastructural features. — Aust. J. Plant Physiol. 2: 111–128, 1975.

Hatfield, J. L., Carlson, R. E.: Photosynthetically active radiation, $CO_2$ uptake, and stomatal diffusive resistance profiles with soybean canopies. — Agron. J. 70: 592–596, 1978.

Hattersley, P. W., Browning, A. J.: Occurrence of the suberized lamella in leaves of grasses of different photosynthetic types. I. In parenchymatous bundle sheaths and PCR ("Kranz") sheaths. — Protoplasma 109: 371–401, 1981.

Hattersley, P. W., Watson, L.: $C_4$ grasses: an anatomical criterion for distinguishing between

NADP-malic enzyme species and PCK or NAD-malic enzyme species. — Aust. J. Bot. 24: 297–308, 1976.

Haupt, W., Feinleib, M. E. (ed.): Physiology of Movements. (Encyclopedia of Plant Physiology. Vol. 7.) — Springer-Verlag, Berlin–Heidelberg–New York 1979.

Havelange, A.: Ultrastructure des chloroplastes des feuilles au cours de la croissance végétative et de la mise à fleurs de *Sinapis alba* L. — Physiol. vég. 15: 723–734, 1977.

Hawke, J. C., Rumşby, M. G., Leech, R. M.: Lipid biosynthesis in green leaves of developing maize. — Plant Physiol. 53: 555–561, 1974.

Heath, O. V. S., Orchard, B.: Carbon assimilation at low carbon dioxide levels II. The processes of apparent assimilation. — J. exp. Bot. 19: 176–192, 1968.

Heber, U., Krause, G. H.: What is the physiological role of photorespiration? — Trends biochem. Sci. 5: 32–34, 1980.

Hedley, C. L., Harvey, D. M.: The involvement of $CO_2$ uptake in the flowering behaviour of two varieties of *Antirrhinum majus*. — In: Marcelle, R. (ed.): Environmental and Biological Control of Photosynthesis. Pp. 149–160. Dr. W. Junk, B. V. Publ., The Hague 1975.

Hedley, C. L., Rowland, A. O.: Changes in the activities of some respiratory and photosynthetic enzymes during the early leaf development of *Antirrhinum majus* L. — Plant Sci. Lett. 5: 119 to 126, 1975.

Heichel, G. H.: Genetic control of epidermal cell and stomatal frequency in maize. — Crop Sci. 11: 830–832, 1971a.

Heichel, G. H.: Stomatal movements, frequencies, and resistances in two maize varieties differing in photosynthetic capacity. — J. exp. Bot. 22: 644–649, 1971b.

Hellkvist, J., Hillerdal-Hagströmer, K., Mattson-Djos, E.: Field studies of water relations and photosynthesis in Scots pine using manual techniques. — Ecol. Bull. (Stockholm) 32 [Persson, T. (ed.): Structure and Function of Northern Coniferous Forests — An Ecosystem Study]: 183–204, 1980.

Hendry, G. A. F., Stobart, A. K.: Glycine metabolism in etiolated barley leaves on exposure to light. — Phytochemistry 17: 69–72, 1978.

Henningsen, K. W., Boardman, N. K.: Development of photochemical activity and the appearance of the high potential form of cytochrome *b*-559 in greening barley seedlings. — Plant Physiol. 51: 1117–1126, 1973.

Henningsen, K. W., Boynton, J. E.: Macromolecular physiology of plastids. IX. Development of plastid membranes during greening of dark-grown barley seedlings. — J. Cell Sci. 15: 31–55, 1974.

Henriques, F., Park, R. B.: Development of the photosynthetic unit in lettuce. — Proc. nat. Acad. Sci. USA 73: 4560–4564, 1976.

Hernández-Gil, R., Schaedle, M.: Functional and structural changes in senescing *Populus deltoides* (Bartr.) chloroplasts. — Plant Physiol. 51: 245–249, 1973.

Hernández-Gil, R., Schaedle, M.: Changes in chloroplast fine structure during the senescence of *Populus deltoides* leaves. — Acta cient. venezolana 25: 13–18, 1974.

Hetherington, S. E., Smillie, R. M.: Humidity-sensitive degreening and regreening of leaves of *Borya nitida* Labill. as followed by changes in chlorophyll fluorescence. — Aust. J. Plant Physiol. 9: 587–599, 1982.

Heyes, J. K., Dale, J. E.: A virescent mutant of *Phaseolus vulgaris*: photosynthesis and metabolic changes during leaf development. — New Phytol. 70: 415–426, 1971.

Hieke, B.: Die Hill-Aktivität isolierter Chloroplasten von *Triticum aestivum* L. und *Phaseolus vulgaris* L. bei Einsatz des künstlichen Elektronendonators DPC (Diphenylcarbazid). — Biol. Rundschau 12: 195–197, 1974.

Hill, R.: Oxygen evolved by isolated chloroplasts. — Nature 139: 881–882, 1937.

Hill, R., Bendall, F.: Function of the two cytochrome components in chloroplasts: a working hypothesis. — Nature 186: 136–137, 1960.

Hiroi, T., Monsi, M.: Dry-matter economy of *Helianthus annuus* communities grown at varying densities and light intensities. — J. Fac. Sci. Univ. Tokyo, Sec. III, 9: 241–285, 1966.

Ho, L. C.: Effects of $CO_2$ enrichment on the rates of photosynthesis and translocation of tomato leaves. — Ann. appl. Biol. 87: 191–200, 1977.

Ho, L. C., Shaw, A. F.: Carbon economy and translocation of [14]C in leaflets of the seventh leaf of tomato during leaf expansion. — Ann. Bot. 41: 833–848, 1977.

Ho, L. C., Shaw, A. F.: Net accumulation of minerals and water and the carbon budget in an expanding leaf of tomato. — Ann. Bot. 43: 45–54, 1979.

Hodáňová, D: Photosynthetic rate and chlorophyll content as related to leaf age and ontogenesis of sugar-beet plants. — In: Productivity of Terrestrial Ecoystems. Production Processes. Czechosl. nat. Comm. IBP, PT-PP Report No. 1. Pp. 175–177. Praha 1970.

Hodáňová, D.: Structure and development of sugar beet canopy. III. Chlorophyll characteristics. — Photosynthetica 7: 338–344, 1973.

Hodáňová, D.: Specific leaf weight and photosynthetic rate in sugar beet leaves of different age. — Biol. Plant. 17: 314–317, 1975.

Hodáňová, D.: Sugar beet canopy photosynthesis as limited by leaf age and irradiance. Estimation by models. — Photosynthetica 13: 376–385, 1979.

Hodáňová, D.: Photosynthetic capacity, irradiance and sequential senescence of sugar beet leaves. — Biol. Plant. 23: 58–67, 1981.

Hodge, A. J., McLean, J. D., Mercer, F. V.: Ultrastructure of the lamellae and grana in the chloroplasts of *Zea mays* L. — J. biophys. biochem. Cytol. 1: 605–613, 1955.

Hodgkinson, K. C.: Influence of partial defoliation on photosynthesis, photorespiration and transpiration by lucerne leaves of different ages. — Aust. J. Plant Physiol. 1: 561–578, 1974.

Hofäcker, W.: Untersuchungen über den Einfluss wechselnder Bodenwasserversorgung auf die Photosyntheseintensität und den Diffusionswiderstand bei Rebblättern. — Vitis 15: 171–182, 1976.

Hoffmann, P.: Zur Physiologie der Photosynthese bei höheren Pflanzen. — Bot. Stud. (Jena) 18: 1–151, 1968.

Hoffmann, P.: Pigmentphysiologische Untersuchungen an isolierten Kinetin-behandelten Weizenprimärblättern. — Wiss. Z. Ernst-Moritz-Univ. Greifswald, math.-nat. Reihe 19: 55–68, 1970.

Hoffmann, P., Michaelis, G.: Physiologische Gradienten in Primärblättern von *Triticum aestivum* L. — Wiss. Z. Humboldt-Univ. Berlin, math.-nat. Reihe 25: 787–795, 1976.

Hoffmann, P., Plescher, A., Meinl, G.: Pigment- und N-stoffwechselphysiologische Grundlagen im Verlauf der Lagerung von Kopfkohl (*Brassica oleracea* L. var. *capitata*) unter besonderer Berücksichtigung resistenz-physiologischer Aspekte. — Arch. Phytopathol. Pflanzenschutz 13: 61–78, 1977.

Höfner, W., Orlovius, K.: Einfluss der N-Düngung auf den [14]C-Einbau in die Komponenten der äthanollöslichen Fraktion von Sommerweizen verschiedener Entwicklungsstadien. — Z. Pflanzenernähr. Bodenkunde 140: 491–504, 1977.

Hofstra, G., Hesketh, J. D.: The effects of temperature and $CO_2$ enrichment on photosynthesis in soybean. — In: Marcelle, R. (ed.): Environmental and Biological Control of Photosynthesis. Pp. 71–80. Dr. W. Junk b. v. Publ., The Hague 1975.

Holaday, A. S., Bowes, G.: $C_4$ acid metabolism and dark $CO_2$ fixation in a submersed aquatic macrophyte (*Hydrilla verticillata*). — Plant Physiol. 65: 331–335, 1980.

Holmgren, P., Jarvis, P. G.: Carbon dioxide efflux from leaves in light and darkness. — Physiol. Plant. 20: 1045–1051, 1967.

Holmgren, P., Jarvis, P. G., Jarvis, M. S.: Resistance to carbon dioxide and water vapour transfer in leaves of different plant species. — Physiol. Plant. 18: 557–573, 1965.

Holowinsky, A. W., Moore, P. B., Torrey, J. G.: Regulatory aspects of chloroplast growth in leaves of *Xanthium pensylvanicum* and etiolated red kidney bean seedling leaves. — Protoplasma 60: 94–110, 1965.

Homann, P. H.: Carbon dioxide exchange of young tobacco leaves in light and darkness. – In:

Marcelle, R. (ed.): Environmental and Biological Control of Photosynthesis. Pp. 183–190. Dr. W. Junk b. v. Publ., The Hague 1975.

Hong, Y. - N., Schopfer, P.: Control by phytochrome of urate oxidase and allantoinase activities during peroxisome development in the cotyledons of mustard (*Sinapis alba* L.) seedlings. – Planta 152: 325–335, 1981.

Hopkins, W. G., Hayden, D. B., Walden, D. B.: Analysis of greening in virescent mutants of maize by *in vivo* spectrophotometry. — Can. J. Bot. 53: 2720–2724, 1975.

Hopkins, W. G., Walden, D. B.: Temperature sensivitity of virescent mutants of maize. — J. Heredity 68: 283–286, 1977.

Hopkinson, J. M.: Studies on the expansion of the leaf surface. IV. The carbon and phosphorus economy of a leaf. – J. exp. Bot. 15: 125—137, 1964.

Hopkinson, J. M.: Studies on the expansion of the leaf surface. VI. Senescence and the usefulness of old leaves. — J. exp. Bot. 17: 762–770, 1966.

Horak, A., Zalik, S.: Development of photoreductive activity in plastids of a virescent mutant of barley. — Can. J. Bot. 53: 2399–2404, 1975.

Horie, T.: Simulation of sunflower growth. I. Formulation and parametrization of dry matter production, leaf photosynthesis, respiration and partitioning of photosynthates. — Bull. nat. Inst. agr. Sci., Ser. A (Tokyo) 24: 45–70, 1977.

Horie, T., de Wit, C. T., Goudriaan, J., Bensink, J.: A formal template for the development of cucumber in its vegetative stage I, II, III. — Proc. kon. nederl. Akad. Wetenschapp. 82C: 433–479, 1979.

Horie, T., Udagawa, T.: Canopy photosynthesis of sunflower plants. Its measurements and modeling. — Bull. nat. Inst. agr. Sci. (Japan), Ser, A, 18: 1–56, 1971.

Horváth, I., Szalay, L., Szász, K., Raafat, A.: The effect of the spectral composition of light on the metabolism. Utilization of light and the chlorophylls of *Sinapis alba*. – Acta biochim. biophys. Acad. Sci. hung. 8: 161–169, 1973.

Hoshino, M., Matsumoto, F., Okubo, T.: [Studies on the assimilation and translocation of $^{14}CO_2$ in ladino clover. V. Relation between leaf age and ability of assimilation and translocation.] — Proc. Crop Sci. Soc. Jap. 40: 468–473, 1971.

Howard, J. A.: Spectral energy relations of isobilateral leaves. — Aust. J. biol. Sci. 19: 757–766, 1966.

Huber, D. J., Newman, D. W.: Relationships between lipid changes and plastid ultrastructural changes in senescing and regreening soybean cotyledons. — J. exp. Bot. 27: 490–511, 1976.

Huber, W., Sankhla, N.: $C_4$ pathway and regulation of the balance between $C_4$ and $C_3$ metabolism. — In: Lange, O. L., Kappen, L., Schulze, E.-D. (ed.): Water and Plant Life. Pp. 335–363. Springer-Verlag, Berlin–Heidelberg–New York 1976.

Hudák, J.: Plastid senescence. 1. Changes of chloroplast structure during natural senescence in cotyledons of *Sinapis alba* L. — Photosynthetica 15: 174–178, 1981.

Hudák, J., Herich, R.: Effect of boron on the ultrastructure of sunflower chloroplasts. — Photosynthetica 10: 463–465, 1976.

Humphries, E. C.: Leaf growth of white mustard (*Sinapis alba*) in different environments. — Planta 72: 223–231, 1967.

Hunt, L. A., Christie, B. R.: Determination of stomatal numbers, stomatal lengths, and dry weight increments of detached bromegrass leaves. — Can. J. Plant Sci. 49: 597–602, 1969.

Hunt, W. F., Loomis, R. S.: Respiration modelling and hypothesis testing with a dynamic model of sugar beet growth. — Ann. Bot. 44: 5–17, 1979.

Hurkman, W. J.: Ultrastructural changes of chloroplasts in attached and detached, aging primary wheat leaves. — Amer. J. Bot. 66: 64–70, 1979.

Hurkman, W. J., Kennedy, G. S.: Development and cytochemistry of the thylakoidal body in tobacco chloroplasts. — Amer. J. Bot. 64: 86–95, 1977.

Huzulák, J., Eliáš, P.: Within-crown pattern of ecophysiological features in leaves of *Acer campestre* and *Carpinus betulus*. — Folia geobot. phytotax. 10: 337–350, 1975.

Hygen, G.: Studies in plant transpiration I. — Physiol. Plant. 4: 57–183, 1951.
Hygen, G.: Studies in plant transpiration II. — Physiol. Plant. 6: 106–133, 1953.

Ikeda, T., Ueda, R. R.: Light and electron microscopical studies on the senescence of chloroplasts in *Elodea* leaf cells. — Bot. Mag. (Tokyo) 77: 336–341, 1964.
Imai, K., Murata, Y.: Changes in apparent photosynthesis, $CO_2$ compensation point and dark respiration of leaves of some *Poaceae* and *Cyperaceae* species with senescence. — Plant Cell Physiol. 20: 1653–1658, 1979.
Imai, K., Ogura, F., Murata, Y.: Photosynthesis and respiration of papaya (*Carica papaya* L.) leaves. — Acta oecol., Oecol. Plant. 3: 399–407, 1982.
Imamaliev, A. I., Zikiryaev, A.: Vliyanie defoliantov na ATF-aznuyu aktivnost' list'ev khlopchatnika. [Effect of defoliants on ATPase activity in cotton leaves.] — Dokl. Akad. Nauk. uzb. SSR 29 (4): 57–58, 1972.
Impens, I. I., Stewart, D. W., Allen, L. H., Jr., Lemon, E. R.: Diffusive resistances at, and transpiration rates from leaves *in situ* within the vegetative canopy of a corn crop. — Plant Physiol. 42: 99–104, 1967.
Inada, K.: Action spectra for photosynthesis in higher plants. — Plant Cell Physiol. 17: 355–365, 1976.
Inada, K.: Effects of leaf color and the light quality applied to leaf-developing period on the photosynthetic response spectra in crop plants. – Jap. J. Crop Sci. 46: 37–44, 1977.
Intykbaeva, B. B., Sokolova, V. S.: Opticheskie svoĭstva list'ev i posevov ozimoĭ pshenitsy. [Optical properties of leaves and canopies of winter wheat.] — In: Fiziologiya Ozimoĭ Pshenitsy na Yugo--Vostoke Kazakhstana. Pp. 96–104. Nauka kaz. SSR, Alma-Ata 1974.
Ïordanov, I., Popov, K.: Vliyanie fiziologicheskogo sostoyaniya rasteniĭ na intensivnost' fotosinteza list'ev fasoli i na fotokhimicheskuyu aktivnost' izolirovannykh khloroplastov. [Effect of physiological state of plants on photosynthetic rate in *Phaseolus* leaves and on photochemical activity of isolated chloroplasts.] — Stud. biophys. 5: 123–130, 1967.
Ïordanov, I., Todorova-Trifonova, A., Zeĭnalov, Yu.: Vliyanie na khloramfenikola i tetratsiklina v"rkhu s"d"rzhanieto, efektivnostta i spektralnite izmeneniya na plastidnite pigmenti. [Effect of chloramphenicol and tetracycline on the content, efficiency, and spectral changes in plastid pigments.] — Nauch. Tr. Plovdiv. Univ. „Paisiĭ Khilendarski“, Biol. 11 (3): 139–145, 1973.
Ïordanov, I. T.: Fotosinteticheskaya aktivnost' razni kh list'ev rasteniĭ fasoli. [Photosynthetic activity in various leaves of the bean plants.] — Sel'skokhoz. Biol. 14: 655–660, 1979.
Isebrands, J. G., Larson, P. R.: Anatomical changes during leaf ontogeny in *Populus deltoides*. — Amer. J. Bot. 60: 199–208, 1973.
Ishihara, K.: Diurnal course of stomatal aperture of leaf blades in rice plants. — Jap. agr. Res. Quart. 13 (2): 85–89, 1979.
Ishihara, K., Ebara, H., Hirasawa, T., Ogura, T.: [The relationship between environmental factors and behaviour of stomata in the rice plants. VII. The relation between nitrogen content in leaf blades and stomatal aperture.] — Jap. J. Crop Sci. 47: 664–673, 1978a.
Ishihara, K., Ishida, Y., Ogura, T.: [The relationship between environmental factors and behaviour of stomata in the rice plant. 2. On the diurnal movement of the stomata.] — Proc. Crop Sci. Soc. Jap. 40: 497–504, 1971a.
Ishihara, K., Ishida, Y., Ogura, T.: [The relationship between environmental factors and behaviour of stomata in the rice plant. 3. On the aperture of the stomata and their diurnal movement in the leaf at different position on the stem.] — Proc. Crop Sci.Soc. Jap. 40: 505–512, 1971b.
Ishihara, K., Sago, R., Ogura, T.: [The relationship between environmental factors and behaviour of stomata in the rice plants. V. Effects of partial excision of root system on diurnal course of stomatal aperture.] — Jap. J. Crop Sci. 47: 499–505, 1978b.
Ishihara, K., Sago, R., Ogura, T.: [The relationship between environmental factors and behaviour of stomata in the rice plants. VI. Comparison between the diurnal course of stomatal aperture

of rice plants grown in the border and interior of paddy fields.] — Jap. J. Crop Sci. 47: 515–528, 1978c.

Ito, K.: Studies on photosynthesis in sugar beet plant. II. The difference in photosynthetic and respiratory abilities and response to temperature and light between leaves of different leaf position. — Proc. Crop Sci. Soc. Jap. 33: 487–491, 1965.

Ito, T.: Absorption and distribution of radioactive phosphorus in tomato plant with respect to the carbon dioxide concentration in the atmosphere. — Techn. Bull. Fac. Horticult. Chiba Univ. 18: 21–28, 1970.

Ito, T.: [Plant growth and physiology of vegetable plants as influenced by carbon dioxide environment.] — Trans. Fac. Hort., Chiba Univ. 1973 (7): 1–134, 1973.

Ivanchenko, V. M., Legenchenko, B. I., Urbanovich, T. A., Kruchinina, S. S., Marshakova, M. I.: O vremennoĭ izmenchivosti strukturno-funktsional'nykh kharakteristik fotosinteticheskogo apparata. [Time-dependent changes in structural and functional parameters of the photosynthetic apparatus.] — Fiziol. Rast. 26: 28–34, 1979.

Iwakiri, S., Inayama M.: [Studies on the canopy photosynthesis of the horticultural crops in controlled environment. (4) Photosynthetic characteristics of single cucumber leaves.] — J. agr. Meteorol. 30: 161–166, 1975.

Izawa, S., Good, N. E.: Effect of salts and electron transport on the conformation of isolated chloroplasts. II. Electron microscopy. — Plant Physiol. 41: 544–552, 1966.


Jacoby, B., Tirosh, T., Plessner, O. E.: Relationship between age of bean leaves, sodium export, and permeability of leaf tissue. — Bot. Gaz. 134: 46–49, 1973.

Jarvis, P. G.: The estimation of resistances to carbon dioxide transfer. — In: Šesták, Z., Čatský, J., Jarvis, P. G. (ed.): Plant Photosynthetic Production: Manual of Methods. Pp. 566–631. Dr. W. Junk N. V. Publ., The Hague 1971.

Jarvis, P. G., Morison, J. I. L.: The control of transpiration and photosynthesis by the stomata. — In: Jarvis, P. G., Mansfield, T. A. (ed.): Stomatal Physiology. Pp. 247–279. Cambridge Univ. Press, Cambridge –London–New York–New Rochelle–Melbourne–Sydney 1981.

Jenkins, G. I., Baker, N. R., Bradbury, M., Woolhouse, H. W.: Photosynthetic electron transport during senescence of the primary leaves of Phaseolus vulgaris L. III. Kinetics of chlorophyll fluorescence emission from intact leaves. — J. exp. Bot. 32: 999–1008, 1981a.

Jenkins, G. I., Baker, N. R., Woolhouse, H. W.: Changes in chlorophyll content and organization during senescence of the primary leaves of Phaseolus vulgaris L. in relation to photosynthetic electron transport. — J. exp. Bot. 32: 1009–1020, 1981b.

Jenkins, G. I., Woolhouse, H. W.: Photosynthetic electron transport during senescence of the primary leaves of Phaseolus vulgaris L. I. Non-cyclic electron transport. — J. exp. Bot. 32: 467–478, 1981a.

Jenkins, G. I., Woolhouse, H. W.: Photosynthetic electron transport during senescence of the primary leaves of Phaseolus vulgaris L. II. The activity of photosystems one and two, and a note on the site of reduction of ferricyanide. — J. exp. Bot. 32: 989–997, 1981b.

Jewiss, O. R., Woledge, J.: The effect of age on the rate of apparent photosynthesis in leaves of tall fescue (Festuca arundinacea Schreb.). — Ann. Bot. 31: 661–671, 1967.

Jodo, S.: [Stomatal movement and water relations in crops. 2. Stomatal behavior of tobacco leaves of different ages and the influence of soil water shortage.] — Proc. Crop Sci. Soc. Jap. 42: 123–130, 1973.

Jones, H. G.: Crop characteristics and the ratio between assimilation and transpiration. — J. appl. Ecol. 13: 605–622, 1976.

Jones, H. G.: Aspects of the water relations of spring wheat (Triticum aestivum L.) in response to induced drought. — J. agr. Sci. 88: 267–282, 1977.

Jones, H. G., Osmond, C. B.: Photosynthesis by thin leaf slices in solution. I. Properties of leaf slices and comparison with whole leaves. — Aust. J. biol. Sci. 26: 15–24, 1973.

Jones, H. G., Slatyer, R. O.: Estimation of the transport and carboxylation components of the intracellular limitation to leaf photosynthesis. — Plant Physiol. 50: 283–288, 1972.

Jones, J. W., Hesketh, J. D.: Predicting leaf expansion. — In: Hesketh, J. D., Jones, J. W. (ed.): Predicting Photosynthesis for Ecosystem Models. Vol. II. Pp. 85–122. CRC Press, Boca Raton 1980.

Jones, M. B., Mansfield, T. A.: A circadian rhythm in the level of carbon dioxide compensation in *Bryophyllum fedtschenkoi* with zero values during the transient. — Planta 103: 134–146, 1972.

Jones, M. M., Rawson, H. M.: Influence of rate of development of leaf water deficits upon photosynthesis, leaf conductance, water use efficiency, and osmotic potential in sorghum. — Physiol. Plant. 45: 103–111, 1979.

Jones, R., Buchanan, I. C., Wilkins, M. B., Fewsen, C. A., Malcolm, A. D. B.: Phosphoenolpyruvate carboxylase from the crassulacean plant *Bryophyllum fedtschenkoi* Hamet et Perrier. Activity changes and kinetic behaviour in crude extracts. — J. exp. Bot. 32: 427–441, 1981.

Jordan, W. R., Brown, K. W., Thomas, J. C.: Leaf age as a determinant in stomatal control of water loss from cotton during water stress. — Plant Physiol. 56: 595–599, 1975.

Jouy, M.: Effect of age of etiolated leaves of *Phaseolus vulgaris* on the 695 nm fluorescence kinetics during first irradiation. — Photosynthetica 16: 234–238, 1982.

Judel, G. K., Linser, H., Zeid, F. A.: Kupfer, Reinprotein und Phenoloxidase in der Blattfolge. von *Helianthus annuus* im Verlaufe der Vegetationsperiode. — Z. Pflanzenern. Bodenk. 1975 (1): 39–48, 1975.

Jurik, T. W., Chabot, J. F., Chabot, B. F.: Ontogeny of photosynthetic performance in *Fragaria virginiana* under changing light regimes. — Plant Physiol. 63: 542–547, 1979.

Kabaki, N., Saka, H., Akita, S.: [Effects of nitrogen, phosphorus and potassium deficiencies on photosynthesis and RuBP carboxylase-oxygenase activities in rice plants.] — Jap. J. Crop Sci. 48: 378–384, 1979.

Kahn, J. S., Chang, I. C.: A soluble protein-chlorophyll complex from spinach chloroplasts. III. Determination of molecular weight and comparison of complex isolated from different sources. — Photochem. Photobiol. 4: 733–738, 1965.

Kakie, T.: Physical properties of starch granules of tobacco leaves during maturity stage. — Soil Sci. Plant Nutr. 18: 7–14, 1972.

Kaler, V. L., Akulovich, E. M.: Izmenenie pigmentnoĭ sistemy zhasmina obvyknovennogo zolotistogo. [Changes in pigment system of jasmine.] — In: Fiziologo-Biokhimicheskie Issledovaniya Rasteniĭ. Pp. 26–32. Nauka i Tekhnika, Minsk 1965.

Kaler, V. L., Fridlyand, L. E.: Ontogeneticheskaya adaptatsiya fotosinteza rasteniĭ kak sledstvie kolichestvennykh izmeneniĭ osnovnykh ėlementov fotosinteticheskogo apparata. Teoreticheskoe rassmotrenie. [Ontogenetic adaptation of plant photosynthesis as a result of quantitative changes in the main elements of the photosynthetic apparatus: a theoretical consideration.] — Fiziol. Rast. 25: 664–670, 1978.

Kanemasu, E. T., Chen, A. - J., Powers, W. L., Teare, I. D.: Stomatal resistance as an indicator of water stress. — Trans. Kansas Acad. Sci. 76 (2): 159–166, 1973/74.

Kanemasu, E. T., Tanner, C. B.: Stomatal diffusion resistance of snap beans. II. Effect of light. — Plant Physiol. 44: 1542–1546, 1969.

Kaniuga, Z., Sochanowicz, B., Ząbek, J., Krzystyniak, K.: Photosynthetic apparatus in chilling-sensitive plants I. Reactivation of Hill reaction activity inhibited on the cold and dark storage of detached leaves and intact plants. — Planta 140: 121–128, 1978.

Kannangara, C. G.: The formation of ribulose diphosphate carboxylase protein during chloroplast development in barley. — Plant Physiol. 44: 1533–1537, 1969.

Kannangara, C. G., Woolhouse, H. W.: Changes in the enzyme activity of soluble protein fractions in the course of foliar senescence in *Perilla frutescens* (L.) Britt. — New Phytol. 67: 533–542, 1968.

Kar, M., Mishra, D.: Catalase, peroxidase, and polyphenoloxidase activities during rice leaf senescence. — Plant Physiol. 57: 315–319, 1976.

Karanov, E., Nikolov, K., Popov, K. Y.: Influence of some growth regulators on the aging and state of plastide pigments in leaves of different age. — Dokl. bolg. Akad. Nauk 23: 1155–1158, 1970.

Karpilova, I. F., Chugunova, N. G., Bil', K. Ya., Chermnykh, L. N.: Ontogeneticheskie izmeneniya ul'trastruktury khloroplastov, produktov fotosinteza i ikh ottoka iz lista ogurtsov pri ponizhennoĭ nochnoĭ temperature. [Ontogenetic changes in chloroplast ultrastructure, photosynthate pattern and transport from the cucumber leaf under low night temperature.] — Fiziol. Rast. 29: 113–120, 1982.

Kasanaga, H., Monsi, M.: On the light-transmission of leaves, and its meaning for the production of matter in plant communities. — Jap. J. Bot. 14: 304–324, 1954.

Kasemir, H.: Mini review. Control of chloroplast formation by light. — Cell Biol. int. Rep. 3: 197–214, 1979.

Kataoka, K., Oohara, H.: [On the phenotypic plasticity of some photosynthesis releted characters for the changes of temperature in rice varieties.] — Bull. Fac. Agr., Tamagawa Univ. 20: 32–39, 1980.

Kato, S., Hozyo, Y., Shimotsubo, K.: [Translocation of $^{14}$C-photosynthates from the leaves at different stages of development in Ipomoea grafts.] — Jap. J. Crop Sci. 38: 254–259, 1979.

Kaval'chuk, R. A., Vechar, A. S.: Lipidny sastaŭ khlaraplastaŭ prarostkaŭ zhyta. [Lipid composition of chloroplasts of rye seeedlings.] — Vestsi Akad. Navuk belarus. SSR, Ser. biyal. Navuk 1973 (4): 20–28, 1973.

Kawashima, N., Mitake, T.: Studies on protein metabolism in higher plats. Part VI. Changes in ribulose diphosphate carboxylase activity and fraction 1 protein content in tobacco leaves with age. — Agr. biol. Chem. 33: 539–543, 1969.

K"drev, T., Georgieva, M.: Vliyanie na nedostiga na magneziĭ v"rkhu s"d"rzhanieto na galaktolipidite i pigmentite v otdelni lista na tsarevichni rasteniya. [Effect of magnesium deficiency on the contents of galactolipids and pigments in individual leaves of maize plants.] — Fiziol. Rast. (Sofiya) 1 (3): 10–16, 1975.

Keay, J., Turton, A. G., Campbell, N. A.: Some effects of nitrogen and phosphorus fertilization of Pinus pinaster in Western Australia. — Forest Sci. 14: 408–417, 1968.

Keck, R. W., Boyer, J. S.: Chloroplast response to low leaf water potentials. III. Differing inhibition of electron transport and photophosphorylation. — Plant Physiol. 53: 474–479, 1974.

Keith, B.: Mass transfer and $^{14}$C translocation in detached maize leaves. — Can. J. Bot. 57: 657–665, 1979.

Kennedy, R. A.: Relationship between leaf development, carboxylase enzyme activities and photorespiration in the $C_4$-plant Portulaca oleracea L. — Planta 128: 149–154, 1976.

Kennedy, R. A., Johnson, D.: Changes in photosynthetic characteristics during leaf development in apple. — Photosynthesis Res. 2: 213–223, 1981.

Kennedy, R. A., Laetsch, W. M.: Relationship between leaf development and primary photosynthetic products in the $C_4$ plant Portulaca oleracea L. — Planta 115: 113–124, 1973.

Keresztes, Á., Faludi-Dániel, Á.: Ultrastructure, pigment content and photosynthetic activity of the normal and mutant chloroplasts in developing Tradescantia leaves. — Acta biol. Acad. Sci. hung. 24: 175–189, 1973.

Keys, A. J., Bird, I. F., Cornelius, M. J.: Possible use of chemicals for the control of photorespiration. - In: McLaren, J. S. (ed.): Chemical Manipulation of Crop Growth and Development. Pp. 39–53. Butterworth Sci., London–Boston–Durban–Singapore–Sydney–Toronto–Wellington 1982.

Keys, A. J., Whittingham, C. P.: Photorespiratory carbon dioxide loss. — In: Johnson, C. B. (ed.): Physiological Processes Limiting Plant Productivity. Pp. 137–145. Butterworths, London – Boston–Sydney–Wellington–Durban–Toronto 1981.

Khavari-Nejad, R. A.: Growth of tomato plants in different oxygen concentrations. — Photosynthetica 14: 326–336, 1980.

Khavkin, E. E.: Formirovanie Metabolicheskikh Sistem v Rastushchikh Kletkakh Rasteniĭ.

[Formation of Metabolic Systems in Growing Cells of Plants.] — Nauka, Novosibirsk 1977.

Khlyastikov, G. P.: Fotosintez i metabolizm fosfora u fasoli pri razlichnykh urovnyakh azotnogo i fosfornogo pitaniya. [Photosynthesis and phosphorus metabolism in *Phaseolus* under different levels of nitrogen and phosphorus nutrition.] — In: Produktivnost' Nazemnykh Fotosinteziruyushchikh Sistem v Ėkstremal'nykh Usloviyakh. Pp. 91–99, 188. Sib. Otd. Akad. Nauk SSSR, Buryat. Filial, Ulan Udé 1977.

Khodasevich, Ė. V., Arnautova, A. I., Godnev, T. N.: Formirovanie i sostoyanie fonda pigmentov i plastid v ontogeneze lista khvoïnykh. [Formation and state of pigments and plastids during ontogenesis of needles.] — In: Metabolizm i Stroenie Fotosinteticheskogo Apparata. Pp. 152–163. Nauka i Tekhnika, Minsk 1970.

Khodasevich, Ė. V., Arnautova, A. I., Myshkovets, E. N.: Ul'trastrukturnaya organizatsiya khloroplastov v svyazi s obratimoï degradatsieï fonda pigmentov u khvoïnykh. [The structural organization of chloroplasts related to reversible degradation of the pigment pool in conifers.] — Fiziol. Rast. 25: 810–814, 1978.

Khodasevich, Ė. V., Arnautova, A. I., Gvardiyan, V. N., Myshkovets, E. N.: Strukturnaya organizatsiya khloroplasta i fotosinteticheskaya funktsiya pri dlitel'noï vegetatsii lista. [Structural organization of the chloroplast and photosynthetic function during prolonged leaf life span.] — Zh. obshch. Biol. 40: 603–609, 1979.

Khodasevich, Ė. V., Lis, P. I.: Issledovanie spektrov nizkotemperaturnoï fluorestsentsii pigmentov v ontogeneze khvoi. [Study of spectra of low-temperature fluorescence of pigments during needle ontogeny.] — Dokl. Akad. Nauk belorus. SSR 24: 182–185, 1980.

Khodasevich, Ė.V., Mel'nikova, L. M., Arnautova, A. I., Godnev, T. N.: Sostoyanie pigmentbelkovogo kompleksa u khvoïnykh v svyazi s obratimoï degradatsieï fonda khlorofillov *a* i *b*. [State of pigment-protein complex in conifers in relation to reversible degradation of the fund of chlorophylls *a* and *b*.] — Dokl. Akad. Nauk beloruss. SSR 17: 80–83, 1973.

Khodzhaev, A. S., Meïstrik, I. A., Rakhmankulova, M. E.: O vzaimosvyazi chisla khloroplastov v kletke s ikh aktivnost'yu v ontogeneze lista khlopchatnika. [Correlation between the number of chloroplasts in the cell and their activity in the course of cotton leaf ontogenesis.] — Fiziol. Rast. 25: 541–546, 1978.

Khristin, M. S., Akulova, E. A.: Obnaruzhenie ferredoksinov i plastotsianina v rannem ontogeneze i pri zelenenii ėtiolirovannykh prorostkov gorokha. [Detection of ferredoxins and plastocyanin in the early ontogenesis and in greening etiolated pea sprouts.] — Dokl. Akad. Nauk SSSR 223: 758–761, 1975.

Kimmerer, T. W., Kozlowski, T. T.: Stomatal conductance and sulfur uptake of five clones of *Populus tremuloides* exposed to sulfur dioxide. — Plant Physiol. 67: 990–995, 1981.

Kimura, M., Yokoi, Y., Hogetsu, K.: Quantitative relationships between growth and respiration II. Evaluation of constructive and maintenance respiration in growing *Helianthus tuberosus* leaves. — Bot. Mag. (Tokyo) 91: 43–56, 1978.

Kirby, E. J. M., Appleyard, M., Fellowes, G.: Effect of sowing date on the temperature response of leaf emergence and leaf size in barley. — Plant Cell Environm. 5: 477–484, 1982.

Kirchanski, S. J.: The ultrastructural development of the dimorphic plastids of *Zea mays* L. — Amer. J. Bot. 62: 695–705, 1975.

Kirita, H., Hozumi, K.: Estimation of the total chlorophyll amount and its seasonal change in a warm-temperate evergreen oak forest at Minamate, Japan. – Jap. J. Ecol. 23: 195–200, 1973.

Kirk, J. T.O.: Chloroplast structure and biogenesis. — Annu. Rev. Biochem. 40: 161–196, 1971.

Kirk, J. T.O.: The relation of chlorophyll synthesis to protein synthesis in the growing thylakoid membrane. — Portug. Acta Biol., Ser. A 14: 127–152, 1974.

Kirk, J. T. O., Goodchild, D. J.: Relationship of photosynthetic effectiveness of different kinds of light to chlorophyll content and chloroplast structure in greening wheat and in ivy leaves. — Aust. J. biol. Sci. 25: 215–241, 1972.

Kirk, J. T. O., Tilney-Bassett, R. A. E.: The Plastids. Their Chemistry, Structure, Growth and

Inheritance. 2$^{nd}$ Ed. — Elsevier/North-Holland Biomed. Press, Amsterdam–New York–Oxford 1978.

Kisaki, T., Hirabayashi, S., Yano, N.: Effect of the age of tobacco leaves on photosynthesis and photorespiration. — Plant Cell Physiol. 14: 505–514, 1973.

Kislyakova, T. E., Golubkova, B. M., Bogacheva, I. I.: Vzaimosvyaz' struktury i funktsii fotosinteticheskogo apparata v ontogeneze kartofelya. [Relation of structure and function of the photosynthetic apparatus in potato ontogenesis.] — Fiziol. Rast. 14: 5–14, 1967.

Kisser, J.: Untersuchungen über den Einfluss der Nährsalze auf die Wasserabgabe, Wasseraufnahme, relative Spross- und Wurzelmasse und die Blattstruktur. II. Teil: Veränderungen der Blattstruktur unter dem Einflusse der Nährsalze. — Planta 3: 578–596, 1927.

Klee, R., Steubing, L.: Studien über das Interzellularvolumen von Laubblättern. — Ber. deut. bot. Ges. 80:,416–425, 1967.

Klein, S.: Diversity of chloroplast structure. — In: Schiff, J. A., Lyman, H. (ed.): On the Origins of Chloroplasts. Pp. 35–53. Elsevier/North Holland, New York –Amsterdam–Oxford 1982.

Klimov, S. V., Bocharov, E. A., Dzhanumov, D. A.: Svyaz' formativnykh protsessov s fotosintezom i poslesvecheniem u prorostkov ozimoï pshenitsy. [Correlation between formative processes, photosynthesis and delayed light emission in winter wheat seedlings.] — Fiziol. Rast. 25: 106–112, 1978.

Klockare, B.: Far-red induced changes of the protochlorophyllide components in wheat leaves. — Physiol. Plant. 48: 104–110, 1980.

Kluge, M.: The flow of carbon in Crassulacean Acid Metabolism (CAM). — In: Gibbs, M., Latzko, E. (ed.): Photosynthesis II. Pp. 113–125. Springer-Verlag, Berlin–Heidelberg–New York 1979.

Kluge, M., Ting, I. P.: Crassulacean Acid Metabolism. Analysis of an Ecological Adaptation. — Springer-Verlag, Berlin–Heidelberg–New York 1978.

Klyuchareva, E. A.: Izmenenie fotokhimicheskoï aktivnosti i kharaktera fraktsionirovaniya pigment-belkovogo fonda v plastidakh zeleneyushchikh list'ev. [Changes in photochemical activity and character of fractionation of the pigment-protein fund in plastids of greening leaves.] — In: Biologiya i Nauchno-Tekhnicheskiï Progress. Pp. 48–50. Akad. Nauk SSSR, Pushchino 1974.

Knecht, G. N., Orton, E. R., Jr.: Stomate density in relation to winter hardiness of *Ilex opaca* Ait. — J. amer. Soc. hort. Sci. 95: 341–345, 1970.

Knipling, E. B.: Physical and physiological basis for the reflectance of visible and near-infrared radiation from vegetation. — Remote Sens. Environ. 1: 155–159, 1970.

Kobayashi, S.: Growth analysis of plant as an assemblage of internodal segments — a case of sunflower plants in pure stands. — Jap. J. Ecol. 25: 61–70, 1975.

Koch, H.: Zebrastreifung, eine Chlorose keimender Monokotylen, verursacht durch Belichtung und Temperatur. — Angew. Bot. 50: 233–250, 1976.

Koch, W.: Blattfarbstoffe von Fichte (*Picea abies* (L.) Karst.) in Abhängigkeit vom Jahresgang, Blattalter und -typ. — Photosynthetica 10: 280–290, 1976.

Koehler, P. G.: The roles of cell division and cell expansion in the growth of alfalfa leaf mesophyll. — Ann. Bot. 37: 65—68, 1973.

Korkes, S.: Carbohydrate metabolism. — Annu. Rev. Biochem. 25: 685–734, 1956.

Korn, R. W.: Arrangement of stomata on the leaves of *Pelargonium zonale* and *Sedum stahlii*. — Ann. Bot. 36: 325–333, 1972.

Körner, C.: Stomatal behaviour and water potential in apricot with symptoms of wilt disease. — Angew. Bot. 55: 469–476, 1981.

Kowallik, K. V., Herrmann, R. G.: Chloroplasts. — Progr. Bot. 39: 1–17, 1977.

Kozhushko, N. N., Chernysheva, S. V.: Sostoyanie pigmentnogo kompleksa plastidnogo apparata list'ev pshenitsy v reproduktivnyï period razvitiya. [State of the pigment complex of plastid apparatus of wheat leaves in the reproductive period of development.] — Byull. vses. nauch. issled. Inst. Rastenievod. Im. N. I. Vavilova 63: 15—18, 1976.

Krasichkova, G. V., Chandylova, L. V., Giller, Yu. E.: Issledovanie fotokhimicheskoï aktivnosti khloroplastov gibridnykh form khlopchatnika. [Photochemical activity of chloroplasts of cotton hybrid forms.] — Dokl. Akad. Nauk tadzh. SSR 20 (9): 54–56, 1977.

Krenzer, E. G., Jr., Moss, D. N., Crookston, R. K.: Carbon dioxide compensation points of flowering plants. — Plant Physiol. 56: 194–206, 1975.

Kreutz, W.: On the structural arrangement of light reaction centres I and II in the photosynthetic membrane. — In: Metzner, H. (ed.): Photosynthetic Oxygen Evolution. Pp. 77–90. Academic Press, London–New York –San Francisco 1978.

Kriedemann, P. E.: Photosynthesis in vine leaves as a function of light intensity, temperature, and leaf age. — Vitis 7: 213–220, 1968.

Kriedemann, P. E.: Photosynthesis and transpiration as a function of gaseous diffusive resistances in orange leaves. — Physiol. Plant. 24: 218–225, 1971.

Kriedemann, P. E., Kliewer, W. M., Harris, J. M.: Leaf age and photosynthesis in Vitis vinifera L. — Vitis 9: 97–104, 1970.

Kriedemann, P. E., Loveys, B. R., Possingham, J. V., Satoh, M.: Sink effects on stomatal physiology and photosynthesis. — In: Wardlaw, I. F., Passioura, J. B. (ed.): Transport and Transfer Processes in Plants. Pp. 401–414. Academic Press, New York–San Francisco–London 1976.

Krizek, D. T., Milthorpe, F. L.: Effect of photoperiodic induction on the transpiration rate and stomatal behaviour of debudded Xanthium plants. — J. exp. Bot. 24: 76–86, 1973.

Ku, S. B., Hunt, L. A.: Effects of temperature on the morphology and photosynthetic activity of newly matured leaves of alfalfa. — Can. J. Bot. 51: 1907–1916, 1973.

Kubelka, P., Munk, F.: Ein Beitrag zur Optik der Farbanstriche. — Z. tech. Phys. 11: 593–601, 1931.

Kufner, R., Czygan, F.-C., Schneider, L.: Veränderungen des Pigmentgehalts und der Ultrastruktur bei den Plastiden der Nadelblätter von Taxus baccata (L.) während ihrer Entwicklung. — Ber. deut. bot. Ges. 91: 325–337, 1978.

Kufner, R. B.: The biological degradation of chlorophyll in senescent tissues. — In: Czygan, F.-C. (ed.): Pigments in Plants. 2nd Ed. Pp. 308–313. G. Fischer Verlag, Stuttgart–New York 1980. Akademie-Verlag, Berlin 1981.

Kulandaivelu, G., Daniell, H.: Dichlorophenyl dimethylurea (DCMU) induced increase in chlorophyll a fluorescence intensity — an index of photosynthetic oxygen evolution in leaves, chloroplasts and algae. — Physiol. Plant. 48: 385–388, 1980.

Kumar, H. D., Singh, H. N.: Plant Metabolism. — MacMillan Press Ltd., London–Basingstoke 1979.

Kumar, N. C., Mukherjee, K. L.: Physiology of host parasite relation in Dendrophthoe falcata infection: respiratory studies with the leaf discs of host and parasite. — Indian Phytopathol. 22: 215–220, 1969.

Kumura, A.: [Studies on dry matter production of soybean plant. IV. Photosynthetic properties of leaf as subsequently affected by light conditions.] — Proc. Crop Sci. Soc. Jap. 37: 583–588, 1968.

Kumura, A.: [Studies on dry matter production in soybean plant. V. Photosynthetic system of soybean plant population.] — Proc. Crop Sci. Soc. Jap. 38: 74–90, 1969.

Kumura, A., Naniwa, I.: [Studies on dry matter production of soybean plant I. Ontogenic changes in photosynthetic- and respiratory capacity of soybean plant and its parts.] — Proc. Crop Sci. Soc. Jap. 33: 467–472, 1965.

Kuno, H.: [Effects of photochemical oxidant on the growth of poplar cuttings II. Effects of photochemical oxidant on chlorophyll contents, photosynthetic and dark respiratory rates, soluble carbohydrate and nitrogen contents in leaves of different ages.] — Taiki Osen Gakkaishi 15: 155–162, 1980.

Kupka, J., Truong Quang Tan: Fotosyntéza a obsah chlorofylu v ontogenezi kukuřice. [Photosynthesis and chlorophyll content in maize ontogenesis.] — Rostlinná Výroba 21: 403–408, 1975.

Kursanov, A. L., Paramonova, N. V.: On the state of membranes in mesophyll cells of Beta *vulgaris* in terms of assimilate transport. — In: Jacques, R. (ed.): Etudes de Biologie Végétale. Hommage au Professeur Pierre Chouard. Pp. 509–519. Paris 1976.

Kutík, J.: The relationships between quantitative characteristics of stomata and epidermal cells of leaf epidermis. — Biol. Plant. 15: 324–328, 1973.

Kutík, J., Beneš, K.: The anatomical study of heterotrophic starch formation in leaf segments of maize and pea. — Biol. Plant. 23: 52–57, 1981.

Kutík, J., Šesták, Z., Volfová, A.: Ontogenetic changes in the internal limitations to bean–leaf photosynthesis. 8. Primary leaf blade characteristics and chloroplast number, size and ultrastructure. — Photosynthetica 18: 1–8, 1984.

Kutík, J., Tageeva, S. V., Popov, V. I.: The development of chloroplast ultrastructure during the ontogeny of *Tilia cordata* Mill. leaves. — Photosynthetica 15: 261–263, 1981.

Kuznetsova, G. K., Khazanov, V. S., Shishov, D. M., Zaïko, L. N., Nakaidze, A. Kh., Geïer, N. I., Mukhina, V. I.: Fiziologicheskie osobennosti krestovnika ploskolistnogo i soderzhanie v nem alkaloidov. [Physiological characteristics of *Senecio platyphylloides* Somm. et Lév. and its content of alkaloids.] — Rastitel'. Resursy 10: 82–87, 1974.


Ladenburger, K., Albert, R.: Stoffwechselphysiologische Untersuchungen an verschieden alten Blättern einiger Halophyten und Glykophyten. — Z. Pflanzenphysiol. 102: 303–314, 1981.

Laetsch, W. M.: The $C_4$ syndrome: a structural analysis. — Annu. Rev. Plant Physiol. 25: 27–52, 1974.

Laetsch, W. M., Price, I.: Development of the dimorphic chloroplasts of sugar cane. — Amer. J. Bot. 56: 77–87, 1969.

Laïsk, A.: Svetovye krivye fotosinteza dlya opticheski tolstykh list'ev. [Light curves of photosynthesis considering light profile in the leaf.] — In: Voprosy Éffektivnosti Fotosinteza. Pp. 93 to 116. Inst. Fiz. Astron., Akad. Nauk eston. SSR, Tartu 1969.

Laïsk, A., Oya, V., Rakhi, M.: Diffuzionnye soprotivleniya list'ev v svyazi s ikh anatomieï. [Diffusion resistances of leaves in connection with their anatomy.] — Fiziol. Rast. 17: 40–48, 1970.

Laïsk, A. Kh.: Kinetika Fotosinteza i Fotodykhaniya $C_3$-Rastenii. [Kinetics of Photosynthesis and Photorespiration in $C_3$-Plants.] — Nauka, Moskva 1977.

Laison, R. A., Thornley, J.H.M.: A model for leaf expansion in cucumber. — Ann. Bot. 50: 407–425, 1982.

Lambers, H., Szaniawski, R. K., de Visser, R.: Respiration for growth, maintenance and ion uptake. An evaluation of concepts, methods, values and their significance. — Physiol. Plant. 58: 556—563, 1983.

Lamoreaux, R. J., Chaney, W. R., Brown, K. M.: The plastochron index: A review after two decades of use. — Amer. J. Bot. 65: 586–593, 1978.

Lancer, H. A., Cohen, C. E., Schiff, J. A.: Changing ratios of phototransformable protochlorophyll (Pchl) and protochlorophyllide (Pchlide) in leaves of bean seedlings developing in the dark. — Plant Physiol. 56 (Suppl.): 33, 1975.

Lancer, H. A., Cohen, C. E., Schiff, J. A.: Changing ratios of phototransformable protochlorophyll and protochlorophyllide of bean seedlings developing in the dark. — Plant Physiol. 57: 369–374, 1976.

Landi, R., Antongiovanni, M.: Contributo allo studio del carotene contenuto nelle piante di sorgo: influenza di alcuni fattori biologici ed agronomici. [Contribution to the study of carotene contained in *Sorghum* plants: the influence of some biological and agronomic factors.] — Maydica 18: 50–62, 1973.

Landsberg, J. J., Cutting, C. V. (ed.): Environmental Effects on Crop Physiology. — Academic Press, London –New York–San Francisco 1977.

Lane, H. C., Hesketh, J. D.: Cotyledon photosynthesis during seedling growth of cotton, *Gossypium hirsutum* L. — Amer. J. Bot. 64: 786–790, 1977.

Lane, H. C., Thompson, A. C.: Morphological and physiological differences between the cotyledons of intact and debudded cotton plants. — Bot. Gaz. 139: 207–210, 1978.

322

Larcher, W.: Physiological Plant Ecology. 2nd Ed. — Springer-Verlag, Berlin–Heidelberg–New York 1980.

Larson, P. R., Gordon, J. C.: Leaf development, photosynthesis, and $C^{14}$ distribution in *Populus deltoides* seedlings. — Amer. J. Bot. 56: 1058–1066, 1969.

Larson, P. R., Isebrands, J. G., Dickson, R. E.: Fixation patterns of $^{14}C$ within developing leaves of eastern cottonwood. — Planta 107: 301–314, 1972.

Lasley, S. E., Garber, M. P.: Photosynthetic contribution of cotyledons to early development of cucumber. — HortScience 13: 191–193, 1978.

Latimer, P.: Light scattering *vs.* microscopy for measuring average cell size and shape. — Biophys. J. 27: 117–126, 1979.

Laudi, G.: Ricerche comparate sulla morfologia e sulla fisiologia di *Larix* e di *Picea*. Infrastruttura dei cloroplasti dei cotiledoni di plantule mantenute al buio ed esposte alla luce. [Comparative research on morphology and physiology of *Larix* and *Picea*. Ultrastructure of chloroplasts of cotyledons on seedlings maintained in darkness and exposed to light.] — G. bot. ital. 71: 177–182, 1964.

Lawlor, D. W., Milford, G. F. J.: The control of water and carbon dioxide flux in water-stressed sugar beet. — J. exp. Bot. 26: 657–665, 1975.

Lechowski, Z.: Chloroplast arrangement as a factor of photosynthesis in multilayered leaves. — Acta Soc. Bot. Pol. 43: 531–540, 1974.

Lee, R., Gates, D. M.: Diffusion resistance in leaves as related to their stomatal anatomy and microstructure. — Amer. J. Bot. 51: 963–975, 1964.

Leech, R. M.: Studies of plastid membrane development using suspensions of isolated proplastids and isolated etioplasts. — Portug. Acta Biol., Ser. A 14: 429–450, 1974.

Leech, R. M.: Plastid development in isolated etiochloroplasts and isolated etioplasts. — In: Sunderland, N. (ed.): Perspectives in Experimental Biology. Vol. 2. Botany. Pp. 145–162. Pergamon Press, Oxford–New York 1976.

Leech, R. M., Rumsby, M. G., Thomson, W. W.: Plastid differentiation, acyl lipid, and fatty acid changes in developing green maize leaves. — Plant Physiol. 52: 240–245, 1973.

Leech, R. M., Rumsby, M. G., Thomson, W. W., Crosby, W., Wood, P.: Lipid changes during plastid differentiation in developing maize leaves. — In: Forti, G., Avron, M., Melandri, A. (ed.): Photosynthesis, Two Centuries after Its Discovery by Joseph Priestley. Vol. 3. Pp. 2479–2488. Dr. W. Junk N. V. Publ., The Hague 1972.

Leese, B. M., Leech, R. M.: Biosynthesis and structure of leaf lipids: Lipid changes during plastid and leaf development. — Biochem. Soc. Trans. 5: 1266–1269, 1977.

Lehner, M. I.: The application of Zalenski's law to certain submerged aquatics. — Papers Michigan Acad. Sci., Arts Lett. 32: 91–97, 1946.

Lemon, E. R. (ed.): $CO_2$ and Plants. The Response of Plants to Rising Levels of Atmospheric Carbon Dioxide. — Westview Press, Boulder 1983.

Leopold, A. C.: Aging and senescence in plant development. — In: Thimann, K. V. (ed.): Senescence in Plants. Pp. 1–12. CRC Press, Boca Raton 1980.

Lerman, J. C., Deleens, E., Nato, A., Moyse, A.: Variation in the carbon isotope composition of a plant with Crassulacean acid metabolism. — Plant Physiol. 53: 581–584, 1974.

Leverenz, J. W., Jarvis, P. G.: Photosynthesis in Sitka spruce (*Picea sitchensis* (Bong.) Carr.) IX. The relative contribution made by needles at various positions on the shoot. — J. appl. Ecol. 17: 59–68, 1980.

Lewandowska, M., Hart, J. W., Jarvis, P. G.: Photosynthetic electron transport in shoots of Sitka spruce from different levels in a forest canopy. — Physiol. Plant. 41: 124–128, 1977.

Lewandowska, M., Jarvis, P. G.: Changes in chlorophyll and carotenoid content, specific leaf area and dry weight fraction in Sitka spruce, in response to shading and season. — New Phytol. 79: 247–256, 1977.

Lewandowska, M., Jarvis, P. G.: Quantum requirements of photosynthetic electron transport in Sitka spruce from different light environments. — Physiol. Plant. 42: 277–282, 1978.

Lewandowska, M., Öquist, G.: Structural and functional relationships in developing *Pinus silvestris* chloroplasts. — Physiol. Plant. 48: 39–46, 1980.

Lewington, R. J., Simon, E. W.: The effect of light on the senescence of detached cucumber cotyledons. — J. exp. Bot. 20: 138–144, 1969.

Li, X., Wu, G.: [Changes of ribulose bisphosphate carboxylase-oxygenase and malate dehydrogenase activities with leaf age in *Vicia faba*.] — Acta Phytophysiol. sin. 8: 197–203, 1982.

Liang, G. H., Dayton, A. D., Chu, C. C., Casady, A. J.: Heritability of stomatal density and distribution on leaves of grain sorghum. — Crop Sci. 15: 567–570, 1975.

Lichtenthaler, H. K.: Verbreitung und relative Konzentration der lipophilen Plastidenchinone in grünen Pflanzen. — Planta 81: 140–152, 1968a.

Lichtenthaler, H. K.: Plastoglobuli and the fine structure of plastids. — Endeavour 27: 144–149, 1968b.

Lichtenthaler, H. K.: Plastoglobuli und Lipochinongehalt der Chloroplasten von *Cereus peruvianus* (L.) Mill. — Planta 87: 304–310, 1969a.

Lichtenthaler, H. K.: Die Bildung überschüssiger Plastidenchinone in den Blättern von *Ficus elastica* Roxb. — Z. Naturforsch. 24b: 1461–1466, 1969b.

Lichtenthaler, H. K.: Plastoglobuli und Lipochinongehalt der Chloroplasten von *Cereus peruvianus* (L.) Mill. — Planta 87: 304–310, 1969c.

Lichtenthaler, H. K.: Die unterschiedliche Synthese der lipophilen Plastidenchinone in Sonnen- und Schattenblättern von *Fagus silvatica* L. — Z. Naturforsch. 26b: 832–842, 1971.

Lichtenthaler, H. K.: Zur Synthese fettlöslicher Vitamine und Lipochinone in pflanzlichen Geweben. – Deut. med. Wochenschr. 2: 105–106, 1972.

Lichtenthaler, H. K.: Regulation of prenylquinone synthesis in higher plants. — In: Tevini, M., Lichtenthaler, H. K. (ed.): Lipids and Lipid Polymers in Higher Plants. Pp. 231–258. Springer-Verlag, Berlin–Heidelberg –New York 1977.

Lichtenthaler, H. K.: Adaptation of leaves and chloroplasts to high quanta fluence rates. — In: Akoyunoglou, G . (ed.): Photosynthesis. Vol. VI. Pp. 273–287. Balaban Int. Sci. Serv., Philadelphia 1981.

Lichtenthaler, H. K., Burkard, G., Kuhn, G., Prenzel, U.: Light-induced accumulation and stability of chlorophylls and chlorophyll-proteins during chloroplast development in radish seedlings. — Z. Naturforsch. 36c: 421–430, 1981a.

Lichtenthaler, H. K., Buschmann, C., Döll, M., Fietz, H.-J., Bach, T., Kozel, U., Meier, D., Rahmsdorf U.: Photosynthetic activity, chloroplast ultrastructure, and leaf characteristics of high-light and low-light plants and of sun and shade leaves. — Photosynthesis Res. 2: 115–141, 1981b.

Lichtenthaler, H. K., Buschmann, C., Rahmsdorf, U.: The importance of blue light for the development of sun-type chloroplasts. — In: Senger, J. (ed.): The Blue Light Syndrome. Pp. 485–494. Springer-Verlag, Berlin–Heidelberg–New York 1980.

Lichtenthaler, H. K., Peveling, E.: Plastoglobuli und osmiophile cytoplasmatische Lipideneinschlüsse in grünen Blättern von *Hoya carnosa* R. Br. — Z. Pflanzenphysiol. 56: 153–165, 1967.

Lichtenthaler, H. K., Weinert, H.: Die Beziehungen zwischen Lipochinonsynthese und Plastoglobulibildung in den Chloroplasten von *Ficus elastica* Roxb. — Z. Naturforsch. 25b: 619–623, 1970.

Lieth, H., Whittaker, R. H. (ed.): Primary Productivity of the Biosphere. — Springer-Verlag, Berlin–Heidelberg–New York 1975.

Liljenberg, C.: Characterization and properties of a protochlorophyllide ester in leaves of dark grown barley with geranylgeraniol as esterifying alcohol. — Physiol. Plant. 32: 208–213, 1974.

Lin, Z. F., Ehleringer, J.: Effects of leaf age on photosynthesis and water use efficiency of papaya. — Photosynthetica 16: 514–519, 1982.

Linder, S.: Seasonal variation of pigments in needles. A study of Scots pine and Norway spruce seedlings grown under different nursery conditions. — Stud. forest. suec. 100: 1–37, 1972.

Linder, S.: Chlorophyll as an indicator of nitrogen status of coniferous seedlings. – New Zeal. J. Forest. Sci. 10: 166–175, 1980.

Linder, S., Troeng, E.: Photosynthesis and transpiration of 20-year-old Scots pine. — Ecol. Bull. (Stockholm) 32 [Persson, T. (ed.): Structure and Function of Northern Coniferous Forests — An Ecosystem Study]: 165–181, 1980.

Lipskaya, G. A.: Ab spetsyfitsy dzeyannya adnol'kavaï kantsêntratsyi kobal'tu na fotasintêtychny aparat roznykh raslin. [Characteristics of the effect of the same concentration of cobalt on the photosynthetic apparatus of different plants.] — Vestsi Akad. Navuk belarus. SSR, Ser. biyal. Navuk 1971 (1): 14–20, 133–134, 1971.

Lipskaya, G. A.: Kobal't i Strukturno-Funktsional'naya Organizatsiya Lista. [Cobalt and Structural and Functional Organization of a Leaf.] — Izd. BGU, Minsk 1980.

Little, C. H. A., Loach, K.: Effect of gibberellic acid on growth and photosynthesis in Abies balsamea. — Can. J. Bot. 53: 1805–1810, 1975.

Littleton, E. J., Dennett, M. D., Elston, J., Monteith, J. L.: The growth and development of cowpeas (Vigna unguiculata) under tropical field conditions 1. Leaf area. — J. agr. Sci. 93: 291–307, 1979.

Littleton, E. J., Dennett, M. D., Elston, J., Monteith, J. L.: The growth and development of cowpeas (Vigna unguiculata) under tropical field conditions. 3. Photosynthesis of leaves and pods. — J. agr. Sci. 97: 539–550, 1981.

Ljubešić, N.: Feinbau der Chloroplasten während der Vergilbung und Wiederergrünung der Blätter. — Protoplasma 66: 369–379, 1968.

Lloyd, E. J.: The influence of shading on enzyme activity in seedling leaves of barley. — Z. Pflanzenphysiol. 78: 1–12, 1976.

Lloyd, E. J.: The effects of leaf age and senescence on the distribution of carbon in Lolium temulentum. — J. exp. Bot. 31: 1067–1079, 1980.

Lloyd, N. D. H., Woolhouse, H. W.: Leaf resistances in different populations of Sesleria caerulea (L.) Ard. — New Phytol. 80: 79–85, 1978.

Lof, H.: Water use efficiency and competition between arid zone annuals especially the grasses Phalaris minor and Hordeum murinum. — Agr. Res. Rep. (Wageningen) 853: 1–107, 1976.

Lott, J. N. A.: Changes in the cotyledons of Cucurbita maxima during germination. III. Plastids and chlorophylls. — Can. J. Bot. 48: 2259–2265, 1970.

Lovell, P. H., Moore, K. G.: A comparative study of cotyledons as assimilatory organs. — J. exp. Bot. 21: 1017–1030, 1970.

Lozova, G. I.: Stan pigmentiv plastyd kukurudzy riznogo genetychnogo pokhodzhennya. [State of plastid pigments of maize of various genetic origins.] — In: Pytannya Fiziologiï, Tsitoémbriologiï i Flory Ukraïny. Pp. 37–45. Kyïv 1963.

Ludlow, M. M.: Effect of oxygen concentration on leaf photosynthesis and resistances to carbon dioxide diffusion. — Planta 91: 285–290, 1970.

Ludlow, M. M.: Effect of water stress on the decline of leaf net photosynthesis with age. — In: Marcelle, R. (ed.): Environmental and Biological Control of Photosynthesis. Pp. 123–134. Dr. W. Junk, The Hague 1975.

Ludlow, M. M.: Measurement of stomatal conductance and plant water status. — In: Coombs, J., Hall, D. O. (ed.): Techniques in Bioproductivity and Photosynthesis. Pp. 44–57. Pergamon Press, Oxford–New York–Toronto–Sydney–Paris–Frankfurt 1982.

Ludlow, M. M., Jarvis, P. G.: Methods for measuring photorespiration in leaves. — In: Šesták, Z., Čatský, J., Jarvis, P. G. (ed.): Plant Photosynthetic Production. Manual of Methods. Pp. 294–315. Dr. W. Junk N. V. Publ., The Hague 1971.

Ludlow, M. M., Wilson, G. L.: Photosynthesis of tropical pasture plants. III. Leaf age. — Aust. J. biol. Sci. 24: 1077–1087, 1971.

Ludwig, L. J., Saeki, T., Evans, L. T.: Photosynthesis in artificial communities of cotton plants in relation to leaf area. I. Experiments with progressive defoliation of mature plants. — Aust. J. biol. Sci. 18: 1103–1118, 1965.

Lugg, D. G., Sinclair, T. R.: Variation in stomatal density with leaf position in field-grown soybeans. — Crop Sci. 19: 407–409, 1979.

Lugg, D. G., Sinclair, T. R.: Seasonal changes in photosynthesis of field-grown soybean leaflets 1. Relation to leaflet dimensions. — Photosynthetica 15: 129–137, 1981.

Lurie, S.: Stomatal development in etiolated *Vicia faba*: Relationship between structure and function. — Aust. J. Plant Physiol. 4: 61–68, 1977.

Lurie, S., Paz, N., Struch, N., Bravdo, B. A.: Effect of leaf age on photosynthesis and photorespiration. —In: Marcelle, R., Clijsters, H., Van Poucke, M. (ed.): Photosynthesis and Plant Development. Pp. 31–38. Dr. W. Junk bv Publ., The Hague–Boston–London 1979.

Lüttge, U., Kramer, D., Ball, E.: Photosynthesis and apparent proton fluxes in intact cells of greening etiolated barley and maize leaves. — Z. Pflanzenphysiol. 71: 6–21, 1974.

Lyubimenko, V.: O prevrashcheniyakh pigmentov plastid v zhivoï tkani rasteniya. [Changes in plastid pigments in the living tissue of plant.] — Zap. imper. Akad. Nauk Petrograd, Ser. 8, Otd. fiz. mat. 33 (12): 1–275, 1916.

Lyubimenko, V. N.: Soderzhanie khlorofilla v khlorofillnom zerne i énergiya fotosinteza. [Chlorophyll content in a chloroplast and energy of photosynthesis.] — Tr. SPb. Obshch. Estestvozn., Ser. 3 — bot. 41: 1–266, 1910.


Ma, P., Hunt, L. A.: Photosynthesis of newly matured leaves during the ontogeny of barley grown at different nutrient levels. — Can. J. Bot. 53: 2389–2398, 1975.

Ma, P. C., Hunt, L. A.: Effects of two photoperiod treatments on leaf photosynthesis throughout ontogeny in "Fergus" barley. — Can. J. Bot. 61: 792—797, 1983.

Maag, H. P., Nösberger, J.: Photosynthetic rate, chlorophyll content and dry matter production of *Trifolium pratense* L. as influenced by nitrogen nutrition. — Angew. Bot. 54: 187–194, 1980.

Maciejewska, U.: The effect of embryonal axis on the development of photosynthetic activity in apple seedlings. — New Phytol. 82: 81–88, 1979.

Mader, P., Nauš, J., Makovec, P., Kupka, J., Novák, V., Gréc, L.: Effect of Nitrogen Nutrition on the Photophysical, Photochemical and Photosynthetic Activities of the Spring Barley Autotrophic Apparatus. — Praha 1978.

Mader, P., Nauš, J., Schmidt, O., Chladová, J., Chlad, F.: Pigment-protein composition and fluorescence activity of the spring barley photosynthetic apparatus. Effect of ontogeny and nitrogen nutrition. — Photosynthetica 15: 61–74, 1981.

Madgwick, H. A. I., Olah, F. D., Burkhart, H. E.: Biomass of open-grown Virginia pine. — Forest Sci. 23: 89–91, 1977.

Madsen, E.: Effect of $CO_2$-Concentration on Morphological, Histological, Cytological and Physiological Processes in Tomato Plants. — State Seed Testing Station, Lyngby 1976.

Mae, T., Ohira, K.: The remobilization of nitrogen related to leaf growth and senescence in rice plants (*Oryza sativa* L.). — Plant Cell Physiol. 22: 1067–1074, 1981.

Maier-Maercker, U.: "Peristomatal transpiration" and stomatal movement: A controversial view. VII. Correlation of stomatal aperture with evaporative demand and water uptake through the roots. — Z. Pflanzenphysiol. 102: 397–413, 1981.

Majid, M. A., Shaikh, M. A. Q., Begum, S., Ahmed, Z. U.: Genotypic variability for frequency, distribution and size of stomata in jute (*Corchorus capsularis* L.). — Beitr. Biol. Pflanz. 54: 399–406, 1978.

Makrides, S. C., Goldthwaite, J.: Biochemical changes during bean leaf growth, maturity, and senescence. Content of DNA, polyribosomes, ribosomal RNA, protein, and chlorophyll. — J. exp. Bot. 32: 725–735, 1981.

Maksymowych, R.: Analysis of Leaf Development. — Cambridge Univ. Press, Cambridge 1973.

Malkin, S., Fork, D. C.: Photosynthetic units of sun and shade plants. — Plant Physiol. 67: 580–583, 1981.

Malkina, I. S.: O stepeni uchastiya palisadnoï parenkhimy v fotosinteze list'ev klena ostrolistnogo i berezy povisloï. [Degree of palisade parenchyma participation in photosynthesis of *Acer platanoides* and *Betula pendula* leaves.] — Lesovedenie 1976 (2): 51–57, 1976a.

Malkina, I. S.: Izmenenie svetovykh krivykh fotosinteza s vozrastom lista klena ostrolistnogo. [Changes in light curves of photosynthesis with ageing of the leaf of Norway maple.] — Fiziol. Rast. 23: 247–253, 1976b.

Malkina, I. S.: Opredelenie intensivnosti fotosinteza v krone vzroslykh derev'ev. [Determination of photosynthetic rate in the crown of mature oak trees.] — Fiziol. Rast. 25: 792–797, 1978a.

Malkina, I. S.: Fotosintez v krone vzroslogo dereva. [Photosynthesis in the crown of a mature tree.] — Lesovedenie 1978 (1): 78–85, 1978b.

Mangat, B. S., Levin, W. B., Bidwell, R. G. S.: The extent of dark respiration in illuminated leaves and its control by ATP levels. — Can. J. Bot. 52: 673–681, 1974.

Manning, C. E., Miller, D. G., Teare, I. D.: Effect of moisture stress on leaf anatomy and water-use efficiency of peas. — J. amer. Soc. hort. Sci. 102: 756–760, 1977.

Mantai, K. E., Wong, J., Bishop, N. I.: Comparison studies on the effects of ultraviolet irradiation on photosynthesis. — Biochim. biophys. Acta 197: 257–266, 1970.

Marcelle, R. (ed.): Environmental and Biological Control of Photosynthesis. — Dr. W. Junk b.v. Publ., The Hague 1975.

Marcelle, R., Clijsters, H., Van Poucke, M. (ed.): Photosynthesis and Plant Development. — Dr. W. Junk b.v. Publ., The Hague–Boston – London 1979.

Marco, G. Di, Grego, S., Pietrosanti, T., Tricoli, D.: Seasonal trends of nitrate reductase, carboxylating enzymes, and water-soluble proteins in two field-grown cultivars of Triticum. — J. exp. Bot. 27: 725–734, 1976.

Marco, G. Di, Grego, S., Tricoli, D.: RuBP carboxylase-oxygenase in field-grown wheat. — J. exp. Bot. 30: 851–861, 1979.

Mares, D. J., Coote, M. A., Possingham, J. V.: Membrane-bound plastid inclusions and chloroplast thylakoid formation in sunflower (Helianthus annuus L.). — Ann. Bot. 43: 191–196, 1979.

Marini, R. P., Barden, J. A.: Seasonal correlations of specific leaf weight to net photosynthesis and dark respiration of apple leaves. — Photosynth. Res. 2: 251–258, 1981.

Mark, H. E., Sweeney, B. M.: Circadian rhythm of chloroplast ultrastructure in Gonyaulax polyedra, concentric organization around a central cluster of ribosomes. — J. Ultrastruct. Res. 50: 347–354, 1975.

Marks, T. C., Taylor, K.: The carbon economy of Rubus chamaemorus L. I. Photosynthesis. — Ann. Bot. 42: 165–179, 1978.

Maróti, I.: Photosynthetical pigments in the spongy and palisade parenchymas and the alternative ways of photosynthesis. — Acta biol. (Szeged) 22 (1–4): 7–14, 1976.

Maróti, J., Gábor, G.: Thylakoid aggregation and pigment ratios in the spongy and palisade parenchymas. — Acta biol. (Szeged) 22 (1–4): 15–27, 1976.

Marshall, B., Biscoe, P. V.: A model for C3 leaves describing the dependence of net photosynthesis on irradiance. I. Derivation . — J. exp. Bot. 31: 29–39, 1980a.

Marshall, B., Biscoe, P. V.: A model for C3 leaves describing the dependence of net photosynthesis on irradiance II. Application to the analysis of flag leaf photosynthesis. — J. exp. Bot. 31: 41–48, 1980b.

Martin, B., Öquist, G.: Seasonal and experimentally induced changes in the ultrastructure of chloroplasts of Pinus silvestris. — Physiol. Plant. 46: 42–49, 1979.

Maslow, M.: Phosphorylation photosynthétique des chloroplastes isolés de Bryophyllum. — Physiol. vég. 2: 209–220, 1964.

Maximov, N. A.: The Plant in Relation to Water. A Study of the Physiological Basis of Drought Resistance. — Allen and Unwin Ltd., London 1929.

McCashin, B. G., Canvin, D. T.: Photosynthetic and photorespiratory characteristics of mutants of Hordeum vulgare L. — Plant Physiol. 64: 354–360, 1979.

McCree, K. J.: An equation for the rate of respiration of white clover plants grown under controlled conditions. — In: Prediction and Measurement of Photosynthetic Productivity. Pp. 221–229. PUDOC, Wageningen 1970.

McCree, K. J.: The action spectrum, absorptance and quantum yield of photosynthesis in crop plants. — Agr. Meteorol. 9: 191–216, 1971/1972.

McCree, K. J.: Maintenance requirements of white clover at high and low growth rates. — Crop Sci. 22: 345–351, 1982a.

McCree, K. J.: The role of respiration in crop production. — Iowa State J. Res. 56: 291–306, 1982b.

McCree, K. J., Davis, S. D.: Effect of water stress and temperature on leaf size and on size and number of epidermal cells in grain sorghum. — Crop Sci. 14: 751–755, 1974.

McCree, K. J., Keener, M. E.: Effect of atmospheric turbidity on the photosynthetic rates of leaves. — Agr. Meteorol. 13: 349–357, 1974.

McLaren, A. D., Luse, R. A.: Mechanism of inactivation of enzyme protein by ultraviolet light. — Science 134: 836, 1961.

McPherson, H. G., Slatyer, R. O.: Mechanisms regulating photosynthesis in *Pennisetum typhoides*. — Aust. J. biol. Sci. 26: 329–339, 1973.

Medeghini-Bonatti, P., Bonetta Conte, M. D.: Ultrastrutture plastidiali in foglioline di gemme di *"Picea excelsa"* e di *"Larix decidua"* nel corso della germogliazione. [Ultrastructural modifications of plastids in leaflets of *Picea excelsa* and *Larix decidua* during sprouting of buds.] — G. bot. ital. 110: 9–20, 1976.

Medina, E.: Relationships between nitrogen level, photosynthetic capacity, and carboxydismutase activity in *Atriplex patula* leaves. — Carnegie Inst. Year Book 69: 655–662, 1971.

Medina, E., Delgado, M.: Photosynthesis and night $CO_2$ fixation in *Echeveria columbiana* v. Poellnitz. — Photosynthetica 10: 155–163, 1976.

Meidner, H.: Water supply, evaporation, and vapour diffusion in leaves. — J. exp. Bot. 26: 666 to 673, 1975.

Meidner, H., Mansfield, T. A.: Stomatal responses to illumination. — Biol. Rev. 40: 483–509, 1965.

Meier, D., Lichtenthaler, H. K.: Ultrastructural development of chloroplasts in radish seedlings grown at high- and low-light conditions and in the presence of the herbicide bentazon. — Protoplasma 107: 195–207, 1981.

Meier, D., Lichtenthaler, H. K., Burkard, G.: Change of chloroplast ultrastructure in radish seedlings under the influence of the photosystem II herbicide bentazon. — Z. Naturforsch. 35c: 656–664, 1980.

Meinl, G., Bellmann, K.: Untersuchungen über die Photosynthese, Respiration und Transpiration des Maises unter Berücksichtigung von Unterschieden zwischen Populationen, Pflanzen, Blättern und Blattabschnitten. — Biol. Plant. 7: 41–57, 1965.

Meister, A.: Messung von Absorptionsspektren *in vivo*. — Kulturpflanze 25: 141–154, 1977.

Menke, W.: Das allgemeine Bauprinzip des Lamellarsystem der Chloroplasten. — Experientia 16: 537—538, 1960.

Meyer, W. S., Walker, S.: Leaflet orientation in water-stressed soybeans. — Agron. J. 73: 1071 to 1074, 1981.

Miedema, P., Sinnaeve, J.: Photosynthesis and respiration of maize seedlings at suboptimal temperatures. — J. exp. Bot. 31: 813–819, 1980.

Migahid, A. M., Abu Raya, M. A.: Studies in stomatal frequency I. Stomatal frequency in relation to position on the leaf. — Bull. Inst. Fouad Ier Désert 2: 40–47, 1952a.

Migahid, A. M., Abu Raya, M. A.: Studies in stomatal frequency II. Stomatal frequency in relation to position of leaf upon the plant. — Bull. Inst. Fouad Ier Désert 2: 48–59, 1952b.

Migahid, A. M., Abu Raya, M. A.: Studies in stomatal frequency III. Analysis of factors affecting the distribution of stomata among the leaves of a plant. — Bull. Inst. Fouad Ier Désert 2: 60–63, 1952c.

Migahid, A. M., Abu Raya, M. A.: Studies in stomatal frequency IV. The significance of variation in stomatal frequency. — Bull. Inst. Fouad Ier Désert 2: 64–71, 1952d.

Migahid, A. M., Abu Raya, M. A.: Studies in stomatal frequency V. The interrelation of stomata frequency and leaf water content. — Bull. Inst. Fouad Ier Désert 2: 72–83, 1952e.

Migus, W. N., Hunt, L . A.: Gas exchange rates and nitrogen concentrations in two winter wheat cultivars during the grain-filling period. — Can. J. Bot. 58: 2110–2116, 1980.

Millar, B. D., Denmead, O. T.: Water relations of wheat leaves in the field. — Agron. J. 68: 303–307, 1976.

Miller, E. C.: Plant Physiology. - McGraw-Hill Book Co., New York — London 1931.

Miller, J. H.: The effect of growth conditions and the stage of leaf development on the Hill reaction in homogenates of *Pisum sativum* leaves. — Amer. J. Bot. 47: 532–540, 1960.

Miller, M. M., Nobel, P. S.: Light-induced changes in the ultrastructure of pea chloroplasts *in vivo*. Relationship to development and photosynthesis. — Plant Physiol. 49: 535–541, 1972.

Millerd, A., Simon, M., Stern, H.: Legumin synthesis in developing cotyledons of *Vicia faba* L. — Plant Physiol. 48: 419–425, 1971.

Milthorpe, F. L. (ed.): The Growth of Leaves. — Butterworths Sci. Publ., London 1956.

Milthorpe, F. L.: Quantitative aspects of leaf growth. — In: Sunderland, N. (ed.): Perspectives in Experimental Biology. Vol. 2. Pp. 33–40. Pergamon Press, Oxford–New York–Toronto–Sydney––Paris–Braunschweig 1976.

Milthorpe, F. L., Penman, H. L.: The diffusive conductivity of the stomata of wheat leaves. — J. exp. Bot. 18: 422–457, 1967.

Miranda, V., Baker, N. R., Long, S. P.: Anatomical variation along the length of the *Zea mays* leaf in relation to photosynthesis. — New Phytol. 88: 595–605, 1981a.

Miranda, V., Baker, N . R., Long, S. P.: Limitations of photosynthesis in different regions of the *Zea mays* leaf. — New Phytol. 89: 179–190, 1981b.

Miskin, K. E., Rasmusson, D. C.: Frequency and distribution of stomata in barley. — Crop Sci. 10: 575–578, 1970.

Mitrofanov, B. A., Gulyaev, B. I., Makhovskaya, M. A., Lavrentovich, D. I., Pochinok, Kh. N., Okanenko, A. S.: Rol' list'ev, steblei i kolos'ev ozimoi pshenitsy v fotosinteze poseva. [Role of leaves, stems and ears in photosynthesis of a winter wheat stand.] — In: Puti Povysheniya Intensivnosti i Produktivnosti Fotosinteza. Vol. 3. Pp. 69–86. Naukova Dumka, Kiev 1969.

Miura, K., Osada, A.: Effect of shading on photosynthesis, respiration, leaf area and corm weight in konjak plants (*Amorphophallus konjac* K. Koch). — Jap. J. Crop Sci. 50: 553–559, 1981.

Miyake, H., Maeda, E.: Starch accumulation in bundle sheath chloroplasts during the leaf development of C3 and C4 plants of the *Gramineae*. — Can. J. Bot. 56: 880–882, 1978.

Młodzianowski, F., Młodzianowska, L.: Chloroplast degeneration and its inhibition by kinetin in detached leaves of *Cichorium intybus* L. — Acta Soc. Bot. Pol. 42: 649–656, 1973.

Mohr, H.: Phytochrome and chloroplast development. — Endeavour, new Ser. 1: 107–141, 1977.

Mohr, H.: Control of chloroplast development by light — some recent aspects. — In: Akoyunoglou, G. (ed.): Photosynthesis. Vol. V. Pp. 869–889. Balaban Int. Sci. Serv., Philadelphia 1981.

Mokronosov, A. T.: Transport assimilyatov kak faktor endogennoi regulyatsii fotosinteza. [Transport of asimilates as a factor in the endogenous regulation of photosynthesis.] — Tr. biol.-pochv. Inst., nov. Ser. 20 (Transport Assimilyatov i Otlozhenie Veshchestv v Zapas u Rastenii): 76–84, 1973.

Mokronosov, A. T.: Mezostruktura i funktsional'naya aktivnost' fotosinteticheskogo apparata. [Mesostructure and functional activity of the photosynthetic apparatus.] — In: Mezostruktura i Funktsional'naya Aktivnost' Fotosinteticheskogo Apparata. Pp. 5–30. Ural'. gos. Univ., Sverdlovsk 1978.

Mokronosov, A. T.: Ontogeneticheskii Aspekt Fotosinteza. [Ontogenetic Aspect of Photosynthesis.] — Nauka, Moskva 1981.

Mokrosonov, A. T., Bagautdinova, R. I.: Dinamika khloroplastov v list'yakh kartofelya. [Dynamics of chloroplasts in potato leaves.] — Fiziol. Rast. 21: 1132–1138, 1974.

Mokronosov. A. T., Bagautdinova, R. I., Bubnova, E. A., Kobeleva, I. V.: Fotosinteticheskii metabolizm v palisadnoĭ i gubchatoĭ tkanyakh lista. [Photosynthetic metabolism in palisade and spongy leaf tissues.] — Fiziol. Rast. 20: 1191–1197, 1973a.

Mokronosov, A. T., Bagautdinova, R. I., Fedoseeva, G. P., Nekrasova, G. F., Borzenkova, R. A., Nazarov, S. K.: Strukturnaya i funktsional'naya dinamika lista v ontogeneze. [Structural and functional dynamics of a leaf during ontogenesis.] — In: Voprosy Regulyatsii Fotosinteza. Vol. 3. Pp. 3—44, 161. Ural'sk. gos. Univ., Sverdlovsk 1973b.

Mokronosov, A. T., Nekrasova, G. F.: Ontogeneticheskiĭ aspekt fotosinteza (na primere lista kartofelya). [Ontogenetic aspect of photosynthesis studied with potato leaf.] — Fiziol. Rast. 24: 458–465, 1977.

Moll, A., Henniger, W.: Genotypische Photosyntheserate von Kartoffeln und ihre mögliche Rolle für die Ertragsbildung. — Photosynthetica 12: 51–61, 1978.

Möller, G., Stamp, P., Geisler, G.: Fotometrische Messung der PEP-Carboxylase-Aktivität in Maisblättern unter Berücksichtigung des Entwicklungszustandes der Pflanze. — Z. Pflanzenernähr. Bodenk. 140: 481–490, 1977.

Molyaka, O. N., Marunchenko, Yu. M., Romodan, V. N.: Biologiya, produktyvnist' ta khimichni osoblyvosti vydiv rodu Typha L. Kremenchuts'kogo vodoĭmyshcha. [Biology, productivity and chemical properties of Typha species from the Kremenchug reservoir.] — In: Roslynni Resursy Ukraïny, Ïkh Vyvchennya ta Ratsional'na Vykorystannya. Pp 78–83, 205. Naukova Dumka, Kyïv 1973.

Monteith, J. L.: Gas exchange in plant communities. — In: Evans, L. T. (ed.): Environmental Control of Plant Growth. Pp. 95–112. Academic Press, New York–London 1963.

Monteith, J. L.: Principles of Environmental Physics. — Edward Arnold, London 1973.

Mooney, H. A., Field, C., Gulmon, S. L., Bazzaz, F. A.: Photosynthetic capacity in relation to leaf position in desert versus old-field annuals. — Oecologia 50: 109–112, 1981.

Moorby, J., Munns, R., Walcott, J.: Effect of water deficit on photosynthesis and tuber metabolism in potatoes. — Aust. J. Plant Physiol. 2: 323–333, 1975.

Moore, K. G.: Changes in leaf composition in Parthenocissus tricuspidata Planch. during growth and senescence of short shoots. — Ann. Bot. 39: 631–637, 1975.

Moore, P.: The varied ways plants tap the Sun. — New Scient. 89: 394–397, 1981.

Moore, R. T., Miller, P. C., Ehleringer, J., Lawrence, W.: Seasonal trends in gas exchange characteristics of three mangrove species. — Photosynthetica 7: 387–394, 1973.

Morgan, D. C., Smith, H.: Non-photosynthetic responses to light quality. — In: Lange, O. L., Nobel, P. S., Osmond, C. B., Ziegler, H. (ed.): Physiological Plant Ecology I. Responses to the Physical Environment. Pp. 109–134. Springer-Verlag, Berlin–Heidelberg–New York 1981.

Morot-Gaudry, J.-F., Bethenod, O. Chartier, M., Chartier, P.: Photosynthèse comparée d'un Maïs normal (W 64 A) et d'un Maïs mutant opaque 2 (W 64 A o₂). — Physiol. vég. 14: 595 to 606, 1976.

Morot-Gaudry, J. F., Farineau, J., Jolivet, E.: Effect of leaf position and plant age on photosynthetic carbon metabolism in leaves of 8·and 16 day-old maize seedlings (W 64 A) with and without the gene opaque 2. — Photosynthetica 13: 365–375, 1979.

Morozov, V. L.: Struktura assimilyatsionnogo apparata dominantov kamchatskogo krupnotrav'ya. [Structure of the assimilatory apparatus of tall herb dominants in Kamchatka.] — Izv. sib. Otd. Akad. Nauk SSSR, Ser. biol. Nauk 1978 (2): 36–42, 1978.

Morozov, V. L.: Produktsionnaya deyatel'nost' kamchatskogo krupnotrav'ya. I. Geometricheskaya struktura i fotosintez dominantov. [The production activity of the Kamchatka tall herbaceous vegetation. I. Geometrical structure and photosynthesis of dominants.] — Byull. mosk. Obshch. Ispyt. Prir., Otd. Biol. 85 (2): 68–78, 1980.

Moss, D. N.: Optimum lighting of leaves. — Crop Sci. 4: 131–136, 1964.

Moss, D. N.: Carbon dioxide compensation in plants with C₄ characteristics. — In: Hatch, M. D., Osmond, C. B., Slatyer, R. O. (ed.): Photosynthesis and Photorespiration. Pp. 120–123. Wiley--Interscience, New York–London–Sydney–Toronto 1971.

Mühlethaler, K.: Introduction to structure and function of the photosynthesis apparatus. — In: Trebst, A., Avron, M. (ed.): Photosynthesis I. Pp. 503–521. Springer-Verlag, Berlin–Heidelberg–New York 1977.

Mühlethaler, K., Frey-Wyssling, A.: Entwicklung und Struktur der Proplastiden. — J. biophys. biochem. Cytol. 6: 507–512, 1959.

Murakami, T.: [Studies on the photosynthetic rate of mulberry plant: I. The variation of photosynthetic rates of leaves with leaf order in various growing periods.] — Bull. sericult. Exp. Sta. (Tokyo) 27: 353–368, 1978.

Murakami, T.: Characteristics of photosynthesis in mulberry leaves. — Jap. agr. Res. Quart. 16: 46–50, 1982.

Murakami, T., Takeda, T.: [Effect of light intensity on the photosynthetic rate in mulberry leaves.] — J. Sericult. Sci. Jap. 42: 417–424, 1973.

Mustárdy, L. A., Brangeon, J.: 3-dimensional chloroplast infrastructure: developmental aspects. — In: Akoyunoglou, G., Argyroudi-Akoyunoglou, J. H. (ed.): Chloroplast Development. Pp. 489–494. Elsevier/North Holland Biomedical Press, Amsterdam–New York–Oxford 1978.

Mutsaers, H. J. W.: Leaf growth in cotton (*Gossypium hirsutum* L.) 1. Growth in area of main-stem and sympodial leaves. — Ann. Bot. 51: 503—520, 1983.

Nadler, K., Granick, S.: Controls of chlorophyll synthesis in barley. — Plant Physiol. 46: 240–246, 1970.

Nadler, K. D., Herron, H. A., Granick, S.: Development of chlorophyll and Hill activity. — Plant Physiol. 49: 388–392, 1972.

Nagarajah, S.: Effect of debudding on photosynthesis in leaves of cotton. — Physiol. Plant. 33: 28–31, 1975a.

Nagarajah, S.: The relation between photosynthesis and stomatal resistance of each leaf surface in cotton leaves. — Physiol. Plant. 34: 62–66, 1975b.

Nagarajah, S.: The effects of increased illumination and shading on the low–light-induced decline in photosynthesis in cotton leaves. — Physiol. Plant. 36: 338–342, 1976.

Naito, K., Iida, A., Suzuki, H., Tsuji, H.: The effect of benzyladenine on changes in nuclease and protease activities in intact bean leaves during ageing. — Physiol. Plant. 46: 50–53, 1979.

Naito, K., Tsuji, H., Hatakeyama, I.: Effect of benzyladenine on DNA, RNA, protein, and chlorophyll contents in intact bean leaves: differential responses to benzyladenine according to leaf age. — Physiol. Plant. 43: 367–371, 1978.

Naito, K., Tsuji, H., Hatakeyama, I., Ueda, K.: Benzyladenine-induced increase in DNA content per cell, chloroplast size, and chloroplast number per cell in intact bean leaves. — J. exp. Bot. 30: 1145–1151, 1979.

Naito, K., Ueda, K., Tsuji, H.: Differential effects of benzyladenine on the ultrastructure of chloroplasts in intact bean leaves according to their age. — Protoplasma 105: 293–306, 1981.

Napp-Zinn, K.: Anatomie des Blattes II. Blattanatomie der Angiospermen. A, 1 (Handbuch der Pflanzenanatomie Vol. VIII). — Gebrüder Borntraeger, Berlin–Stuttgart 1973.

Napp-Zinn, K.: Anatomie des Blattes II. Blattanatomie der Angiospermen. A, 2 (Handbuch der Pflanzenanatomie Vol. VIII). — Gebrüder Borntraeger, Berlin–Stuttgart 1974.

Nasrulhaq-Boyce, A., Jones, O. T. G.: Cytochromes of developing plastids of greening barley: effects of inhibitors of haem synthesis. — Phytochemistry 20: 1197–1199, 1981.

Nátr, L.: Influence du déficit en éléments minéraux sur la production de matière sèche, l'intensité de la photosynthèse et la quantité de N, P et K, dans les plantes. — Physiol. vég. 8: 573–583, 1970.

Nátr, L.: Influence of mineral nutrition on photosynthesis and the use of assimilates. — In: Cooper, J. P. (ed.): Photosynthesis and Productivity in Different Environments. Pp. 537–555. Cambridge Univ. Press, Cambridge–London–New York–Melbourne 1975.

Nátr, L., Kousalová, I., Kopecký, M., Vu Van Vu: Produkční potenciál a akumulační kapacita jarního ječmene. [Production potential and accumulation capacity of spring barley.] — Rostl. Výroba (Praha) 21: 419–427, 1975.

Naylor, D. G., Teare, I. D., Kanemasu, E. T.: Photosynthesis in field-grown sorghum. — Fyton 33: 97–102, 1975.

Neese, P.: Zur Kenntnis der Struktur der Niederblätter und Hochblätter einiger Laubhölzer. — Flora 109: 144–187, 1917.

Nekrasova, G. F.: Formirovanie struktury i fotosinteticheskoĭ funktsii v protsesse rosta lista. [Formation of structure and photosynthetic function during leaf growth.] — In: Mezostruktura i Funktsional'naya Aktivnost' Fotosinteticheskogo Apparata. Pp. 61–73. Ural'. gos. Univ., Sverdlovsk 1978.

Nesterenko, T.V., Sid'ko, F. Ya.: Induktsiya fluorestsentsii list'ev pshenitsy v ikh ontogeneze. [Induction of fluorescence of wheat leaves in their ontogeny.] — Fiziol. Rast. 27: 336–340, 1980.

Nestsyarovich, M. D., Panamarova, A. V., Dzyarugina, T. F.: Zmyanenne anatamichnaĭ budovy khvoi nekatorykh drevavykh parod u zalezhnastsi ad yae ŭzrostu i vyshyni prymatsavannya. [Changes in anatomical structure of needles in some woody plants as affected by age and insertion level.] — Vestsi Akad. Navuk belarus. SSR, Ser. biyal. Navuk 1963 (3): 5–13, 1963.

Nevins, D. J., Loomis, R. S.: Nitrogen nutrition and photosynthesis in sugar beet (Beta vulgaris L.). — Crop Sci. 10: 21–25, 1970.

Ng, P. A. P.: Response of Stomata to Environmental Variables in Pinus sylvestris L. — Ph.D. Thesis. Univ. Edinburgh 1978.

Niemann, G. J., Schulz, T. K. F., van Genderen, H. H., Baas, W. J.: Investigations on Hoya species. IV. Leaf phenolics and leaf-wax triterpenes of Hoya australis R. Br. ex Traill in relation to leaf age. — Z. Pflanzenphysiol. 97: 241–248, 1980.

Nilsen, S., Mortensen, L.: Spectral response of photorespiration effect of plant age and chlorophyll content in spruce. — Z. Pflanzenphysiol. 89: 433–441, 1978.

Nimbalkar, J. D., Joshi, G. V.: Physiological studies in senescent leaves of sugarcane var. Co. 740. — Indian J. exp. Biol. 13: 384–386, 1975.

Nishida, K.: Effects of internal and external factors on photosynthetic $^{14}CO_2$ fixation in general and on formation of $^{14}$C-maltose in Acer leaf in particular. — Physiol. Plant. 15: 47–58, 1962.

Nishida, K.: Effect of leaf age on light and dark $^{14}CO_2$ fixation in a CAM plant, Bryophyllum calycinum. — Plant Cell Physiol. 19: 935–941, 1978.

Nobel, P. S.: Light-induced chloroplast shrinkage in vivo detectable after rapid isolation of chloroplasts from Pisum sativum. — Plant Physiol. 43: 781–787, 1968.

Nobel, P. S.: Introduction to Biophysical Plant Physiology. — W. H. Freeman, San Francisco 1974.

Nobel, P. S., Longstreth, D. J.: Effects of environmental factors on leaf anatomy, mesophyll cell conductance, and photosynthesis. — In: Akoyunoglou, G. (ed.): Photosynthesis. Vol. VI. Pp. 245–254. Balaban Int. Sci. Serv., Philadelphia 1981.

Nobel, P. S., Zaragoza, L. J., Smith W. K.: Relation between mesophyll surface area, photosynthetic rate, and illumination level during development for leaves of Plectranthus parviflorus Henckel. — Plant Physiol. 55: 1067–1070, 1975.

Noodén, L. D.: Regulation of senescence. — In: Corbin, F. T. (ed.): World Soybean Research Conference II. Pp. 139–152. Westview Press, Boulder 1980.

Novitskaya, G. V., Rutskaya, L. A., Molotkovskiĭ, Yu. G.: Vozrastnye izmeneniya lipidnogo sostava i aktivnost' membran khloroplastov bobov. [Chages in lipid composition and activity of chloroplast membranes of broadbean during ageing.] — Fiziol. Rast. 24: 35–43, 1977.

Nunes, M. A.: Water relations in coffee. Significance of plant water deficits to growth and yield: A review. — J. Coffee Res. 6: 4–21, 1976.

Obendorf, R. L., Huffaker, R. C.: Influence of age and illumination on distribution of several Calvin cycle enzymes in greening barley leaves. — Plant Physiol. 45: 579–582, 1970.

Ochiai, H., Shibata, H.: Effect of 4-thiouridine on chloroplast development. — Agr. biol. Chem 34: 1751–1753, 1970.

Ochiai, H., Shibata, H., Suekane, T.: Chloroplast development in 4-thiouridine treated radish cotyledons. — Agr. biol. Chem. 35: 1259–1266, 1971a.

Ochiai, H., Shibata, H., Suekane, T., Kono, Y.: [Effect of 4-thiouridine on chloroplast development in a higher plant.] — Amino Acid nucl. Acid 24: 1–13, 1971b.

Ogawa, M., Tsutsui, Y., Konishi, M.: Effects of illumination on absorption peak shifts in spectra of intact etiolated cotyledons of *Pharbitis nil* II. Effects of leaf age on protochlorophyllide regeneration and the Shibata shift. — Plant Cell Physiol. 19: 127–132, 1978.

Ogura, N.: Studies on chlorophyllase of tea leaves II. Seasonal change of a soluble chlorophyllase. — Bot. Mag. (Tokyo) 82: 392–396, 1969.

Ojakian, G. K., Satir, P.: Particle movements in chloroplast membranes: quantitative measurements of membrane fluidity by the freeze-fracture technique. — Proc. nat. Acad. Sci. USA 71: 2052 to 2056, 1974.

Okubo, T., Kawanabe, S., Hoshino, M.: [Chlorophyll amount for analysis of matter production in forage crops. II. Seasonal variations in maximum crop growth rate and leaf photosynthesis, and their correlations with chlorophyll content in alfalfa and ladino clover.] — J. jap. Soc. Grassland Sci. 21: 124–135, 1975a.

Okubo, T., Kawanabe, S., Hoshino, M.: [Chlorophyll amount for analysis of matter production in forage crops. III. Leaf photosynthesis under dim light, maximum crop growth rate under different plant density, and their dependence on chlorophyll content in alfalfa.] — J. jap. Soc. Grassland Sci. 21: 136–145, 1975b.

Olesen, P.: Structure of chloroplast membranes as revealed by natural and experimental fixation with tannic acid: particles in and on the thylakoid membrane. — Biochem. Physiol. Pflanzen 172: 319–342, 1978.

Ongun, A., Stocking, C. R.: Effect of light on the incorporation of serine into the carbohydrates of chloroplasts and nonchloroplast fractions of tobacco leaves. — Plant Physiol. 40: 819–824, 1965.

Onwueme, I. C., Lawanson, A. O.: Chlorophyll accumulation in cowpea (*Vigna*) leaves and melon (*Colocynthis*) cotyledons as influenced by prior heat stress and seedling age. — Fyton 33: 69–73, 1975.

Öquist, G.: Chloroplast structure and photosynthetic efficiency. — In: Johnson, C. B. (ed.): Physiological Processes Limiting Plant Productivity. Pp. 53–80. Butterworths, London–Boston––Sydney–Wellington–Durban–Toronto 1981.

Öquist, G., Brunes, L., Hällgren, J. - E.: Photosynthetic efficiency during ontogenesis of leaves of *Betula pendula*. — Plant, Cell Environm. 5: 17–21, 1982.

Öquist, G., Liljenberg, C.: Lipid and fatty acid composition of chloroplast thylakoids isolated from *Betula pendula* leaves of different stages of development or acclimated to different quantum flux densities. — Z. Pflanzenphysiol. 104: 233–242, 1981.

Orsenigo, M., Rascio, N.: Chloroplast fine structure in the *japonica-2* maize mutant exposed to continuous illumination. 1. The green tissues. — Cytobios 16: 171–182, 1976.

Osakovskiĭ, V. L., Solomonova, T. N.: Vozrastnye izmeneniya funktsional'noĭ aktivnosti khloroplastov u yarovoĭ pshenitsy. [Growth changes of functional activity of chloroplasts of winter wheat.] — Izv. sib. Otd. Akad. Nauk SSSR, Ser. biol. Nauk 1980 (2): 117–121, 1980.

Osborne, D. J.: Effects of kinetin on protein & nucleic acid metabolism in *Xanthium* leaves during senescence. — Plant Physiol. 37: 595–602, 1962.

Osipova, O.: Ob izvlekaemosti khlorofilla iz zelenykh rasteniĭ. [Extractability of chlorophyll from green plants.] — Dokl. Akad. Nauk SSSR 57: 799–801, 1947.

Osipova, O. P., Ashur, N. I.: Struktura khloroplastov list'ev kukuruzy, vyrosshikh v raznykh usloviyakh osveshcheniya. [Structure of chloroplasts in maize leaves grown at various light conditions.] — Fiziol. Rast. 12: 257–262, 1965.

Osipova, O. P., Nikolaeva, M. K., Kheĭn, Kh. Ya.: K voprosu o deĭstvii khloramfenikola na fotosinteticheskiĭ apparat rasteniĭ. [Action of chloramphenicol on the photosynthetic apparatus of plants.] — Fiziol. Rast. 14: 210–218, 1967.

Osman, A. M., Goodman, P. J., Cooper, J. P.: The effects of nitrogen, phosphorus and potassium on rates of growth and photosynthesis of wheat. — Photosynthetica 11: 66–75, 1977.

Osman, A. M., Milthorpe, F. L.: Photosynthesis of wheat leaves in relation to age, illuminance and nutrient supply II. Results. — Photosynthetica 5: 61–70, 1971.

Osmond, C. B., Björkman, O.: Simultaneous measurements of oxygen effects on net photosynthesis and glycolate metabolism in $C_3$ and $C_4$ species of Atriplex. — Carnegie Inst. Year Book 71: 141–148, 1972.

Osmond, C. B., Holtum, J. A. M.: Crassulacean acid metabolism. — In: Hatch, M. D., Boardman, N. K. (ed.): The Biochemistry of Plants. Vol. 8. Photosynthesis. Pp. 283–328. Academic Press, New York–San Francisco–London 1981.

Osmond, C. B., Troughton, J. H., Goodchild, D. J.: Physiological, biochemical and structural studies of photosynthesis and photorespiration in two species of Atriplex. — Z. Pflanzenphysiol. 61: 218–237, 1969.

Osmond, C. B., Winter, K., Ziegler, H.: Functional significance of different pathways of $CO_2$ fixation in photosynthesis. — In: Lange, O. L., Nobel, P. S. Osmond, C. B., Ziegler, H. (ed.): Physiological Plant Ecology II. Water Relations and Carbon Assimilation. Pp. 479–547. Springer-Verlag, Berlin–Heidelberg–New York 1982.

O'Toole, J. C., Ludford, P. M., Ozbun, J. L.: Gas exchange and enzyme activity during leaf expansion in Phaseolus vulgaris L. — New Phytol. 78: 565–571, 1977.

Outlaw, W. H., Jr., Fisher, D. B.: Compartmentation in Vicia faba leaves. I. Kinetics of $^{14}C$ in the tissues following pulse labeling. — Plant Physiol. 55: 699—703, 1975a.

Outlaw, W. H., Jr., Fisher, D. B.: Compartmentation in Vicia faba leaves. III. Photosynthesis in the spongy and palisade parenchyma. — Aust. J. Plant Physiol. 2: 435–439, 1975b.

Outlaw, W. H., Jr., Fisher, D. B., Christy, A. L.: Compartmentation in Vicia faba leaves II. Kinetics of $^{14}C$-sucrose redistribution among individual tissues, following pulse labeling. — Plant Physiol. 55: 704–711, 1975.

Outlaw, W. H., Jr., Schmuck, C. L., Tolbert, N. E.: Photosynthetic carbon metabolism in the palisade parenchyma and spongy parenchyma of Vicia faba L. – Plant Physiol. 58: 186–189, 1976.

Overdieck, D.: $CO_2$-Gaswechsel und Transpiration von Sonnen- und Schattenblättern bei unterschiedlichen Strahlungsqualitäten. — Ber. deut. bot. Ges. 91: 633–644, 1978.

Oya, V. M., Laïsk, A. Kh.: Adaptatsiya fotosinteticheskogo apparata k profilyu sveta v liste. [Adaptation of the photosynthetic apparatus to the profile of light in the leaf.]— Fiziol. Rast. 23: 445–451, 1976.

Ozbun, J. L., Volk, R. J., Jackson, W. A.: Effects of potassium deficiency on photosynthesis, respiration and the utilization of photosynthetic reductant by mature bean leaves. — Crop Sci. 5: 497–500, 1965.


Pallett, K. E., Dodge, A. D.: Modifications of chloroplasts of flax cotyledons treated with monuron: myelinoid figures formed under low light conditions. — Plant, Cell Environ. 3: 183–188, 1980.

Pandey, R. M., Farmahan, H. L.: Changes in rate of photosynthesis and respiration in leaves and berries of Vitis vinifera grapevines at various stages of berry development. — Vitis 16: 106–111, 1977.

Paolillo, D. J. Jr.: The three-dimensional arrangement of intergranal lamellae in chloroplasts. — J. Cell Sci. 6: 243—255, 1970.

Paolillo, D. J. Jr., MacKay, N. C., Graffius, J. R.: The structure of grana in flowering plants. — Amer. J. Bot. 56: 344–347, 1969.

Parkhurst, D. F.: Stereological methods for measuring internal leaf structure variables. — Amer. J. Bot. 69: 31–39, 1982.

Parkhurst, D. F., Loucks, O. L.: Optimal leaf size in relation to environment. — J. Ecol. 60: 505–537, 1972.

Parlange, J. Y., Waggoner, P. E.: Stomatal dimensions and resistance to diffusion. — Plant Physiol. 46: 337–342, 1970.

Paromenskaya, L. N., Mikhaïlova, L. D., Bagiyan, L. G.: Vliyanie gerbitsidov 2M-4X i 2M-4XM na fotosinteticheskie pigmenty v rasteniyakh gorokha. [Effect of herbicides 2M-4X and 2M-4XM on photosynthetic pigments in pea plants.] — Fiziol. Rast. 22: 421–423, 1975.

Parshina, Z. S., Bedenko, V. P., Makarova, S. M.: Sezonnaya dinamika khlorofilla v list'yakh ozimoĭ pshenitsy v svyazi s sortovymi osobennostyami i usloviyami vozdelivaniya. [Seasonal dynamics of chlorophyll in winter wheat leaves in connection with varietal peculiarities and cultivation conditions.] — Fiziol. Biokhim. kul't. Rast. 4: 396–401, 1972.

Parshina, Z. S., Bedenko, V. P., Makarova, S. M.: O dissimetrii listovoĭ plastinki pshenitsy po soderzhaniyu khlorofilla. [Dissymetry of the leaf blade in wheat with regard to chlorophyll content.] — Sel'skokhoz. Biol. 9: 385–388, 1974a.

Parshina, Z. S., Makarova, S. M., Aleksandrova, N. Ya.: Pigmenty listovogo apparata. [Pigments of the leaf apparatus.] — In: Fiziologiya Ozimoĭ Pshenitsy na Yugo-Vostoke Kazakhstana. Pp. 149–179. Nauka kaz. SSR, Alma-Ata 1974b.

Parthier, B₁: Licht-induzierte Transformation von Proplastiden zu Chloroplasten in Algen und höheren Pflanzen. — In: Nover, L., Luckner, M., Parthier, B. (ed.): Zelldifferenzierung. Molekulare Grundlagen und Probleme. Pp. 210–259. VEB G. Fischer Verlag, Jena 1978.

Parthier, B.: The equivocal role of phytohormones (cytokinins) in chloroplast development. — Biochem. Physiol. Pflanz. 174: 173–214, 1979.

Passera, C.: Meccanismo di assimilazione della CO₂ e livello di fotorespirazione in piante di orzo di differente età. [Photosynthetic CO₂ fixation mechanisms and photorespiration level in barley plants at different age.] — Riv. Agron. 9: 56–60, 1975.

Passera, C.: Effetto della carenza di zolfo sulla fotosintesi e fotorespirazione di Hordeum vulgare L., cv. "Astrix". [Effect of sulfur deficiency on photosynthesis and photorespiration of barley leaves, cv. "Astrix".] — Riv. Agron. 12: 113–118, 1978.

Passera, C., Albuzio, A.: Source of glycolate and cyclic changes in photosynthetic and photorespiratory activity during the development of barley leaves. — Biol. Plant. 19: 448–452, 1977.

Pasternak, D., Wilson, G. L.: Illuminance, stomatal opening, and photosynthesis in sorghum and cotton. — Aust. J. agr. Res. 24: 527–532, 1973.

Patel, J. S.: Photosynthesizing leaves and nodulated roots as donors of carbon to protein of the shoot of the field pea (Pisum arvense L.). — Ann. Bot. 30: 93–109, 1966.

Patra, H. K., Mishra, D.: ATPase activity during leaf development and senescence. — Photosynthetica 15: 80–86, 1981.

Patterson, T. G., Brun, W. A.: Influence of sink removal in the senescence pattern of wheat. — Crop Sci. 20: 19–23, 1980.

Patterson, T. G., Moss, D. N.: Senescence in field-grown wheat. — Crop Sci. 19: 635–640, 1979.

Patterson, T. G., Moss, D. N., Brun, W. A : Enzymatic changes during the senescence of field-grown wheat. — Crop Sci. 20: 15–18, 1980.

Pazourek, J.: The symmetry of the lateral leaflets of Medicago sativa L. – Biol. Plant. 7: 261–269, 1965.

Pazourek, J.: Anatomical gradients. — Acta Univ. Carolinae, Biol. Suppl. 1966 (1/2): 19–25, 1966.

Pazourek, J.: Anatomical gradients of stomatal apparatus in leaves of Hordeum distichon L. — Advan. Front. Plant Sci. 23: 9–18, 1969.

Pazourek, J.: The effect of light intensity on stomatal frequency in leaves of Iris hollandica hort., var. Wedgwood. – Biol. Plant. 12: 208–215, 1970.

Pazourek, J.: The density of stomata in leaves of two ecotypes of Phragmites communis Trin. in Southern Bohemia. — Preslia 45: 242–249, 1973a.

Pazourek, J.: The density of stomata in leaves of different ecotypes of Phragmites communis. — Folia geobot. phytotax. 8: 15–21, 1973b.

Pazourek, J.: The effect of light intensity on some anatomical characteristics in leaves of *Xiphium holandicum* hort., var. Wedgwood. — Acta Univ. Carolinae, Biol. 1971: 211–221, 1973c.

Pazourek, J., Nátr, L.: Changes in the anatomical structure of the first two leaves of barley caused by the absence of nitrogen or phosphorus in the nutrient medium. — Biol. Plant. 23: 296–301, 1981.

Pearce, R. B., Brown, R. H., Blaser, R. E.: Photosynthesis of alfalfa leaves as influenced by age and environment. — Crop Sci. 8: 677–680, 1968.

Pearce, R. B., Lee, D. R.: Photosynthetic and morphological adaptation of alfalfa leaves to light intensity at different stages of maturity. — Crop Sci. 9: 791–794, 1969.

Pearcy, R. W., Björkman, O.: Biochemical characteristics. — Carnegie Inst. Year Book 69: 632–640, 1971.

Pearcy, R. W., Tumosa, N., Williams, K.: Relationships between growth, photosynthesis and competitive interactions for a $C_3$ and $C_4$ plant. — Oecologia 48: 371–376, 1981.

Peat, W. E.: Relationships between photosynthesis and light intensity in the tomato. — Ann. Bot. 34: 319–328, 1970.

Peisker, M.: Ein Modell der Sauerstoffabhängigkeit des photosynthetischen $CO_2$-Gaswechsels von $C_3$-Pflanzen. — Kulturpflanze 24: 221–235, 1976.

Peisker, M., Apel, P.: Influence of oxygen on photosynthesis and photorespiration in leaves of *Triticum aestivum* L. 2. Response of $CO_2$ gas exchange to oxygen at various leaf ages and its variability. — Photosynthetica 10: 140–146, 1976.

Peisker, M., Tichá, I., Čatský, J.: Ontogenetic changes in the internal limitations to bean-leaf photosynthesis. 7. Interpretation of the linear correlation between $CO_2$ compensation concentration and $CO_2$ evolution in darkness. — Photosynthetica 15: 161–168, 1981.

Penman, H. L., Schofield, R. K.: Some physical aspects of assimilation and transpiration. — Symp. Soc. exp. Biol. 5: 115–129, 1951.

Penning de Vries, F. W. T.: Respiration and growth. — In: Rees, R. A., Cockshull, K. E., Hand, D. W., Hurd, R. G. (ed.): Crop Processes in Controlled Environments. Pp. 327–347. Academic Press, London–New York 1972.

Penning de Vries, F. W. T.: Substrate utilization and respiration in relation to growth and maintenance in higher plants. — Neth. J. agr. Sci. 22: 40–44, 1974.

Penning de Vries, F. W. T.: Use of assimilates in higher plants. — In: Cooper, J. P. (ed.): Photosynthesis and Productivity in Different Environments. Pp. 459–480. Cambridge Univ. Press, Cambridge–London–New York–Melbourne 1975a.

Penning de Vries, F. W. T.: The cost of maintenance processes in plant cells. — Ann. Bot. 39: 77–92, 1975b.

Penning de Vries, F. W. T., Brunsting, A. H. M., van Laar, H. H.: Products, requirements and efficiency of biosynthetic processes: a quantitative approach. — J. theor. Biol. 45: 339–377, 1974.

Penning de Vries, F. W. T., Witlage, J. M., Kremer, D.: Rates of respiration and of increase in structural dry matter in young wheat, ryegrass and maize plants in relation to temperature, to water stress and to their sugar content. — Ann. Bot. 44: 595—609, 1979.

Peoples, M. B., Dalling, M. J.: Degradation of ribulose-1,5-bisphosphate carboxylase by proteolytic enzymes from crude extracts of wheat leaves. — Planta 138: 153–160, 1978.

Pereira, J. S., Kozlowski, T. T.: Leaf anatomy and water relations of *Eucalyptus camadulensis* and *E. globosus* seedlings. — Can. J. Bot. 54: 2868–2880, 1976.

Pethő, M.: Dry matter accumulation in the stalk- and leaf-levels of maize (*Zea mays* L.). — Acta agron. Acad. Sci. hung. 16: 139–146, 1967.

Phan, C. T., Brach, E. J., Jasmin, J. J.: Studies on the detection of lettuce maturity: anatomical observations and reflectance measurements in the visible range (350–650 nm). — Can. J. Plant Sci. 59: 1067–1075, 1979.

Pieters, G. A.: Some aspects of the problem of sun and shade leaves. — In: Proc. XVIth Int. Hort. Congress. Pp. 393–399. Brussels 1962.

Pieters, G. A.: The growth of sun and shade leaves of *Populus euramericana* "*robusta*" in relation to age, light intensity and temperature. — Med. Landbouwhogesch. Wageningen 74 (11): 1–106, 1974.

Platt-Aloia, K. A., Thomson, W. W.: Chloroplast development in young sesame plants. — New Phytol. 78: 599–605, 1977.

Plesničar, M., Bendall D. S.: The development of photochemical activities during greening of etiolated barley. — In: Forti, G., Avron, M., Melandri, A. (ed.): Photosynthesis, Two Centuries after Its Discovery by Joseph Priestley. Vol. 3. Pp. 2367–2374. Dr. W. Junk N. V. Publ., The Hague 1972.

Plesničar, M., Bendall, D. S.: The photochemical activities and electron carriers of developing barley leaves. — Biochem. J. 136: 803–812, 1973.

Plesničar, M., Bogdanović, M.: Fotohemijske aktivnosti hloroplasta crnog bora. [Photochemical activities of the black pine chloroplasts.] — Acta bot. croat. 35: 71–75, 1976.

Poincelot, R. P.: Carbonic anhydrase. — In: Gibbs, M., Latzko, E. (ed.): Photosynthesis II. Pp. 230–238. Springer-Verlag, Berlin–Heidelberg–New York 1979.

Pollock, C. J.: Changes in the activity of sucrose-synthesizing enzymes in developing leaves of *Lolium temulentum*. — Plant Sci. Lett. 7: 27–31, 1976.

Ponomareva, A. N.: Aktivnost' adenozintrifosfatazy pshenitsy pri prorastanii. [Acivity of adenosine triphosphatase of wheat during germination.] — Dokl. Akad. Nauk SSSR 121: 515–518, 1958.

Popov, K., Bakardshijeva, N.: Über den Zustand und die Extrahierbarkeit der Plastidenpigmente. — Stud. biophys. 5: 51–58, 1967.

Popov, K., Bakardzhieva, N.: Vliyanie na ultravioletovoto obl"chvane i nyakoi mikroelementi v"rkhu pigmentniya rezhim na listata. [Effect of ultraviolet radiation and some microelements on the pigment regime of leaves.] — Izv. Inst. Fiziol. Rast. "Metodiĭ Popov" b"lg. Akad. Nauk 16: 65–81, 1970.

Popov, K., Tsoneva, P.: Sezonni i v"zrastovi izmeneniya v s"d"rzhanieto i s"stoyanieto na pigmentite v listata na *Pinus silvestris* i *Taxus baccata*. [Seasonal and ontogenetic changes in the content and state of pigments in leaves of *Pinus silvestris* and *Taxus baccata*.] — God. sofiisk. Univ. biol. Fak., Kn. 2. Bot. Mikrobiol. Fiziol. Biokhim. Rast. 61: 147–164, 1966/7.

Popov, V. I., Kaurov, B. S., Finakov, G. Z., Gulyaev, B. A., Kukarskikh, G. P., Allakhverdov, B. L., Tageeva, S. V.: Chloroplast membrane structure. Ultrastructural and functional organization of digitonin-fractionated pea subchloroplast fragments. — Photosynthetica 14: 343–354, 1980.

Porokhnevich, N. V.: Izmenenie tsitologicheskikh pokazateleĭ fotosinteticheskogo apparata l'na v ontogeneze pri razlichnom urovne snabzheniya rasteniĭ tsinkom. [Changes of cytological characteristics of the photosynthetic apparatus of *Linum* during ontogeny at different Zn supply of the plants.] — Nauch. Dokl. vyssh. Shkoly, biol. Nauki 1972(6): 68–74, 1972.

Poskuta, J., Parys, E., Ostrowska, E.: Growth, $CO_2$ exchange rates and yield of pea (*Pisum sativum* L.) cv. "Bordi" in the field conditions after pretreatment of seeds with gibberellic acid ($GA_3$). — Biul. warzyw. 18: 197–206, 1975.

Pospíšilová, J., Solárová, J.: Ontogenetic changes in response of conductances of adaxial and abaxial epidermis to water stress. — Biol. Plant. 26: 49–55, 1984.

Pospíšilová, J., Zima, J., Šesták, Z.: Effect of hydration level in primary bean leaves on the activity of photosystems 1 and 2 in isolated chloroplasts. — Biol. Plant. 18: 473–479, 1976.

Possingham, J. V., Saurer, W.: Changes in chloroplast number per cell during leaf development in spinach. — Planta 86: 186–194, 1969.

Possingham, J. V., Smith, J. W.: Factors affecting chloroplast replication in spinach. — J. exp. Bot. 23: 1050–1059, 1972.

Powles, S. B., Osmond, C. B.: Photoinhibition of intact attached leaves of $C_3$ plants illuminated in the absence of both carbon dioxide and of photorespiration.—Plant Physiol. 64: 982–988, 1979.

Prenzel, U., Lichtenthaler, H. K., Meier, D.: Level of chlorophyll *b* and the light harvesting chlorophyll-protein complex in *Raphanus* seedlings grown at different light quanta fluence rates. —

In: Mazliak, P., Benveniste, P., Costes, C., Douce, R. (ed.): Biogenesis and Function of Plant Lipids. Pp. 369–372. Elsevier/North-Holland Biomedical Press, Amsterdam 1980.

Prioul, J. L.: Réactions des feuilles de *Lolium multiflorum* à l'éclairement pendant la croissance et variation des résistances aux échanges gazeux photosynthétiques. — Photosynthetica 5: 364–375, 1971.

Prioul, J. L.: Eclairement de croissance et infrastructure des chloroplastes de *Lolium multiflorum* Lam. Relation avec les résistances au transfert de $CO_2$. — Photosynthetica 7: 373–381, 1973.

Prioul, J. L., Brangeon, J., Reyss, A.: Interaction between external and internal conditions in the development of photosynthetic features in a grass leaf. I. Regional responses along a leaf during and after low-light or high-light acclimation. — Plant Physiol. 66: 762–769, 1980a.

Prioul, J. L., Brangeon, J., Reyss, A.: Interaction between external and internal conditions in the development of photosynthetic features in a grass leaf. II. Reversibility of light-induced responses as a function of developmental stages. — Plant Physiol. 66: 770–774, 1980b.

Prioul, J. L., Chartier, P.: Partitioning of transfer and carboxylation components of intracellular resistance to photosynthetic $CO_2$ fixation: A critical analysis of the methods used.—Ann. Bot. 41: 789–800, 1977.

Quarrie, S. A., Henson, I. E.: Abscisic acid accumulation in detached cereal leaves in response to water stress. II. Effects of leaf age and leaf position. — Z. Pflanzenphysiol. 101: 439–446, 1981.

Raafat, A., Gausz, J., Szalay, L., Horváth, I.: Photobiology of aging bean leaves *in vivo*. — Acta biochim. biophys. Acad. Sci. hung. 4: 403–410, 1969.

Raafat, A., Höfner, W.: Effects of age on the fixation of $^{14}CO_2$ in sugars, organic acids and amino acids of bean leaves. — Phytochemistry 10: 2373–2381, 1971.

Raafat, A., Höfner, W., Linser, H.: $^{14}CO_2$ assimilation during photosynthesis of ageing bean seedlings. — Z. Pflanzenphysiol. 64: 22–33, 1971.

Raafat, A., Stur, J., Sipos, M., Marek, N.: Some aspects of oxidative-reductive changes in chloroplast suspensions during the process of ageing. — Acta biochim. biophys. Acad. Sci. hung 5: 265–272, 1970a.

Raafat, A., Szász, K., Horváth, I.: Effect of leaf age on the chlorophyll fractions extracted with two different acetone concentrations. — Acta bot. Acad. Sci. hung. 16: 187–191, 1970b.

Raafat, A., Szász, K., Horváth, I.: Effect of leaf age on the chlorophyll fractions extracted with two different acetone concentrations. — Agrochimica 14: 392–397, 1970c.

Rabinowitch, E. I.: Photosynthesis and Related Processes. Vol. II/1. — Interscience Publishers, New York 1951.

Rabinowitch, E. I., Govindjee: Photosynthesis. — John Wiley & Sons, New York–London–Sydney–Toronto 1969.

Radin, J. W.: Water relations of cotton plants under nitrogen deficiency. IV. Leaf senescence during drought and its relation to stomatal closure. — Physiol. Plant. 51: 145–149, 1981.

Radin, J. W., Parker, L. L.: Water relations of cotton plants under nitrogen deficiency. I. Dependence upon leaf structure. — Plant Physiol. 64: 495–498, 1979.

Raggi, V.: The $CO_2$ compensation point, photosynthesis and respiration in rust infected bean leaves. — Physiol. Plant Pathol. 13: 135–139, 1978a.

Raggi, V.: Fotorespirazione, respirazione, fotosintesi e loro influenza sul punto di compensazione per la $CO_2$ in piante di fagiolo affette de leggeri attacchi di "ruggine". [Photorespiration, respiration, photosynthesis and their correlation with the $CO_2$ compensation point in French bean leaves mildly infected by rust.] — Phytopathol. mediter. 17: 105–109, 1978b.

Raggi, V.: Correlation of $CO_2$ compensation point ($\Gamma$) with photosynthesis and respiration and $CO_2$-sensitive $\Gamma$ in rust-affected bean leaves. — Physiol. Plant Pathol. 16: 19–24, 1980.

Raghavendra, A. S.: Characteristics of plant species intermediate between $C_3$ and $C_4$ pathways of photosynthesis: Their focus of mechanism and evolution of $C_4$ syndrome. — Photosynthetica 14: 271–283, 1980.

Raghavendra, A. S., Rajendrudu, G., Das, V. S. R.: Simultaneous occurrence of $C_3$ and $C_4$ photosynthesis in relation to leaf position in *Mollugo nudicaulis*. — Nature 273: 143–144, 1978.

Raïtsina, G. I., Lebedeva, T. D., Vecher, A. S.: Fotokhimicheskaya aktivnost' khloroplastov izolirovannykh iz list'ev gorokha v protsesse zeleneniya. [Photochemical activity of chloroplasts isolated during greening of pea leaves.] — Dokl. Akad. Nauk belorus. SSR 12: 731–734, 1968.

Rakhi, M.: Ob anatomicheskikh parametrakh lista v svyazi s diffuzionnymi soprotivleniyami. [Characteristics of leaf anatomy and diffusion resistances.] — Izv. Akad. Nauk eston. SSR, Biol. 20: 84–94, 1971.

Raper, C. D., Downs, R. J.: Factors affecting the development of flue-cured tobacco grown in artificial environments: IV. Effects of carbon dioxide depletion and light intensity. — Agron. J. 65: 247–252, 1973.

Rascio, N., Casadoro, G.: Sunflower etioplast membranes. — J. Ultrastruct. Res. 68: 325–327, 1979.

Rascio, N., Mariani Colombo, P., Orsenigo, M.: The ultrastructural development of plastids in leaves of maize plants exposed to continuous illumination. — Protoplasma 102: 131–139, 1980.

Rascio. N., Orsenigo, M., Arboit, D.: Prolamellar body transformation with increasing cell age in the maize leaf. — Protoplasma 90: 253–263, 1976.

Rathnam, C. K. M., Chollet, R.: Photosynthetic carbon metabolism in $C_4$ plants and $C_3$–$C_4$ intermediate species. — In: Reinhold, L., Harborne, J. B., Swain, T. (ed.): Progress in Phytochemistry. Vol. 6. Pp. 1–48. Pergamon Press, New York–Oxford–Frankfurt–Paris 1980.

Rathnam-Chagúturu: $C_3$–$C_4$ intermediate species. — What's New Plant Physiol. 12: 21–24, 1981.

Raven, C. W.: Synthesis of chlorophyll at different intensities of monochromatic light. — In: Forti, G., Avron, M., Melandri, A. (ed.): Photosynthesis, Two Centuries after Its Discovery by Joseph Priestley. Vol. 3. Pp. 2325–2332. Dr. W. Junk N. V. Publ., The Hague 1972.

Raven, C. W.: Chlorophyll formation and phytochrome. — Meded. Landbouwhogesch. Wageningen 73 (9): 1–100, 1973.

Raven, J. A., Glidewell, S. M.: Processes limiting photosynthetic conductance. — In: Johnson, C. B. (ed.): Physiological Processes Limiting Plant Productivity. Pp. 109–136. Butterworths, London–Boston–Sydney–Wellington–Durban–Toronto 1981.

Rawson, H. M.: Vertical wilting and photosynthesis, transpiration, and water use efficiency of sunflower leaves. — Aust. J. Plant Physiol. 6: 109–120, 1979.

Rawson, H. M., Constable, G. A.: Carbon production of sunflower cultivars in field and controlled environments. I Photosynthesis and transpiration of leaves, stems and heads. — Aust. J. Plant Physiol. 7: 555–573, 1980.

Rawson, H. M., Constable, G. A., Howe, G. N.: Carbon production of sunflower cultivars in field and controlled environments. II Leaf growth. — Aust. J. Plant Physiol. 7: 575–586, 1980.

Rawson, H. M., Craven, C. L.: Stomatal development during leaf expansion in tobacco and sunflower. — Aust. J. Bot. 23: 253–261, 1975.

Rawson, H. M., Gifford, R. M., Bremner, P. M.: Carbon dioxide exchange in relation to sink demand in wheat. — Planta 132: 19–23, 1976.

Rawson, H. M., Hackett, C.: An exploration of the carbon economy of the tobacco plant. III. Gas exchange of leaves in relation to position on the stem, ontogeny and nitrogen content. — Aust. J. Plant Physiol. 1: 551–560, 1974.

Rawson, H. M., Hindmarsh, J. H.: Effects of temperature on leaf expansion in sunflower. — Aust. J. Plant Physiol. 9: 209–219, 1982.

Rawson, H. M., Hofstra, G.: Translocation and remobilization of [14]C assimilated at different stages by each leaf of the wheat plant. — Aust. J. biol. Sci. 22: 321–331, 1969.

Rawson, H. M., Turner, N. C.: Recovery from water stress in five sunflower (*Helianthus annuus* L.) cultivars. II. The development of leaf area. — Aust. J. Plant Physiol. 9: 449–460, 1982.

*339*

Rawson, H. M., Woodward, R. G.: Photosynthesis and transpiration in dicotyledonous plants. I. Expanding leaves of tobacco and sunflower. — Aust. J. Plant Physiol. 3: 247–256, 1976.

Rebeiz, C. A., Yaghi, M., Abou-Haidar, M., Castelfranco, P. A.: Protochlorophyll biosynthesis in cucumber (*Cucumis sativus*, L.) cotyledons. — Plant Physiol. 46: 57–63, 1970.

Reed, H. S., Hirano, E.: The density of stomata in *Citrus leaves*. — J. agr. Res. 43: 209–222, 1931.

Reinert, J. (ed.): Chloroplasts. — Springer-Verlag, Berlin–Heidelberg–New York 1980.

Rejmánková, E.: Chlorophyll content in leaves of *Phragmites communis* Trin. — In: Hejný, S. (ed.): Ecosystem Study on Wetland Biome in Czechoslovakia. Czechosl. IBP/PT-PP Report No. 3. Pp. 143–145. Třeboň 1973.

Repka, J., Jureková, Z.: Heterogeneity of the maize leaf blade in photosynthetic characteristics, respiration, mineral nutrient contents, and growth substances. — Biol. Plant. 23: 145–155, 1981.

Reynolds, P. E., Raigosa, J., Trip, P.: Qualitative and quantitative changes in first products of photosynthesis in *Zea mays* as related to age. — Plant Physiol. 1974 (Suppl.): 62, 1974.

Reyss, A., Prioul, J. L.: Carbonic anhydrase and carboxylase activities from plants (*Lolium multiflorum*) adapted to different light regimes. — Plant Sci. Lett. 5: 189–195, 1975.

Rhoades, M. M., Carvalho, A.: The function and structure of the parenchyma sheath plastids of the maize leaf. — Bull. Torrey bot. Club 71: 335–346, 1944.

Rhodes, M. J. C., Yemm, E. W.: The development of chloroplasts and photosynthetic activities in young barley leaves. — New Phytol. 65: 331–342, 1966.

Richter, G.: Plant Metabolism. Physiology and Biochemistry of Primary Metabolism. — G. Thieme, Stuttgart 1978.

Robards, A. W.: General and molecular cytology. — Progr. Bot. 43: 1–12, 1981.

Röbbelen, G.: Gestörte Thylakoidbildung in Chloroplasten einer *Xantha*-Mutante von *Arabidopsis thaliana* (L.) Hyenh. — Planta 69: 1–26, 1966.

Robberecht, R., Caldwell, M. M.: Leaf epidermal transmittance of ultraviolet radiation and its implications for plant sensitivity to ultraviolet-radiation induced injury. — Oecologia 32: 277–287, 1978.

Roberts, S. W., Miller, P. C., Valamanesh, A.: Comparative field water relations of four co-occurring chaparral shrub species. — Oecologia 48: 360–363, 1981.

Robertson, D., Laetsch, W. M.: Structure and function of developing barley plastids. — Plant Physiol. 54: 148–159, 1974.

Robson, M. J., Deacon, M. J.: Nitrogen deficiency in small closed communities of S24 ryegrass. II. Changes in the weight and chemical composition of single leaves during their growth and death. – Ann. Bot. 42: 1199–1213, 1978.

Rogan, P. G., Smith, D. L.: Rates of leaf initiation and leaf growth in *Agropyron repens* (L.) Beauv. – J. exp. Bot. 26: 70–78, 1975.

Rook, D. A., Corson, M. J.: Temperature and irradiance and the total daily photosynthetic production of the crown of a *Pinus radiata* tree. — Oecologia 36: 371–382, 1978.

Rosinski, J., Rosen, W. G.: Chloroplast development: fine structure and chlorophyll synthesis. — Quart. Rev. Biol. 47: 160–191, 1972.

Rübel, E.: Experimentelle Untersuchungen über die Beziehungen zwischen Wasserleitungsbahnen und Transpirationsverhältnissen bei *Helianthus annuus* L. — Beih. bot. Centralbl., Abt. I, 37: 1–62, 1920.

Rudenko, T. I., Shmeleva, V. L., Makarov, A. D.: Izuchenie konformatsionnykh kolebaniǐ khloroplastov iz rasteniǐ gorokha, vyrashchennykh pri razlichnoǐ intensivnosti osveshcheniya. [A study of conformational oscillations of chloroplasts from pea plants grown under various light intensities.] — Fiziol. Rast. 26: 1150–1155, 1979.

Rühle, W., Wild, A.: The intensification of absorbance changes in leaves by light-dispersion. Differences between high-light and low–light leaves. — Planta 146: 551–557, 1979.

Rumi, C. P., Carpinetti, R. M.: Effect of sunlight on the development of *Tropaeolum majus* L. 1. Leaf growth. — Fyton 35: 137–143, 1977.

Rumi, C. P., Carpinetti, R. M.: Effect of sunlight on the development of *Tropaeolum majus* L. II. Leaf development. — Fyton 41: 129–137, 1981.

Rychnovská, M.: A contribution to the autecology of *Phragmites communis* Trin. I. Physiological heterogeneity of leaves. — Folia geobot. phytotaxon. 2: 179–188, 1967.

Ryle, G. J. A., Cobby, J. M., Powell, C. E.: Synthetic and maintenance respiratory losses of $^{14}CO_2$ in uniculm barley and maize. — Ann. Bot. 40: 571–586, 1976.

Saakov, V. S., Baranov, A. A., Hoffmann, P.: Derivativ-spektroskopische Charakteristik des pigmentphysiologischen Zustandes des Photosyntheseapparates unter besonderer Berücksichtigung der Temperatur. — Studia biophys. 70: 163–173, 1978.

Saeki, T.: Variation of photosynthetic activity with aging of leaves and total photosynthesis in a plant community. — Bot. Mag. (Tokyo) 72: 404–408, 1959.

Sagromsky, H.: Weitere Beobachtungen zur Bildung des Spaltöffnungsmusters in der Blattepidermis. Zur Frage der Gruppenbildung. — Z. Naturforsch. 4 B: 360–367, 1949.

Sălăgeanu, N.: On light absorbed, reflected and passing through leaves and on photosynthesis efficiency in some species. — Rev. roum. Biol. — Ser. Bot. 10: 393–402, 1965.

Salema, R., Abreu, I.: Fine structure of green and white leaves of *Phyllanthus nivosus* Bull. — Port. Acta biol., Sér. A 12: 255—266, 1972.

Salema, R., Brandão, I.: Development of microtubules in chloroplasts of two halophytes forced to follow Crassulacean acid metabolism. — J. Ultrastruct. Res. 62: 132–136, 1978.

Salin, M. L., Homann, P. H.: Changes of photorespiratory activity with leaf age. — Plant Physiol. 48: 193–196, 1971.

Salin, M. L., Homann, P. H.: Glycolate metabolism in young and old tobacco leaves, and effects of $\alpha$-hydroxy-2-pyridinemethanesulfonic acid. — Can. J. Bot. 51: 1857–1865, 1973.

Salisbury, E. J.: On the causes and ecological significance of stomatal frequency, with special reference to the woodland flora. — Phil. Trans. roy. Soc. London B 216: 1–65, 1927.

Sambo, E. Y., Moorby, J., Milthorpe, F. L.: Photosynthesis and respiration of developing soybean pods. — Aust. J. Plant Physiol. 4: 713–721, 1977.

Sams, C. E., Flore, J. A.: The influence of age, position, and environmental variables on net photosynthetic rate of sour cherry leaves. — J. amer. Soc. hort. Sci. 107: 339–344, 1982.

Samsuddin, Z., Impens, I.: Relationship between leaf age and some carbon dioxide exchange characteristics of four *Hevea brasiliensis* Muell. Arg. clones. — Photosynthetica 13: 208–210, 1979a.

Samsuddin, Z., Impens, I.: The development of photosynthetic rate with leaf age in *Hevea brasiliensis* Muell. Arg. clonal seedlings. — Photosynthetica 13: 267–270, 1979b.

Samsuddin, Z., Rahman, M. K. A., Impens, I.: Development of leaf blade class concept for the characterisation of *Hevea brasiliensis* Muell. Arg. leaf age. — J. Rubb. Res. Inst. Malaysia 26: 1–5, 1978.

Sanada, Y., Nishida, K.: The presence of pyruvate, orthophosphate dikinase in CAM plants. — Z. Pflanzenphysiol. 105: 189–192, 1982.

Sandanam, S., Gee, G. W., Mapa, R. B.: Leaf water diffusion resitance in clonal tea (*Camellia sinensis* L.): Effects of water stress, leaf age and clones. — Ann. Bot. 47: 339–349, 1981.

Sandhu, B. S., Horton, M. L.: Response of oats to water deficit. I. Physiological characteristics. — Agron. J. 69: 357–360, 1977.

Sane, P. V.: The topography of the thylakoid membrane of the chloroplast. — In: Trebst, A., Avron, M. (ed.): Photosynthesis I. Pp. 522–542. Springer-Verlag, Berlin–Heidelberg–New York 1977.

Sant, F. I.: A comparison of the morphology and anatomy of seedling leaves of *Lolium multiflorum* Lam. and *L. perenne* L. — Ann. Bot. 33: 303–313, 1969.

Sarda, C., Prioul, J.-L., Moyse, A.: Structure du limbe et accumulation d'amidon dans les chloroplastes des feuilles d'une Crassulacée, *Bryophyllum daigremontianum* Berger. — Physiol. vég. 13: 563–577, 1975.

Sasahara, T.: Changes in size and number of mesophyll cells, nitrogen content and photosynthesis with leaf order in *Brassica* spp. — Ann. Bot. 50: 379–383, 1982.

Sasahara, T., Tsunoda, S.: Genetic variations in cell and tissue forms in relation to plant growth. I. Relationship between growth rates of the *Brassica* species and cell sizes of the shoot apex and palisade parenchyma. — Jap. J. Breed. 21: 1–8, 1971.

Sato, K., Kim, J. M.: [Relationships between environmental conditions and production- and consumption activities of individual leaves in the population of rice plant in a paddy field. I. Changes in photosynthesis and dark respiration of individual leaves under field conditions.] — Jap. J. Crop Sci. 49: 243–250, 1980a.

Sato, K., Kim, J. M.: [Relationships between environmental conditions and production- and consumption activities of individual leaves in the population of rice plant in a paddy field. II. Effects of temperature on photosynthesis and dark respiration of individual leaf of different position.] — Jap. J. Crop Sci. 49: 251–256, 1980b.

Sato, K., Kim, J. M.: [Relationships between environmental conditions and production- and consumption activities of individual leaves in the population of rice plant in a paddy field. III. Effects of shading on photosynthesis and dark respiration of individual leaf of different position.] —Jap. J. Crop Sci. 49: 257–262, 1980c.

Sato, K., Kim, J. M.: [Relationships between environmental conditions and production- and consumption activities of individual leaves in the population of rice plant in a paddy field. IV. Leaf positional and seasonal changes in the rates of net photosynthesis and dark respiration in paddy fields of different plant spacing and fertilization.] – Jap. J. Crop Sci. 49: 262–269, 1980d.

Sato, T., Kawai, M., Fukuyama, T.: [Studies on matter production of taro plant (*Colocasia esculenta* Schott) I. Changes with growth in photosynthetic rate of single leaf.] — Jap. J. Crop Sci. 47: 425–430, 1978.

Satoh, M.: Studies on photosynthesis and translocation of photosynthate in mulberry tree. III. Translocation of $^{14}C$-photosynthetic product from leaves of different ages. — Proc. Crop Sci. Soc. Jap. 43: 99–104, 1974.

Satoh, M., Hazama, K.: [Studies on photosynthesis and translocation of photosynthate in mulberry tree. I. Photosynthetic rate of remained leaves after short pruning.] — Proc. Crop Sci. Soc. Jap. 40: 7–11, 1971.

Satoh, M., Kriedemann, P. E., Loveys, B. R.: Changes in photosynthetic activity and related processes following decapitation in mulberry trees. — Physiol. Plant. 41: 203–210, 1977.

Saurer, W., Possingham, J. V.: Studies on the growth of spinach leaves (*Spinacea oleracea*). — J. exp. Bot. 21: 151–158, 1970.

Savchenko, G. E.: Chaĭka, M. T.: Issledovanie kinetiki i temnovogo nakopleniya protokhlorofillida na raznykh stadiyakh razvitiya list'ev yachmenya. [Kinetics of dark protochlorophyllide accumulation in barley leaves at various developmental stages.] — In: Shlyk, A. A. (ed.): Biosintez i Sostoyanie Khlorofillov v Rastenii. Pp. 83–103, 245. Nauka i Tekhnika, Minsk 1975.

Sawada, S.: An ecophysiological analysis of the difference between the growth rates of young wheat seedlings grown in various seasons. – J. Fac. Sci., Univ. Tokyo, Sec. III, 10: 233–263, 1970.

Sawada, S., Igarashi, T., Miyachi, S.: Effects of nutritional levels of phosphate on photosynthesis and growth studied with single, rooted leaf of dwarf bean. — Plant Cell Physiol. 23: 27—33, 1982.

Sawyer, W. H., Jr.: Stomatal apparatus of the cultived cranberry *Vaccinium macrocarpon*. – Amer. J. Bot. 19: 508–513, 1932.

Schaedle, M.: Tree photosynthesis. — Annu. Rev. Plant Physiol. 26: 101–115, 1975.

Schäfer, K., Tirtapradja, H.: Nettoassimilation, Dunkelatmung, Stomataanzahl und Chlorophyllgehalt von Blattabschnitten bei einigen Sorten von *Dactylis glomerata* L. — Z. Acker- Pflanzenbau 132: 320–339, 1970.

Schiff, J. A.: Development, inheritance, and evolution of plastids and mitochondria. — In: Tolbert, N. E. (ed.): Biochemistry of Plants. Vol. 1. Pp. 209–272. Academic Press, New York–London 1980.

Schiff, J. A., Lyman, H. (ed.): On the Origins of Chloroplasts. — Elsevier/North-Holland, New York–Amsterdam–Oxford 1982.

Schimper, A. F.W.: Untersuchungen über die Chlorophyllkörper und die ihnen homologen Gebilde. — Jahrb. wiss. Bot. 16: 1–247, 1885.

Schlesinger, W. H., Chabot, B. F.: The use of water and minerals by evergreen and deciduous shrubs in Okefenokee swamp. — Bot. Gaz. 138: 490–497, 1977.

Schnepf, E.: Types of plastids: Their development and interconversions. — In: Reinert, J. (ed.): Chloroplasts. Pp. 1–27. Springer-Verlag, Berlin–Heidelberg–New York 1980.

Schoch, P. G.: Influence of air temperature and humidity during the vegetative growth on some structural characteristics of the leaf of *Capsicum annuum* L. — In: Proc. IX[th] Congr. Caribbean Food Soc. Pp. 1–19. Guyane 1971.

Schoch, P. G.: Effects of shading on structural characteristics of the leaf and yield of fruit in *Capsicum annuum* L. — J. amer. Soc. hort. Sci. 97: 461–464, 1972a.

Schoch, P. G.: Variation de la densité stomatique de *Capsicum annuum* L. en fonction du rayonnement global. — Compt. rend. Acad. Sci. Paris, Sér. D 274: 2496–2498, 1972b.

Schoch, P. G., Candelario, L. S.: Croissance des feuilles de *Vigna sinensis*. Bilan individuel de la productivité foliaire lors des phases diurnes et nocturnes. — Oecol. Plant. 8: 301–308, 1973.

Schoch, P. G., Candelario, L. S.: Influencia de la sombra sobre el crecimiento y la productividad de las hojas de *Vigna sinensis* L. [The influence of shade on growth and productivity of leaves of *Vigna sinensis* L.] – Turrialba 24: 84–89, 1974.

Schulze, E.-D., Fuchs, M. I., Fuchs, M.: Spacial distribution of photosynthetic capacity and performance in a mountain spruce forest of Northern Germany. I. Biomass distribution and daily $CO_2$ uptake in different crown layers. — Oecologia 29: 43–61, 1977a.

Schulze, E.-D., Fuchs, M., Fuchs, M. I.: Spacial distribution of photosynthetic capacity and performance in a mountain spruce forest of Northern Germany III. The significance of the evergreen habit. — Oecologia 30: 239–248, 1977b.

Schwarz, Z.: Das Verhalten der NAD- und der NADP-abhängigen Glycerinaldehyd-Phosphat--Dehydrogenase im Verlaufe der Entwicklung des Photosyntheseapparates bei *Phaseolus vulgaris* L. – Biol. Rundschau 9: 333–335, 1971.

Scott, N. S., Possingham, J. V.: Leaf development. — In: Smith, H., Grierson, D. (ed.): The Molecular Biology of Plant Development. Pp. 223—255. Blackwell Sci. Publ., Oxford — London — Edinburgh — Boston — Melbourne 1982.

Senser, M., Beck, E.: Photochemically active chloroplasts from spruce (*Picea abies* (L.) Karst.). — Photosynthetica 12: 323–327, 1978.

Senser, M., Beck, E.: Kälteresistenz der Fichte. II. Einfluß von Photoperiode und Temperatur auf die Struktur und photochemischen Reaktionen von Chloroplasten. — Ber. deut. bot. Ges. 92: 243–259, 1979.

Senser, M., Schötz, F., Beck, E.: Seasonal changes in structure and function of spruce chloroplasts. — Planta 126: 1–10, 1975.

Šesták, Z.: Changes in the chlorophyll content as related to photosynthetic activity and age of leaves. – Photochem. Photobiol. 2: 101–110, 1963a.

Šesták, Z.: On the question of the quantitative relation between the amount of chlorophyll, its forms, and the photosynthetic rate. — In: La Photosynthèse. Coll. int. du C. N.R.S. No. 119. Pp. 343—356. Édit. C.N.R.S., Paris 1963b.

Šesták.: Age- and chlorophyll-dependent changes in the photosynthetic activity of leaves. — In: 4th Int. Photobiology Congress. Authors' Abstracts. P. 134. Oxford 1964.

Šesták, Z.: Limitations for finding a linear relationship between chlorophyll content and photosynthetic activity. — Biol. Plant. 8: 336–346, 1966a.

Šesták, Z.: Leaf ageing, chlorophyll content and photosynthetic rate. — Acta Univ. Carolinae, Biol. 1966 (Suppl. 1/2): 115–118, 1966b.

Šesták, Z.: Ratio of photosystem 1 and 2 particles in young and old leaves of spinach and radish. —Photosynthetica 3: 285—287, 1969.

Šesták, Z.: Photosystem 1 and 2 particles from leaves of diverse ages. — Carnegie Inst. Year Book 68: 572–574, 1970.

Šesták, Z.: Leaf age and the shape of the red absorption band of maize chloroplasts. — Photosynthetica 6: 75–79, 1972.

Šesták, Z.: Photosynthetic characteristics during ontogenesis of leaves. 1. Chlorophylls. — Photosynthetica 11: 367–448, 1977a.

Šesták, Z.: Photosynthetic characteristics during ontogenesis of leaves. 2. Photosystems, components of electron transport chain, and photophosphorylation. — Photosynthetica 11: 449–474, 1977b.

Šesták, Z.: Photosynthetic characteristics during ontogenesis of leaves. 3. Carotenoids. — Photosynthetica 12: 89–109, 1978.

Šesták, Z.: Changes in activities of photosystems during leaf ontogenesis. — In: Marcelle, R., Cli) sters, H., Van Poucke, M. (ed.): Photosynthesis and Plant Development. Pp. 21–29. Dr. W. Junk bv. Publ., The Hague–Boston–London 1979.

Šesták, Z.: Leaf ontogeny and photosynthesis. — In: Johnson, C. B. (ed.): Physiological Processes Limiting Plant Productivity. Pp. 147–158. Butterworths, London–Boston–Sydney–Wellington––Durban–Toronto 1981.

Šesták, Z.: Chlorophylls during leaf ontogeny. — In: Poskuta, J. (ed.): Photosynthetic Solar Energy Conversion and Storages. Pp. 55—82. Warsaw University Press, Warsaw 1983.

Šesták, Z.: Effects of leaf age on protochlorophyllide and chlorophyllide formation (a review). — In: Sironval, C., Brouers, M. (ed.): Protochlorphyllide Reduction and Greening. Pp. 365—375. M. Nijhoff/Dr W. Junk Publ. The Hague — Boston — Lancaster 1984.

Šesták, Z., Bartoš, J.: Vliv snížení obsahu chlorofylu na intenzitu fotosyntézy u kukuřice. [Effect of decline in chlorophyll content on the photosynthetic rate in maize.]—Rostl. Výr. 9: 119–134, 1963.

Šesták, Z., Čatský, J.: Intensity of photosynthesis and chlorophyll content as related to leaf age in *Nicotiana sanderae* hort. — Biol. Plant. 4: 131–140, 1962.

Šesták, Z., Čatský, J.: Fotosyntetická heterogenita rostliny. [Photosynthetic heterogenity of plants.] — Studijní Inform. ÚVTI MZLVH, základní pomocné Vědy Zemědělství (Praha) 1966 (4–5): 1–141, 1966.

Šesták, Z., Čatský, J.: Chlorophyll and photosynthesis in fodder cabbage plants growing in controlled conditions. — Stud. biophys. 5: 91–96, 1967a.

Šesták, Z., Čatský, J.: Heterogeneity in photosynthetic characteristics of fodder cabbage plants grown in controlled conditions. — In: Photochemistry and Photobiology in Plant Physiology. European Photobiol. Symp. Hvar. Book of Abstracts. Pp. 93–96. Hvar 1967b.

Šesták, Z., Čatský, J.: Sur les relations entre le contenu en chlorophylle et l'activité photosynthétique pendant la croissance et le vieillissement des feuilles. — In: Sironval, C. (ed.): Le Chloroplaste, Croissance et Vieillissement. Pp. 213–262. Masson et Cie, Paris 1967c.

Šesták, Z., Čatský, J., Solárová, J., Strnadová, H., Tichá, I.: Carbon dioxide transfer and photochemical activities as factors of photosynthesis during ontogenesis of primary bean leaves. — In: Nasyrov, Yu. S., Šesták, Z. (ed.): Genetic Aspects of Photosynthesis. Pp. 159–166. Dr. W. Junk b.v. Publ., The Hague 1975.

Šesták, Z., Demeter, S.: Changes in circular dichroism and P700 content of chloroplasts during leaf ontogenesis. – Photosynthetica 10: 182–187, 1976.

Šesták, Z., Solárová, J., Zima J., Václavík, J.: Effect of growth irradiance on photosynthesis and transpiration in *Phaseolus vulgaris* L. – Biol. Plant. 20: 234–238, 1978a.

Šesták, Z., Václavík, J.: Relationship between chlorophyll content and photosynthetic rate during the vegetation season in maize grown at different constant soil water levels. — In: Slavík, B. (ed.): Water Stress in Plants. Pp. 210–218. Publ. House czechosl. Acad. Sci., Praha, Dr. W. Junk N.V. Publ., The Hague 1965.

Šesták, Z., Zima, J., Strnadová, H.: Ontogenetic changes in the internal limitations of bean-leaf photosynthesis. 2. Activities of photosystems 1 and 2 and non-cyclic photophosphorylation and their dependence on photon flux density. — Photosynthetica 11: 282–290, 1977.

Šesták, Z., Zima, J., Wilhelmová, N.: Ontogenetic changes in the internal limitations to bean-leaf photosynthesis. 4. Effect of pH of the isolation and/or reaction medium on the activities of photosystems 1 and 2. — Photosynthetica 12: 1–6, 1978b.

Shabel'skaya, É. F., Gvardiyan, V. N.: Vliyanie prodolzhitel'nogo polnogo zatemneniya na soderzhanie fotolabil'noï formy khlorofilla v list'yakh rasteniï. [Effect of prolonged full darkening on the content of photolabile form of chlorophyll in leaves of plants.] — Vestsi Akad. Navuk belarus. SSR, Ser. biyal. Navuk 1978 (1): 115–116, 1978.

Shakhov, A. A., Golubkova, B. M.: Struktura khloroplastov v ontogeneze rasteniï. [Structure of chloroplasts in plant ontogeny.] — Bot. Zh. 49: 503–510, 1964.

Shakhov, A. A., Golubkova, B. M.: Struktura khloroplastov i lipoidnaya globulyatsiya. [Structure of chloroplasts and lipoid globulation.] — Bot. Zh. 51: 551–553, 1966.

Sharma, G. K., Dunn, D. B.: Effect of environment on the cuticular features in Kalanchoe fedschenkoi. — Bull. Torrey bot. Club 95: 464–473, 1968.

Sharma, G. K., Dunn, D. B.: Environmental modifications of leaf surface traits in Datura stramonium. — Can. J. Bot. 47: 1211–1216, 1969.

Sharpe, P. J. H., DeMichele, D. W.: A morphological and physiological model of the leaf. — Trans. ASAE 17: 355–359, 1974.

Shatilov, I. S., Sharov, A. F.: Fotosinteticheskiï potentsial, intensivnost' fotosinteza i rol' otdel'nykh organov rasteniï v formirovanii biologicheskogo urozhaya ozimoï pshenitsy na raznykh agrofonakh. [Photosynthetic potential, photosynthetic rate and the role of separate plant organs in the formation of the photosynthetic surface of winter wheat grown under different soil fertility conditions.] — Sel'skokhoz. Biol. 13: 36–43, 1978.

Shatkovskiï, T. A.: Soderzhanie pigmentov i intensivnost' fotosinteza v list'yakh yabloni po dline pobega. [Pigment content and photosynthetic rate in apple tree leaves along the shoot.] — In: Fotosintez Odnoletnikh i Mnogoletnikh Rasteniï. Pp. 72–95, 140. Shtiintsa, Kishinev 1972.

Sheehy, J. E.: Some optical properties of leaves of eight temperate forage grasses. — Ann. Bot. 39: 377–386, 1975.

Shiba, Y.: Seasonal-changes in rates of photosynthesis and respiration of three poplar clones in relation to leaf age. — In: Monsi, M., Saeki, T. (ed.): Ecophysiology of Photosynthetic Productivity. JIBP Synthesis Vol. 19. Pp. 67–72. Univ. Tokyo Press, Tokyo 1978.

Shibata, H., Ochiai, H.: [Studies on δ-amino levulinic acid dehydratase during chloroplast development in radish cotyledons.] — Amino Acid nucl. Acid 32: 16–24, 1975.

Shibata, H., Ochiai, H.: Studies on δ-amino levulinic acid dehydratase in radish cotyledons during chloroplast development. — Plant Cell Physiol. 17: 281—288, 1976.

Shiryaev, A. I.: Submikroskopicheskaya i Makromolekulyarnaya Organizatsiya Khloroplastov. [Submicroscopic and Macromolecular Organization of Chloroplasts.] — Naukova Dumka, Kiev 1978.

Shishkanu, G. V.: Kontsentratsiya pigmentov v list'yakh yabloni. [Concentration of pigments in apple tree leaves.] — In: Fotosintez Sel'skokhozyaïstvennykh Rasteniï Moldavii v Svyazi s Usloviyami Proizrastaniya. Pp. 49–67. Red.-izd. Otdel Akad. Nauk mold. SSR, Kishinev 1970.

Shishkanu, G. V., Grozov, D. N., Grati, M. I.: Fotosinteticheskaya raznokachestvennost' krony u yabloni. [Photosynthetic heterogeneity of apple tree crown.] — In: Fotosinteticheskaya Deyatel'nost' Yabloni i Slivy v Usloviyakh Moldavii. Pp. 3–39. Kishinev 1970.

Shlyk, A. A., Averina, N. G.: Stimulyatsiya kinetinom temnovogo nakopleniya protokhlorofillida v normal'nykh zelenykh list'yakh. [Kinetin stimulated dark accumulation of protochlorophyllide in normal green leaves.] — Dokl. Akad. Nauk SSSR 186: 1209–1212, 1969.

Shlyk, A. A., Averina, N. G.: O kharaktere vliyaniya kinetina na protsess nakopleniya protokhlorofillida v étiolirovannykh i zelenykh list'yakh yachmenya. [Character of kinetin effect on the process of protochlorophyllide accumulation in etiolated and green leaves of barley.] — Dokl. Akad. Nauk SSSR 213: 235–238, 1973.

Shlyk, A. A., Fradkin, L. I.: Izotopno-kineticheskiĭ analiz vozmozhnosti posledovatel'nogo biosinteza khlorofillov *b* i *a*. [Isotope-kinetic analysis of the possibility of sequential biosynthesis of chlorophylls *b* and *a*.] — Biofizika 6: 424–435, 1961.

Shlyk, A. A., Fradkin, L. I.: O skorosti metabolizma khlorofilla v zelenom rastenii. [Rate of chlorophyll metabolism in a green plant.] — Biofizika 7: 281–291, 1962.

Shlyk, A. A., Gaponenko, V. I., Nikolaeva, G. N., Stanishevskaya, E. M., Shevchuk, S. N., Lositskaya, T. V., Mikhaĭlova, S. A.: Metabolicheskie proyavleniya geterogennosti khlorofillov *a* i *b* v zavisimosti ot osveshchennosti. [Metabolic manifestation of heterogeneity of chlorophylls *a* and *b* in dependence on illuminance.] — Dokl. Akad. Nauk SSSR 193: 487–490, 1970.

Shlyk, A. A., Gaponenko, V. I., Prudnikova, I. V., Kukhtenko, T. V., Lyakhnovich, Ya. P., Kaler, V. L.: Sravnitel'noe issledovanie obnovleniya khlorofilla v raznykh chastyakh rasteniya. [Comparative study of chlorophyll renewal in various parts of the plant.]—Fiziol. Rast. 7: 625–637, 1960.

Shlyk, A. A., Kostyuk, N. N.: Vliyanie δ-aminolevulinovoĭ kisloty na vyzvannoe khloramfenikolom ingibirovanie temnovogo nakopleniya protokhlorofillida v zelenykh list'yakh yachmenya. [Effect of δ-aminolevulinic acid on chloramphenicol induced inhibition of dark accumulation of protochlorophyllide in green barley leaves.] — Dokl. Akad. Nauk SSSR 202: 707–710, 1972.

Shlyk, A. A., Kukhtenko, T. V.: Kinetika udel'nykh aktivnosteĭ ugleroda fitola i forbina khlorofillov *a* i *b* v protsesse obnovleniya. [Kinetic of specific activities of carbon of the phytol and phorbine chlorophylls *a* and *b* in the process of renewal.] — Fiziol. Rast. 8: 526–535, 1961.

Shlyk, A. A., Mikhaĭlova, S. A.: Lipofil'nye i gidrofil'nye formy khlorofillov *a* i *b* i osobennosti ikh metabolizma u sinkhronnoĭ kul'tury khlorelly. [Lipophilic and hydrophylic forms of chlorophylls *a* and *b* and peculiarities of their metabolism in a synchronous culture of *Chlorella*.] — Dokl. Akad. Nauk SSSR 177: 236–239, 1967.

Shlyk, A. A., Savchenko, G. E.: Fraktsionirovanie nesushchikh protokhlorofillid chastits khloroplastov zelenykh list'ev yachmenya. [Fractionation of protochlorophyllide-bearing particles of chloroplasts from green barley leaves.] — Dokl. Akad. Nauk SSSR 176: 1437–1440, 1967.

Shlyk, A. A., Savchenko, G. E.: Metabolizm protokhlorofillida v zelenykh list'yakh. [Protochlorophyllide metabolism in green leaves.] — In: Andreenko, S. S. (ed.): Fiziologiya i Biokhimiya Zdorovogo i Bol'nogo Rasteniya. Pp. 185–197. Izdat. mosk. Univ., Moskva 1970.

Shlyk, A.A., Savchenko, G. E., Stanishevskaya, E. M., Shevchuk, S. N., Gaponenko, V. I., Gatikh, O. A.: Rol' fitokhroma v metabolizme khlorofilla v zelenom rastenii. [Role of phytochrome in chlorophyll metabolism in a green plant.] — Dokl. Akad. Nauk SSSR 171: 1443–1446, 1966.

Shmat'ko, I. G., Gulyaev, B. I., Shvedova, O. E., Golik, K. N., Latashenko, O. P.: Parametry vodnogo rezhima i gazoobmena sortov ozimoĭ pshenitsy pri ukhudshenii vodoobespechennosti. [Parameters of water relations and gas exchange in winter wheat varieties under limited water supply.] — Fiziol. Biokhim. kul't. Rast. 11: 312–317, 332, 1979.

Shomer-Ilan, A., Beer, S., Waisel, Y.: *Suaeda monoica*, a C₄ plant without typical bundle sheaths. — Plant Physiol. 56: 676–679, 1975.

Shul'gin, I. A.: Solnechnaya Radiatsiya i Rastenie. [Solar Radiation and Plant.] — Gidrometeoizdat, Leningrad 1967.

Shul'gin, I. A.: Rastenie i Solntse. [The Plant and Sun.] — Gidrometeoizdat, Leningrad 1973.

Shul'gin, I. A., Nichiporovich, A. A., Klimov, S. V., Mureĭ, I. A.: K strukturnoĭ organizatsii lista kak optikofotosinteziruyushcheĭ sistemy. [Structural organization of the leaf as an optical photosynthetic system.] — Fiziol. Rast. 24: 684–690, 1977.

Shumway, L. K., Kleinhofs, A.: Aspects of the biochemistry and ultrastructure of a cytoplasmically inherited plastid defect (Dpl) of tobacco. — Biochem. Genet. 8: 271–280, 1973.

Signol, M.: Modifications morphologiques, pigmentaires et cytologiques induites chez divers groupes de végétaux par quelques substances antichlorophylliennes. — Rev. gén. Bot. 72: 417–573, 1965.

Silaeva, A. M.: Rol' svetovogo faktora v organizatsii struktury khloroplastov. [Role of light factor in organization of chloroplast structure.] — Fiziol. Biokhim. kul't. Rast. 10: 563–572, 1978a.

Silaeva, A. M.: Struktura Khloroplastov i Faktory Sredy. [Chloroplast Structure and Environmental Factors.] — Naukova Dumka, Kiev 1978b.

Silaeva, A. M., Silaev, A. V.: Metody kolichestvennogo analiza élektronno-mikroskopicheskikh izobrazheniĭ khloroplastov. [Methods for quantitative analysis of electron-microscopic images of chloroplasts.] — Fiziol. Biokhim. kul't. Rast. 11: 547–562, 1979.

Silsbury, J. H.: Leaf growth in pasture grasses. — Trop. Grasslands 4: 17–36, 1970.

Silvius, J. E., Johnson, R. R., Peters, D. B.: Effect of water stress on carbon assimilation and distribution in soybean plants at different stages of development. — Crop Sci. 17: 713–716, 1977.

Silvius, J. E., Kremer, D. F., Lee, D. R.: Carbon assimilation in soybean leaves at different stages of development. — Plant Physiol. 62: 54–58, 1978.

Simola, L. K.: Development of chloroplasts in intact Atropa belladonna and in stem callus cultures during greening and leaf differentiation. — Ann. Acad. Sci. fenn., Ser. A IV. Biol. 196: 1–10, 1973.

Simonis, W.: CO₂-Assimilation und Stoffproduktion trocken gezogener Pflanzen. — Planta 35: 188–224, 1947.

Sinclair, J., Garland, S., Arnason, T., Hope, P., Granville, M.: Polychlorinated biphenyls and their effects on photosynthesis and respiration. — Can. J. Bot. 55: 2679–2684, 1977.

Sinclair, T. R., Schreiber, M. M., Hoffer, R. M.: Diffuse reflectance hypothesis for the pathway of solar radiation through leaves. — Agron. J. 65: 276–283, 1973.

Singer, S. J., Nicolson, G. L.: The fluid mosaic model of the structure of cell membranes. — Science 175: 720–731, 1972.

Sionit, N., Kramer, P. J.: Water potential and stomatal resistance of sunflower and soybean subjected to water stress during various growth stages. — Plant Physiol. 58: 537–540, 1976.

Sirohi, G. S., Ghildiyal, M. C.: Varietal differences in photosynthetic carboxylases & chlorophylls in wheat varieties. — Indian J. exp. Biol. 13: 42–44, 1975.

Sirohi, G. S., Shrivastava, A. K.: Carbon dioxide compensation concentration and its relationship to photorespiration and net carbon exchange — a review. — Indian J. Plant Physiol. 21: 70–89, 1978.

Sironval, C. (ed.): Le Chloroplaste, Croissance et Vieillissement. — Masson et Cie, Paris 1967.

Sirotkin, Yu. D., Anufrieva, V. G.: Osobennosti sezonnogo rosta sosny i eli v smeshannykh lesnykh kul'turakh. [Features of seasonal growth of pine and fir in mixed forests.] — Lesovedenie les. Khoz. 1973 (7): 50–57, 1973.

Sisler, E. C., Klein, W. H.: The effect of age and various chemicals on the lag phase of chlorophyll synthesis in dark grown bean seedlings. — Physiol. Plant. 16: 315–322, 1963.

Sisson, W. B.: Photosynthesis, growth, and ultraviolet irradiance absorbance of Cucurbita pepo L. leaves exposed to ultraviolet-B radiation (280–315 nm). — Plant Physiol. 67: 120–124, 1981.

Sisson, W. B., Caldwell, M. M.: Atmospheric ozone depletion: Reduction of photosynthesis and growth of a sensitive higher plant exposed to enhanced u.v. - B. radiation. — J. exp. Bot. 28: 691–705, 1977.

Skene, D. S.: Chloroplast structure in mature apple leaves grown under different levels of illumination and their response to changed illumination. — Proc. roy. Soc. London B 186: 75–78, 1974.

Skośkiewicz, K.: Stomatal movements in summer rape Bronowski IHAR (Brassica napus L. ssp. oleifera (Metzg.) Sinsk. f. annua Thel.) in dependence on the age of the leaf, water deficit, light intensity and CO₂ concentration. — Hodowla Rośl., Aklimat. Nasienn. 17: 359–386, 1973.

Slatyer, R. O.: Comparative photosynthesis, growth and transpiration of two species of Atriplex. — Planta 93: 175–189, 1970.

Slatyer, R. O., Bierhuizen, J. F.: The influence of several transpiration suppressants on transpiration, photosynthesis and water-use efficiency of cotton leaves. — Aust. J. biol. Sci. 17: 131–146, 1964.

Slavík, B.: Grafické stanovení intensity průduchové a kutikulární složky transpirace rostlin. [Graphic determination of stomatal and cuticular components of transpiration rate in plants.] — Českoslov. Biol. 7: 347–352, 1958a.

Slavík, B.: The influence of water deficit on transpiration. — Physiol. Plant. 11: 524–536, 1958b.

Slavík, B.: The distribution pattern of transpiration rate, water saturation deficit, stomata number and size, photosynthetic and respiration rate in the area of the tobacco leaf blade. — Biol. Plant. 5: 143–153, 1963.

Slavík, B.: Methods of Studying Plant Water Relations. — Academia, Praha 1974. Springer-Verlag, Berlin–Heidelberg–New York 1974.

Slavík, B.: Water stress, photosynthesis and the use of photosynthates. — In: Cooper, J. P. (ed.): Photosynthesis and Productivity in Different Environments. Pp. 511–536. Cambridge Univ. Press, Cambridge–London–New York–Melbourne 1975.

Smelyanskaya, E. P.: Osobennosti anatomii tkaneĭ lista sakharnoĭ svekly v svyazi s predposevnoĭ obrabotkoĭ semyan. [Anatomical peculiarities of leaf tissue of sugar beet with respect to pre-sowing seed management.] — In: Fotosintez i Produktivnost' Rasteniĭ. Pp. 72–81. Naukova Dumka, Kiev 1965.

Smillie, R. M.: Photosynthetic & respiratory activities of growing pea leaves. — Plant Physiol. 37: 716–721, 1962.

Smillie, R. M., Krotkov, G.: Enzymic activities of subcellular particles from leaves. IV. Photosynthetic phosphorylation and photosynthesis by isolated chloroplasts from pea leaves. — Can. J. Bot. 37: 1217–1225, 1959.

Smirnova, N. P.: K voprosu o svyazi khlorofilla s lipo–proteidnym kompleksom inbrednoĭ, gibridnoĭ i svobodno opylyaemoĭ kukuruzy. [Binding of chlorophyll with the lipo-protein complex of inbred, hybrid and free–pollinated maize.] — Tr. gorsk. sel'.-khoz. Inst. 27: 160—163, 1967.

Smith, E. W., Tolbert, N. E., Ku, H. S.: Variables affecting the $CO_2$ compensation point. — Plant Physiol. 58: 143–146, 1976.

Smith, J. H. C.: The relationship of plant pigments to photosynthesis. — J. chem. Educ. 26: 631–638, 1949.

Soikkeli, S.: Seasonal changes in mesophyll ultrastructure of needles of Norway spruce (Picea abies). — Can. J. Bot. 56: 1932–1940, 1978.

Soikkeli, S.: Ultrastructure of the mesophyll in Scots pine and Norway spruce: seasonal variation and molarity of the fixative buffer. — Protoplasma 103: 241–252, 1980.

Solárová, J.: Stomata reactivity in leaves at different insertion level during wilting. — In: Slavík, B. (ed.): Water Stress in Plants. Pp. 147–154. Publ. House czechosl. Acad. Sci., Praha 1965.

Solárová, J.: Changes in minimal diffusive resistances of leaf epidermes during ageing of primary leaves of Phaseolus vulgaris L. — Biol. Plant. 15: 237–240, 1973.

Solárová, J.: Diffusive conductances of adaxial (upper) and abaxial (lower) epidermes: Response to quantum irradiance during development of primary Phaseolus vulgaris L. leaves. — Photosynthetica 14: 523–531, 1980.

Solárová, J., Pospíšilová, J.: Photosynthetic characteristics during ontogenesis of leaves 8. Stomatal diffusive conductance and stomata reactivity. — Photosynthetica 17: 101–151, 1983.

Solárová, J., Pospíšilová, J.: The effect of water stress during ontogeny of primary bean leaves on the light-induced stomatal opening. — Biol. Plant. 26: 56—61, 1984.

Solárová, J., Pospíšilová, J., Tichá, I., Čatský, J., Pleskanka, J.: Modifikace závislosti epidermální vodivosti na kvantové ozářenosti biologickými a ekologickými faktory. [Relationship between epidermal conductance and quantum irradiance modified by biological and ecological factors.] — In: Dny Rostlinné Fyziologie II. Pp. 375–379. Vysoká škola Zemědělská, Brno 1980.

Soldatini, G. F.: Changes of glycolate oxidase activity with leaf age in Zea mays L. - Z. Pflanzenphysiol. 94: 267–271, 1979.

Specht-Jürgensen, I.: Untersuchungen über Stickstoffverbindungen und Chlorophyll während des Vergilbens der Laubblätter von *Ginkgo biloba*. II. Isolierte Laubblätter. — Flora A 157: 471–502, 1967.

Spruit, C. J. P., Raven, C. W.: Regeneration of protochlorophyll in dark grown seedlings following illumination with red and far red light. — Acta bot. neerl. 19: 165–174, 1970.

Srivastava, B. I. S., Atkin, R. K.: Effect of second-leaf removal on the nucleic acid metabolism of senescing first seedling leaf of barley. — Biochem. J. 107: 361–366, 1968.

Srivastava, H. S., Jolliffe, P. A., Runeckles, V. C.: Inhibition of gas exchange in bean leaves by NO₂. — Can. J. Bot. 53: 466–474, 1975.

Staehelin, L. A.: Freeze-fracture studies of green plant and *Prochloron* thylakoids. A status report. — In: Akoyunoglou, G. (ed.): Photosynthesis. Vol. III. Pp. 3–14. Balaban Int. Sci. Services, Philadelphia 1981.

Stamp, P.: Der Chlorophyllgehalt, die PEP- und RuDP-Carboxylase-Aktivitäten während der Blattentwicklung einer ergrünenden Chlorophyllmutante und einer normalen Linie von *Zea mays* L. - Z. Pflanzenphysiol. 86: 395–404, 1978.

Stamp, P.: Pigmentgehalte und Aktivitäten photosynthetisch wirksamer Enzyme in Blättern junger Maispflanzen in Abhängigkeit von der Temperatur zur Zeit der Kornausreife. — Z. Acker- Pflanzenbau 148: 230–238, 1979.

Starzecki, W.: The role of the palisade and spongy parenchymas of leaves in photosynthesis. — Acta Soc. Bot. Pol. 31: 419–436, 1962.

Stearns, M. E., Wagenaar, E. B.: Ultrastructural changes in chloroplasts of autumn leaves. — Can. J. gen. Cytol. 13: 550–560, 1971.

Steer, B. T.: The dynamics of leaf growth and photosynthetic capacity in *Capsicum frutescens* L. — Ann. Bot. 35: 1003–1015, 1971.

Steer, B. T.: Leaf-growth parameters associated with photosynthetic capacity in expanding leaves of *Capsicum frutescens* L. – Ann. Bot. 36: 377–384, 1972.

Steer, B. T.: Control of ribulose-1,5-diphosphate carboxylase activity during expansion of leaves of *Capsicum frutescens* L. – Ann. Bot. 37: 823–829, 1973.

Steer, B. T., Darbyshire, B.: Some aspects of carbon metabolism and translocation in onions. — New Phytol. 82: 59–68, 1979.

Steffens, G. L.: Chlorophyll of "chloroplasts" from Connecticut shade tobacco leaves as they mature. — Tobacco 151 (25): 20–23, 1960. Tobacco Sci. 4: 234–237, 1960.

Stevenson, K. R., Shaw, R. H.: Diurnal changes in leaf resistance to water vapor diffusion at different heights in a soybean canopy. — Agron. J. 63: 17–19, 1971.

Stigter, C. J.: The epidermal resistance to diffusion of water vapour: an improved measuring method and field results in Indian corn (*Zea mays*). — Agr. Res. Rep. (Wageningen) 831: 1–25, 1974.

Stigter, C. J., Goudriaan, J., Bottemanne, F. A., Birnie, J., Lengkeek, J. G., Sibma, L.: Experimental evaluation of a crop climate simulation model for Indian corn (*Zea mays* L.). — Agr. Meteorol. 18: 163–186, 1977.

Stigter, C. J., Lammers, B.: Leaf diffusion resistance to water vapour and its direct measurement. III. Results regarding the improved diffusion porometer in growth rooms and fields of Indian corn (*Zea mays*). — Med. Landbouwhogesch. Wageningen 74–21: 1–76, 1974.

Stoddart, J. L., Thomas, H.: Leaf senescence. — In: Boulter, D., Parthier, B. (ed.): Nucleic Acids and Proteins in Plants I. Structure, Biochemistry and Physiology of Proteins. Pp. 592–636. Springer-Verlag, Berlin–Heidelberg–New York 1982.

Strnadová, H., Šesták, Z.: Reliability of methods used for determining ontogenetic changes in Hill reaction rate. — Photosynthetica 8: 130–133, 1974.

Sud'ïna, O. G., Dovbysh, K., Golod, M. G., Fomishyna, R. M.: Do pytannya pro stan khlorofilazy ta ïogo minlyvist'. [Chlorophyllase state and its variability.] — Ukr. bot. Zh. 32: 330—334, 397, 1975.

Sud'ïna, O. G., Golod, M. G., Dovbysh, K. P., Baïdulova-Babko, T. Yu.: Dynamika vmistu pigmentiv ta khlorofilaznoĭ aktyvnosti riznykh bilkovykh fraktsiĭ v ontogenezi lystka. [Dynamics of pigment content and chlorophyllase activity of different protein fractions in leaf ontogenesis.] — Ukr. bot. Zh. 33: 132–136, 1976.

Sundqvist, C., Björn, L. O., Virgin, H. I.: Factors in chloroplast differentiation. — In: Reinert, J. (ed.): Chloroplasts. Pp. 201–224. Springer-Verlag, Berlin –Heidelberg–New York 1980.

Sundqvist, C., Odengård, B., Persson, G.: Light-stimulated accumulation of protochlorophyllide in leaves of different ages treated with δ-aminolevulinic acid. — Plant Sci. Lett. 4: 89–96, 1975.

Suzuki, S.: The role of cotyledons in growth and photosynthesis of radish plants. — J. jap. Soc. hort. Sci. 45: 275–282, 1976.

Sveshnikova, I. N., Kulaeva, O. N., Bolyakina, Yu. P.: Obrazovanie lamell i gran v khloroplastakh zheltykh list'ev pod deĭstviem 6-benzilaminopurina. [Lamellae and grana formation in chloro- plasts of yellow leaves induced by 6-benzylaminopurine.] — Fiziol. Rast. 13: 769–774, 1966.

Swanson, C. A., Hoddinott, J.: Effect of light and ontogenetic stage on sink strength in bean leaves. — Plant Physiol. 62: 454–457, 1978.

Sýkorová, M.: Intensity of photosynthesis of the leaves and of the bracts of maize (Zea mays L.) when measured by the gravimetric method. — Acta Fac. Rerum nat. Univ. comen,. Physiol. Plant. 11: 61–67, 1976.

Symonides, E.: Changes in the chlorophyll a and b content in different-aged leaves of Spergula vernalis Willd. — Acta Soc. Bot. Pol. 43: 235–241, 1974.

Syvertsen, J. P., Cunningham, G. L.: Rate of leaf production and senescence and effect of leaf age on net gas exchange in creosotebush. — Photosynthetica 11: 161–166, 1977.

Syvertsen, J. P., Smith, M. L., Jr., Allen, J. C.: Growth rate and water relations of citrus leaf flushes. — Ann. Bot. 47: 97–105, 1981.

Szczepański, A.: Chlorophyll in the assimilation parts of helophytes. — Pol. Arch. Hydrobiol. 20: 67–71, 1973.


Tabentskiĭ, A. A.: K voprosu ob upravlenii protsessami obrazovaniya zelenykh plastid. [Control of processes of green plastid formation.] — Izv. Akad. Nauk SSSR, Ser. biol. 1953: 71–95, 1953.

Tageeva, S. V., Generozova, I. P., Derevyanko, V. G., Ladygin, V. G., Semenova, G. A.: Razno- obrazie ul'trastrukturnoĭ organizatsii khloroplastov i zavisimosti ot funktsional'nogo sostoyaniya tkaneĭ rastenii, geneticheskogo faktora i svetovykh uslovii. [Ultrastuctural organization of chloro- plasts as affected by the functional state of plant tissues and organs, genetic factors and light conditions.] — In: Zalenskiĭ, O. V. (ed.): Fotosintez i Ispol'zovanie Solnechnoĭ Ėnergii. Pp. 126–144. Nauka, Leningrad 1971.

Tageeva, S. V., Kutík, J., Popov, V. I.: The development of chloroplast ultrastructure during the ontogeny of Larix europaea LAM. et DC. needles. — Photosynthetica 15: 258–260, 1981.

Tageeva, S. V., Savchenko, G. E., Semenova, G. A., Shlyk, A. A.: Dinamika organizatsii khloro- plastov i metabolizm protokhlorofillida v raznykh zonakh lista kukuruzy pri zatemnenii i osvesh- chenii. [Dynamics of chloroplast organization and metabolism of protochlorophyllide in various parts of a maize leaf in darkness and light.] — Fiziol. Rast. 16: 581–593, 1969.

Takami, S., Turner, N. C., Rawson, H. M.: Leaf expansion of four sunflower (Helianthus annuus L.) cultivars in relation to water deficits. I. Patterns during plant development. — Plant, Cell Environm. 4: 399–407, 1981.

Takano, Y., Tsunoda, S.: Light reflection, transmission and absorption rates of rice leaves in rela- tion to their chlorophyll and nitrogen contents. — Tohoku J. agr. Res. 21: 111–117, 1970.

Takaoki, T.: [A simple volumetric method for measuring photosynthesis and respiration rates in higher plants (2).] — Bot. Mag. (Tokyo) 82: 244–252, 1969.

Takeda, G., Udagawa, T.: [Ecological studies on the photosynthesis of winter cereals III. Changes of the photosynthetic ability of various organs with growth.] — Proc. Crop Sci. Soc. Jap. 45: 357–368, 1976.

Tamàs, I. A., Atkins, B. D., Ware, S. M., Bidwell, R. G. S.: Indoleacetic acid stimulation of phosphorylation and bicarbonate fixation by chloroplast preparations in light. — Can. J. Bot. 50: 1523–1527, 1972.

Tan, G. - Y., Dunn, G. M.: Stomatal length, frequency, and distribution in Bromus inermis Leyss. — Crop Sci. 15: 283–286, 1975.

Tanaka, A.: Studies on the nutrio-physiology of leaves of rice plants. — J. Fac. Agr. Hokkaido Univ. 51: 449–550, 1961.

Tanaka, A., Fujita, K.: Growth, photosynthesis and yield components in relation to grain yield of the field bean. — J. Fac. Agr. Hokkaido Univ. 59: 145–238, 1979.

Tanaka, A., Fujita, K., Kikuchi, K.: Nutrio-physiological studies on the tomato plant. III. Photosynthetic rate of individual leaves in relation to the dry matter production of plants. — Soil Sci. Plant Nutr. 20: 173–183, 1974.

Tanaka, A., Kikuchi, K.: [Nutrio-physiological studies on field beans (Phaseolus vulgaris L.). 3. Changes in the photosynthetic rate of individual leaves during growth of the determinate and the semi-determinate.] – J. Sci. Soil Manure Jap. [Nippon Dojo-Hiryogaku Zasshi] 47: 506–510, 1976.

Tanaka, A., Yamaguchi, J.: Dry matter production, yield components and grain yield of the maize plant. — J. Fac. Agr., Hokkaido Univ. 57: 71–132, 1972.

Tanaka, K., Sugahara, K.: Role of superoxide dismutase in defense against $SO_2$ toxicity and an increase in superoxide dismutase activity with $SO_2$ fumigation. — Plant Cell Physiol. 21: 601–611, 1980a.

Tanaka, K., Sugahara, K.: Role of superoxide dismutase in the defense against $SO_2$ toxicity and induction of superoxide dismutase with $SO_2$ fumigation. — Res. Rep. nat. Inst. environ. Stud. 11 (Studies on the Effects of Air Pollutants on Plants and Mechanisms of Phytotoxicity): 155–164, 1980b.

Tanaka, S., Tatemichi, Y.: [Simulating dry matter production in various light conditions in the tobacco field.] — Bull. Hatano Tobacco exp. Sta. 69: 1–37, 1971.

Tanaka, T.: [Studies on the light-curves of carbon assimilation of rice plants — The interrelation among the light-curves, the plant type and the maximizing yield of rice.] — Bull. nat. Inst. agr. Sci. (Jap.), Ser. A 1972 (19): 1–100, 1972.

Tanaka, T., Matsushima, S.: [Analysis of yield-determining process and its application to yield prediction and culture improvement of lowland rice. XCIV. Relation between the light intensity on both sides and the amount of carbon assimilation in each side of a single leaf-blade.] —Proc. Crop Sci. Soc. Jap. 39: 325–329, 1970.

Tanaka, T., Mizuno, M.: [Effect of age and part of the leaf on photosynthetic rate of Ophiopogon chekiangensis.] — Shoyakugaku Zasshi 35: 169–172, 1981.

Tarnowska, K.: Badania nad vpływem niektórych czynników na tworzenie się karotenu w liściach pomidorów. [Effect of some factors on carotene formation in tomato leaves.] — Roczn. Nauk roln. 89 (A-4): 645–655, 1964.

Tatemichi, Y.: [Studies on the photosynthesis of tobacco plants. I. An apparatus for the measurement of photosynthesis in plant, and changes of photosynthesis and respiration in the course of development.] — Proc. Crop Sci. Soc. Jap. 37: 129–134, 1968.

Taylor, A. O., Rowley, J. A., Jepsen, N. M.: Factors regulating the growth rate of Lolium perenne L. cv. "Grasslands Ruanui" and L. multiflorum Lam. cv. "Grasslands Tama" a tetraploid. 1. Seeds, photosynthetic rates, photosynthetic products, translocation, and proportion of plant parts. — New Zeal. J. Bot. 9: 504–518, 1971.

Taylor, F. J.: Some aspects of the development of mango (Mangifera indica L.) leaves I. Leaf area, dry weight and water content. — New Phytol. 69: 377–394, 1970.

Taylor, R. J., Pearcy, R. W.: Seasonal patterns of the $CO_2$ exchange characteristics of understory plants from a deciduous forest. — Can. J. Bot. 54: 1094–1103, 1976.

Tenhunen, J. D., Hesketh, J. D., Gates, D. M.: Leaf photosynthesis models. — In: Hesketh,

J. D., Jones, J. W. (ed.): Predicting Photosynthesis for Ecosystem Models. Vol. I. Pp. 123–181. CRC Press, Boca Raton 1980a.

Tenhunen, J. D., Meyer, A., Lange, O. L., Gates, D. M.: Development of a photosynthesis model with an emphasis on ecological applications. V. Test of the applicability of a steady-state model to description of net photosynthesis of *Prunus armeniaca* under field conditions. — Oecologia 45: 147–155, 1980b.

Teramura, A. H., Biggs, R. H., Kossuth, S.: Effects of ultraviolet-B irradiances on soybean. II. Interaction between ultraviolet-B and photosynthetically active radiation on net photosynthesis, dark respiration, and transpiration. — Plant Physiol. 65: 483–488, 1980.

Terry, N.: Limiting factors in photosynthesis I. Use of iron stress to control photochemical capacity *in vivo*. — Plant Physiol. 65: 114–120, 1980.

Terry, N., Mortimer, D. C.: Estimation of the rates of mass carbon transfer by leaves of sugar beet. — Can. J. Bot. 50: 1049–1054, 1972.

Tetley, R. M., Thimann, K. V.: The metabolism of oat leaves during senescence. I. Respiration, carbohydrate metabolism, and the action of cytokinins. — Plant Physiol. 54: 294–303, 1974.

Tezuka, T., Sekiya, H., Ohno, H.: Physiological studies on the action of CCC in Kyoko grapes. — Plant Cell Physiol. 21: 969–977, 1980.

Thiagarajah, M. R., Hunt, L. A., Hunter, R. B.: Effects of short-term temperature fluctuations on leaf photosynthesis in corn (*Zea mays*). — Can. J. Bot. 57: 2387–2393, 1979.

Thiagarajah, M. R., Hunt, L. A., Mahon, J. D.: Effects of position and age on leaf photosynthesis in corn (*Zea mays*). — Can. J. Bot. 59: 28–33, 1981.

Thimann, K. V. (ed.): Senescence in Plants. — CRC Press, Boca Raton 1980.

Thimann, K. V., Tetley, R. R., Thanh, T. V.: The metabolism of oat leaves during senescence. II. Senescence in leaves attached to the plant. — Plant Physiol. 54: 859–862, 1974.

Thomas, H., Stoddart, J. L.: Leaf senescence. — Annu. Rev. Plant Physiol. 31: 83–111, 1980.

Thomas, J. R., Myers, V. I., Hielman, M. D., Wiegand, C. L.: Factors affecting light reflectance in cotton. — In: Proceedings of the 4th Symposium on Remote Sensing of Environment. Pp. 305–312. Inst. Sci. Technol., Univ. Michigan, Ann Arbor 1966.

Thomas, J. R., Namken, L. N., Oerther, G. F., Brown, R. G.: Estimating leaf water content by reflectance measurements. — Agron. J. 63: 845–847, 1971.

Thomas, J. R., Oerther, G. F.: Estimating nitrogen content of sweet pepper leaves by reflectance measurements. — Agron. J. 64: 11–13, 1972.

Thomas, J. R., Wiegand, C. L., Myers, V. I.: Reflectance of cotton leaves and its relation to yield. — Agron. J. 59: 551–554, 1967.

Thomas, S. M., Hall, N. P., Merrett, M. J.: Ribulose 1,5-bisphosphate carboxylase/oxygenase activity and photorespiration during the ageing of flag leaves of wheat. — J. exp. Bot. 29: 1161–1168, 1978.

Thomas, S. M., Thorne, G. N.: Effect of nitrogen fertilizer on photosynthesis and ribulose 1,5-diphosphate carboxylase activity in spring wheat in the field. — J. exp. Bot. 26: 43–51, 1975.

Thompson, A., Vogel, J., Lee, R. E.: Carbon dioxide uptake in relation to a plastid inclusion body in the succulent *Kalanchoë pinnata* Persoon. — J. exp. Bot. 28: 1037–1041, 1977.

Thornley, J. H. M.: Respiration, growth and maintenance in plants. — Nature 227: 304–305, 1970.

Thornley, J. H. M.: Energy, respiration, and growth in plants. — Ann. Bot. 35: 721–728, 1971.

Thornley, J. H. M.: Mathematical Models in Plant Physiology. A Quantitative Approach to Problems in Plant and Crop Physiology. — Academic Press, London–New York–San Francisco 1976.

Thornley, J. H. M.: Growth, maintenance and respiration: a re-interpretation. — Ann. Bot. 41: 1191–1203, 1977.

Thornley, J. H. M., Hurd, R. G., Pooley, A.: A model of growth of the fifth leaf of tomato. — Ann. Bot. 48: 327–340, 1981.

Tichá, I.: Einige quantitative anatomische Merkmale von Blättern verschiedener Insertionshöhe bei *Potamogeton perfoliatus* L. und *P. lucens* L. — Biol. Plant. 6: 108–116, 1964.

Tichá, I.: Heterogenita některých anatomických znaků a ukazatelů vodní bilance na rostlině ve vztahu k fotosyntéze. [Heterogeneity of some anatomical parameters and indexes of water balance in plants as related to photosynthesis.] — Stud. Inform. ÚVTI MZLH, zákl. pom. Vědy Zeměděl. (Praha) 1966 (6-7-8): 1–207, 1966.

Tichá, I.: Ontogenetische Veränderungen der Photosyntheseintensität bei Blättern verschiedener Insertionshöhe von Plectranthus fructicosus L'Hérit.-Pflanzen aus zwei künstlichen Klimas. — Photosynthetica 2: 167–171, 1968.

Tichá, I.: Anatomická a fyziologická heterogenita listů na rostlině. [Anatomical and physiological heterogeneity of leaves on the plant.] — Thesis. Inst. exp. Bot. czechosl. Acad. Sci., Praha 1970a.

Tichá, I.: Insertion gradients of leaf area and photosynthetic rate in plants of different age. — In: Dykyjová, D. (ed.): Productivity of Terrestrial Ecosystems. Production Processes. Pp. 185–186. Czechosl. Acad. Sci., Praha 1970b.

Tichá, I.: Blattflächenbestimmung bei Plectranthus fructicosus L'Hérit. auf Grund von linearen Messungen und die Veränderungen im Laufe der Blatt- und Pflanzenentwicklung. — Biol. Plant. 16: 152–155, 1974.

Tichá, I.: Der Beitrag einzelner Blätter an der Pflanze zur Photosynthese des gesamten Blattapparates im Laufe der Pflanzenentwicklung. — Biol. Plant. 18: 237–240, 1976a.

Tichá, I.: Photosynthesis of plants in two controlled environments. — In: Doklady i Tezisy Dokladov Koordinatsionnogo Soveshchaniya SĖV, Fotosintez I-18.3. Pp. 32–46. Szeged 1976b.

Tichá, I.: Photosynthetic characteristics during ontogenesis of leaves. 7. Stomata density and sizes. — Photosynthetica 16: 375–471, 1982.

Tichá, I.: Quantitative Veränderungen der Kenngrösse Nettophotosyntheserate mit Rücksicht auf Blattontogenese und -insertion. — Wiss. Z. Humboldt—Univ. (Berlin), math.-nat. R. 33: 300–302, 1984.

Tichá, I., Čatský, J.: Ontogenetic changes in the internal limitations to bean-leaf photosynthesis 3. Leaf mesophyll structure and intracellular conductance for carbon dioxide transfer. — Photosynthetica 11: 361–366, 1977.

Tichá, I., Čatský, J.: Photosynthetic characteristics during ontogenesis of leaves 5. Carbon dioxide compensation concentration. — Photosynthetica 15: 401–428, 1981.

Tichá, I., Čatský, J.: Regulation des photosynthetischen $CO_2$-Transportes durch Aussenluft-$CO_2$-Konzentration im Laufe der Blattontogenese. — In: Hoffmann, P., Hieke, B. (ed.): Photosynthese: Regulation und Evolution. Colloquia Pflanzenphysiol. Nr. 5. Pp. 244–246. Humboldt — Univ., Berlin 1982.

Tichá, I., Čatský, J., Peisker, M.: Ontogenetic changes in the internal limitations to bean-leaf photosynthesis. 6. $CO_2$ dependence of net photosynthetic rate. — Photosynthetica 14: 489–496, 1980.

Tichá, I., Čatský, J., Peisker, M., Kaše, M.: The ontogenetic pattern of leaf photosynthesis as affected by irradiance, carbon dioxide concentration and temperature. — In: Sybesma, C. (ed.): Advances in Photosynthesis Research. Vol. IV. Pp. 255—258. Martinus Nijhoff/Dr W. Junk Publ., The Hague — Boston — Lancaster 1984.

Tinus, R. W.: Impact of the $CO_2$ requirement on plant water use. — Agr. Meteorol. 14: 99–112, 1974.

Todd, G. W., Basler, E.: Fate of various protoplasmic constituents in droughted wheat plants. — Fyton 22: 79–85, 1965.

Tolbert, N. E.: Glycolate metabolism by higher plants and algae. — In: Gibbs, M., Latzko, E. (ed.): Photosynthesis II. Pp. 338–352. Springer-Verlag, Berlin–Heidelberg–New York 1979.

Toyama, S.: Electron microscope studies on the morphogenesis of plastids. X. Ultrastructural changes of chloroplasts in morning glory leaves exposed to ethylene. — Amer. J. Bot. 67: 625–635, 1980.

Trachtenberg, C. H., McCloud, D. E.: Net photosynthesis of peanut leaves at varying light intensities and leaf ages. — Soil Crop Sci. Soc. Florida Proc. 35: 54–55, 1975.

Trebst, A.: Organization of the photosynthetic electron transport system of chloroplasts in the thylakoid membrane. — In: Schäfer, G., Klingenberg, M. (ed.): Energy Conservation in Biological Membranes. Pp. 84–95. Springer-Verlag, Berlin–Heidelberg –New York 1978.

Treffry, T.: Developmental changes in the surface properties of chloroplast membranes. — Planta 126: 11–17, 1975.

Treffry, T.: Biogenesis of the photochemical apparatus. — Int. Rev. Cytol. 52: 159–196, 1978.

Treharne, K. J., Cooper, J. P., Taylor, T. H.: Growth response of orchard-grass (Dactylis glomerata L.) to different light and temperature environments. II. Leaf age and photosynthetic activity. — Crop Sci. 8: 441–445, 1968.

Treharne, K. J., Eagles, C. F.: Effect of temperature on photosynthetic activity of climatic races of Dactylis glomerata L. — Photosynthetica 4: 107–117, 1970.

Tretyak, T. V., Okanenko, A. S.: Osobennosti anatomicheskoĭ struktury i vodnogo rezhima u poliploidnykh form sakharnoĭ svekly. [Some features of the anatomical structure and water regime in polyploid sugar beet.] — Fiziol. Rast. 13: 469–478, 1966.

Tripodi, G.: Negative staining of chloroplast lamellae in Acanthus leaves below the compensation point. — Protoplasma 103: 163–168, 1980.

Troeng, E., Linder, S.: Gas exchange in a 20-year-old stand of Scots pine. II. Variation in net photosynthesis and transpiration within and between trees. — Physiol. Plant. 54: 15–23, 1982.

Troughton, J. H., Slatyer, R. O.: Plant water status, leaf temperature, and the calculated mesophyll resistance to carbon dioxide of cotton leaves. — Aust. J. biol. Sci. 22: 815–827, 1969.

Tschakalova, E.: Struktur-Funktionsbeziehungen bei Phaseolus vulgaris L. unter besonderer Berücksichtigung der Photosynthese I. Untersuchungen der Spaltöffnungen und des Interzellularvolumens von Phaseolus vulgaris-Blättern im Verlauf der Ontogenese. — Godish. sofiĭ. Univ., biol. Fak., Kn. 2, 68: 1–10, 1976.

Tschakalova, E., Hoffmann, P.: Strukturelle und funktionelle Grundlagen des photosynthetischen Gaswechsels bei Triticum aestivum L. – Wiss. Z. Humboldt-Univ. Berlin, math.-naturwiss. Reihe 25: 723–736, 1976.

Tschakalova, E. S., Hoffmann, P. E.: Das Interzellularvolumen in den oberirdischen Organen der Weizenkeimpflanze im Verlauf der Entwicklung. — Dokl. bolg. Akad. Nauk 27: 533–535, 1974.

Tsel'niker, Yu. L.: Ritmy rosta tkaneĭ, khloroplastov i determinatsiya priznakov svetovoĭ i tenevoĭ struktury lista u klena ostrolistnogo. [Growth rhythms of tissues and chloroplasts and determination of characteristics of light and dark leaf structure in Norway maple.] — Fiziol. Rast. 20: 1182–1190, 1973.

Tsel'niker, Yu. L.: Replikatsiya khloroplastov, ee regulyatsiya i znachenie dlya fotosinteza. [Chloroplasts replication, its regulation and significance for photosynthesis.] — In: Mezostruktura i Funktsional'naya Aktivnost' Fotosinteticheskogo Apparata. Pp. 31–45. Ural'skiĭ Gosudarstvennyĭ Universitet, Sverdlovsk 1978.

Tsel'niker, Yu. L., Maĭ, V. V., Andreeva, T. F.: Sootnoshenie aktivnosti ribulozodifosfatkarboksilazy i intensivnosti fotosinteza u list'ev osiny. [Relationship between RuBPC activity and photosynthetic rate in aspen leaves.] — Fiziol. Rast. 28: 953–961, 1981.

Tsoneva, P. N.: Sezonni i v"zrastovi izmeneniya v aktivnosta na khlorofilazata v nyakoi iglolistni i shirokolistni rasteniya. [Seasonal and age changes in chlorophyllase activity in some conifers and broad-leaved plants.] — God. sofiĭ. Univ., biol. Fak., Kn. 2 — Bot., Mikrobiol., Fiziol., Biokhim. Rast. 64: 185–193, 1969/70.

Tsoneva, P. N.: Izmeneniya v aktivnostta na khlorofilazata i pigmentnoto s"d"rzhanie v listata na Pinus silvestris prez proletta. [Changes in chlorophyllase activity and pigment content in Pinus silvestris needles.] — God. sofiĭ. Univ., Biol. Fak. 67 (2): 139–147, 1972/73.

Tsuji, H., Isa, Y., Hatakeyama, I.: Changes in two parameters characterizing the light-photosynthesis curve of growing bean leaves. — In: Monsi, M., Saeki, T. (ed.): Ecophysiology of Photosynthetic Productivity. JIPB Synthesis. Vol. 19. Pp. 46–54. Univ. Tokyo Press, Tokyo 1978a.

354

Tsuji, H., Naito, K., Hatakeyama, I.: Effect of benzyladenine on the changes in two parameters of the light-photosynthesis curve of bean leaves during aging. — In: Monsi, M., Saeki, T. (ed.): Ecophysiology of Photosynthetic Productivity. JIBP Synthesis. Vol. 19. Pp. 55–58. Univ. Tokyo Press, Tokyo 1978b.

Tsuno, Y.: [Some characteristics of the photosynthesis of the potato plant.] — Bull. Fac. Agr., Tottori Univ. 29: 89–95, 1977.

Tucker, C. J., Garratt, M. W.: Leaf optical system modeled as a stochastic process. — Appl. Optics 16: 635–642, 1977.

Tumanow, J. J.: Ungenügende Wasserversorgung und das Welken der Pflanzen als Mittel zur Erhöhung ihrer Dürreresistenz. — Planta 3: 391–480, 1927.

Tuquet, C., Newman, D. W.: Aging and regreening in soybean cotyledons. 1. Ultrastructural changes in plastids and plastogloguli. — Cytobios 29: 43–59, 1980.

Turgeon, R., Webb, J. A.: Leaf development and phloem transport in Cucurbita pepo: Carbon economy. — Planta 123: 53–62, 1975.

Turner, N. C.: Stomatal behavior and water status of maize, sorghum, and tobacco under field conditions. II. At low soil water potential. — Plant Physiol. 53: 360–365, 1974a.

Turner, N. C.: Stomatal response to light and water under field conditions. — In: Bieleski, R. L., Ferguson A. R., Cresswell, M. M. (ed.): Mechanisms of Regulation of Plant Growth. Roy. Soc. New Zeal. Bull. 12. Pp. 423–432. Roy. Soc. New Zeal., Wellington 1974b.

Turner, N. C.: Concurrent comparisons of stomatal behavior, water status, and evaporation of maize in soil at high or low water potential. — Plant Physiol. 55: 932–936, 1975.

Turner, N. C., Begg, J. E.: Stomatal behavior and water status of maize, sorghum, and tobacco under field conditions. I. At high soil water potential. — Plant Physiol. 51: 31–36, 1973.

Turner, N. C., Heichel, G. H.: Stomatal development and seasonal changes in diffusive resistance of primary and regrowth foliage of red oak (Quercus rubra L.) and red maple (Acer rubrum L.). — New Phytol. 78: 71–81, 1977.

Turner, N. C., Incoll, L. D.: The vertical distribution of photosynthesis in crops of tobacco and sorghum. — J. appl. Ecol. 8: 581–591, 1971.

Turrell, F. M.: The area of the internal exposed surface of dicotyledon leaves. — Amer. J. Bot. 23: 255–264, 1936.

Turrell, F. M.: Internal surface—intercellular space relationships and the dynamics of humidity maintenance in leaves. — In: Wexler, A. (ed.): Humidity and Moisture. Vol. II. Applications. Pp. 39–53. Reinhold Book Co., New York 1965.

Uchida, N., Itoh, R., Murata, Y.: [Studies on the changes in the photosynthetic activity of a crop leaf during its development and senescence I. Changes in the developmental stage of a rice leaf.] — Jap. J. Crop Sci. 49: 127–134, 1980.

Urmantsev, Yu. A.: Zasukho- i zharoustoïchivost' D i L list' ev pervogo yarusa u fasoli (gipoteza o prichinakh razlichnoï vstrechaemosti D, L, DL bioob'ektov v prirode). [Drought- and heat resistance in right (D) and left (L) bean leaves of the first insertion level (hypothesis on the causes of various occurrences of D, L, and DL bioobjects in the nature).] — Fiziol. Rast. 17: 937–944, 1970.

Václavík, J.: Effect of different leaf age on the relationship between the $CO_2$ uptake and water vapour efflux in tobacco plants. — Biol. Plant. 15: 233–236, 1973.

Václavík, J.: $CO_2$ and water vapor exchange through adaxial and abaxial surfaces of tobacco leaves of different insertion level. — Biol. Plant. 16: 389–394, 1974.

Václavík, J.: Comparison of the changes in net photosynthetic $CO_2$ uptake and water vapour efflux during leaf ontogenesis with the differences between the leaves according to their descending insertion level. — Biol. Plant. 17: 411–415, 1975.

Václavík, J.: Distribution pattern of gas exchange in the area of maize leaf blades during the generative phase. — Biol. Plant. 19: 457–461, 1977.

Václavík, J.: Photosynthesis, transpiration and stomatal conductance in *Zea mays* leaves as affected by their irradiance at normal and inverse position. — Photosynthetica 14: 482–488, 1980.

Valanne, N., Aro, E.-M., Repo, E.: Changes in photosynthetic capacity and activity of RuBPC-ase and glycolate oxidase during the early growth of moss protonemata in continuous and rhythmic light. — Z. Pflanzenphysiol. 88: 123–131, 1978.

Valanne, N., Pennanen, A., Vapaavuori, E.: Comparison between the preservation of chloroplast structure in the dark and the turnover rate of chlorophyll-protein complexes in a moss and two varieties of pea. — Plant Cell Physiol. 20: 1511–1522, 1979.

Valanne, N., Valanne, T., Niemi, H., Aro, E.-M.: The development of the photosynthetic apparatus during leaf opening in silver birch (*Betula pendula* Roth.). — In: Akoyunoglou, G. (ed.): Photosynthesis. Vol. V. Pp. 397–406. Balaban Int. Sci. Serv., Philadelphia 1981.

Valcke, R.: Influence of relative humidity and age on the development of the photosynthetic apparatus in barley (*Hordeum vulgare* c. v. Union). — In: Akoyunoglou, G., Argyroudi-Akoyunoglou, J. H. (ed.): Chloroplast Development. Pp. 871–874. Elsevier/North-Holland Biomedical Press, Amsterdam–New York–Oxford 1978.

Van, T. K., Garrard, L. A., West, S. H.: Effects of UV-B radiation on net photosynthesis of some crop plants. — Crop Sci. 16: 715–718, 1976.

Van den Driessche, R., Connor, D. J., Tunstall, B. R.: Photosynthetic response of brigalow to irradiance, temperature and water potential. — Photosynthetica 5: 210–217, 1971.

Van Steveninck, M. E., Van Steveninck, R. F. M.: Plastids with densely staining thylakoid contents in *Nymphoides indica*. I. Plastid development. — Protoplasma 103: 333–342, 1980a.

Van Steveninck, M. E., Van Steveninck, R. F. M.: Plastids with densely staining thylakoid contents in *Nymphoides indica*. II. Characterization of stainable substance. — Protoplasma 103: 343–360, 1980b.

Varlet·Grancher, C., Bonhomme, R., Chartier, M., Artis, P.: Évolution de la réponse photosynthétique des feuilles et efficience théorique de la photosynthèse brute d'une culture de canne à sucre (*Saccharum officinarum* L.). — Agronomie 1: 473–481, 1981.

Vechar, A. S., Troitskaya, T. M., Masny, M. M.: Ab adroznennyakh nekatorykh biyakhimichnykh pakazchykaǔ u dy- i tetraploidnykh tsukrovykh burakoǔ. [Differences in some biochemical characteristics in di- and tetraploid sugar beets.] — Vestsi Akad. Navuk belarus. SSR, Ser. biyal. Navuk 1970 (1): 24–29, 138, 1970.

Vecher, A. S., Lebedeva, T. D.: Aktivnost' reaktsii Khilla s raznymi okislitelyami v protsesse razvitiya list'ev [Rate of Hill reaction with various oxidants during leaf ontogenesis.] — Dokl. Akad. Nauk belorus. SSR 11: 727–730, 1967.

Vecher, A. S., Lebedeva, T. I., Raïtsina, G. I.: Aktivnost' reaktsii Khilla i fotosinteticheskogo fosforilirovaniya v izolirovannykh khloroplastakh list'ev gorokha. [Rates of Hill reaction and photosynthetic phosphorylation in isolated chloroplasts of pea leaves.] — Dokl. Akad. Nauk belorus. SSR 11: 451–454, 1967.

Vecher, A. S., Nenadovich, R. A., Mas'ko, A. A., Reshetnikov, V. N: Lipidy i plastokhinony khloroplastov kartofelya i rzhi. [Lipids and plastoquinones of potato and rye chloroplasts.] — Fiziol. Biokhim. kul't. Rast. 10: 269–275, 1978.

Vecher, A. S., Predkel', K. I.: Porfirinsoderzhayushchie pigmenty i aktivnost' nekotorykh gemoproteidnykh oksidoreduktaz v list'yakh i plastidakh. [Porphyrin containing pigments and activity of some haemoproteid oxidoreductases in leaves and plastids.] — In: Shlyk, A. A. (ed.): Khlorofill. Pp. 313–319, 412. Nauka i Tekhnika, Minsk 1974.

Vecher, A. S., Raïtsina, G. I.: Izuchenie ATF-aznoï aktivnosti v khloroplastakh izolirovannykh iz list'ev razlichnogo vozrasta. [ATPase activity in chloroplasts isolated from leaves of different ages.] — Dokl. Akad. Nauk belorus. SSR 11: 544–547, 1967.

Verbelen, J. P., de Greef, J. A.: Leaf development of *Phaseolus vulgaris* L. in light and in darkness. — Amer. J. Bot. 66: 970–976, 1979.

Vietor, D. M., Ariyanayagam, R. P., Musgrave, R. B.: Photosynthetic selection of *Zea mays* L. I.

Plant age and leaf position effects and a relationship between leaf and canopy rates. — Crop Sci. 17: 567–573, 1977.

Vignes, D., Calmés, J.: Quelques modifications physico-chimiques et physiologiques liées à la sénescence des feuilles. — Physiol. Plant. 33: 188–193, 1975.

Virgin, H. I.: On the formation of protochlorophyll in normal green wheat leaves of varying age. — Physiol. Plant. 14: 384–392, 1961.

Virgin, H. I.: Carotenoid synthesis in leaves of wheat after irradiation by red light. — Physiol. Plant. 19: 40–46, 1966.

Virgin, H. I.: In vivo absorption spectra of protochlorophyll$_{650}$ and protochlorophyll$_{636}$ within the region 530–700 nm. — Photosynthetica 9: 84–92, 1975.

Viro, M., Kloppstech, K.: Differential expression of the genes for ribulose-1,5-bisphosphate carboxylase and light-harvesting chlorophyll $a/b$ protein in the developing barley leaf. — Planta 150: 41–45, 1980.

Vlasova, M. P., Osipova, O. P.: Vliyanie intensivnosti sveta na tonkuyu strukturu khloroplastov rastenii Vicia faba. [Effect of irradiance on chloroplast ultrastucture of Vicia faba.] — Fiziol. Rast. 20: 742—746, 1973.

Volodarskiï, N. I., Bystrykh, E. E.: Funktsional'naya aktivnost' fotosinteticheskogo apparata rastenii podsolnechnika pod vliyaniem zasukhi. [Functional activity of photosynthetic apparatus of sunflower plants during drought.] — Sel'skokhoz. Biol. 10: 716–721, 1975.

Volodarskiï, N. I., Bystrykh, E. E.: Funktsional'naya aktivnost' fotosinteticheskogo apparata pri narushenii vodnogo rezhima podsolnechnika. [Functional activity of the photosynthetic apparatus in the sunflower during disturbance of the water state.] — Fiziol. Rast. 23: 497–501, 1976.

Volodarskiï, N. I., Bystrykh, E. E., Nikolaeva, E. K.: Fotosinteticheskaya aktivnost' verkhnego lista pshenitsy u sortov razlichnoï produktivnosti. [Photosynthetic activity of the upper leaf of wheat in cultivars with different productivity.] — Sel'skokhoz. Biol. 13: 703–710, 1978.

Vorob'eva, L. M., Krasnovskiï, A. A.: Khlorofillid v khloroplastakh i gomogenatakh list'ev. [Chlorophyllide in chloroplasts and leaf homogenates.] — Fiziol. Rast. 13: 929–936, 1966.

Voskresenskaya, N. P.: Fotoregulyatornye Aspekty Metabolizma Rastenii. [Photoregulatory Aspects of Plant Metabolism.] — In: Timiryazevskie Chteniya. Vol. 38. Pp. 1—48. Nauka, Moskva 1979.

Voznesenskaya, E. V.: Razvitie khloroplastov v khlorenkhime Haloxylon persicum (Chenopodiaceae). [The development of chlorenchyma chloroplasts of Haloxylon persicum (Chenopodiaceae).] — Bot. Zh. 66: 98–101, 1981.

Wada, Y.: [Changes of photosynthetic and respiratory activities and of chlorophyll content in growing leaves of some tobacco varieties.] — Bot. Mag. (Tokyo) 81: 25–32, 1968.

Wada, Y.: Changes in activities of ribulosediphosphate carboxylase and phosphopyruvate carboxylase during leaf growth of tobacco. — Bot. Mag. (Tokyo) 84: 159–168, 1971.

Wada, Y., Kuroda, S.: [Changes in the photosynthetic activity during aging in different parts of intact tobacco leaves.] — Bot. Mag. (Tokyo) 81: 226–231, 1968.

Wada, Y., Watanabe, S., Kuroda, S.: [Changes in the photosynthetic activity and chlorophyll contents of growing tobacco leaves.] — Bot. Mag. (Tokyo) 80: 123–129, 1967.

Walker, J. M., Waygood, E. R.: Ecology of Phragmites communis. I. Photosynthesis of a single shoot in situ. — Can. J. Bot. 46: 549–555, 1968.

Walles, B., Hudák, J.: A comparative study of chloroplast morphogenesis in seedlings of some conifers (Larix decidua, Pinus sylvestris and Picea abies). — Stud. forest. suec. 127: 1–22, 1975.

Wardlaw, I. F.: The effect of water stress on translocation in relation to photosynthesis and growth. II. Effect during leaf development in Lolium temulentum L. — Aust. J. biol. Sci. 22: 1–16, 1969.

Watanabe, I.: Mechanism of varietal differences in photosynthetic rate of soybean leaves. II. Varietal differences in the balance between photochemical activities and dark reaction activities. — Proc. Crop Sci. Soc. Jap. 42: 428–436, 1973.

Watson, R. L., Landsberg, J. J.: The photosynthetic characteristics of apple leaves (cv. Golden Delicious) during their early growth. — In: Marcelle, R., Clijsters, H., Van Poucke, M. (ed.): Photosynthesis and Plant Development. Pp. 39–48. Dr. W. Junk b.v. - Publ., The Hague–Boston –London 1979.

Watts, W. R., Neilson, R. E., Jarvis, P. G.: Photosynthesis in Sitka spruce (*Picea sitchensis* (Bong.) Carr.) VII. Measurements of stomatal conductance and $^{14}CO_2$ uptake in a forest canopy. — J. appl. Ecol. 13: 623–638, 1976.

Weier, T. E.: The ultramicro structure of starch-free chloroplasts of fully expanded leaves of *Nicotiana rustica*. — Amer. J. Bot. 48: 615—630, 1961.

Weiland, R. T., Noble, R. D., Crang, R. E.: Photosynthetic and chloroplast ultrastructural consequences of manganese deficiency in soybean. — Amer. J. Bot. 62: 501–508, 1975.

Weiss, A.: Untersuchungen über die Zahlen- und Grössenverhältnisse der Splatöffnungen. — Jahrb. wiss. Bot. 4: 125–196, 1865.

Wellburn, A. R., Hampp, R.: Appearance of photochemical function in prothylakoids during plastid development. — Biochim. biophys. Acta 547: 380–397, 1979.

Wells, R., Liebhardt, W. C., Svec, L. V., Frick, H.: Photosynthetic capacity in potassium stressed soybean: Comparison of $CO_2$ fixation and $O_2$ evolution assays. — J. Plant Nutr. 1: 283–293, 1979.

West, D. W., Black, J. D. F.: Irrigation timing — its influence on the effects of salinity and waterlogging stresses in tobacco plants. — Soil Sci. 125: 367–376, 1978.

Wettstein, D. von: Chloroplast and nucleus: concerted interplay between genomes of different cell organelles. The Emil Heitz lecture. — In: Schweiger, H. G. (ed.): International Cell Biology 1980–1981. Pp. 250–272. Springer-Verlag, Berlin–Heidelberg–New York 1981.

Whatley, J. M.: Ultrastructural changes in chloroplasts of *Phaseolus vulgaris* during development under conditions of nutrient deficiency. — New Phytol. 70: 725–742, 1971.

Whatley, J. M .: Chloroplast development in primary leaves of *Phaseolus vulgaris* L. — New Phytol. 73: 1097–1110, 1974.

Whatley, J. M.: Variations in the basic pathway of chloroplast development. — New Phytol. 78: 407–420, 1977a.

Whatley, J. M.: The effect of cotyledons on chloroplast development in primary leaves of *Phaseolus vulgaris*. — New Phytol. 79: 55–60, 1977b.

Whatley, J. M.: A suggested cycle of plastid developmental interrelationships. — New Phytol. 80: 489–502, 1978.

Whatley, J. M.: Plastid development in the primary leaf of *Phaseolus vulgaris*: variations between different types of cell. — New Phytol. 82: 1–10, 1979.

Whatley, J. M.: Plastid growth and division in *Phaseolus vulgaris*. — New Phytol. 86: 1–16, 1980.

Whatley, J. M., John, P., Whatley, F. R.: From extracellular to intracellular: the establishment of mitochondria and chloroplasts. — Proc. roy. Soc. London B 204: 165–187, 1979.

Whittaker, R. H., Likens, G. E.: The biosphere and man. — In: Lieth, H., Whittaker, R. H. (ed.): Primary Productivity of the Biosphere. Pp. 305–328. Springer-Verlag, Berlin–Heidelberg– New York 1975.

Więckowski, S.: Changes in the content of carotenoids in growing bean leaves. — Bull. Acad. pol. Sci., Sér. Sci. biol. 9: 325–332, 1961.

Więckowski, S.: Photosynthesis in the growing leaf of *Phaseolus vulgaris*. — Acta Soc. Bot. Pol. 35: 437–443, 1966.

Więckowski, S.: Chloroplasts in growing bean leaf. — Acta Soc. Bot. Pol. 36: 161–169, 1967a.

Więckowski, S.: Ultrastructure and activity of the photosynthetic apparatus in growing bean leaf. — Stud. biophys. 5: 77–84, 1967b.

Więckowski, S.: Daily changes in the photosynthetic rate and chloroplast ultrastructure in growing bean leaf. — Photosynthetica 2: 172–177, 1968.

Więckowski, S.: Studies on the activity and ultrastucture of the photosynthetic apparatus in the earliest stages of primary bean leaves development. — Acta Soc. Bot. Pol. 38: 103–114, 1969.

Więckowski, S.: Photosynthetic activity of bean leaves and the relative content of chlorophyll in particles isolated from chloroplasts after irradiation of etiolated seedlings. — Photosynthetica 5: 44–49, 1971.

Więckowski, S.: The Hill reaction activity at different phases of chlorophyll accumulation in primary bean leaves. — Bull. Acad. pol. Sci., Sér. Sci. biol. 20: 425–429, 1972.

Więckowski, S.: Metabolism of chloroplast pigments and photosynthetic activity at different phases of leaf ontogenesis. — Pol. ecol. Stud. 1: 33–40, 1975.

Wignarajah, K., Jennings, D. H., Handley, J. F.: The effect of salinity on growth of *Phaseolus vulgaris* L. I. Anatomical changes in the first trifoliate leaf. — Ann. Bot. 39: 1029–1038, 1975.

Wild, A.: Physiologie der Photosynthese höherer Pflanzen. Die Anpassung an die Lichtbedingungen. — Ber. deutsch. bot. Ges. 92: 341–364, 1979.

Wild, A., Belz, J., Rühle, W.: Cyclic and noncyclic photophosphorylation during the ontogenesis of high-light and low-light leaves of *Sinapis alba*. — Planta 153: 308–311, 1981a.

Wild, A., Forschner, W., Zerbe, R., Rühle, W.: The effect of kinetin on the transpiration and the photosynthetic capacity of primary leaves of *Sinapis alba*. — Z. Pflanzenphysiol. 105: 93–96, 1981b.

Wild, A., Höhler, T.: Die Wirkung unterschiedlicher Lichtintensitäten während der Anzucht auf die $CO_2$-Kompensationslage, die Glykolsäure-Oxidase- und Ribulose-biphosphat-Carboxylase-Aktivitäten bei *Sinapis alba*. — Z. Pflanzenphysiol. 87: 413–428, 1978.

Wild, A., Holzapfel, A.: The effect of blue and red light on the content of chlorophyll, cytochrome *f*, soluble reducing sugars, soluble proteins and the nitrate reductase during growth of the primary leaves of *Sinapis alba*. — In: Senger, H. (ed.): The Blue Light Syndrome. Pp. 444–451. Springer-Verlag, Berlin–Heidelberg–New York 1980.

Wild, A., Wolf, G.: The effect of different light intensities on the frequency and size of stomata, the size of cells, the number, size and chlorophyll content of chloroplasts in the mesophyll and the guard cells during the ontogeny of primary leaves of *Sinapis alba*. — Z. Pflanzenphysiol. 97: 325–342, 1980.

Wild, A., Zerbe, R., Kind, S.: The influence of indole-3-acetic acid on the photosynthetic apparatus of *Sinapis alba* plants grown under high light conditions. — In: Akoyunoglou, G. (ed.): Photosynthesis. Vol. VI. Pp. 339–348. Balaban Int. Sci. Serv., Philadelphia 1981c.

Wilhelm, W. W., Nelson, C. J.: Leaf growth, leaf aging, and photosynthetic rate of tall fescue genotypes. — Crop Sci. 18: 769–772, 1978.

Willert, D. J. von: Vorkommen und Regulation des CAM bei Mittagsblumengewächsen (*Mesembryanthemaceae*). — Ber. deut. bot. Ges. 92: 133–144, 1979.

Willert, D. J. von, Curdts, E., Willert, K. von: Veränderung der PEP-Carboxylase während einer durch NaCl geförderten Ausbildung eines CAM bei *Mesembryanthemum crystallinum*. — Biochem. Physiol. Pflanzen 171: 101–107, 1977.

Willert, D. J. von, Treichel, S. , Kirst, G. O., Curdts, E.: Environmentally controlled changes in phosphoenolpyruvate carboxylases in *Mesembryanthemum*. — Phytochemistry 15: 1435–1436, 1976.

Willert, D. J. von, Willert, K. von: Light modulation of the activity of the PEP-carboxylase in CAM-plants in the *Mesembryanthemaceae*. — Z. Pflanzenphysiol. 95: 43–49, 1979.

Williams, L. E., Kennedy, R. A.: Relationship between early photosynthetic products, photorespiration and stage of leaf development in *Zea mays*. — Z. Pflanzenphysiol. 81: 314–322, 1977.

Williams, R. F.: The Shoot Apex and Leaf Growth. A Study in Quantitative Biology. — Cambridge Univ. Press, Cambridge 1975.

Willstätter, R., Stoll, A.: Untersuchungen über Chlorophyll. — Berlin 1913.

Willstätter, R., Stoll, A.: Untersuchungen über die Assimilation der Kohlensäure. — Berlin 1918.

Wilson, D.: Effect of selection for stomatal length and frequency on theoretical stomatal resistance to diffusion in *Lolium perenne* L. — New Phytol. 71: 811–817, 1972.

Wilson, D.: Variation in leaf respiration in relation to growth and photosynthesis of *Lolium*. — Ann. appl. Biol. 80: 323–338, 1975.

Wilson, D., Cooper, J. P.: Assimilation of *Lolium* in relation to leaf mesophyll. — Nature 214: 989–992, 1967.

Wilson, D., Cooper, J. P.: Apparent photosynthesis and leaf characters in relation to leaf position and age, among contrasting *Lolium* genotypes. — New Phytol. 68: 645–655, 1969.

Wilson, G. L., Ludlow, M. M.: Bean leaf expansion in relation to temperature. — J. exp. Bot. 19: 309–321, 1968.

Wilson, G. L., Ludlow, M. M.: Net photosynthetic rates of tropical grass and legume leaves. — In: Proceedings of the XI[th] International Grassland Congress. Pp. 534–538. Univ. Queensland Press, St. Lucia 1970.

Wilson, J. R.: Variation of leaf characteristics with level of insertion on a grass tiller. II. Anatomy. — Aust. J. agr. Res. 27: 355–364, 1976.

Wilson, J. R.: Variation of leaf characteristics with level of insertion on a grass tiller. III. Tissue water relations. — Aust. J. Plant Physiol. 4: 733–743, 1977.

Winter, K., Greenway, H.: Phosphoenolpyruvate carboxylase from *Mesembryanthemum crystallinum*: Its relation and inactivation *in vitro*. — J. exp. Bot. 29: 539–546, 1978.

Winzeler, H., Nösberger, J.: Carbon dioxide exchange of spring-wheat in relation to age and photon-flux density at different growth temperatures. — Ann. Bot. 46: 685–693, 1980.

Wirth, E., Kelly, G. J., Fischbeck, G., Latzko, E.: Enzyme activities and products of $CO_2$ fixation in various photosynthetic organs of wheat and oat. — Z. Pflanzenphysiol. 82: 78–87, 1977.

Wise, R. R., Harris, J. B.: Thylakoid-dense chloroplasts of *Callisia fragrans*. — Cytologia 45: 113–126, 1980.

Wittenbach, V. A.: Ribulose bisphosphate carboxylase and proteolytic activity in wheat leaves from anthesis through senescence. — Plant Physiol. 64: 884–887, 1979.

Woledge, J.: The effect of light intensity during growth on the subsequent rate of photosynthesis of leaves of tall fescue (*Festuca arundinacea* Schreb.). — Ann. Bot. 35: 311–322, 1971.

Woledge, J.: The effect of shading on the photosynthetic rate and longevity of grass leaves. — Ann. Bot. 36: 551–561, 1972.

Woledge, J.: The photosynthesis of ryegrass leaves grown in a simulated sward. — Ann. appl. Biol. 73: 229–237, 1973.

Woledge, J.: The effects of shading and cutting treatments on the photosynthetic rate of ryegrass leaves. — Ann. Bot. 41: 1279–1286, 1977.

Woledge, J.: The effect of shading during vegetative and reproductive growth on the photosynthetic capacity of leaves in a grass sward. — Ann. Bot. 42: 1085–1089, 1978.

Woledge, J.: Effect of flowering on the photosynthetic capacity of ryegrass leaves grown with and without natural shading. — Ann. Bot. 44: 197–207, 1979.

Woledge, J., Jewiss, O. R.: The effect of temperature during growth on the subsequent rate of photosynthesis in leaves of tall fescue (*Festuca arundinacea* Schreb.). — Ann. Bot. 33: 897–913, 1969.

Wolf, F. T.: Effects of chemical agents in inhibition of chlorophyll synthesis and chloroplast development in higher plants. — Bot. Rev. 43: 395–425, 1977.

Wolińska, D.: Functional and structural changes in chloroplasts of senescent tobacco leaves. — Acta Soc. Bot. Pol. 45: 341–352, 1976.

Wood, G. B.: Spatial variation in leaf chlorophyll within the crown of a radiata pine sapling. — Aust. Forest Res. 6 (4): 5–14, 1974.

Woodman, J. N.: Variation of net photosynthesis within the crown of a large forest-grown conifer. — Photosynthetica 5: 50–54, 1971.

Woodward, R. G.: Photosynthesis and expansion of leaves of soybean grown in two environments. — Photosynthetica 10: 274–279, 1976.

Woodward, R. G., Rawson, H. M.: Photosynthesis and transpiration in dicotyledonous plants. II. Expanding and senescing leaves of soybean. — Aust. J. Plant Physiol. 3: 257–267, 1976.

Woolhouse, H. W.: The nature of senescence in plants. — In: Aspects of the Biology of Ageing. (Symp. Soc. exp. Biol. Vol. 21.) Pp. 179–213. Univ. Press, Cambridge 1967.

Woolhouse, H. W.: Leaf age and mesophyll resistance as factors in the rate of photosynthesis. — Hilger J. 11: 7–12, 1967/8.

Woolhouse, H. W.: Longevity and senescence in plants. — Sci. Progr. (Oxford) 61: 123–147, 1974.

Woolhouse, H. W.: Light-gathering and carbon assimilation processes in photosynthesis; their adaptive modifications and significance for agriculture. — Endeavour N. S. 2: 35–46, 1978.

Woolhouse, H. W.: Leaf senescence. — In: Smith, H., Grierson, D. (ed.): The Molecular Biology of Plant Development. Pp. 256—281. Blackwells, Oxford 1982.

Woolhouse, H. W., Batt, T.: The nature and regulation of senescence in plastids. — In: Sunderland, N. (ed.): Perspectives in Experimental Biology. Vol. 2. Pp. 163–175. Pergamon Press, Oxford–New York 1976.

Woolley, J. T.: Reflectance and transmittance of light by leaves. — Plant Physiol. 47: 656–662, 1971.

Woolley, J. T.: Change of leaf dimensions and air volume with change in water content. — Plant Physiol. 51: 815–816, 1973.

Woolley, J. T.: Refractive index of soybean leaf cell walls. — Plant Physiol. 55: 172–174, 1975.

Wort, D. J.: Mechanism of plant growth stimulation by naphthenic acid. II. Enzymes of $CO_2$ fixation, $CO_2$ compensation point, bean embryo respiration. — Plant Physiol. 58: 82–86, 1976.

Woźny, A., Szweykowska, A.: Effect of cytokinins and antibiotics on chloroplast development in cotyledons of Cucumis sativus. — Biochem. Physiol. Pflanzen 168: 195–209, 1975.

Wróblewska, H.: Influence of water deficit and age of plant on the intensity of photosynthesis and air passage capacity in leaves of Nicotiana rustica L. — Hodowla Rośl., Aklimat., Nasienn. 17: 387–411, 1973.

Wu Guang-yao, Deng Yue-fen, Lu Chong-en, Pu Zong-shi, Xu Shi-rong: [On the change of photosynthetic characteristics in Setaria italica.] — Acta Phytophysiol. sin. 8: 111–116, 1982.

Wu, J. H.: Retardation of ultraviolet light accelerated leaf senescence by a cytokinin: $N^6$ benzyl-adenine. — Photochem. Photobiol. 13: 179–181, 1971.


Yakshina, A. M., Malkina, I . S.: Uglekislotnyï gazoobmen duba chereshchatogo i ego vliyanie na gazovyï sostav atmosfery. [Carbon dioxide exchange of Quercus pedunculata Ehrh. and its effect on the gas composition of the atmosphere.] — Zh. obshch. Biol. 40: 926–930, 1979.

Yamaguchi, J.: Respiration and the growth efficiency in relation to crop productivity. — J. Fac. Agr. Hokkaido Univ. 59: 59–129, 1978.

Yamaguchi, T., Friend, D. J. C.: Effect of leaf age and irradiance on photosynthesis of Coffea arabica. — Photosynthetica 13: 271–278, 1979.

Yamashita, T., Tsutsumi, M., Yoshinari, S.: [Changes in the activities of enzyme related to carbon dioxide fixation and glycolate pathway with the age of mulberry leaves.] — J. Sericult. Sci. Jap. 44: 1–6, 1975.

Yates, D. J.: Effect of the angle of incidence of light on the net photosynthesis rates of Sorghum almum leaves. — Aust. J. Plant Physiol. 8: 335–346, 1981.

Yordanov, I.: Influence of the physiological state of the leaves on the photophosphorylating capacity of the chloroplasts isolated from them. — Dokl. bolg. Akad. Nauk 23: 999–1002, 1970.

Yordanov, I. T.: Photophosphorylation activity in vivo and intensity of the photosynthesis of bean leaves of different physiological state. — Dokl. bolg. Akad. Nauk 24: 1551–1554, 1971.

Yoshida, T.: Effect of stomatal frequency on photosynthesis and its use for breeding in barley. — Bull. Kyushu nat. agr. exp. Sta. 20: 129–193, 1978.

Youn, K. B., Ota, Y.: [Changes in the chlorophyll content and chlorophyll retention of leaf segments according to the growth of various leaf blades in rice plant.] — Proc. Crop Sci. Soc. Jap. 42: 6–12, 1973.

Yukimoto, M., Ishitani, A., Yoshida, K., Kobayashi, N.: [Phytotoxicity of MBCP (4-bromo-2,5-dichlorophenyl methyl phenylphosphonothionate) to Chinese cabbage seedlings (*Brassica pekinensis* Rupr.).] — Nihon Noyakugaku Kaishi [J. Pestic. Sci.] 3: 243–247, 1978.

Yusufov, A. G., Ashurova, O. B.: Starenie ukorenennykh list'ev. [Aging of rooted leaves.] — Fiziol. Rast. 22: 741–746, 1975.

Zaïtseva, N. A.: Reaktsiya Khilla v khloroplastakh zdorovykh i khloroznykh rastenii. [Hill reaction in chloroplasts of healthy and chlorotic plants.] — Fizol. Biokhim. kul' t. Rast. 2: 64–67, 1970.

Zakaryan, N. E.: Metabolizm khlorofilla v list'yakh tabaka v zavisimosti ot ikh yarusnogo raspolozheniya. [Chlorophyll metabolism in tobacco leaves in dependence on their insertion.] — Biol. Zh. Armenii 19 (12): 70–76, 1966.

Zalenskiĭ, V.: Materialy k kolichestvennoĭ anatomii razlichnykh list'ev odnikh i tekhzhe rastenii. [Materials for the study of the quantitative anatomy of different leaves of the same plant.] — Mem. politekh. Inst. Kiev 4: 1–203, 1904.

Żelawski, W., Gowin, T.: O rozmieszczeniu aparatów szparkowych na powierzchni igieł sosny pospolitej (*Pinus silvestris* L.). [Distribution of stomata on the surface of the needles of Scots pine (*Pinus silvestris* L.).] — Folia forest. polon., Ser. A 1967 (13): 111–117, 1967a.

Żelawski, W., Gowin, T.: Badania cech strukturalnych igliwia sosny (*Pinus silvestris* L.) w nasłonecznionej i ocienionej strefie korony drzewa. [Investigation of the structure of Scots pine (*Pinus silvestris* L.) needles in insolated and shadowed parts of the tree crown.] — Folia forest. polon., Ser. A 1967 (13): 119–126, 1967b.

Zelenskiĭ, M. I., Mogileva, F., Shitova, I., Fattakhova, F.: Hill reaction of chloroplasts from some species, varieties and cultivars of wheat. — Photosynthetica 12: 428–435, 1978.

Zelitch, I.: Photosynthesis, Photorespiration, and Plant Productivity. — Academic Press, New York–London 1971.

Zelitch, I.: Pathways of carbon fixation in green plants. — Annu. Rev. Biochem. 44: 123–145, 1975.

Zelitch, I.: Photorespiration: Studies with whole tissues. — In: Gibbs, M., Latzko, F. (ed.): Photosynthesis II. Pp. 353–367. Springer-Verlag, Berlin–Heidelberg–New York 1979.

Zerbe, R., Wild, A.: The effect of kinetin on the photosynthetic apparatus of *Sinapis alba*. — Photosynthesis Res. 1: 53–64, 1980a.

Zerbe, R., Wild, A.: The effect of indole-3-acetic-acid on the photosynthetic apparatus of *Sinapis alba*. — Photosynthesis Res. 1: 71–81, 1980b.

Zerbe, R., Wild, A.: The effects of gibberellic acid and kinetin on fresh weight, dry weight, leaf area, contents of soluble reducing sugars, soluble proteins, chlorophylls and cytochrome *f*. — In: Akoyunoglou, G. (ed.): Photosynthesis. Vol. VI. Pp. 349–357. Balaban Int. Sci. Serv., Philadelphia 1981.

Zima, J., Šesták, Z.: Photosynthetic characteristics during ontogenesis of leaves. 4. Carbon fixation pathways, their enzymes and products. — Photosynthetica 13: 83–106, 1979.

Zima, J., Šesták, Z., Čatský, J., Tichá, I.: Ontogenetic changes in leaf $CO_2$ uptake as controlled by photosystem and carboxylation activities and $CO_2$ transfer. — In: Akoyunoglou, G. (ed.): Photosynthesis. Vol. VI. Pp. 23–31. Balaban Int. Sci. Serv., Philadelphia 1981.

Zirkle, C.: Development of normal and divergent plastid types in *Zea mays*. — Bot. Gaz. 88: 186–203, 1929.

Zobel, D. B., Liu, V. T.: Leaf-conductance patterns of seven palms in a common environment. — Bot. Gaz. 141: 283–289, 1980.

Zrůst, J., Smolíková, A.: Rozdíly v rychlosti fotosyntézy u kříženců a některých rodičovských odrůd brambor. [Differences in photosynthetic rate in potato hybrids and in some parental cultivars.] — Rostl. Výr. (Praha) 23: 723–732, 1977.

Zuo Buo-yu, Duan Xu-chuang: [Ultrastructure and function of chloroplasts from the winter wheat leaves at different ranks of attachment to the main stem.] — Acta bot. sin. 20: 223–228, 1978.

Zurzycki, J.: The influence of chloroplast displacements on the optical properties of leaves. — Acta Soc. Bot. Pol. 30: 503–527, 1961.

Zurzycki, J., Gabryś, H.: Changes in light absorption by the chloroplast, related to its structural transformations. — Acta Soc. Bot. Pol. 46: 369–380, 1977.

Zurzycki, J., Gierka, A.: The effect of light intensity and spectral region on the structure of thylakoid membranes of *Phaseolus vulgaris*. — Acta Physiol. Plant. 1: 27–34, 1978.

# AUTHORS' INDEX

The numbers in italics refer to the list of references.

# SUBJECT INDEX

The terms development, insertion, leaf, ontogeny, and photosynthesis are not included (with the exception of common terms). As concerns chemical substances, their contents, activities, and effects are mostly included in single item; the environmental factors are treated similarly. Abbreviations are only explained, while the reference figures accompany the fully expressed items.

plastoquinone 111, 128—9, 131—2, 144
plastosome 64
pollution, atmospheric 68, 73, 95, 242
polyribosomes 86, 146, 268
postillumination burst 251, 253—7
potassium nutrition 143, 204, 261
PP = palisade parenchyma
Pr = phytochrome (red form)
primordia 16, 21, 62—3, 172, 270
prolamellar body 54, 56, 59, 60, 62, 65, 68—9, 74, 81
proplastid 51, 55—6, 60, 62, 64
proteins 14, 73, 111, 129—30, 155, 209, 268, 270
prothylakoid vesicles 81
protochlorophyll(ide) 71, 77—81, 100
PS = Photosystem
PSU = photosynthetic unit
pyruvate orthophosphate dikinase 153

$q$ = photosynthetic requirement
$Q$, electron acceptor 128—9, 137
quantum yield, requirement 137—9, 141, 183, 187—8
quercetin 105

$r_a$ = boundary layer resistance
$r_i$ = intercellular resistance
$r_M$ = intracellular (mesophyll) resistance
$r_s$ = stomatal (epidermal) resistance
R = reflectance
$R_{abs}$ = absorbed radiation
$R_D$ = „dark" respiration rate
$R_G$ = growth respiration
$R_L$ = photorespiration rate
$R_M$ = maintenance respiration
radiation absorbed, incident, reflected, scattered, transmitted 107
– wavelengths 72, 88, 101—2, 107, 109—111, 116—7, 121, 125, 133, 185, 189, 277
reaction centres 190—10, 268
reflectance 107, 115, 121—4
reflection coefficient 107, 119—20, 122—5
–, leaf 112—6, 118—21, 125—7
reflectivity 107, 122—4
refraction index 113—4, 119
resistance to $CO_2$ and water vapour transfer see conductance...
respiration, growth, maintenance 205—7, 209, 277, 279—80
– rate, dark 157, 166—7, 190, 198, 200, 205—
—210, 215, 252, 275—9, 283

respiration rate in light 191, 194, 276—7
reticulum, endoplasmic 60, 66—7
–, peripheral 54, 70
review papers on photosynthesis topics 14, 16, 42, 53, 56—7
RH = relative humidity
ribonucleases 73
ribonucleic acid 86, 111, 209
ribose-5-phosphate isomerase 151—2, 274
ribosome 53—4, 63—5, 72—3, 86, 268
ribulose-1,5-bisphosphate 160, 192, 203, 250
– carboxylase (fraction 1 protein) 53—5, 70, 111, 145—53, 156, 158, 160, 171, 192, 198, 245—6, 270, 273—4
ribulose-1,5-bisphosphate carboxylase/oxygenase 157, 199, 203, 250, 259—62, 284
RNA = ribonucleic acid
rooted leaves 85, 171
roughness, leaf 220
RuBP = ribulose-1,5-bisphosphate
RuBPC = ribulose-1,5-bisphosphate carboxylase
RuBPCO = ribulose-1,5-bisphosphate carboxylase/oxygenase

S = scattering
saccharides (fructose, glucose, raffinose, ribose, saccharose, stachyose, starch, trisaccharides) and their phosphates 153—6, 160, 193, 209, 214, 277
saccharose phosphate synthetase 277
salinity 18, 27, 32, 34, 55, 70, 120, 239
scattering of radiation 114, 119—24, 126
seeds 214, 277
Shibata shift 101
SI = Stomatal Index
sieve effect 112, 119
sink-source 214, 231, 290
SLW = Specific Leaf Mass
$SO_2$ 239
SP = spongy parenchyma
Specific Leaf Mass 210—3
starch grains (inclusions) 55, 59, 65—70, 155, 269, 277
stomata density 24—34, 49, 50, 158, 222—3, 227, 231, 266
– functions, aperture, reactivity 161, 217, 221—3, 233—5, 237—8, 240, 266, 274, 276
– guard and mother cells 110, 227, 231, 274, 276
– number per leaf 35, 50

stomata size, dimensions 25—7, 30—1, 33—5, 49, 50, 221—2, 227, 263, 266
Stomatal Index 35—6, 222
stroma 42, 52, 54, 57—60, 63, 65—9, 217, 269
stromacentrum 55, 60
succinate 79, 155
succulence index 154
sucrose synthetase 153
superoxide dismutase 140
surfaces, leaf 23—34, 49, 115, 117, 179, 182, 185—7, 192, 222—3, 227, 232—3, 236, 239, 248, 282
surgical treatments, debudding, decapitation, defoliation, ear removal, pruning 48, 73—4, 87, 137—8, 144, 147, 183, 212, 231, 261—2, 268, 271

T = transmittance
$T$ = lamina thickness growth
tannins 63, 68
tannins 63, 68
tartaric acid 155
temperature 13, 17—8, 20, 23, 28—9, 32, 48—9, 68, 72, 91, 96, 112, 120, 137, 142, 158—9, 170—2, 187—8, 193—4, 200—2, 209, 212—3, 239, 243, 268, 271, 281, 285
terminology 12—4
thickness, epidermis 23, 50
—, leaf 17, 20—1, 39, 40, 101, 114, 116, 166, 212, 266
—, mesophyll 38—9
2-thiouracil 73
thylakoid 50, 52—75, 81, 111, 217, 248, 268—70, 272, 284
– fracture faces 55
thylakoids, appressed see grana
—, non-appressed 52, 54, 64, 66—8, 74
tip (apex), leaf 32—4, 40—1, 47, 49, 80, 97—8, 100, 125, 137, 139, 141, 143, 149, 153, 181—2, 187, 199, 212, 214, 233, 277
tonoplast 52
transketolase 152
translocation, transport, export 109, 209—10, 212, 214—5, 277—80, 284
transmission, leaf 112—6, 118—20, 125—7
– coefficient 107, 119, 122—5
transmissivity 107, 122—4
transmittance 107, 115, 121—4

transpiration 20, 24, 110, 159, 200, 223, 227, 231, 236, 258, 266, 274—6, 281—3
– ratio 159
tree crown 47—8, 64, 88—9, 91, 93, 95, 97, 139
2, 3', 6-trichlorophenol indophenol 134
trichomes, hairs, pubescence 22—4, 49, 113, 115, 117, 125, 220
trifluralin 137
TPIP = 2,3',6-trichlorophenol indophenol
triosephosphate isomerase 152
TRIS = tris (hydroxymethyl) aminomethane
tris(hydroxymethyl)aminomethane 137
*Triton X-100* 80, 100

$u$ = wind speed
UDP-galactose-diglyceride galactosyl transferase 130
ultraviolet radiation 108, 110—2, 125
uncouplers 135, 144
unifacial leaf 117, 187
UV = ultraviolet radiation

vacuole 38, 52, 67, 124
vascular bundles 36, 38, 49, 70
vernalization 242
violaxanthin 77, 90—1, 96—9

$W$ = dry matter; leaf width
Warburg-Dickens-Horecker pathway 208
Warburg effect 159
water content, deficit, potential, relations, stress of leaf 31, 35, 42, 88, 102, 116—7, 120, 124, 126, 137, 140, 161, 175, 182, 200, 202, 213, 232, 235—9, 242, 281—3, 285
– supply, soil moisture, irrigation 18—9, 31, 34, 42, 172, 202, 215, 227, 229, 234, 237—8, 242—3
– use efficiency 20, 24, 112, 158—9, 199, 281—3
waxes, epidermal 23, 108, 111, 117
wilting 31, 202—3
wind speed 219—20
WUE = water use efficiency

xanthophylls 76—7, 88, 90—1, 96—9
xeromorphy, xerophytes 49—50, 118—9, 125

$Y_G$ = true growth yield
Zalenskiï law 23—4, 49—50
Z-scheme 128

# PLANT INDEX

Scientific (Latin) names of plant genera are accompanied by those English names (in brackets) which are used in the text.